Communications in Computer and Information Science 2440

Rationale

The CCIS series is devoted to the publication of proceedings of computer science conferences. Its aim is to efficiently disseminate original research results in informatics in printed and electronic form. While the focus is on publication of peer-reviewed full papers presenting mature work, inclusion of reviewed short papers reporting on work in progress is welcome, too. Besides globally relevant meetings with internationally representative program committees guaranteeing a strict peer-reviewing and paper selection process, conferences run by societies or of high regional or national relevance are also considered for publication.

Topics

The topical scope of CCIS spans the entire spectrum of informatics ranging from foundational topics in the theory of computing to information and communications science and technology and a broad variety of interdisciplinary application fields.

Information for Volume Editors and Authors

Publication in CCIS is free of charge. No royalties are paid, however, we offer registered conference participants temporary free access to the online version of the conference proceedings on SpringerLink (http://link.springer.com) by means of an http referrer from the conference website and/or a number of complimentary printed copies, as specified in the official acceptance email of the event.

CCIS proceedings can be published in time for distribution at conferences or as post-proceedings, and delivered in the form of printed books and/or electronically as USBs and/or e-content licenses for accessing proceedings at SpringerLink. Furthermore, CCIS proceedings are included in the CCIS electronic book series hosted in the SpringerLink digital library at http://link.springer.com/bookseries/7899. Conferences publishing in CCIS are allowed to use Online Conference Service (OCS) for managing the whole proceedings lifecycle (from submission and reviewing to preparing for publication) free of charge.

Publication process

The language of publication is exclusively English. Authors publishing in CCIS have to sign the Springer CCIS copyright transfer form, however, they are free to use their material published in CCIS for substantially changed, more elaborate subsequent publications elsewhere. For the preparation of the camera-ready papers/files, authors have to strictly adhere to the Springer CCIS Authors' Instructions and are strongly encouraged to use the CCIS LaTeX style files or templates.

Abstracting/Indexing

CCIS is abstracted/indexed in DBLP, Google Scholar, EI-Compendex, Mathematical Reviews, SCImago, Scopus. CCIS volumes are also submitted for the inclusion in ISI Proceedings.

How to start

To start the evaluation of your proposal for inclusion in the CCIS series, please send an e-mail to ccis@springer.com

Saurabh Singh · Karm Veer Arya ·
Ciro Rodriguez Rodriguez · Altaf Osman Mulani
Editors

Emerging Trends in Artificial Intelligence, Data Science and Signal Processing

First International Conference, AIDSP 2023
Kanpur, India, October 20–21, 2023
Proceedings, Part II

Springer

Editors
Saurabh Singh
Galgotias University
Noida, Uttar Pradesh, India

Karm Veer Arya
ABV-Indian Institute of Information Technology and Management
Gwalior, Madhya Pradesh, India

Ciro Rodriguez Rodriguez
Universidad Nacional Mayor de San Marcos
Lima, Peru

Altaf Osman Mulani
SKN Sinhgad College of Engineering
Solapur, Maharashtra, India

ISSN 1865-0929 ISSN 1865-0937 (electronic)
Communications in Computer and Information Science
ISBN 978-3-031-88761-1 ISBN 978-3-031-88762-8 (eBook)
https://doi.org/10.1007/978-3-031-88762-8

This Springer imprint is published by the registered company Springer Nature Switzerland AG
The registered company address is: Gewerbestrasse 11, 6330 Cham, Switzerland

Preface

The rapid evolution of technology in recent decades has propelled humanity into an era dominated by data and intelligence-driven decision-making. This transformation has been marked by the convergence of cutting-edge technologies, with Artificial Intelligence (AI), Data Science, and Signal Processing standing at the forefront. These fields are not only reshaping traditional industries but are also giving rise to entirely new sectors of innovation. AI, with its ability to simulate human-like cognition, is enabling breakthroughs in areas such as robotics, natural language processing, and autonomous systems. Data Science, leveraging the power of big data and analytics, provides the tools needed to extract valuable insights from complex datasets, driving informed decisions and predictive capabilities. Signal Processing, the backbone of modern communication and multimedia systems, enhances the ability to analyze, interpret, and transform diverse signals. Together, these disciplines are paving the way for solutions to challenges that once seemed insurmountable, fostering a future of unprecedented possibilities.

With this objective in mind, the *International Conference on* Emerging Trends in Artificial Intelligence, Data Science, and Signal Processing was organized by Maharana Pratap Engineering College, Kothi, Mandhana, Kanpur, India, during October 20–21, 2023. We are thankful to our contributors and participants who supported this event wholeheartedly.

This edited book showcases the outcome of a thorough and transparent review process conducted using an open peer review model, where reviewers' identities were known. Out of around 371 submissions received, 260 papers were selected for review, each evaluated by an average of three reviewers. The process, managed through the EasyChair platform, was enhanced by its efficient communication features. Ultimately, 26 full papers and 11 short papers were accepted, reflecting the highest standards of research. The review involved dedicated program committee members and external reviewers, with each handling an average of five papers. This rigorous process ensured the quality and credibility of the work presented in this volume.

The conference was organized with four parallel technical sessions besides inaugural and valedictory sessions as the most suitable tracks capable of serving the electronics, IT, and software industries, to be specific. Few presentations by Indian and international industry specialists took place in these sessions. This was done with a view to establishing a connection between academia and industry, so that each of them could get fruitful ideas from the other. We are really thankful to all our overseas as well as Indian participants and academic experts who either joined us physically or in online mode.

Saurabh Singh
Karm Veer Arya
Ciro Rodriguez Rodriguez
Altaf Osman Mulani

Acknowledgement

I express my heartfelt gratitude to Mr. Gaurav Bhadauria, Honorable Joint Secretary of the Maharana Pratap Group of Institutions, for his unwavering support in making this event a grand success.

I am deeply grateful to Prof. (Dr.) Karm Veer Arya, Chief Technical Advisor to the conference, from ABV-Indian Institute of Information Technology, Gwalior, for his invaluable technical guidance and support, from ideation to the successful execution of AIDSP-2023.

A special note of appreciation goes to our reviewers for dedicating their valuable time and effort to evaluate over 350 submissions. Their commitment to our shared vision is truly commendable.

Lastly, I am thankful to the organizing committee, jointly coordinated by Mr. Milind Bhatt and Mr. Subhash Chandra Maurya, along with all the faculty members of the CSE Department, Maharana Pratap Engineering College, Kanpur, for their active involvement and dedication to AIDSP-2023.

Saurabh Singh

Organization

Program Committee Chairs

Singh, Saurabh	Maharana Pratap Engineering College, India
Arya, K. V.	ABV-Indian Institute of Information Technology, Gwalior, India
Rodriguez, Ciro	Universidad Nacional Mayor de San Marcos, Peru
Mulani, Altaf Osman	SKN Sinhgad College of Engineering, Pandharpur, India

Program Committee Members

García Vázquez, Mireya Saraí	IPN Tijuana, Mexico
Singh, Raghuraj	Harcourt Butler Technological University, India
Singhal, Abhishek	Amity University, Noida, India
Maurya, Subhash Chandra	Maharana Pratap Engineering College, India
Bhatt, Milind	Maharana Pratap Engineering College, India
Singh, Pooja	Galgotias University, India

Reviewers

Singhal, Abhishek	Amity University, Noida, India
Joshi, Hem Dutt	Thapar University, India
Srivastava, Prakash	Graphic Era University, India
Bajaj, Rakesh	Jaypee Institute of Information Technology, Noida, India
Shukla, Varun	Pran Veer Singh Institute of Technology, India
Narwal, Bhawna	Indira Gandhi National Technical University for Women, India
Chaudhury, Meeta	KIET, Noida, India
Rastogi, Rahul	KIET, Noida, India
Chaudhary, Tanu Priya	Symbiosis University, India
Jha, Pooja	Amity University, Ranchi, India
Mishra,Apoorv	Maharana Pratap Engineering College, India
Khan, Farah	Maharana Pratap Engineering, India
Kuashwah, Sushil	Maharana Pratap Engineering College, India

Penubaka, Balaji	K L University, India
Gangwar, Ashok	Amity University, Gwalior, India
Indian, Ajay	Central University of Rajasthan, India

Contents – Part II

Contents – Part I

Impulsive Transaction Prediction for People with Bipolar Disorder Using Supervised Learning Algorithms

Parth Kulkarni(✉)

Department of Information Technology, SCTR's Pune Institute of Computer Technology, Pune, India
parthkulkarni998@gmail.com

Abstract. Predictive Analytics have been widely used in Payment and Banking applications for transaction processing. Companies are using various algorithms to determine the credit limit and spending capacity for an individual. With the rise in data collection and data analytics advancements, this process needs to be more inclusive for people with mental disorders such as Bipolar Disorder, which can help them get control over their finances. In order to conduct this research analysis, a real-world anonymized dataset is used with few synthetic fields. To improve the effectiveness of different algorithms, concept of Dimensionality Reduction is utilized. To determine the accuracy, various Supervised Machine Learning classification algorithms have been implemented. Findings revealed that Random Forest was most accurate with an accuracy of 0.951 using the RFCEV method, while the Artificial Neural Network gave the best results with PCA of 0.913. By deploying the solution on public cloud network, it can be used as Mobile first application to help users with mental disorders monitor, analyze and control their spending in real-time. This will help in making banking and payment industry more inclusive and improve accessibility.

Keywords: Impulsive Spending Behavior · Bipolar Disorder · Machine Learning

1 Introduction

Credit Card Companies, Payment Startup Companies and Internet Banking has revolutionized the way we transact. Spread of Internet and availability of mobile devices has made it simpler, quicker, and rewarding for everyone to perform transaction on the go [1]. With the ease of the transaction, comes the challenge of overspending. Lots of people end up buying things they rarely use, or they end up in the basement filled with dust. This happens with everyone, some, or the other time. But for people with Bipolar Disorder this is a routine thing [2]. They purchase these things and then they experience depressive lows which lead to loss of motivation and enthusiasm towards life. On the other hand, they experience euphoria moments or manic, which leads them to state of

S. Singh et al. (Eds.): AIDSP 2023, CCIS 2440, pp. 1–12, 2025.
https://doi.org/10.1007/978-3-031-88762-8_1

high enthusiasm and excessive energy [3]. Manic episodes can vary from person to person, but a common challenge that most feel is the problem of involuntary rash spending. This overspending disturbs their financial plan and in turn worsened their anxiety or in some cases added to their anxiety [4]. Traditionally there have been approaches designed to determine the credit card limits for the customers [5, 6]. However, these studies never considered Impulsive Spending Disorder suffered by people with Bipolar Disorder. The goal of this research work is determining how predictive modelling techniques can be used to determine whether a transaction should be refrained from by a user considering his available past transaction data, health data. This research work has potential to improve their lives by giving financial freedom.

2 Literature Survey

Over the years, Credit Card companies have resorted to using traditional financial factors such as the individual's current earnings, assets, future growth potential, number of other credit cards possessed, for how long the user owns the card, credit card spending history etc. for determining the credit limit [7, 8]. Also, digital payment wallets use more standard ways for setting the spending limit. The user is asked to set spending limit based on their requirement. User receives alerts if the balance in their wallet falls below certain user defined threshold or if spending crosses users credit limit [9].

The challenge with the above approaches is that it is assumed that user of the application/card would be in a composed state of mind while setting their spending limit. However, this assumption does not always hold true for people with mental disorders such as Bipolar Disorder. For these individuals, factors such as mindfulness and impulsivity rather play a more critical role. Psychological factors are having high correlation with financials difficulties for people with Bipolar disorder [10, 11]. Factors such as the current stress level, anxiety level, any recent emotional impact or historic mood levels [12] also need to be considered when determining the credit limit for people with mental disorders.

Currently people dealing with mental disorders depend on their guardians, partners, or their caretakers for managing their finances. This dependency leads to tension and has an impact on their personal relationships [13]. A commonly used medical terminology for measuring self-stress test is Perceived Stress Scale [14]. Through this research work, the objective is to use technology to overcome the challenge faced by the people with Bipolar Disorder or other mental disorders to manage their finances. This will help them gain confidence to manage their finances independently in a regulated and a thoughtful manner by levering Supervised Learning algorithms.

3 Methodology

The correlation between various factors would be investigated using logical and well-defined approach mentioned ahead. Supervised Learning algorithms are used to support these objectives. Figure 1 showcase the data flow of the proposed methodology. It shows steps involved and different processes that occur in sequential manner involving steps such as Data Cleaning, Data preparation, Exploratory Data Analysis, Machine Learning prediction and visualization steps.

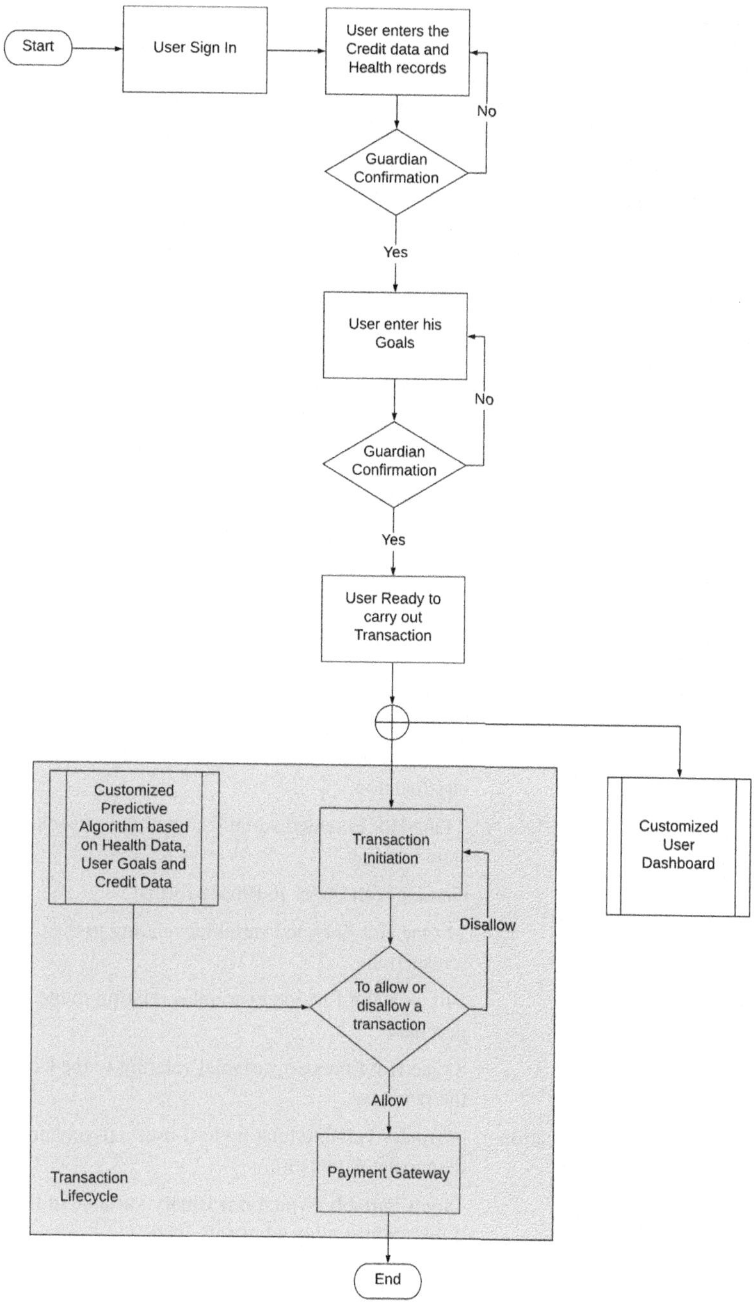

Fig. 1. System Flow Diagram

3.1 Description of Dataset

The problem statement used for this research is novel. The dataset used is a collection of Real and Synthetic records. There are 4940 records and 217 attributes in this dataset, which represents both the financial and the health aspects of an individual. The dataset consists of Categorical, Continuous, Nominal and Discrete variables. Table 1 shows the description of all the variables.

Table 1. Characteristics of Dataset

Features	Explanation
Self_rating_on_personal_health	Self-Assessed Health Score of User
Age	Normalized age of the User
Total_credit_card_limit	Normalized total credit limit of the User
Current_available_limit	Normalized total available limit to the User currently
Total_credit_liability	Normalized total credit liability of the User
Date	Date of the transaction
Amount	Normalized value of transaction
Perceived_stress_level	3 levels of Stress perceived presently by the User
Branch_number_1-23	23 One Hot Encoded variables relating to user's current bank branch
City_1-10	10 One Hot Encoded variables relating to user's Branch city location
State_1-2	2 One Hot Encoded variables relating to user's present state location
Gender	Gender of the User in Binary format
Category_expense_1-21	21 One Hot Encoded variables relating to the Category of transaction
Purchase_City_1-140	140 One Hot Encoded variables relating to the city of the purchase
purchase_country_1-8	8 One Hot Encoded variables relating to the Country of the purchase
psychiatric_diagnosis_medication	A Binary variable relating to if user's psychiatric diagnosis/ medication
Response	Target Variable, which is a Binary variable indicating if transaction is allowed or refrained

3.2 Data Preprocessing

Many factors affect the success of a ML application. The algorithms used do play a significant role in determining the effectiveness, but more important than that is the process of Data Preprocessing. Data Preprocessing involves various processes like Data Cleaning, Normalization, Transformation, Selection, etc. As its rightly said "Garbage In, Garbage Out", if we want our model to provide good result, we need to ensure the data provided to it is free from noise, redundancy, missing values and is accurately scaled.

In this step, the dataset from Brazilian Credit Card Spending is cleaned to handle missing values. Missing values is one the biggest data processing challenge that occurs more frequently than others. In real-world data, it's not always that data is available in its entirety. Some values might not be collected, missed, ignored. The missing data points in the dataset will be imputed.

3.3 Data Normalization

Data Normalization is critical to make the features dimensionless and bring them in same scale. When the features in a dataset have a varied range of values, some features might get undue importance in the model because of their large values. This provides all the features an equal importance as input and thereby allowing the model to give more importance to the features based on their variance [15]. Ex. If the value of one of the features is 0.001 and that of another feature as 1000. To bring both values in same scale so that model doesn't give unfair advantage to the later feature as compared to the prior, Normalization is performed.

3.4 Methods of Dimensionality Reduction

Dimensionality Reduction is a technique of adjusting the number of available variables in order to perform efficient modelling. Dimensionality Reduction methods can be broadly categorized into two categories – Feature Extraction and Feature Selection. Dimensionality Reduction is useful for improving the model training time and tuning model accuracy [16]. Two of the popular Dimensionality Reduction methods used for this research are Principal Component Analysis and Recursive Feature Elimination and Cross Validation. The two methods are discussed ahead [17].

Principal Component Analysis

Principal Component Analysis or PCA is used to generate a set of principal components from a set of features. These principal components are then used as input to the machine learning model as they have a higher variance ratio. More the variance, more the information it carries, and more is it relevant for the prediction task [18].

Variance shown by the principal components is highest for the first set of components and it goes down from there. PCAs represent the maximum variance in the dataset. Scholars like [19] and [20] have conducted the research and deduced that PCA is effective when used along with supervised algorithms.

Recursive Feature Elimination and Cross Validation

The Recursive Feature Elimination and Cross Validation or RFECV is a technique to remove insignificant features. Features are ranked based on feature_importance_ attribute. In every iteration a certain number of features are eliminated thereby removing collinearity. Subset of features which show more variance as compared to other subsets is selected [21].

3.5 Supervised Learning Algorithms

Different classification algorithms have been implemented on the dataset. Classification algorithms are subpart of Supervised learning algorithms where the variable i.e. the label, which is getting predicted is categorical in nature. Classification Algorithms are again subdivided into two categories – Binary Classification and Multiclass Classification. In Binary Classification there are two resultant label classes while in Multiclass Classification there are more than two resultant classes [22, 23].

As in this case, we have two classes as 'Approve' and 'Disapprove' as the output labels, hence Binary Classification Algorithms have been used. This section covers different algorithms that have been implemented on the dataset to achieve the resultant predictive model.

Decision Tree

A Decision Tree is a supervised learning algorithm that infers the dataset and generates a tree shaped decision structure using straight forward decision rules. It is used for Regression and Classification algorithms, but most used for Classification algorithms. A decision tree builds a hierarchy of decision structure based on straight-forward if-then-else logic [24].

The biggest advantage of Decision Tree is its interoperability as it works for both classification and aggression set of problems. It is a white-box technique, it has a clear binary logic representation which can be easily interpreted by humans. It works well on the large datasets and do not need primary data preparation. It inherently does feature selection, as the hierarchy of attributes reflect the importance of the attributes. The features that are on the top are the features that are more important [25].

Random Forest Classifier

An ensemble or a collection of decision trees is known as Random Forest Classifier. In case of Random Forest, a "Forest" consisting of multiple trees is generated using "bagging method" [26, 27]. Bagging improves accuracy by combining multiple models thereby increasing the overall accuracy.

Random Forest unlike Decision trees considers random feature set for classifying the datasets rather than using the most optimal one, thereby providing a more diverse range of decision trees. Random Forest during the training phase gives random importance to the features and does not give high importance to a particular set of features unlike Decision Trees [28, 29].

Logistic Regression

Logistic Regression, despite the word regression in its name, is used for classification problems. To predict the possibility of a class being one of the two values or one of the many values, Logistic Regression is used. Logistic Regression used Logistic function in its basic form to classify a dependent variable as "0" or "1".

In Binary Logistic Regression, the dependent variable has two levels which are categorical while for Multinomial Logistic Regression more than two values are modelled [30].

K Nearest Neighbor

K Nearest Neighbor in its true form is a non-parameterized classification algorithm. The Nearest Neighbor class of algorithms work very well in many classification scenarios like Dense data points present in case of Financial Datapoints [31, 32]. The principle behind K-Nearest Neighbor algorithm is that it finds a fix number of data points that are close in distance to the point under consideration. The distance can range from simple Euclidean distance to more sophisticated distances such as cosine similarity measure, Minkowsky, correlation, and Chi square [33].

Support Vector Machine

Support Vector Machine builds the model for each data-point by building a n-dimensional space. SVM does the mapping of the training examples to widen the gap between different classes thus forming a hyperplane. The examples are then mapped onto the same plane and based on what side of hyperplane they lie; their class is determined [34].

Linear Classification SVMs have two sub-categories based on the margin that divides the classes. If each datapoint impacts the hyperplane in that case, it's called as Hard Margin SVM while if some datapoints can be misclassified it's called Soft Margin SVM. Hard Margin overfits the model as each data point impacts the hyperplane whereas in case of Soft Margin, a loss function is used such as Hinge Loss to determine the hyperplane. Soft Margin SVMs are more useful in real world scenarios as it doesn't get impacted by the outliers [35].

4 Experimental Results

4.1 Data Preprocessing

Dataset had a total of 4942 records and 217 attributes which were processed using Python programming language for data preparation step.

4.2 Processing Missing Data

Features having more than 35% of data absent will be removed from analysis because of sufficient datapoints available. For other features such as purchase_city, purchase_country that have number of missing fields lesser than this will need to be treated for imputation by handling missing data.

Missing data is handled using various techniques based on the type of the missing data. Missing data can be categorical or continuous in nature. In case of categorical data, datapoints can be substituted with most common feature. Or if number of missing fields being significant, a newer category for such data points can be created. On the other hand, for continuous variables, techniques such as mean or median value replacement is utilized. Random number generation within a certain range is also considered in certain scenarios where computation of mean or median is not a viable option.

4.3 Dashboard for Visual Analytics

Microsoft Power BI is used for getting dynamic insights and visualizations. The cleaned dataset had 4942 records and 217 attributes. Microsoft Power BI provides features to import unclean dataset and perform data cleaning operation before starting with analytics using visualization. However, in current setup Data Cleanup is been performed before hand separately via Python programming and resultant cleaned dataset is imported.

Figure 2 shows different interactive charts that have been generated for a user with mental disorders such as Bipolar Disorder based on his previous spending and health history. Dashboard helps user to quickly grasp their spending patterns and take actions in their spending habits. Different charts such as Geographical map, Bar chart, Line chart and Tree map chart are used to denote spending trend, spending based on category, spending based on health and stress levels etc. for the user. This makes user take better control over his finances and manage it in a systematic way.

Currently, the dashboard is visualized in Microsoft Power BI from a browser instance, but similar visualization can be generated for a mobile first application.

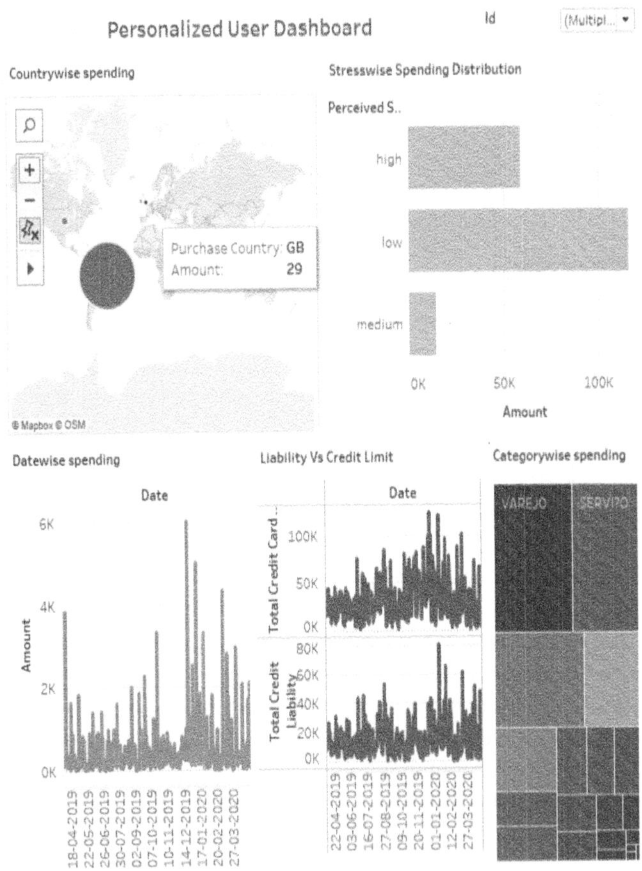

Fig. 2. User Dashboard for Analytics

4.4 Comparison Between Different Predictive Algorithms

The tests were implemented in Jupyter Notebook using Python 3 kernel. All the 217 features and 4942 records were used for implementation purpose after data preprocessing steps discussed earlier. As the greater number of features in a large dataset could lead to variety of challenges such as overfitting, large processing time and huge amount of memory requirement, Dimensionality Reduction techniques are better suited to make the model more generalized and more efficient.

All these subcategories have been tested with different supervised learning algorithms. The accuracy score for different classification algorithms was recorded. The accuracy scores were recorded in Table 2.

PCA feature extraction works in a way by assigning ranks to all the 217 features available in the dataset. The technique works by combining these features in a way that they are in a better place to predict the output label. The number of principal components to be selected was decided by taking standard deviation of feature set into consideration. A total of 50 attributes were retained by setting the cut-off threshold of 0.6.

Following this the Recursive Feature Elimination method was used to perform feature selection from the available features using Random Forest Classifier as the estimator. The validation is performed using Stratified K-fold Cross Validation, which ensures that the distribution of classes remain uniform.

The maximum accuracy score of 0.951 was achieved using RFECV method. Using Decision Tree and Artificial Neural Network, accuracy score of 0.943 and 0.938 respectively was achieved using RFECV.

However, using PCA, highest accuracy score of 0.913 was achieved using ANN. Without performing Dimensionality Reduction there is risk of overfitting. Highest score of 0.937 was achieved for Random Forest followed by Decision Tree score of 0.924 without performing Dimensionality Reduction.

Table 2. Comparison of accuracies for different Supervised Learning Algorithms

Algorithms	Accuracy Measures		
	Without Dimensionality Reduction	With Dimensionality Reduction	
	N = 217	PCA	RFECV
Decision Tree	0.924	0.841	0.943
Random Forest	0.937	0.889	0.951
Logistic Regression	0.872	0.865	0.862
K Nearest Neighbor	0.855	0.889	0.886
Support Vector Machine	0.816	0.831	0.766
Artificial Neural Network	0.921	0.913	0.938

5 Conclusions

Improving the accessibility for people with special needs and disorders is a big challenge in front of financial and banking organizations. This research work has tremendous potential to act as a baseline to start addressing this challenge for people with mental disorders who find it difficult to manage their finances by effectively utilizing predictive algorithms. Also, predictive analytics have been used successfully by credit card companies in the past, which can itself be extended further to the use of people with mental disorders such as Bipolar Disorder. This will help in truly building an inclusive and conducive environment which caters to the requirements of people with special needs thereby improving accessibility.

The data obtained from Brazilian Credit Card Spending dataset was preprocessed using Python programming language. The features that had more than 35% of misplaced data values were removed from the study.

The research showed the use of dimensionality reduction and its benefit by selecting only the important features thereby improving the accuracy, training time, and reducing challenge of over-fitting. 50 attributes were identified using PCA while 11 attributes were identified using RFECV.

The models were evaluated based on how effective they are in correct classification using K-Fold (10) Cross Validation. Findings suggested that Random Forest had the highest accuracy of 0.951, for the RFCEV method. While Decision Tree achieved a score of 0.943 using RFECV method. Conversely for the PCA method, Artificial Neural Network, Random Forest and KNN achieved the highest accuracy score of 0.913, 0.889 and 0.889 respectively. Ultimately, it can be witnessed from these results that Machine Learning algorithms are effective in this use-case and can be used for managing the finances and controlling the spending of people with mental disorders by considering various health and financial factors.

As a next step to this work, deeper analysis needs to be done with the guidance of a medical practitioner and psychiatrist community. The solution including the predictor system and the visualization dashboard currently runs locally, which can be deployed on a Public Cloud platform for universal access and all other benefits that cloud provides. It has a potential to provide social and financial benefits of the banking and payment industry to these individuals.

References

1. Bona, J.T.C.: Factors affecting the spending behavior of college students. J. Fundam. Appl. Sci. **10**(3S), 142–152 (2018)
2. Lin, H.F.: An empirical investigation of mobile banking adoption: the effect of in-novation attributes and knowledge-based trust. Int. J. Inf. Manage. **31**(3), 252–260 (2011)
3. Belmaker, R.H.: Bipolar disorder. N. Engl. J. Med. **351**(5), 476–486 (2004)
4. Kersten, M., Cox, C.R., Van Enkevort, E.A., Arrowood, R.B.: The influence of money-related metaphors on financial anxiety and spending. Metaphor. Symb. **34**(4), 229–242 (2019)
5. Licari, J., Loiseau-Aslanidi, O.L.G.A., Tolstova, V., Sadat, M.: Determining the optimal dynamic credit card limit (2021)

6. Ru, H., Schoar, A.: Do credit card companies screen for behavioral biases? (No. w22360). National Bureau of Economic Research (2016)
7. Teoh, W.M.Y., Chong, S.C., Yong, S.M.: Exploring the factors influencing credit card spending behavior among Malaysians. Int. J. Bank Mark. (2013)
8. Gan, C.E., Cohen, D.A., Hu, B., Tran, M.C., Dong, W., Wang, A.: The relationship between credit card attributes and the demographic characteristics of card users in China. Int. J. Bank Mark. (2016)
9. Upadhayaya, A.: Electronic Commerce and E-wallet. Int. J. Recent Res. Rev. **1**(1), 37–41 (2012)
10. Richardson, T., Jansen, M., Fitch, C.: Financial difficulties in bipolar disorder part 1: longitudinal relationships with mental health. J. Ment. Health **27**(6), 595–601 (2018)
11. Richardson, T., Jansen, M., Fitch, C.: Financial difficulties in bipolar disorder part 2: psychological correlates and a proposed psychological model. J. Ment. Health (2019)
12. Mansell, W., Brawn, P., Griffiths, R., Silver, B., Tai, S.J.: Bipolar disorders and problematic mood swings. In: The Complete CBT Guide for Depression and Low Mood (2015)
13. Birmaher, B., et al.: Comparison of manic and depressive symptoms between children and adolescents with bipolar spectrum disorders. Bipolar Disord. **11**(1), 52–62 (2009)
14. Denovan, A., Dagnall, N., Dhingra, K., Grogan, S.: Evaluating the perceived stress scale among UK university students: implications for stress measurement and management. Stud. High. Educ. **44**(1), 120–133 (2019)
15. Jo, J.M.: Effectiveness of normalization pre-processing of big data to the machine learning performance. J. Korea Inst. Electron. Commun. Sci. **14**(3), 547–552 (2019)
16. Ghodsi, A.: Dimensionality reduction a short tutorial. Department of Statistics and Actuarial Science, Univ. of Waterloo, Ontario, Canada, vol. 37, no. 38, p. 2006 (2006)
17. Khalid, S., Khalil, T., Nasreen, S.: A survey of feature selection and feature extraction techniques in machine learning. In: 2014 Science and Information Conference, pp. 372–378. IEEE (2014)
18. Taguchi, Y., Iwadate, M., Umeyama, H.: Principal component analysis-based unsupervised feature extraction applied to in silico drug discovery for posttraumatic stress disorder-mediated heart disease. BMC Bioinf. **16**, 1–26 (2015)
19. Shi, X., Guo, Z., Nie, F., Yang, L., You, J., Tao, D.: Two-dimensional whitening reconstruction for enhancing robustness of principal component analysis. IEEE Trans. Pattern Anal. Mach. Intell. **38**, 2130–2136 (2016)
20. Yi, S., Lai, Z., He, Z., Cheung, Y.M., Liu, Y.: Joint sparse principal component analysis. Pattern Recogn. **61**, 524–536 (2017)
21. Misra, P., Yadav, A.S.: Improving the classification accuracy using recursive feature elimination with cross-validation. Int. J. Emerg. Technol **11**(3), 659–665 (2020)
22. Umadevi, S., Marseline, K.J.: A survey on data mining classification algorithms. In: 2017 International Conference on Signal Processing and Communication (ICSPC), pp. 264–268. IEEE (2017)
23. Sen, P.C., Hajra, M., Ghosh, M.: Supervised classification algorithms in machine learning: a survey and review. In: Emerging Technology in Modelling and Graphics, pp. 99–111. Springer, Singapore (2020)
24. Charbuty, B., Abdulazeez, A.: Classification based on decision tree algorithm for machine learning. J. Appl. Sci. Technol. Trends **2**(01), 20–28 (2021)
25. Madeh, P.S., El-Diraby, T.E.: Role of data analytics in infra-structure asset management: overcoming data size and quality problems. J. Transp. Eng. Part B: Pavements **146**(2), 04020022 (2020). https://doi.org/10.1061/JPEODX.0000175
26. Shaik, A.B., Srinivasan, S.: A brief survey on random forest ensembles in classification model. In: International Conference on Innovative Computing and Communications, pp. 253–260. Springer, Singapore (2019)

27. Ali, J., Khan, R., Ahmad, N., Maqsood, I.: Random forests and decision trees. Int. J. Comput. Sci. Issues (IJCSI) **9**(5), 272 (2012)
28. Resende, P.A.A., Drummond, A.C.: A survey of random forest based methods for intrusion detection systems. ACM Comput. Surv. (CSUR) **51**(3), 1–36 (2018)
29. Dastile, X., Celik, T., Potsane, M.: Statistical and machine learning models in cred-it scoring: a systematic literature survey. Appl. Soft Comput. **91**, 106263 (2020)
30. Hussein, A.S., Khairy, R.S., Najeeb, S.M.M., ALRikabi, H.T.: Credit card fraud detection using fuzzy rough nearest neighbor and sequential minimal optimization with logistic regression. Int. J. Interact. Mob. Technol. **15**(5) (2021)
31. Abdelmoula, A.K.: Bank credit risk analysis with k-nearest-neighbor classifier: case of Tunisian banks. Account. Manage. Inf. Syst. **14**(1), 79 (2015)
32. Khodabakhshi, M., Fartash, M.: Fraud detection in banking using kNN (k-nearest neighbor) algorithm. In: International Conference on Research in Science and Technology (2016)
33. Chomboon, K., Chujai, P., Teerarassamee, P., Kerdprasop, K., Kerdprasop, N.: An empirical study of distance metrics for k-nearest neighbor algorithm. In: Proceedings of the 3rd International Conference on industrial Application Engineering, pp. 280–285 (2015)
34. Pisner, D.A., Schnyer, D.M.: Support vector machine. In: Machine Learning, pp. 101–121. Academic Press (2020)
35. Cervantes, J., Garcia-Lamont, F., Rodríguez-Mazahua, L., Lopez, A.: A compre-hensive survey on support vector machine classification: applications, challenges and trends. Neurocomputing **408**, 189–215 (2020)

A Novel Transformer-Based Approach for Portfolio Allocation Using Attention Mechanism

Laksheeta Pandit(✉), Abhishek Bhatt, and Shubhangi Tidake

Symbiosis Skills and Professional University, Kiwale, Pune 412101, India
laksheetapandit@gmail.com

Abstract. Ever since data science is growing so has its application. The portfolio allocation problem can be solved with ease using the Data Science framework. As there is no generalized method for portfolio allocation, the Modern Portfolio Theory (MPT) formulated by Markowitz has been a benchmark and widely used model (mean-variance). This model limits due to the highly stochastic nature of the market, whereas the existing model such as deep deterministic policy, random forest and other machine learning-based models lack in adapting to the stochastic market nature. In this study, a novel technique is proposed which is inspired by the success of the attention mechanism in natural language processing. The proposed model can quickly adapt to the changing market with its full encoder-decoder architecture, encoded layers, and gated components. It is trained and tested on three datasets i.e., Stock, ETFs & commodities. The accuracy of the model is checked on the widely used and industry go-to parameter Sharpe ratio, which measures the returns on risk. It has been found that using the proposed approach the sharp ratio, return and positive return % have been improved.

Keywords: Portfolio Allocation · Modern Portfolio Theory · Attention Mechanism · Transformer · Stock Price

1 Introduction

Portfolio allocation is an investment strategy that will allow an individual to balance risk with good returns by dividing investments among different assets such as stocks, bonds, metals, etc. The is no standard method to do portfolio allocation. Most individuals follow the "60/40" rule, a well-known fixed allocation strategy where 60% of the portfolio is for equities (stocks) and 40% for bonds. Another popular method is equal weights in which all assets in consideration are given equal importance. This method is simple and reliable, yet portfolio allocation has room for improvement. Markowitz's portfolio allocation theory, Modern Portfolio Theory (MPT) [1], is a go-to tool for most industry practitioners. However, the popular theory has its drawback when applied in practice.

The advancement in technology and wide availability of data has allowed us to use machine learning for portfolio allocation and optimization techniques. For example, random forests [2] gradient-boosted trees [3], the XGBoost model [4] and many more.

S. Singh et al. (Eds.): AIDSP 2023, CCIS 2440, pp. 13–22, 2025.
https://doi.org/10.1007/978-3-031-88762-8_2

The growing utilization of artificial intelligence and deep learning methodologies has brought about significant advancements, resulting in state-of-the-art outcomes across diverse domains. As a result, the number of deep learning applications for portfolio allocation has skyrocketed [5]. Specifically, researchers have integrated recurrent neural networks (RNNs) [6], long short-term memory (LSTM) networks [7], and deep deterministic policy gradient (DDPG) [8] into portfolio selection models, enabling improved portfolio allocation and optimization. The primary objective of any technique is to optimize the Sharpe ratio [9], which is a commonly employed measure of risk-adjusted return in practical applications. Advanced techniques like a transformer, GRN, etc. can be implemented for portfolio allocation.

Portfolio transformer [10] is the latest technique, a variation of the transformer architecture commonly used in natural language processing tasks. The result of the Portfolio Transformer demonstrates that it outperforms several other methods. In this paper, the technique used a novel attention mechanism with Bidirectional LSTM and GRU as a combination. The output of the model gives portfolio weights which are used to calculate the Sharpe ratio, which measures the returns on risk on three different datasets.

2 Background and Related Work

2.1 Time2Vec Embedding

The Time2Vec Embedding [11] was introduced to address the problem of learning with time to capture the underlying patterns and dependencies in time-series data. Time2Vec propose to represent time as a vector in a continuous vector space. The Portfolio Transformer implements time encoding, which takes the following form

$$Time2Vec(t)[i] = \begin{cases} w_i t + \varphi_i & if\ i - 0 \\ \sin(w_i t + \varphi_i) & if\ 1 \leq i \leq k \end{cases} \tag{1}$$

Equation (1) decomposes the temporal signal denoted by t into various frequencies ω and corresponding phase shifts φ_i. This time decomposition method shares a strong connection with Fourier transforms. However, unlike Fourier transforms that rely on pre-determined values, this technique utilizes learnable parameters to capture all frequencies and phase shifts. The time decomposition method is strongly linked to Fourier transforms; however, unlike Fourier transforms which use a predetermined set of values, this technique employs learnable parameters to capture all frequencies and phase shifts.

2.2 Transformer Model

The Transformer architecture [12] was introduced to overcome the limitations of LSTM in capturing significant dependencies. This model has emerged as the state-of-the-art method in numerous natural language processing tasks. In the field of finance, researchers have utilized the Transformer model to enhance various applications. For instance, the Momentum Transformer [13] has demonstrated superior performance compared to the LSTM model, while the Relation-Aware Transformer [14] leverages the Transformer model for portfolio policy learning within a reinforcement learning framework (Fig. 1).

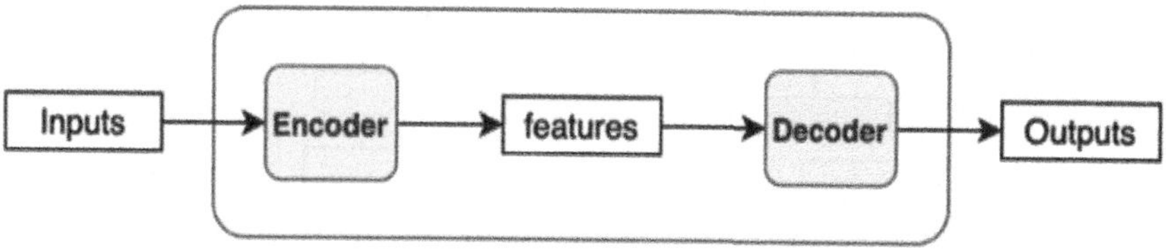

Fig. 1. The Transformer

2.3 Attention Mechanician

At the core of the Transformer architecture lies a pivotal mechanism known as "self-attention" [15]. In the context of Portfolio Transformer, the implementation involves "scale dot-product attention" [12], which can be represented as follows:

$$Attention(Q, K, V) = softmax\left(\frac{QK^T}{\sqrt{d_{model}}} + M\right)V \quad (2)$$

Value matrices $V \in R^{\tau \times d_v}$ in Eq. (2) are weighted using "scores" derived from softmax operations. These scores determine the level of emphasis each time step from the key matrix $K \in R^{\tau \times d_k}$ should receive during the encoding of sequence positions from the query matrix $Q \in R^{\tau \times d_k}$.. By dividing the dot product of Q and K by the square root of the encoding dimension (d_{model}), minor gradient issues are mitigated. Additionally, each decoder layer begins by masking the attention block using a matrix $M \in R^{\tau \times \tau}$.. By masking the previous time steps, the model is able to remain autoregressive. A multi-headed attention (MHA) method repeats the operation defined in Eq. (2) h times. MHA allows the model to capture information from multiple representations simultaneously, which helps improve its performance. Finally, the output of the MHA is concatenated and used as input to a feed-forward network.

$$MHA(Q, K, V) = Concatenate(head_1, \ldots.., head_h)W^O \quad (3)$$

$$head_i = Attention\left(QW_i^Q, KW_i^K, VW_i^V\right) \quad (4)$$

It allows the model to extract information from multiple representation subspaces. A set of linear projection matrices is learned for each attention head, denoted $head_i$: $W_i^Q \in R^{d_{model} \times d_k}$, $W_i^K \in R^{d_{model} \times d_k}$ and $W_i^V \in R^{d_{model} \times d_v}$. In order to obtain the final value, all attention heads' outputs are concatenated and linearly projected using a learned parameter matrix $W^O \in R^{hd_v \times d_{model}}$.

The attention mechanism helps in finding the small pattern in the data which are overlooked by other models. Hence attention mechanism is more accurate than the other model (Fig. 2).

2.4 Bidirectional LSTM

A simple Feed-forward network follows each attention layer in the original Transformer model [12]. However, in the Portfolio Transformer [10] they use a gated residual network

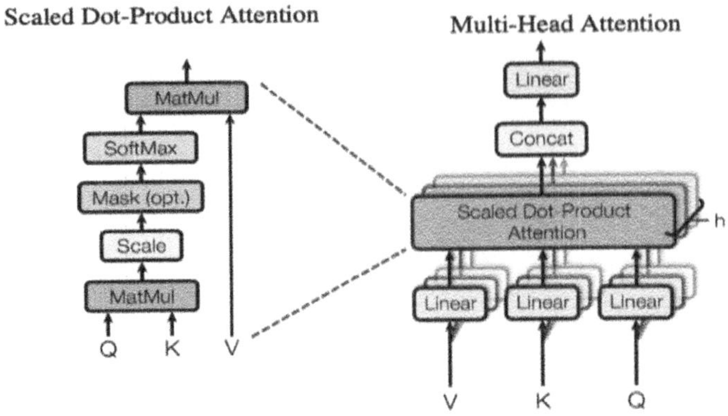

Fig. 2. Attention Mechanician

[16] instead. Moreover, in the Portfolio Transformer, the encoder layer incorporates a Bidirectional LSTM [17]. The Bidirectional LSTM is a variant of the LSTM model that processes data in both the forward and backward directions. Unlike a regular LSTM, which can only process input in a single direction, the Bidirectional LSTM allows for bidirectional information flow. However, in a bi-directional LSTM, the input can be processed in both directions simultaneously, allowing the model to capture and preserve information from both the past and future contexts

$$BLSTM\,(z) = Bi(LSTM\,(z)) \tag{5}$$

$$X = LayerNorm(z + BLSTM\,(z)) \tag{6}$$

In the above input is given by the vector z and the output is stored in the vector X. This will enable simultaneous comprehension of past and future information.

2.5 GRU

The gated residual network (GRU) is used at the decoder layer as the gate mechanism responsible for determining the amount of non-linear processing required for a given portfolio. This functionality is represented by the equation:

$$GRU\,(z) = LayerNorm(z + GRU\,(g_1)) \tag{7}$$

$$g_1 = W_1 g_2 + b_1 \tag{8}$$

$$g_2 = ELU\,(W_2 z + b_2) \tag{9}$$

The input to the Gated Residual Network (GRN) is represented by the vector z, g_1 and g_2 correspond to intermediate layers. The activation function Exponential Linear Unit (ELU) [18] is used here. A gated linear unit (GLU) [19] is used to filter out non-linear contributions. The GLU allows the Portfolio Transformer to adjust the level of non-linear processing, allowing for a simpler model when dealing with scenarios such as small or noisy datasets.

3 Methodology

3.1 Dataset

The model is trained on three datasets that include daily price data. The first dataset includes four Exchange Traded Funds (ETFs) [6] used in the LSTM-based experiment: Aggregate Bond Index (AGG), Commodity Index (DBC), Volatility Index (VIX), and US Stock Index (VTI). The second data set consists of commodities such as metals, agricultural products and energy commodities [7] such as oil and natural gas. The final dataset includes 500 US-based stocks listed on NASDAQ[8]. It should be noted that all datasets are from 2006.

3.2 Training and Model Calibration

The dataset is divided into a training set and a testing set based on a temporal split: data before 2018 is used for training, while data from 2018 onwards is used for testing. The portfolio positions are adjusted daily, taking into account a transaction cost rate of 2 basis points in the valuation. The Portfolio Transformer is trained using a mini-batch stochastic gradient descent, with the batch size specified in the hyperparameters. The Adam optimizer [20] is used for model optimization. To reduce overfitting and facilitate model calibration, 10% of each training segment is reserved as a separate validation dataset. The Portfolio Transformer was developed with TensorFlow.

3.3 Portfolio Transformer Architecture

The Portfolio Transformer [4] proposed by Kisiel D. and Gorse D. is shown in Fig. 3. It consists of four main building blocks: input (Time2Vec embedding, Sect. 2.1), encoder and decoder (Sect. 2.2, Sect. 2.5) where it uses GRN layer instead of feed-forward network and output layer. Keeping the architecture, the same instead of GRN in the encoder layer use Bidirectional LSTM (Sect. 2.4) and added Time Distributed layer [21] before the output layer as shown in Fig. 4. Each of the blocks will be discussed below.

Input Layer: In Fig. 4, each sequence position represented by the vector x_t contains the chained returns of all N assets for a given portfolio on day t. For both encoder and decoder, τ vectors are stacked to form the input matrix X with dimensions (τ, N). To capture time-coded cross-sectional features, the Portfolio Transformer integrates Time2Vec embedding.

Encoder Layer: The model employs four identical encoder layers. Within each layer, the time-coded input is subjected to multi-head attention, where the number of attention heads, denoted h, is determined as a hyperparameter. Bidirectional LSTM is then applied to extract information from both directions. To facilitate learning, a residual connection [22] is utilized, followed by layer normalization [23], for both the multi-head attention and Bidirectional LSTM sub-layers. This structure is illustrated in Fig. 4.

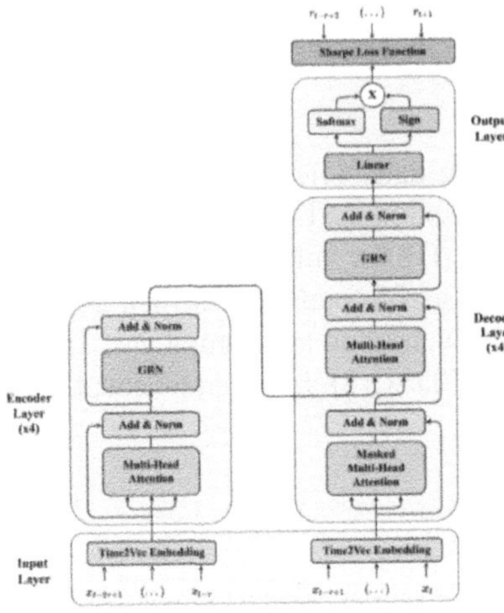

Fig. 3. Original Portfolio Transformer

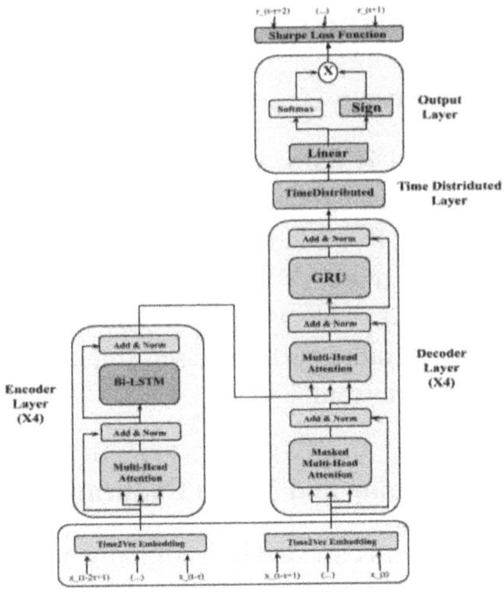

Fig. 4. Optimized Portfolio Transformer

Decoder Layer: The decoder block is composed of four identical layers, each containing two multi-head attention modules. The initial approach involves the application of masking, which guarantees that the predictions made by the Portfolio Transformer solely depend on information from the previous time steps. The second one allows the decoder to process the output from the encoder stack, which provides a much more nuanced representation of the data. A series of GRN modules is applied to determine the ideal amount of non-linearity and it is also used to remove any unnecessary complexity. Following every sub-layer, a residual connection is employed, which is then succeeded by layer normalization.

Time Distributed Layer: The TimeDistributed layer[9] allows us to apply a specific layer, such as a Dence layer or an LSTM layer, to each time step the output sequence. This helps the decoder to model that temporal dependencies in the output sequence and generate more accurate predictions.

Output Layer: The output layer employed in the model is the same as that used in the Portfolio Transformer [10]. It implements one of Zhang et al. proposed composite function [24]. The output of the time-shared layer is first routed through a fully connected layer for processing. The resulting vector $(s_{i,t})$ is then used to calculate the final portfolio weights, as shown in the following equation:

$$w_{i.t} = \sin(s_{i,t}) \times softmax(s_{i,t}) \triangleq \sin(s_{i,t}) \times \frac{e^{s_{i,t}}}{\sum_{j=0}^{N} e^{s_{j,t}}} \tag{10}$$

The softmax operator used in the above equation allows the weight to be positive and negative. The sum of their absolute values is always equal to one.

Sharpe Loss Function: Implementing customized loss function as we are to determine the max returns with low risk, for this purpose we use sharpe ratio[9] which is described as follows:

$$SR = \frac{E(R_P)}{\sqrt{E\left({R_P}^2\right) - (E(R_P))^2}} \tag{11}$$

where, the shaper ratio (SR) is expected returns ($E(R_p)$) divided by its volatility, R_P is portfolio return. The expected returns are computed by averaging all portfolio returns over a trading period of length τ:

$$E(R_P) = \frac{1}{\tau}\sum\nolimits_{t=1}^{\tau} R_{P,t} \tag{12}$$

To mitigate the impact of transaction costs on allocation strategies and reduce when turnover is high, the Portfolio Transformer considers cost-adjusted portfolio returns. This adjustment is represented by the following expression:

$$R_{P,t} = \sum\nolimits_{i}^{N} w_{i,t-1} \times r_{i,t} - C \times \sum\nolimits_{i}^{N} \left|w_{i,t-1} - w_{i,t-2}\right| \tag{13}$$

In the given expression, C represents a constant rate of cost, usually set to a realistic value of two basis points (2 basis points). The weight of asset i on day t-1, denoted by $w_{i,t-1}$ and $r_{i,t}$, represents the arithmetic return of asset i on day t-1 to t The calculation of $r_{i,t}$ can be performed using the following equation:

$$r_{i,t} = \frac{P_{i,t} - P_{i,t-1}}{P_{i,t-1}} \tag{14}$$

The portfolio positions are constrained within the range of $w_{i,t} \in [-1,1]$. Additionally, the sum of the absolute values of all portfolio weights, $\sum_{i}^{N}\left|w_{i,t}\right| = 1$, These constraints are enforced by utilizing the compound function described in Eq. (10).

4 Results

The Portfolio transformer was compared to the benchmark algorithms such as Mean-Variance [3], XGBoost, LSTM, etc. from which the Portfolio Transformer outperform all of them. Hence comparing our Portfolio Transformer (encoder with BiLSTM and decoder with GRU) (PT2) to the Portfolio Transformer (encoder-decoder with GRU) (PT1). The performance of the model is evaluated based on three parameters: portfolio risk level measured by annualized volatility (Vol.), profitability assessed by annualized returns (Returns) and percentage of positive returns (%). In addition, the risk-adjusted performance is evaluated using the annualized Sharpe Ratio. Table 1 presents the results obtained from various algorithms and datasets.

It has been observed from Fig. 5 that proposed approaches outperforms Portfolio Transformer (PT1) for ETFs, Commodities and Stocks dataset in terms of Returns [0.015, 0.019, 0.037], + Returns (%) [0.06, 0.06, 0.062], Sharpe Ratio [0.277, 0.166, 0.22] and Volatility in Commodities dataset [0.012]. Except for Volatility for ETFs [−0.007, −0.017] where Volatility should be minimum.

Table 1. Accuracy Table

	Vol	Returns	+ Returns (%)	Sharpe Ratio
ETFs				
PT1 [4]	**0.067**	0.138	0.548	2.252
PT2 [Proposed Approach]	0.074	**0.153**	**0.608**	**2.529**
Commodities				
PT1[4]	0.108	0.174	0.543	1.506
PT2 [Proposed Approach]	**0.096**	**0.193**	**0.603**	**1.672**
Stocks				
PT1 [4]	0.130	0.334	0.566	2.001
PT2 [Proposed Approach]	0.147	**0.371**	**0.628**	**2.221**

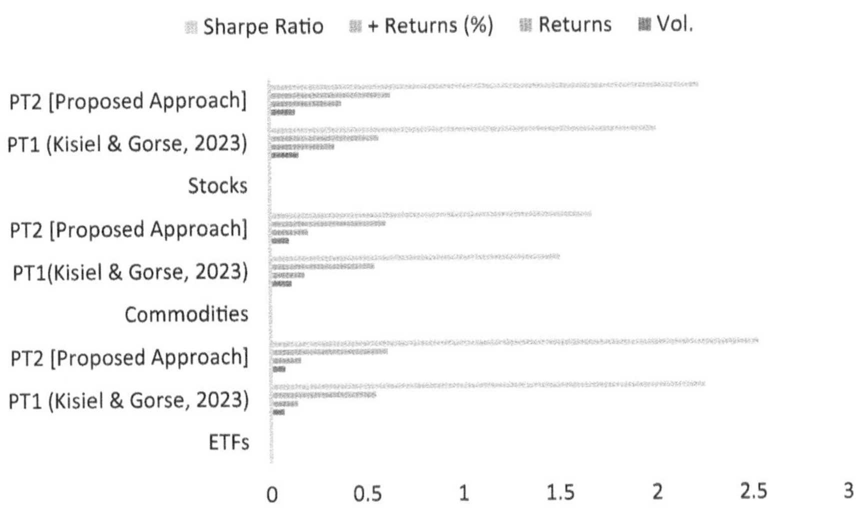

Fig. 5. Accuracy Assessment

5 Conclusions

This paper introduces the Portfolio Transformer (PT2), an attention-based architecture with Bidirectional LSTM, GRU networks, and time-encoding layers. The PT2 model directly optimizes risk-adjusted returns and incorporates transaction costs into the loss function to account for trading cost constraints. The experimental results show that PT2 achieves superior risk-adjusted performance as compared to PT1. The Proposed approach (PT2) outperforms benchmark algorithms on different datasets with different portfolio sizes. This improvement is due to the Portfolio Transformer's full encoder-decoder configuration and attention mechanism, which allows it to learn long-term dependencies and respond more effectively to changing market conditions.

6 Future Work

To further enhance the current model, future research could explore alternative attention mechanisms beyond the scaled dot-product attention presently employed. Additionally, it is worth noting that while the Sharpe ratio is widely used by industry professionals, there are numerous other objective features that could be tweaked to assess portfolio performance. Subsequent studies will study alternative metrics and their effect on overall portfolio performance. Furthermore, exploring alternative network architectures other than Bidirectional LSTM/GRU could be beneficial in optimizing the model.

References

1. Markowitz, H.: Portfolio selection. J. Financ. **7**(1), 77 (1952). https://doi.org/10.2307/2975974
2. Pinelis, M., Ruppert, D.: Machine learning portfolio allocation. J. Financ. Data Sci. **8**, 35–54 (2022). https://doi.org/10.1016/j.jfds.2021.12.001
3. Chen, W., Zhang, H., Mehlawat, M.K., Jia, L.: Mean–variance portfolio optimization using machine learning-based stock price prediction. Appl. Soft Comput. **100** (2021). https://doi.org/10.1016/j.asoc.2020.106943
4. Kisiel, D., Gorse, D.: A meta-method for portfolio management using machine learning for adaptive strategy selection. In: 2021 the 4th International Conference on Computational Intelligence and Intelligent Systems, pp. 67–71. ACM, New York (2021). https://doi.org/10.1145/3507623.3507635
5. Durall, R.: Asset allocation: from markowitz to deep reinforcement learning. SSRN Electron. J. (2022). https://doi.org/10.2139/ssrn.4148379
6. Lin, C.M., Huang, J.J., Gen, M., Tzeng, G.H.: Recurrent neural network for dynamic portfolio selection. Appl. Math. Comput. **175**(2), 1139–1146 (2006). https://doi.org/10.1016/j.amc.2005.08.031
7. Zhang, Z., Zohren, S., Roberts, S.: Deep learning for portfolio optimization. J. Financ. Data Sci. **2**(4), 8–20 (2020). https://doi.org/10.3905/jfds.2020.1.042
8. Chaouki, A., Hardiman, S., Schmidt, C., Sérié, E., de Lataillade, J.: Deep deterministic portfolio optimization. J. Financ. Data Sci. **6**, 16–30 (2020). https://doi.org/10.1016/j.jfds.2020.06.002
9. Sharpe, W.F.: The Sharpe Ratio. J. Portfolio Manage. **21**(1), 49–58 (1994). https://doi.org/10.3905/jpm.1994.409501
10. Kisiel, D., Gorse, D.: Portfolio transformer for attention-based asset allocation, pp. 61–71 (2023). https://doi.org/10.1007/978-3-031-23492-7_6
11. Kazemi, S.M., et al.: Time2Vec: learning a vector representation of time (2019). http://arxiv.org/abs/1907.05321
12. Vaswani, A., et al.: Attention is all you need (2017). https://doi.org/10.48550/arXiv.1706.03762
13. Wood, K., Giegerich, S., Roberts, S., Zohren, S.: Trading with the momentum transformer: an intelligent and interpretable architecture (2021). http://arxiv.org/abs/2112.08534
14. Xu, K., Zhang, Y., Ye, D., Zhao, P., Tan, M.: Relation-aware transformer for portfolio policy learning. In: Proceedings of the Twenty-Ninth International Joint Conference on Artificial Intelligence, California: International Joint Conferences on Artificial Intelligence Organization, pp. 4647–4653 (2020). https://doi.org/10.24963/ijcai.2020/641
15. Sukhbaatar, S., Grave, E., Lample, G., Jegou, H., Joulin, A.: Augmenting self-attention with persistent memory (2019). http://arxiv.org/abs/1907.01470

16. Lim, B., Arik, S.O., Loeff, N., Pfister, T.: Temporal fusion transformers for interpretable multi-horizon time series forecasting (2019)
17. Schuster, M., Paliwal, K.K.: Bidirectional recurrent neural networks. IEEE Trans. Signal Process. **45**(11), 2673–2681 (1997). https://doi.org/10.1109/78.650093
18. Clevert, D.-A., Unterthiner, T., Hochreiter, S.: Fast and accurate deep network learning by exponential linear units (ELUs) (2015)
19. Dauphin, Y.N., Fan, A., Auli, M., Grangier, D.: Language modeling with gated convolutional networks (2016)
20. Kingma, D.P., Ba, J.: Adam: a method for stochastic optimization (2014). http://arxiv.org/abs/1412.6980
21. Graves, A., Fernández, S., Gomez, F., Schmidhuber, J.: Connectionist temporal classification. In: Proceedings of the 23rd international conference on Machine learning - ICML 2006, pp. 369–376. ACM Press, New York (2006). https://doi.org/10.1145/1143844.1143891
22. He, K., Zhang, X., Ren, S., Sun, J.: Deep residual learning for image recognition. In: 2016 IEEE Conference on Computer Vision and Pattern Recognition (CVPR), pp. 770–778. IEEE (2016). https://doi.org/10.1109/CVPR.2016.90
23. Ba, J.L., Kiros, J.R., Hinton, G.E.: Layer Normalization (2016)
24. Zhang, C., Zhang, Z., Cucuringu, M., Zohren, S.: A universal end-to-end approach to portfolio optimization via deep learning (2021)

Query Expansion Using Pseudo-relevance Feedback on Ad Hoc

Parul Kalra[1(✉)], Deepti Mehrotra[1], and Abdul Wahid[2]

[1] Amity University Uttar Pradesh, Noida, India
parulkalra18@gmail.com
[2] Maulana Azad National Urdu University, Gachibowli, Hyderabad, India

Abstract. A well-established and extensively used method in the field of information retrieval is query expansion (QE). By updating the query definition, it improves the relevant documents that are returned. Some data requirements are challenging to meet because users' queries are frequently brief and imprecise due to dialect ambiguity. In order to address this problem, query restructure and expansion techniques have been created. The weighted Divergence from Randomness (DFR) models were compared and analyzed by the authors using Terrier 4.1 on the TREC ad hoc datasets. According to the results, the relevant documents that were located using the QE approaches had substantially higher recall and precision than the DFR models used as a baseline.

Keywords: Query expansion · Terrier · Information retrieval · DFR · Pseudo relevance feedback · Recall · Precision · Automatic methods · TREC

1 Introduction

Information Retrieval (IR) systems are designed to give users relevant information based on their needs, which are often expressed through a query. The two main stages of the query formulation process are indexing, which analyzes the documents using IR techniques, and matching, which further associates the phrases and keywords in the indexed articles with the queries.

As stated, objective of the IR techniques is to rank the relevant documents as per the users' need. Some of the retrieval models are Boolean, language, statistical and probabilistic models. The probabilistic models are of two types: weighing models and field models. The Divergence from Randomness (DFR) models are based on probabilistic weighing models which gives high precision and recall as compared to baseline models. A search engine maps the queries and the documents based on the similarities between them. In the term weight IR techniques recall and the precision is generally very low as the query formed by the user is short and vague [1].

To face this challenge, there are several strategies that the IR system considers to enhance the query. One of the most effective methods is to elaborate the query which is known as local feedback or pseudo feedback. It was introduced by Attar and Frankel. In this approach top document retrieved on submission of query to search engine are

S. Singh et al. (Eds.): AIDSP 2023, CCIS 2440, pp. 23–32, 2025.
https://doi.org/10.1007/978-3-031-88762-8_3

extracted and are considered as input feedback. More keywords are extracted from these documents and based on their significance measure they added back to query. Various query expansion approaches are proposed in literature by carrying the ranking and selection of keywords from document considering for feedback [2]. The terms from the top-ranking papers in the first ranking results are chosen by the local feedback, and the resulting phrases are then added to the initial search together with a weight that has been allocated to them. The Rocchio algorithm is employed in this method to determine how much weight each phrase in the expands [3].

Query Expansion (QE) techniques are merged with probabilistic models which give the improved results in IR system. In this paper effect of different distribution based selection approached mainly: Kullack-Leibler (KL), Boss-Einstein (Bo1, Bo2) and Binomial Approximation (BA) when used with 10 popular DFR models is studied considers probability distribution [4].

This paper pursues two objectives: Finding the best QE model based on a thorough review of the data is the first goal. Comparing the baseline (DFR without QE models) versus DFR with QE models is the second goal [4, 6].

The rest of the parts of the document are arranged in the following way: There is a literature review in Sect. 2. Present the evaluation framework and QE techniques in Sect. 3. Results are covered in Sect. 4 before the conclusion is presented in Sect. 5.

2 Literature Review

QE is an efficient way for analysis of a document and collection which in turn retrieves top most relevant terms from the corpus. Xu and Croft categorized QE in two different ways, first is local method and the second is global method. Local method is the study of a document from the initial query [5, 7]. The global method maps the relationship between the term and the corpus. Global method uses term clustering, latent semantic analysis and similarity terms.

The following steps make up the query transformation and ranking technique:

1. Based on the top n documents in the TF-IDF ranking, determine the expanded query terms T_q.
2. Calculate the enlarged query QE = W_1, W_2, W_3,…,W_q using T_q, where W_i is a weighting factor corresponding to the similarity between the original query Q and the term T.
3. Use QE to conduct the TF-IDF search.

Pseudo Relevance Feedback (PRF) was first suggested by Rocchio 1971. Rocchio method has been derived from vector space model (VSM). The main motivation of QE is to add relevant terms which helps the user to eliminate the ambiguity of the human language. This method re-computes the weighted terms that occurs in highly relevant documents which give positive response but for the original query as the weight of the terms have not yet been computed, it gives non-relevant documents. Thus, it is important to know initially whether QE would be efficient in our considered scenario. PRF is often simple but it is not efficient for user based feedback [8, 9].

By recommending the idea of proximity-based phrase frequency, which emphasizes the close proximity of terms instead of positioning information, Miao et al. offered amplification of the Rocchio's approach. The following steps make up the query transformation and ranking technique:

Several weighting models have been proposed by many researchers in IR. Since each query behaves differently the optimal parameter setting is different for each of them. These models assume the in formativeness of the terms which are more relevant in corpus. Kullback-Leibler divergence, the binomial distribution, and the Bose-Einstein 1 and 2 concepts are instances of DFR models [10, 11]. While performing QE in the DFR, the proposed keywords are arranged accordingly to their contents of information provided in the query Q. Table 1 explains the 11-probabilistic index term weighting models and their description for the frequency distributions of (query) terms.

Table 1. The 11-probabilistic index term weighting models and their description for the frequency distributions of (query) terms

Weighted DFR Model	Description
DPH	Hypergeometric distribution
XSqrA_M	the multinomial M-based XSqrA_M weighting model, which calculated the inner product of Pearson's X2 and the information growth
DFReeKLIM	DFR model free from parameters. DFReeKLIM is the inner product of two KL Information Measures
Js_KLs	The Jefreys' divergence and the Kullback Leibler's divergence combine to form the Js_KLs weighting model
BB2	With the Bernoulli after-effect and normalization, the Bernoulli-Einstein model
In_expB2	Model for inverse expected document frequency with normalization and the Bernoulli after-effect 2
InB2	Method for inverse document frequency with standardization and the Bernoulli after-effect 2
IFB2	Inverse Term Frequency model with Bernoulli after-effect and normalization 2
DFRee	Hypergeometric distribution
In_expC2	Inverse Expected Document Frequency model incorporating Bernoulli after-effect and normalization 2. Alternatively referred to as the In_expB2 weighting method
DLH13	Hypergeometric distribution

Through our experiments, we find the optimal values for normalization parameters by selecting the one that gives the highest Precision@0 (p@n). Table 2 explains the strengths of Query Expansion models [13,14].

Table 2. Query Expansion Models with their strengths

Query Expansion Model	Description	Strength
Kullback Leibler's (KL)	The Kullback-Leibler divergence for query expansion	Kullback-Leibler is the most effective for achieving a good early precision
Binomial Approximation (BA)	To weight terms from queries for search expanding, the Kullback-Leibler divergence is used to estimate the binomial distribution	Binomial expansions are most effective for achieving a good early precision
Bose-Eninstein 1 (Bo1)	The Bose-Einstein dsitribution where the mean is given by the Poisson model	Bose-Einstein statistics are good choices to improve the Mean Average Precision
Bose-Eninstein 2 (Bo2)	The Bose-Einstein statistics where the mean is given by the Bernoulli process	Bose-Einstein statistics are good choices to improve the Mean Average Precision

3 Evaluation

3.1 Data Collection

The TREC and Green Mining stack overflow data set have been used for analysis in this research. The details of the data set used in Terrier IR framework are cited in Table 2. The TREC Disk1 collection includes abstract of news articles from the Associated Press, Federal Register, Ziff-Daviswas, Wall Street Journal, Department of Energy abstracts, and Green Mining stack overflow dataset for software engineering used for the evaluation [17, 18] (Table 3).

Table 3. TREC Volume 1 Collection Analysis

Attributes	Statistics
Volume of the Collection	1.17 GB
The total number of text documents	84678
Number of Unique terms	138809
Number of Pointers	13384542
Number of Tokens	22290723

Below is a data set template that has been used for the analysis (Table 4):

Table 4. TREC and Green mining Software engineering Dataset templates

```
<DOC>
<DOCNO> AP890101-0001 </DOCNO>
<FILEID>AP-NR-01-01-89 2358EST</FILEID>
<HEAD>You Don't Need a Weatherman To Know '60s Films Are
Here</HEAD>
<BYLINE>Associated Press Writer</BYLINE>
<DATELINE>NEW YORK (AP) </DATELINE>
<TEXT>
   The celluloid torch has been passed to a new generation:
filmmakers who grew up in the 1960s.    ``Platoon,'' ``Running
on Empty,'' ``1969'' and ``Mississippi Burning'' are among
the movies released in the past two years from writers and
directors who brought their own experiences of that turbulent
decade to the screen.
</TEXT>
</DOC>
```

3.2 Evaluation Measures

In order to evaluate results, trec_eval is used for which the system parameters have been mentioned in Table 5.

Table 5. System parameters with the corresponding values

System Parameter Name	Modalities
Retrieving Model	TFIDF, DPH, IFB2, DFRee, InexpC2, HiemstraLM, XSqrAM, DLH, DFRBM25, DirichletLM, InexpB2, LGD, PL2, JsKLs, DFI0, InL2, InB2, BM25, DLH13, LemurTFIDF, BB2,IFB2, XSqrA_M
Trec Query Tags Process	TOP, NUM, TITLE
Parameter Free Expansion	FALSE
Query Expansion (QE) Model	KL, BA, Bo2, Bo1
Number of Documents for QE	100
Number of terms for QE	10
Minimum number of documents in which terms should occur	2
Rocchio Beta	0.5d

The results can be obtained through several measures related to information retrieval like Mean Average Precision (MAP), R-precision, average precision (AP)etc.In this research the authors have used average precision to compare different DFR-QE models (Eq. (1), (2), (3), (4)) with baseline.[15, 16] In this evaluation, accuracy scores across the rank instances were a relevant document was obtained are averaged:

$$AP = \sum_{rank=1}^{N} \frac{(Precision(rank) \times relevance(rank)}{C_{rank}}$$

where,

C_{rank} is number of relevant documents.

The average precision value gives the highest precision which is considered as the optimal model for the dataset. Accuracy refers to the portion of the documents that are pertinently returned in response to a user's query. It can be evaluated the cutoff rank which give the top most results to the user. This measure is called *precision at n* or *P@n*.

The experiments are performed on Terrier IR system which provides indexing of the terms. Figure 1 shows the flow of the query expansion process on DFR models.

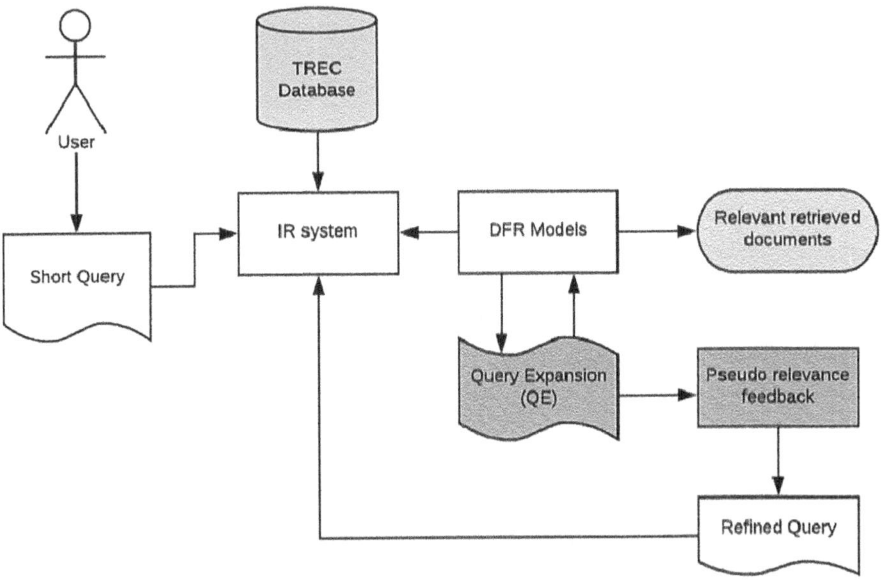

Fig. 1. QE Process

The results file (.res) is generated by the topic file. After that the authors have used the standard TREC and Green Mining QREL files to give the evaluation of the queries and provide the average precision of the result file [17].

The authors have applied Terrier DFR based weighing models with QE techniques namely, KL, Bo1, Bo2 and BA (discussed below) which refines the query given by the user and retrieves top relevant documents. These PRF methods have been proposed by Xu and Croft which are co-occurrence models.

1. KullbackLeibler (KL)

The basic KL function is to extract highly weighted terms from the expected top rank documents.

$$KL(t) = E_x.log_2[\frac{len_x}{E_c}]$$

where,

- $E_x = tf_x / len_x$
- tf_x is the phrase that describes the frequency of the search for the appropriate records.
- len_x is The approximate amount of projected (exp_rel) documents that are pertinent.

2. Bo1

Bo1 query expansion model is defined from the combination of Bose Einstein Statistic and weighted term *t* from highly relevant documents.

$$Bo1(x) = tf_x.log_2\left(\frac{1 + \frac{TF}{doc\,x}}{\frac{TF}{doc\,x}}\right) + log_2(1 + \frac{TF}{doc_x})$$

where,

- tf_x is the phrase that describes the frequency of the search for the appropriate records.
- TF is the phrase appearance.
- doc_x is the number of documents in the corpus.

3. Bo2

Bo2 is the extended version of Bo1. Mathematically, it is given by Eq. (3).

$$Bo2(x) = tf_x.log_2\left(\frac{1 + \frac{TF}{doc\,y}}{\frac{TF}{doc\,y}}\right) + log_2(1 + \frac{TF}{doc_y})$$

4. Binomial Approximation (BA)

BA is the binomial approximation where *t* is the candidate term that is it occurs frequently in the top most relevant documents.

$$BA(term|query) = B_i(TF_{x't}, TF_x, \frac{TF_x(t)}{TF_x})$$

where,

- TF is the term frequency.

The evaluation has been done for all DFR models amongst which the top 10 for each QE technique will be discussed in the results section.

4 Result and Analysis

The authors have compared and analyzed DFR-QE with baseline models. The DFR and QE models have been described in previous sections.

The authors evaluated all weighted probabilistic models with QE amongst which the top 10 have been shown in Table 6. The evaluation file provides the values of MAP, AP, R-precision and precision at k. Table 6 gives the result of top 10 DFR-QE model at p@0.

The Table 6 has been shown graphically in Fig. 2. IFB2 gives the highest precision at 10 in contrast to which DLH13 gives the lowest precision and BB2.

Table 6. *Precision@0* of top 10 DFR – QE models on TREC Dataset

P@0%	InB2	IFB2	In_expB2	BB2	In_expC2	DPH	XSqrA_M	Js_KLs	DFReeKLIM	DLH
DFR_basline	0.3533	0.3533	0.3533	0.3533	0.3733	0.38	0.3667	0.5	0.3733	0.54
DFR_KL	0.6652	0.6651	0.6623	0.6601	0.66	0.6593	0.6509	0.6498	0.648	0.6457
DFR_BA	0.6671	0.6654	0.6635	0.6598	0.6633	0.6595	0.6529	0.6208	0.6512	0.696
DFR_Bo1	0.6682	0.6451	0.6651	0.6575	0.6557	0.6565	0.647	0.6498	0.6505	0.6392
DFR_Bo2	0.6595	0.6658	0.6587	0.6596	0.6623	0.6496	0.6477	0.6381	0.6479	0.6073
Max_P	0.6682	0.6658	0.6651	0.6601	0.6633	0.6595	0.6529	0.6498	0.6512	0.696

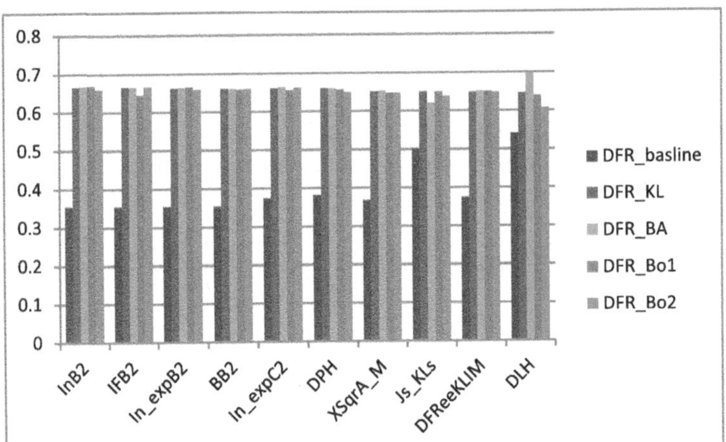

Fig. 2. Comparison of top 10 Baseline with DFR-QE Model at *Precision@0* models on TREC Dataset

The highest AP was obtained by Bo1 with DPH model. This can be seen in Table 6. As per the result, Bo1 and Kl gives the highest AP as compared to Bo2 and BA. The highlighted text in the Table is the top 10% values from the overall values.

The results indicate that highest AP was calculated for Bo1 with value c set to 1.0 but in baseline the result has drastically dropped for all the models as compared to DFR-QE. The KL (with 24.33%) and the Bo1 (with 24.71%) models performed similarly for all the cases.

The authors have noticed a clear degradation of AP for baseline model as compared to DFR with QE. The figure clearly shows that the baseline can be enhanced with the application of QE.

5 Conclusion

The authors have presented the study of baseline models with QE and further gives the distribution analysis of the document. QE is one of the most robust method which enhances the retrieved results of the original query. Query Expansion has improved the average precision of the retrieval to 4 times (4x) the existing retrieved results. The analysis of the results indicate that KL and Bo1 have the best average precision compared to other expansion models. The combination of QE on the baseline enhances the simple baseline model results to a good extent.

The future study aims to project the query formulation that would give optimal value for both long and short queries.

References

1. Colace, F., De Santo, M., Greco, L., Napoletano, P.: Improving relevance feedback-based query expansion by the use of a weighted word pairs approach. J. Am. Soc. Inf. Sci. **66**(11), 2223–2234 (2015)
2. Ermakova, L., Mothe, J., Ovchinnikova, I.: Query expansion in information retrieval: what can we learn from a deep analysis of querie ? In: International Conference on Computational Linguistics - Dialogue 2014, 4–8 June 2014, Moscow, Russian Federation (2014)
3. Ai, Q., O'Connor, B., Croft, W.B.: A neural passage model for ad-hoc document retrieval. In: European Conference on Information Retrieval, pp. 537–543. Springer, Cham (2018)
4. Cieslewicz, A., Dutkiewicz, J., Jedrzejek, C.: Baseline and extensions approach to information retrieval of complex medical data: Poznan's approach to the bioCADDIE 2016. Database **2018** (2018)
5. Ermakova, L., Mothe, J., Ovchinnikova, I.: Query expansion in information retrieval: what can we learn from a deep analysis of queries?. In: International Conference on Computational Linguistics-Dialogue 2014, vol. 20, no. 13, p. 162 (2014)
6. Ermakova, L., Mothe, J., Nikitina, E.: Proximity relevance model for query expansion. In: Proceedings of the 31st Annual ACM Symposium on Applied Computing, pp. 1054–1059. ACM (2016)
7. Ermakova, L., Mothe, J.: Query expansion by local context analysis. In: Conference francophone en Recherche d'Information et Applications (CORIA 2016), pp. p. 235 (2016)
8. Tang, P., Zhao, J., Yu, Z., Wang, Z., Xian, Y.: A method of Chinese and Thai cross-lingual query expansion based on comparable corpus. J. Inf. Process. Syst. **13**(4) (2017)
9. Keikha, A., Ensan, F., Bagheri, E.: Query expansion using pseudo relevance feedback on Wikipedia. J. Intell. Inf. Syst. **50**(3), 455–478 (2017). https://doi.org/10.1007/s10844-017-0466-3
10. Amati, G.: Probabilistic models of information retrieval based on measuring the divergence from randomness [abstract]. The University of Glasgow, Fondazione Ugo Bordoni and CORNELIS JOOST VAN RIJSBERGEN University of Glasgow (n.d.). http://theses.gla.ac.uk/1570/1/2003amatiphd.pdf
11. He, B.: Divergence from randomness (2005). http://ir.dcs.gla.ac.uk/wiki/DivergenceFromRandomness

12. Kalra P., Mehrotra D., Wahid A.: Comparison and analysis of information retrieval DFR models. In: Satapathy S., Bhateja V., Das S. (eds.) Smart Computing and Informatics. Smart Innovation, Systems and Technologies, vol. 78. Springer, Singapore (2018)
13. Gahlawat, T., Lekhwar, S., Kalra, P., Mehrotra, D.: Information retrieval on green mining dataset using divergence from randomness models. In: International Conference on Innovative Computing and Communications, pp. 127–134. Springer, Singapore (2019)
14. Silva, S., Seara Vieira, A., Celard, P., Iglesias, E.L., Borrajo, L.: A query expansion method using multinomial naive bayes. Appl. Sci. **11**, 10284 (2021)
15. Saeed, M.M., Al Aghbari, Z.: ARTC: feature selection using association rules for text classification. Neural Comput. Appl. **34**(24), 22519–22529 (2022)
16. Cambria, E., Mao, R., Han, S., Liu, Q.: Sentic parser: a graph-based approach to concept extraction for sentiment analysis. In: 2022 IEEE International Conference on Data Mining Workshops (ICDMW), pp. 1–8. IEEE (2022)
17. https://trec.nist.gov/data.html
18. http://terrier.org/

Content-Based Linear and Nonlinear Dynamic Texture Synthesis Using Time Series Approach

Premanand Pralhad Ghadekar(✉)

Department of Information Technology, VIT, Pune 411037, Maharshtra, India
ppghadekar@gmail.com

Abstract. Capturing the nonlinear aspects of a dynamic texture is both challenging and tough. The proposed innovative algorithm focuses on capturing both the aspects of dynamic texture that is linear as well as nonlinear motion. The innovative part of the proposed model is computing a seasonal matrix using simple mathematical calculations to predict a new frame as well as insert a predicted frame for reconstructing a dynamic texture from the original ones by using innovative, dynamic texture synthesis models. The proposed approach is used for dynamic texture synthesis with a very fewer amount of model coefficients and very high visual quality. It is useful to create a huge number of frames with a compression ratio of 33% and 90% and PSNR values between 45dB to 65dB. The main advantage over the other widely available methods is that there is no need for transform coding since the algorithm directly applies to the raw video thereby, reducing the model size, model parameters, time, and computational complexity.

Keywords: SVD · HOSVD · YCbCr · linear · nonlinear · time series · seasonal matrix

1 Introduction

In the texture representation certain designs are found to be repeated. Textures are of two categories-static textures and dynamic textures. Image that is formed by spatial redundancy is called as static texture. Dynamic texture is formed by a series of frames that has temporal stationarity in it. Both the features that is spatial and temporal self-similarity is found in a dynamic texture. Flame, grass, waterfall, and flag, etc. are examples of dynamic textures. The procedure of forming dynamic textures artificially from the existing dynamic texture by making use of both similar spatial as well as temporal information as the original one is known as dynamic texture synthesis. The common approaches that are used for dynamic texture synthesis are physics-based, and the image-based [1]. The approach used in the physics-based is to prepare a model from the particles of images with the help of physical laws. Physical laws are helpful to form a model by using particles of images. The two forms of image-based techniques are- parametric and non-parametric. To create a number of output frames in the non-parametric technique different clips are separated from the video, and then all them are mend together to form different parameters that are useful for dynamic texture synthesis in parametric process.

S. Singh et al. (Eds.): AIDSP 2023, CCIS 2440, pp. 33–44, 2025.
https://doi.org/10.1007/978-3-031-88762-8_4

A representation which is made up of linear combination of independent variables is called a linear model. These models are predictable, regular and deterministic. In nonlinear dynamic system the relation between the current and the succeeding state is disconnected. Nonlinear systems show movable action, high variations, but they are moderately deterministic.

The physical model arrangement contains a nonlinear link between images [2]. The connection between linear and nonlinear behavior can be well known by considering an example- A person walking on a road, linear motion can be described using simple mathematical equation. However, the person playing football. In this example motion is nonlinear in nature. It's very difficult to predict the motion as it is nonlinear in nature.

1.1 Literature Survey

Soatto et al. [3] explored a linear dynamic texture synthesis model. In this model, for the process of dynamic texture synthesis the Linear Dynamic System (LDS) is used. The LDS approach is linear and stable. It requires a large amount of model coefficients, and the synthesis value of nonlinear dynamic textures is very poor. The dynamic texture is modelled using subspace mixtures [4, 5] to improve visual quality, the dynamic textures are synthesized with a closed-loop LDS. In the model proposed by Soatto et al. there are two drawbacks: the Principal Component Analysis (PCA) can not be used to capture the non-linear motion of dynamic texture; PCA has its restrictions in synthesizing nonlinear dynamic textures. The Fourier descriptors method is available to make use of the spatial information of dynamic texture [6]. It also requires a huge amount of memory to store its model coefficients. The reason why the computational intricacy of Fourier transform is extremely high because of its complex quality. Ghadekar P et al. presented "the nonlinear dynamic texture synthesis model using discrete wavelet transform and YCbCr model" [7]. The "Higher-Order Singular Value Decomposition" (HOSVD) algorithm [8] is used to synthesize linear dynamic textures. HOSVD approach cannot be used to capture the nonlinear motion of dynamic textures. Ghadekar P et al. proposed the nonlinear dynamic texture synthesis model using hybrid DWT-DCT and kernel PCA with GPU model [9]. It provides a better result with less memory. The dynamic texture synthesis method [10] is used to generate dynamic texture frames with very few number of model coefficient, model size, and less time complexity. It provides better results for linear dynamic textures. The Nonlinear Principal Component Analysis (NLPCA) is a non-linearized form of standard PCA. This method is nonlinear in nature, which captures both linear and nonlinear motion of the effective appearance. It works on the concept of feature extraction. It maps the nonlinear data into a linear form [11, 12]. It requires lots of time for training, its computational and time intricacy required is high. The innovative method [13] of dynamic texture synthesis uses Fourier transform and Dynamic Texture Units (DTUs). In order to store the model coefficients more amount of memory is needed. Kaitai Zhang et al. [14] presented dynamic texture synthesis model to capture the structural and long-range correlation of the reference texture video. Matthew Tesfaldet et al. [15] proposed a dynamic texture synthesis using two-stream based approach. The proposed model is an application based on convolutional networks (ConvNets). Deep Learning based model [16] can also be used to capture linear as well as nonlinear motion of dynamic texture.

The proposed algorithm is based on the time series analysis approach. In the proposed model, innovative models are used to occupy the linear as well as nonlinear motion of a dynamic texture with very few model coefficients, very fewer time complexities, and very high PSNR. The article is structured as follows. After exploring the time series analysis, the working of the suggested algorithm is given in detail. Comparisons are made of the proposed model with different existing models given in the next section. Then finally, it concludes the article.

1.2 Time Series Analysis

The time-series analysis is termed as a nonlinear model. In the previous work of time series analysis, the input time series data is mapped into higher dimensional phase space. It provides better PSNR of reconstructed dynamic textures. This mapping is required to calculate two parameters, namely, the embedding dimension d and lag t [17, 18]. Calculating these two parameters for every frame requires lots of time. Thus, this method is very time-consuming.

Seasonal time series is a concept that is likely to change or pattern in a time series that repeats over one year [19, 20]. Seasonality is the phenomenon that is commonly found in economic, environmental, and other time series.

In the proposed time series analysis model, there is no need to map the original time series into an embedded phase. Initially, the time-series approach was used to predict the non-linearity in the share market. The algorithm was used to predict the future value of a particular share based upon its previous and current value.

1.2.1 Seasonality Adjustment

In-order to remove the Seasonal parameters of a time-series seasonal adjustment methods are used [19, 20]. The suitable technique for seasonal adjustment is selected based on the decomposition of a time-series into components and defined with names like 'cyclic', 'trend', 'seasonal', and 'irregular'. For an example, such parameters may act multiplicatively or additively. Thus, the adjustment technique has two stages, if seasonal element acts multiplicatively or additively: 1) Evaluate the seasonal element of disparity in the time series, typically, in a form that has a zero mean across a series, 2) Then subtract the calculated seasonal parameters from the original time-series parting the seasonally adjusted series [21].

Thus, the same concept of seasonality is used to capture the repetitive nature of dynamic textures. The randomness of a frame is removed by obtaining the mean value for the frame, and it is used to predict the next frame by identifying the trend component, which may be a linear expansion or reduction. In the case of images and videos, this trend component is random. Therefore, the mean value helps to eliminate this randomness and uses this mean value to drive the seasonal matrix that can act as a trend component for the future prediction of values.

1.3 Working in Detail of the Proposed Algorithms

The proposed algorithm is based on time series analysis. In the proposed algorithm, only a part of the time series analysis is implemented. The innovative part of the algorithm lies in obtaining seasonal matrices for intermediate frames. It is achieved by computing the mean value of a frame and then dividing each value in the frame with its mean. Then obtain only the average value of the next frame. Compute the mean value for both the means obtained. A new frame that is predicted and inserted in between these two frames by multiplying the seasonal matrix obtained with the mean of means. Repeat the same process for the remaining frames. Thus, the number of frames is increased at the end of each iteration. The following different approaches are proposed to achieve dynamic texture synthesis.

a) The algorithm using time series analysis by considering 66% of the original frames (i.e., 66% RGB approach).
b) The algorithm using time series analysis by considering 33% of the original frames (i.e., 33% RGB approach).
c) The algorithm using time series analysis and the block-based approach by considering 33% of the original frames (i.e., 33% RGB block-based approach).

The detailed algorithm steps are given below.

1. Find the seasonal matrix and the average of the first image.
2. Find the average of the second image.
3. Generate a new image by considering the average of averages and the seasonal matrix.

Figure 1 shows the proposed dynamic texture synthesis model for an infinite frame sequence with fewer components. The input to the first block is a dynamic texture. The first block, i.e., different colour coding, deals with chromatic redundancy among the pixels of an intra frame. Here, the YCbCr or YIQ colour model is used. The default colour model is RGB. Conversion from RGB to YCbCr or YIQ, and YCbCr to RGB or YIQ to RGB; it changes the values of near about 4–8% pixels of the original frame. The second and third block is for learning (analysis) and synthesizing the dynamic texture. The time series analysis method is used for linear and nonlinear dynamic appearance for composition.

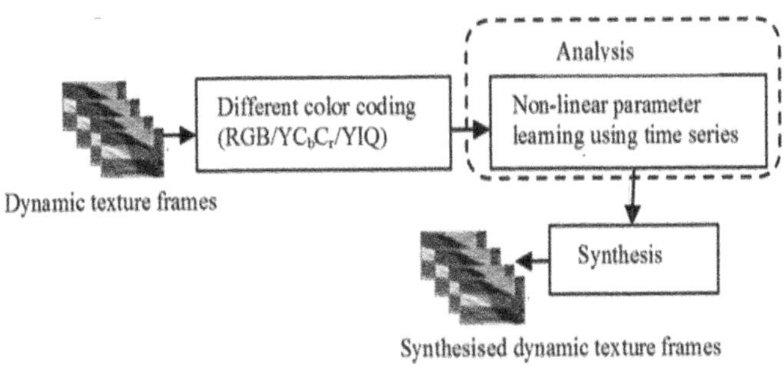

Fig. 1. The proposed dynamic texture synthesis model for an infinite frame sequence with fewer components

In Fig. 2, the proposed dynamic texture synthesis algorithm using time series analysis by considering 66% of the original frames is shown. In this approach, 66% of the original frames mean that alternate frames are considered for analysis and synthesis. The idea is to synthesize frame number 2, from frame numbers 1 and 3. Similarly, frame number 5 is synthesized from frame numbers 4 and 6, and so on. The seasonal matrices of the first frame and the third frame is calculated. It is multiplied by the average of average (average of (M frame) + (M + 2frame))), where M is the frame number to find the skipped frames (frame no 2, 5..., N − 1). The un-skipped frames (frame numbers 1, 3, 4, 6....N − 2, and N) are generated by multiplying its seasonal matrix with the average value. Different colour coding YCbCr/YIQ are used to exploit the chromatic information. This approach requires a fewer number of model coefficients for synthesis.

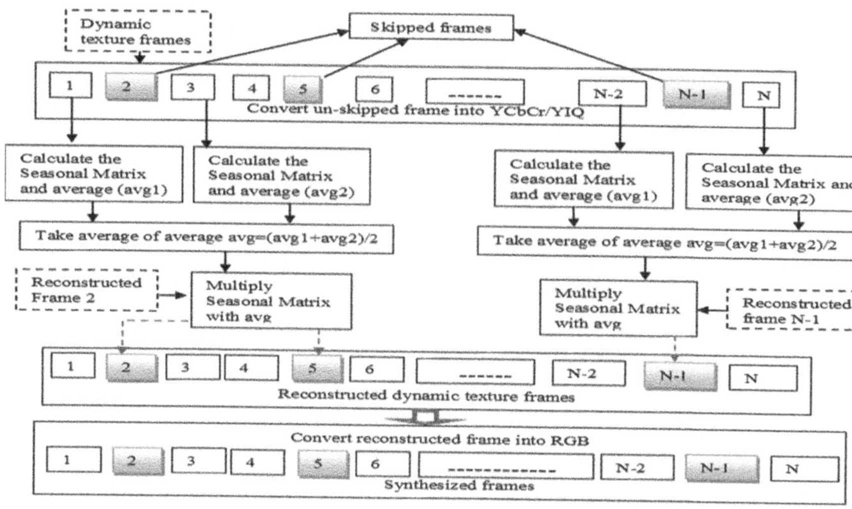

Fig. 2. The proposed dynamic texture synthesis algorithm using time series analysis by considering 66% of the original frames

In Fig. 3, the proposed dynamic texture synthesis algorithm using time series analysis by considering 33% of the original frames is shown. In this approach, only 33% of the original frames are taken into account for synthesis. The idea is to synthesize frame numbers 1 and 3 from frame number 2. Similarly, frame numbers 4 and 6 are synthesized from frame numbers 5, and so on. The seasonal matrices of the retained frames are calculated. The seasonal matrices are then multiplied by the average of a skipped frame to generate frame numbers 1, 3, 4, 6 ..., N − 2, and N. The un-skipped frames (frame numbers 2, 5, 8..., and N − 1) are generated by multiplying its seasonal matrix with the average value. Different colour coding YCbCr/ YIQ are used to exploit the chromatic information. This approach requires a very few number of model coefficients and model size for synthesis.

In Fig. 4 the proposed dynamic texture synthesis algorithm using time series analysis and the block-based approach by considering 33% of the original frames is shown. The proposed block-based approach is useful to occupy nonlinear motion. It is the proposed approach in, which the frame is divided into blocks. The different block sizes can be of 2 by 2, or 4 by 4, or 8 by 8, or 10 by 10 that are used. The seasonal matrix needs

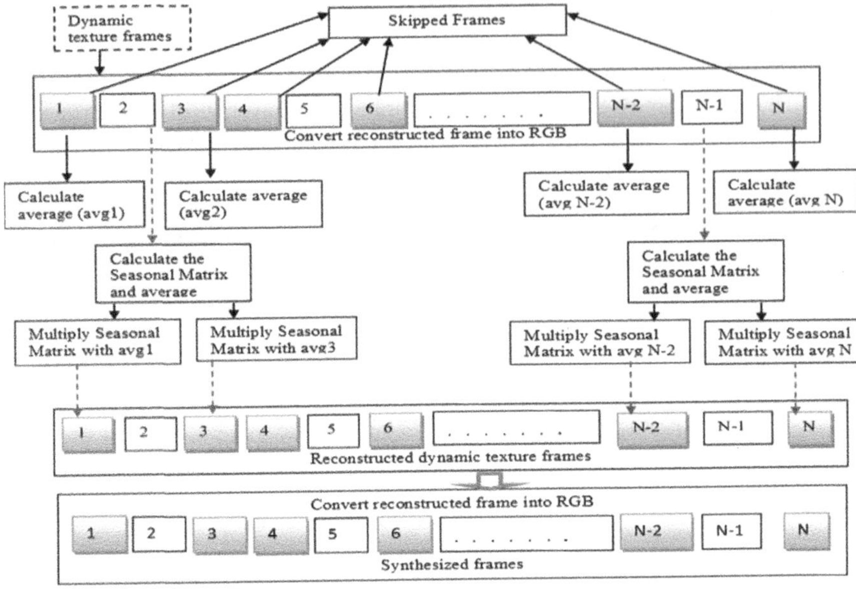

Fig. 3. The proposed dynamic texture synthesis algorithm using time series analysis by considering 33% of the original frames.

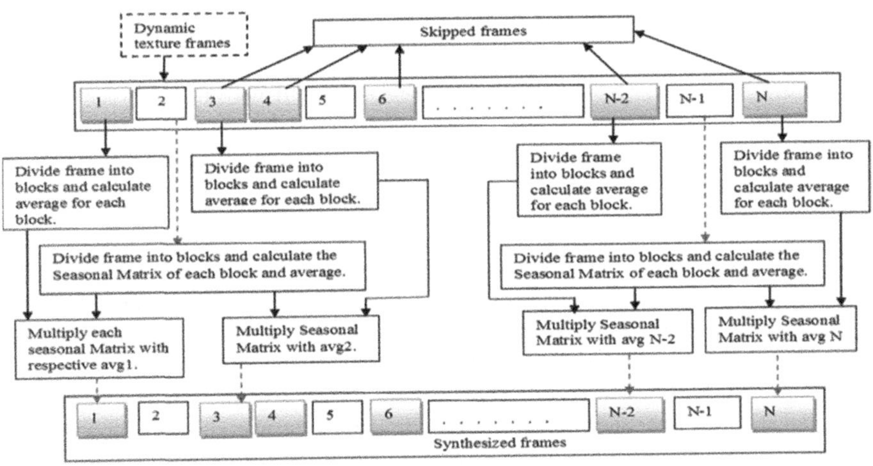

Fig. 4. The proposed dynamic texture synthesis algorithm using time series analysis and the block-based approach by considering 33% of the original frames.

to be calculated for each and every block. The average matrix of the skipped frames is motion of a dynamic texture and provides very high-quality synthesized frames. It has one drawback- blocking artifacts, to remove blocking artifacts, a filter is used. The approach using a block size 2 by 2 provides very high PSNR, as it controls the random variation in the values of the seasonal matrix by minimizing the actual block size of the seasonal matrix.

2 Results

Standard dynamic texture database [22, 23] is used for experiments. The model size is the size of the stored model parameter in the memory for synthesis. The model is said to be a compact model when it has a smaller size than the other models. The stored model parameters in the proposed model are the seasonal matrices of intermediate frames. Table 1 displays the result of the proposed dynamic texture synthesis algorithms using time series analysis by considering model coefficients, self-similarity index, and PSNR parameters. The 66% RGB approach provides very high PSNR, and it needs large numbers of model parameters as compared to a 33% RGB approach and 33% RGB with a 2x2 block-based approach. The 33% RGB approach requires fewer numbers of model coefficients. The 33% RGB with 2x2 block-based approach provides moderate PSNR, and it requires a smaller number of model coefficients as compared to the 66% RGB approach.

Table 1. Shows the result of the proposed dynamic texture synthesis algorithms using time series analysis by considering model coefficients, self-similarity index, and psnr parameters

Dynamic Texture	66% RGB approach			33% RGB approach			33%RGB with a 2×2 block based approach		
	Model Coefficients	Self-similarity index	PSNR (dB)	Model Coefficients	Self-similarity index	PSNR (dB)	Model Coefficients	Self-similarity index	PSNR (dB)
Air Flow	26956800	0.98	78.18	13594122	0.97	56.43	23616000	0.98	62.36
Brownian	26049600	0.85	70.21	13219497	0.70	40.44	22842000	0.91	49.68
Flagl	21643200	0.85	72.63	10887147	0.71	45.26	18954000	0.93	53.17
Flagg	25546752	0.95	74.74	13077873	0.91	49.59	22429440	0.97	56.90
Flame	13824000	0.95	75.49	7142661	0.91	50.99	12153600	0.97	56.47
Grassi	19353600	0.97	78.33	9907569	0.95	56.70	16992000	0.87	54.18
Grass2	6483456	0.88	75.46	3290409	0.76	50.85	5685120	0.87	52.82
Pond	26496000	0.89	74.58	13248450	0.78	49.13	23184000	0.90	51.39
Sparkle	11520000	0.97	79.93	5760450	0.95	59.87	10080000	0.97	56.79
Waterfall	26496000	0.93	76.85	13363713	0.87	53.71	23212800	0.80	49.04

Table 2 shows the proposed dynamic texture synthesis algorithms using time series analysis by considering model size, PSNR, and compression ratio parameters. The 33% RGB approach requires less model size as compared to a 66% RGB approach and 33% RGB with a block-based approach. The 33% RGB with a 2x2 block-based approach provides moderate results.

Table 3 compares the proposed methods with the LDS [3], the HOSVD [8], LDS+FFT [13], and for the flame and flag dynamic textures. The proposed algorithms provide a very high PSNR and compression ratios as compared to LDS [3], HOSVD [8], and

Table 2. Shows the proposed dynamic texture synthesis algorithms using time series analysis by considering model size, psnr, and compression ratio parameters

Dynamic Texture	66% RGB approach			33% RGB approach			33% RGB with a 2 × 2 block-based approach		
	Model Size (MB)	PSNR (dB)	Compression Ratio (%)	Model Size (MB)	PSNR (dB)	Compression Ratio (%)	Model Size (MB)	PSNR (dB)	Compression Ratio (%)
Air Flow	11.33	78.18	70.53	5.75	56.43	85.04	17.54	62.36	54.38
Brownian	10.40	70.21	71.95	5.30	40.44	85.68	14.63	49.68	60.53
Flag1	18.63	72.63	39.71	9.37	45.26	69.65	21.15	53.17	31.55
Flag2	9.04	74.74	75.05	4.65	49.59	87.15	13.54	56.90	62.63
Flame	2.14	75.49	89.04	1.12	50.99	94.26	2.87	56.47	85.31
Grass1	13.79	78.33	49.78	7.06	56.70	74.29	14.28	54.18	47.99
Grass2	4.96	75.46	46.28	2.51	50.85	72.76	5.78	52.82	37.41
Pond	19.81	74.58	47.73	9.82	49.131	74.09	23.28	51.39	38.58
Sparkle	3.20	79.93	80.55	1.65	59.87	89.99	4.26	56.79	74.10
Waterfall	23.24	76.85	38.58	11.71	53.71	69.04	22.69	49.04	40.03

Table 3. Performance comparison between the proposed methods, the lds method [3], lds+fft method [11], and the hosvd method [7]

Dynamic Texture	Parameters	LDS Approach	LDS + FFT Approach	HOSVD Approach	33% RGB Approach	33% RGB 2 × 2 Approach
Flame (240 × 320 × 3 × 89)	Original size MB	19.56	19.56	19.56	19.56	19.56
	Model Size MB	18.59	98.53	7.2	1.12	2.87
	PSNR (dB)	35.53	35.05	35.46	50.99	56.47
	Compression Ratio (%)	4.94	−19.57	63.19	94.26	85.31
Flag1 (180 × 240 × 3 × 250)	Original size MB	30.91	30.91	30.91	30.91	30.91
	Model size MB	23.41	99.6 MB	11.2	9.47	21.15
	PSNR (dB)	21.51	36.4745	21.42	45.26	53.17
	Compression Ratio (%)	24.24	−2.22	63.75	69.65	31.15

LDS+FFT approach [13]. LDS+FFT Memory to store the model components so that its Compression ratio is in a negative form.

Figures 5, and 6 show the average PSNR comparison graphs by using the proposed approaches (i.e., 66% RGB, 33% RGB, 33% RGB with a 2 by 2 block-based, 33% RGB with a 4 by 4 block-based approach, and 33% RGB with a 10 by 10 block-based approach). The proposed 66% RGB approach provides better PSNR.

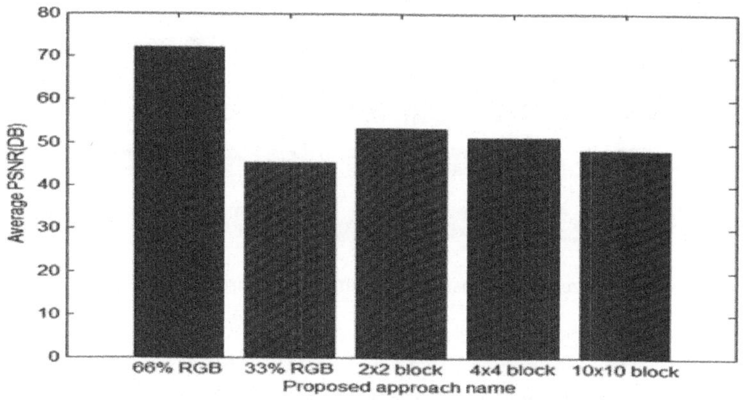

Fig. 5. Average PSNR comparison graph for the flag

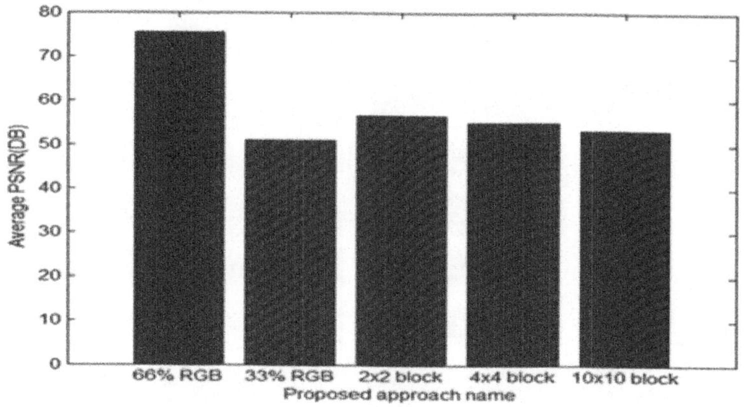

Fig. 6. Average PSNR comparison graph for the flame dynamic texture using the proposed approaches

In Fig. 7 the dynamic texture flag (number of frames-250, height- 180, and width-240) is used. Figure 7 shows the synthesized frames using 250 frames of the original dynamic texture flag. The proposed 66% RGB approach is used to generate an infinite number of frames with better PSNR and a fewer number of model coefficients. The temporal motion of the dynamic texture flag is nonlinear in nature. By using the LDS approach and the HOSVD approach, an infinite number of frames can be generated, but it provides very low PSNR.

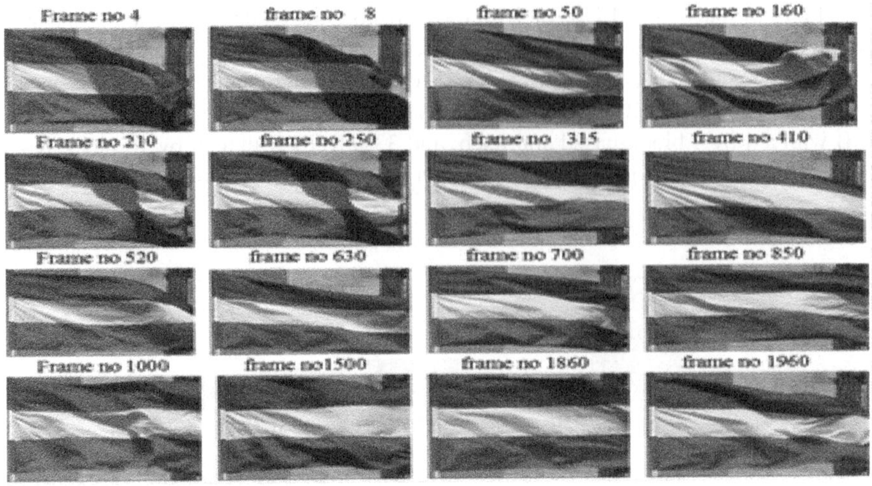

Fig. 7. The synthesized frames using 250 frames of the original dynamic texture flag

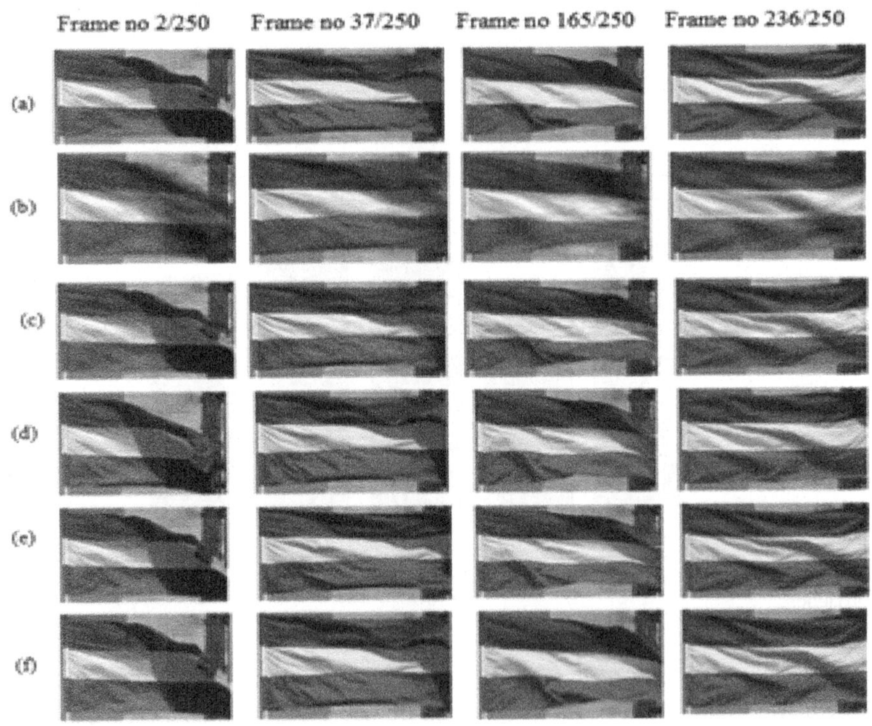

Fig. 8. (a) original flag sequence, (b) LDS approach (c) HOSVD approach (d) 66% RGB approach. (e) 33% RGB approach (f) 33% RGB with a 2x2 block-based approach

Figure 8 represents the visual quality of the reconstructed frames by using the LDS approach [3], HOSVD approach [8], and the proposed approaches (i.e., 66% RGB, 33% RGB, and 33% RGB with a block-based approach). The dynamic texture flag is nonlinear in nature, 33% RGB, and 33% RGB with 2x2 block-based approaches provide better visual quality as compared to the existing LDS [3] and HOSVD approach [8].

3 Conclusion

The proposed algorithms comprise of time series analysis with the block-based approach. Here, innovative, dynamic texture synthesis models have been proposed. The experimental result shows that the proposed approaches are useful to capture linear and also the nonlinear motion of a dynamic texture for synthesis purposes with very fewer model coefficients and better visual quality. The proposed approaches exploits the spatial, temporal, and the chromatic correlation in an excellent way without using the transform-based approach. The synthesis model uses a minimum count of model coefficients with less execution time. The proposed approach delivers the better results as compared to the LDS approach [3], the HOSVD approach [8], and the LDS+FFT approach [13].

References

1. Costantini, R.: Compact representation for static and dynamic texture synthesis. Thesis, Infoscience, pp. 1–160 (2007)
2. Awasthi, I., Elgammal, A.: Learning nonlinear manifolds of dynamic textures. In: Advances in Computer Graphics and Computer Vision Communications in Computer and Information Science, vol. 4, pp. 395–405 (2007)
3. Doretto, G., Chiuso, A., Wu, Y., Soatto, S.: Dynamic textures. Int. J. Comput. Vision **51**(2), 91–109 (2003)
4. Liu, C., Lin, R., Ahuja, N.: Modeling dynamical textures using subspace mixtures. In: Proceedings of International Conference of Multimedia and Expo, pp. 1378–1381 (2005)
5. Yuan, L., Wen, F., Liu, C., Shum, H.: Synthesis dynamic texture with closed-loop linear dynamic system. In: Proceedings of European Conference on Computer Vision (ECCV), pp. 603–616 (2004)
6. Abraham, B., Camps, O., Sznaier, M.: Dynamic texture with Fourier descriptors. In: Proceedings of the 4th International Workshop on Texture Analysis and Synthesis, pp. 53–58 (2005)
7. Ghadekar, P., Chopade, N.: Nonlinear dynamic texture synthesis by using discrete wavelet transform and YCbCr color coding. Int. J. Comput. Sci. Manage. Syst. **3**(1), 47–53 (2011)
8. Costantini, R., Sbaiz, L., Süsstrunk, S.: Higher order SVD analysis for dynamic texture synthesis. IEEE Trans. Image Process. **17**(1), 42–52 (2008)
9. Ghadekar, P., Chopade, N.: Modelling nonlinear dynamic textures using hybrid DWT-DCT and kernel PCA with GPU. J. Inst. Eng. Ser.-B **97**(4), 549–555 (2016)
10. Ghadekar, P., Chopade, N.: Content-based dynamic texture analysis and synthesis based on SPIHT with GPU. J. Inf. Process. Syst. Korea (2016). https://doi.org/10.3745/JIPS.02.009. ISSN 1976-913X (print), ISSN-2092-805X (Electronic)
11. Kramer, M.: NLPCA using ANN mark laboratory for intelligent systems in process engineering. AIChE J. **37**(2), 233–243 (1991)

12. Scholz, M., Fraunholz, M., Selbig, J.: Nonlinear principal component analysis: neural network models and applications. In: Principal Manifolds for Data Visualization and Dimension Reduction Lecture notes in Computational Science and Engineer, vol. 58, pp. 44–67. Springer, Heidelberg (2008)
13. Li, C., Wang, J., Ye, L., Wang, H.: A novel method of dynamic textures analysis and synthesis. In: Proceedings of IEEE International Joint Conference on Computational Sciences and Optimization, pp. 328–332 (2009)
14. Zhang, K., Wang, B., Chen, H.-S., Wang, Y., Mou, S., Jay Kuo, C.-C.: Dynamic texture synthesis by incorporating long-range spatial and temporal correlations. Int. J. Comput. Vision Pattern Recogn. (2021). arXiv:2104.05940v2
15. Tesfaldet, M., Brubaker, M.A., Derpanis, K.G.: Two-stream convolutional networks for dynamic texture synthesis. Int. J. Comput. Vision Pattern Recogn. (2021). arXiv:1706.06982v4
16. Jadhav, R.S., Ghadekar, P.P.: Content based facial emotion recognition model using machine learning algorithm. In: IEEE International Conference on Advanced Computation & Telecommunication (ICACAT-2018), p. 8 (2018)
17. Basharat, A., Shah, M.: Time series prediction by chaotic modelling of nonlinear dynamical systems. In: Proceedings of IEEE 12th International Conference on Computer Vision, University of Central Florida Orlando, FL, USA, pp. 1941–1948 (2009)
18. Cao, L.: Practical method for determining the minimum embedding dimension of a scalar time series. Phys. D: Nonlinear Phenom. **110**, 43–50 (1997)
19. Foldesi, E., Bauer, P., Horvath, B., Urr, B.: Seasonal adjustment methods and practices. European Commission Grant 10300.2005.021-2005.709, Budapest, Final Version 3.1, pp. 1–131 (2007)
20. Koopman, S., Ooms, M., Hindrayanto, I.: Periodic unobserved cycles in seasonal time series with an application to US unemployment. In: Proceedings of Conference on Seasonality, Seasonal Adjustment and Their Implications for Short-Term Analysis and Forecasting, Luxembourg, pp. 1–36 (2006)
21. http://en.wikipedia.org/wiki/Seasonality
22. Péteri, R., Fazekas, S., Huiskes, M.: DynTex: a comprehensive database of dynamic textures. Pattern Recogn. Lett. **31**(12), 1627–1632 (2010)
23. Kwatra, V., Schodl, A., Essa, I., Turk, G., Bobick, A.: Graphcut textures: image and video synthesis using graph cuts. In: Proceedings of SIGGRAPH, GPU Programming guide-GeForce 8 and 9 Series by NVIDIA, pp. 277- 286 (2003)

IoT Based Assistive Device for Speech Impaired People

G. B. Gour(✉), Bheemesh K. Gudur, Sanjana S. Umarani, Shreya B. Bagali, and Sofiya A. Shaikh

BLDEA's V.P. Dr. P.G. Halakatti College of Engineering and Technology, Vijayapura 586103, Karnataka, India
ec.gururaj@bldeacet.ac.in

Abstract. The speech-disabled people are facing difficulties in having normal communication with the world. Communication is the only means by which people with speech impairments can share their thoughts and communicate with others. However, there are several difficulties in communicating with a person who is speech challenged. With the development of wearable technology, it is now possible to put many innovative ideas into practice and benefit humanity in previously unimaginable ways. The previous work was done to develop sensors and a sign language translator as a wearable or assistive device. In order to aid such people, the hand-talking system is proposed in the present work as a wearable electronic device for speech-impaired people. The work consists of simple and friendly communication with speech-impaired people. This is achieved by developing a suitable sign language translator using gestures. The gestures are traced with the help of suitable sensors and accelerometers, which are later converted into recognizable text or speech. The goal of the current work is to develop an assistive device for identifying gesture-based communication that helps normal people and speech-impaired people communicate. The glove will function as a gesture-to-text translator. With the aid of appropriate sensors and accelerometers, gestures are tracked in sign language that can be easily translated into acceptable text or voice.

Keywords: Accelerometer · Hand-talking system · flex sensor · multi-sign language · gesture · Arduino · speech impairment · assistive technology

1 Introduction

1.1 Background

The ability to convey or exchange information through speech, writing, gestures, or another medium is a common definition of communication. Such a definition ignores communication's significance for expressing one's individuality, engaging with friends and family, and its significance for personal growth. It is odd that speech and language impairments are frequently disregarded or not seen as crucial to a person's quality of life when communication is essential to the evolution of Homo sapiens. Speech impediments

S. Singh et al. (Eds.): AIDSP 2023, CCIS 2440, pp. 45–56, 2025.
https://doi.org/10.1007/978-3-031-88762-8_5

are any symptoms that make it difficult to speak or pronounce words. A variety of disorders can lead to speech impairments, which include slurred, fast, stuttering, hoarse, or slowed speech. While some speech problems are temporary, others are caused by an inherent medical issue. When speech impairment occurs suddenly, it is a sign of a dangerous condition, like a heart attack or stroke, and has to be treated very quickly.

According to the National Health Portal, Govt. of India, there are an estimated 360 million individuals worldwide who suffer from moderate to serious hearing loss; 63 million people in India suffer from significant speech impairment, which accounts for 6.3% of the population (a large percentage involves children in the age group of 0–14 years). The previous works focused on the development of an aid for such disabled people by assessing their gestures or sign languages.

The following sections, discuss about the Speech impairment in detail.

1.2 Speech Impairment

The numerous speech, language, and communication impairments have a wide range of medical, surgical, psychological, and environmental causes. Such challenges are likely to have a significant impact on a child or adult with socio-economic limitations. Along with speech impairment, one can also experience many symptoms in addition to the inability to vocally speak clearly. This includes conditions like drooling, weakened face muscles, trouble remembering words, poor expressive language, and sudden vocal muscle contractions [25].

There are different types of speech impairments and dis-orders as follows,

- Aphasia: This condition affects the ability to think of words or pronounce them. It can be brought on by head injuries, brain tumors, and degenerative disorders like alzheimer's disease.
- Dysarthria: Dysarthria is the inability to move the articulators. This occurs frequently in conditions that affect the atrophy of the muscles, including multiple sclerosis, stroke, head injuries, and persistent alcohol use.
- Vocal disorders: These impair one's ability to talk and are brought on by physical trauma, cancer of the throat, polyps or nodules on the vocal cords, and the use of medications like antidepressants.
- Spasmodic dysphonia: This condition causes the voice chords to move uncontrollably. although the specific origin is uncertain, aberrant brain activity has been related to this impairment.

1.3 Diagnosis and Treatment of Speech Impairment

Speech impairment may occur after overusing the voice or during an infection, but persistent impairment calls for attention and diagnostic testing. Following a physical examination, the doctor would assess the patient's symptoms and medical history. Additionally, it requires assessing the extent of communication in a conversation, the level of speaking, and comprehension. Along with it, the common diagnostic tests include blood tests, electrical current tests, radiological tests (CT, MRI, X-ray), and urine tests.

The cause and severity of the speech impairment will determine the treatment. In all cases of speech disability, the actual cause must be treated to recover speech. There are exercises that can be done to help enhance and improve the articulation. ***Assistive communication devices can also be used***. Only in exceptional circumstances may surgery or other invasive procedures be required. If a stroke or heart disease is the cause of the speech impairment, medications as well as surgery may be prescribed [25, 26].

2 Literature Survey

Therefore, by looking at the present health-care facilities, sanitation, and medical facilities, the cost of existing diagnostic tests, and the poor ratio of available doctors' numbers to cases, it is encouraged to take up the present work of developing an assistive device for the speech-impaired community.

A literature survey basically consists of the study of previous work related to the two broad approaches to system development for speech-impaired people, namely video or image processing augmented with deep learning-based and gesture detection-based work.

The wearable electronic glove was developed for speech-impaired people with two-way communication using an LCD display [1]. The typical hand-talking system (HTS) was developed in [2] with multi-lingual (English and Arabic) support. The approach includes speaking in letter-formed words or using hand gestures to symbolize the commonly used words in daily communication. Flex sensors and accelerometers were used in both works. The wearable sign language translation device was developed utilizing strain sensors in [3], which can recognize English alphabets and is designed to comprehend finger and hand gestures.

The previous studies in [4–7, 19] focused on building a translator system supporting Indian sign language (ISL) and Malaysian sign language (MSL) by adopting video and image processing. Some work was trailed with English 26 alphabets and 0–9-digit detection. These attempts were made with deep learning methods with an average accuracy of 90%. Here, actual sign language and gesture actions will be recorded with a camera having enhanced vision capability as the system interprets the data into visual form after analysing the signal. The different algorithms were adopted in recognizing the signs.

Five flex sensors, one on each glove finger, and a three-axis accelerometer were used to develop a portable, affordable, portable speech to sign translator. The eSpeak engine is used to translate the categorized motions from text to speech to work with American Sign Language fingerspelling [8]. The sign language translator was reviewed as an interpreter in [9] for singers and non-singers.

The touchless capacitive sensor was used in the hand gesture detection system in [10]. The work claimed to be low power consumption and low cost, but had a limitation of a 10cm sensing range. This approach was improved by using a flexible strain sensor along with a capacitive sensor in [11]. Here the strain sensor was designed, which improved the accuracy for both static and dynamic gesture recognition.

The home-made flex sensors were adopted along with an on-board gyroscope and accelerometer [12]. The gesture-based recognition system based on hand movements and sensor data for English alphabets was developed in [13–16, 21], as summarized in Table 1. In [14], a different approach was adopted, augmenting the Google Assistant platform for communication in both directions. This improved the response time with the help of a 3-D accelerometer. Different sensors (such as flex, pressure, and gyroscope sensors) were fastened to the gloves in order to create a smart wearable hand gadget. With the help of a CNN, movement characteristics are retrieved and categorized. The ISL was detected with 98% accuracy [15]. In a related device developed in [16], five flex detectors, an inertial measurement unit (IMU), and three force measuring resistors (FSRs) were utilized to calculate the degree of finger bending and hand movement data. The seven greeting gestures were identified for Japanese Sign Language (JSL). The detection accuracy was found to be 82.5% using support vector machines (SVM).

Till now, the work has been able to detect only letters and numbers. The sentence generation using natural language processing (NLP) was shown in [17] for ISL. With a similar concept, a different translator was presented for Hong Kong sign language (HKSL). The accelerometer and gyroscope were employed in conjunction with time, frequency, sensor, and CNN-based characteristics. Two wrist watches were used to detect the 50 commonly spoken sentences in Hong Kong eateries [18].

The mobile application software tool SAHARA was proposed in [19] using video processing for deaf and mute people. Here, the hand's recorded video is processed using hue, saturation, and value (HSV) color space, and skin pixels are segmented. Another different approach was created, namely, the Notifying Glove for Dumb People [20]. A simple and low-cost flex sensor-based glove is intended to alert five messages stored in each of the flex sensors. Smart phones with accelerometers and gyroscopes were used in a unique way to gather sensory data for training and assessment. And it was adopted to present the gesture of hand movements and recognition by Key Intervals in [24]. Using sensory information from accelerometers and gyroscopes, the system can recognize and categorize a series of hand motions. In the series, each hand motion is seen as an object with two significant durations.

The different types of sensors, their comparative study, and rehabilitation and health monitoring were studied in [22, 23]. The relevant summary of the literature review is shown in Table 1.

Table 1. Summary of Literature Survey

Reference	Communication technique	Sensing hardware	Signs implemented
[1]/2022	2-way communication with LCD display	sensors, accelerometer	-
[2]/2020	Hand talking system (HTS)	flex sensors, Arduino, smartphone, and accelerometer	English and Arabic all letters with 98.26% and 99.33% accuracy respectively
[3]/2021	Strain sensors are used in a wearable sign language translating system	Strain sensor	all 26 English alphabets

(continued)

Table 1. (*continued*)

Reference	Communication technique	Sensing hardware	Signs implemented
[4]/2015	Creation of an algorithm to convert the ISL into English	Video processing (1,30,000 videos)	numbers, alphabets, and few phrases, 97.3% accuracy
[5]/2022	Construct A live translator for language signs using Python with TensorFlow object detection	webcam	35 signs consisting of alphabets and 1–9 numbers provided an accuracy between 50% and 80%
[6]/2022	Sign language detection is performed using video input from a smartphone camera that displays the sign language	CNN (Image processing)	There are 300 images in the collection, arranged from A to Z
[7]/2021	Building of a Sign Language Translator for People with Speech Problems	Video processing	The device has been programmed to detect multiple alphabets and figures dependent on the Malaysian Sign Language (MSL)
[8]/2022	Construction of a portable, low-cost sign language to speech interpreter	The system is viewed as gesture detection, gesture categorization, and text-to-speech, and it is powered by a battery bank	The technique is intended to operate with finger-spelling in American Sign Language
[14]/2021	Design and Development of a Communication Device Based on Hand Gestures	A 3-D accelerometer is employed to identify a disabled person's gesture	Less response time with hardware platform-based implantation
[15]/2021	Deep Learning Based Indian Sign Language Words Identification System	An intelligent wireless hand gadget is created with various sensors	98% accurate
[16]/2021	Japanese Sign Language Sensor-Based Hand Gesture Recognition System	To detect the bending and movement data, flex sensors, IMU, and FSRs are employed	7 greeting gestures
[17]/2021	NLP based Sentence Generation for Indian Sign Language	Few Indian words are converted into English sentences, using NLP and bidirectional long short-term memory (BiLSTM) method	95.8% accuracy achieved

(*continued*)

Table 1. *(continued)*

Reference	Communication technique	Sensing hardware	Signs implemented
[18]/2022	A Handheld Sign Language Acquisition and Interpretation Platform Using Smart Watches	A multi-feature framework based on (BLSTM) for precisely performing gesture-based detection with two smart watches	50 frequently utilised statements in Hong Kong eateries
[24]/2022	Key Intervals-Based Sensor-Based Hand Gesture Detection and Recognition	A one-of-a-kind neural network design for sensor-based gesture identification and interpretation	

2.1 Findings

From the discussion of literature survey, the following observations are made,

- Much of the language translation work was done in the English language (letters and numbers).
- Few works were found in specific languages (Arabic, Japanese, Malaysian, Hong Kong).
- Commonly adopted sensors are flex sensors, accelerometers, gyroscopes, strain and capacitor sensors, etc.
- Few have developed their own sensors.

2.2 Objectives with Problem Statement

The objectives of the present work are,

- Adopting domestic sensor development (wearable systems) with the purpose of making the system cost-effective.
- Focusing on using multiple sensors to collect potential gesture data.
- Focusing on developing multi-lingual translators with an emphasis on mother tongue (KANNADA).
- Finally, converting detecting gestures into meaningful communication with the real world.

In this regard, the paper consists of the development of an assistive or wearable device with multiple sensors and a multi-lingual translator (Kannada, English).

3 Methodology

The following methodology, as shown in Fig. 1, is adopted for the present work. The pricey camera might be employed because of its ability to be used in a variety of lighting settings and against a variety of backdrops. However, in order to keep the current system cost-effective, the use of a camera to determine hand position is rejected. Therefore, the glove monitors the alignment of the fingers using flex sensors and the orientation of hands using accelerometer.

Due to affordability, graphite layers are used to make homemade flex sensors. This handmade flex detector is set up by first determining the sensor values based on various finger flex orientations. The values provided by these sensors are then used in the software code. To address the issue of receiving varied values from constructed flex sensors, these sensors are organized in the form of a voltage divider circuit. The connection involves coupling one end of a flex sensor with static resistance (here, a 22-kΩ resistor), which is powered by 5 V from the Arduino. While the other end of the resistor is connected to the detector, the analog input ports or inputs on that end serve as input values to be computed.

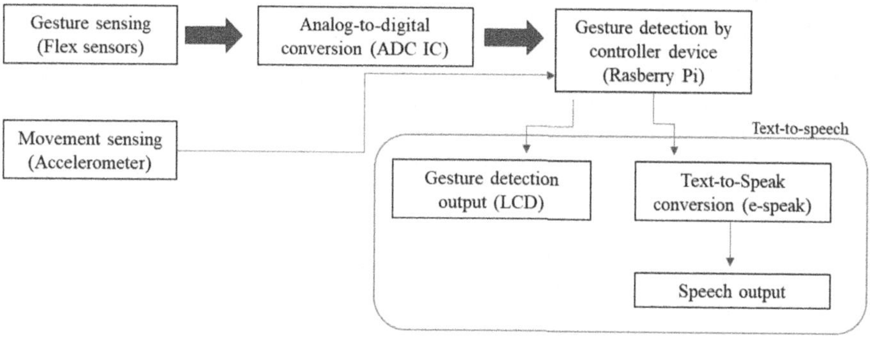

Fig. 1. Methodology of the present work

For the present work, only five flex sensors are used for the representation of five Kannada and English words, alphabets, and numbers. The resistance acquired by the gestures is recorded by the flex sensor. The Arduino then processes the value, sending the appropriate character to the text-to-speech program, which produces the output. Flexes in various topologies create a gesture that can be interpreted into verbal messages or presented as text to assist physically challenged individuals. The data is subsequently examined by the microcontroller Arduino Nano. A Bluetooth module transmits the words connected to the mapped value of the flex sensor to the liquid crystal display panel. Caretakers and medical professionals particularly profit from this strategy because it enables quick and efficient responses.

The present work could be coupled with accelerometer, to identify the direction and speed of the hands dynamically.

3.1 Accelerometer Module

As shown in Fig. 2, the ADXL345 is a tiny accelerometer with spatial resolution in three dimensions and a 13-bit resolution. The digital output of the accelerometer is in the form of 16-bit 2's complement data. It can be linked using either the serial peripheral interface with 3–4 wires or the I2C interface. The module is used to track dynamic as well as static acceleration.

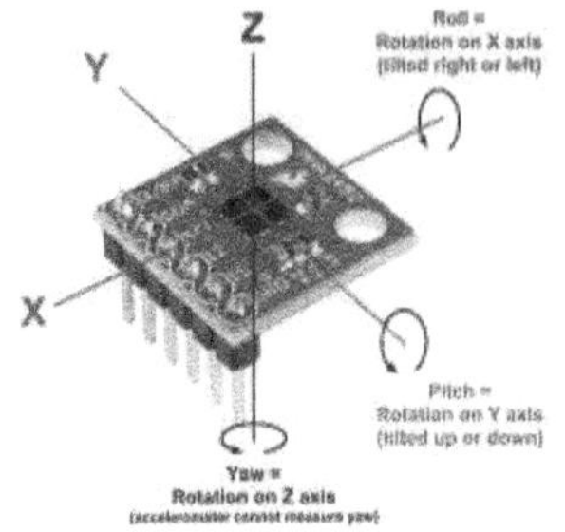

Fig. 2. Tri-axial Accelerometer

3.2 Arduino Module

The Atmega328 is utilized in the present work. It has 32 KB of flash storage for storing code, and 0.5 KB of the corresponding memory is used for the bootloader. It also consists of 2 KB of SRAM and 1 KB of EEPROM. The module works at a clock speed of 16 MHz. The controller samples five ADC values to acquire all the data from the flex sensor. As a result, the MCP3008 ADC IC would be utilized, which enables the conversion of up to 8 analog channels at a time. The sample rate is set to 100 Hz. Because the proposed gesture detection algorithm only considers times when the glove is stationary and does not require the more dynamic portions of the data, this is adequate for the application. The power bank at 5 V is adopted for the system power supply. Because the recommended battery bank has a capacity of 6000 mAh and the typical current usage of a controller is between 120 and 170 mA, This may power the circuit for 35 to 50 h, eliminating the need for frequent charging.

3.3 Bluetooth Module

The handshaking between the Bluetooth and the microcontroller is provided by setting up of HC-05 as shown below,

HC-06>>> Microcontroller; VCC>>3.3v; GND>>GND; TXD>>RXD; RXD>>TXD.

In this case, the HC-5 serves as a serial port, allowing data to be sent and received as well as string transmission to the controller. In this way, the HC-05 can be paired with either a mobile app or Bluetooth device, as shown in Fig. 3.

The typical setup of the complete work is as shown in the Fig. 4. The Table 2 summarizes the system components required for the present work.

In the present work, a low-cost Bluetooth module-HC05 is used. From 3.6 to 6 V can be used to power this module. All the flex sensor data is combined into one form before being delivered. The data is delivered to the microcontroller, Arduino Uno, who suitably decodes it. And does so while displaying the messages or reading the text.

3.4 Translator-to-Speech

The translator has been implemented to convert the detected gestures into the corresponding hand symbols. In the present work, five English words (welcome, tea, hungry, snacks,

Table 2. List of the components which made into the finished system

System aspect	Components
Sensing components	Atmega328 Accelerometer (ADXL345)
Glove	Leather glove
Arm band	Cotton sweat band
ADC chip	MCP3008 ADC IC
Ras berry Pi	Pi
Miscellaneous	Resistors and Capacitors Voltage regulator (7805), transformer, rectifiers, and filters
Power bank	Generic power bank

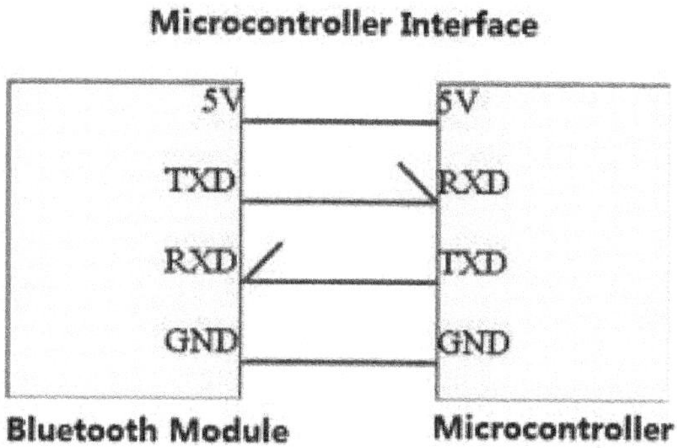

Fig. 3. Configuring HC-05

juice), four Kannada words (banni, uta, dhanyavadagalu, shubhodaya), 26 alphabets, and numbers (0–9) are included. The translator sub-system's identified words, numbers, or letters are conveyed to the speaker circuit. The synthesized speech signal is output via the speaker circuit. A Raspberry Pi GPIO pin is used to deliver the signal in this setup. The circuit has an amplifier and a filter. The filter has a frequency of 20 kHz and is a low-pass filter of second order, which belongs to the higher frequency range of sound that is audible [21]. Before being delivered to the speaker, the signal is amplified with a gain of 20 by the amplification stage.

To evaluate the extent to which the hardware can identify different movements, an analytic study is carried out. Table 3 summarizes the findings of an analysis performed to choose which sensor data ought to be incorporated into the final system. To recognize motions, accelerometer and flex sensor data are combined. This is due to the accelerometer data adding three new dimensions to the readings while enhancing the mean relative standard deviation by 0.03 as compared to the flex sensors alone.

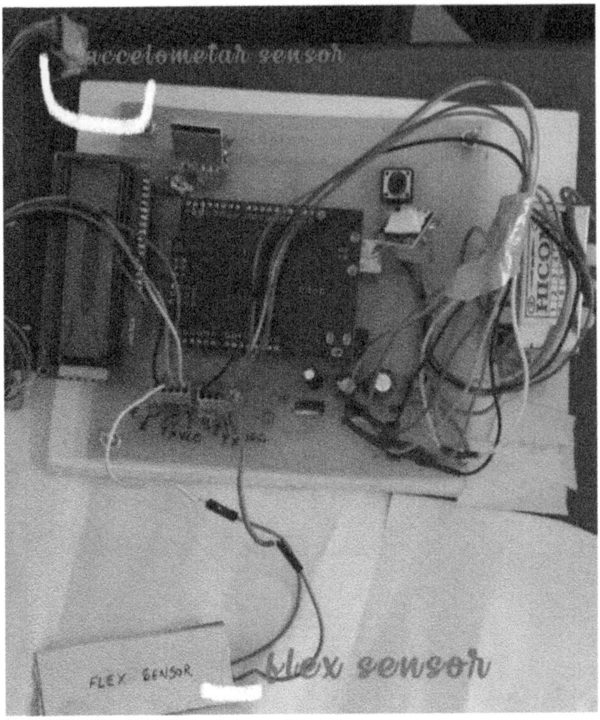

Fig. 4. Complete setup

Table 3. Comparison of sensor data

Sensor type	Mean relative standard deviation
Flex sensor	0.287
Flex sensor with accelerometer	0.317

4 Working

To begin with, five flex sensors, each having three different flex angles, provide a complete representation of 26 alphabets. Hence, out of 243 combinations, these gestures are adopted for alphabets, and the remaining are used for numbers and different words in Kannada and English. This system, when coupled with an accelerometer, can detect the direction and pace of the hand so as to detect dynamic gestures. The combined features and reading are utilized by the Rasberry Pi to detect the gestures, and they are handed over to the text-to-speech converter or translator. Here, the translator is viewed in two different aspects. In one way, the identified gestures are displayed directly onto the LCD display, and in the other way, text is converted into speech as per the detected and mapped literals or words in the translation algorithm.

5 Conclusion and Scope

The paper presents a less expensive, simple, and compact assistive device by adopting the in-house flex sensor and accelerometer as assistive technology. The elimination of vivid cameras and machine learning algorithms would reduce the cost and computation complexity of the system. In the context of the project's overall objectives, the synthesized sensors outperformed store-bought sensors. This is because they are more practical, have more sensitivity and reduced non-linearity, and are less expensive. The temperature drift that is shown by the manufactured flex sensors is their main flaw and has an impact on the system's performance. Moreover, the translator has been implemented to support different Kannada and English routine words, alphabets, and numbers. There are numerous approaches to analyze and account for it further. In a similar light, the brain signals (EEG) could be augmented with sensors for mapping the brain signals with gestures. The brain waves could be mapped in this way, and using machine learning algorithms, they could make an associated curve from an ensemble of several signals for the same motion. This would allow us to create a brain-computer interface system that would let us operate objects simply by thinking about them using electric EEG pulses.

References

1. Banerjee, I., Madhumathy, P.: IoT-based health monitoring system for speech-impaired people using assistive wearable accelerometer. In: Advanced Healthcare Systems: Empowering Physicians with IoT-Enabled Technologies, pp. 81–99 (2022)
2. Baktash, A.Q., Mohammed, S.L., Jameel, H.F.: Multi-sign language glove-based hand talking system. In: IOP Conference Series: Materials Science and Engineering, vol. 1105, no. 1, p. 012078. IOP Publishing (2021)
3. Lee, S., Jo, D., Kim, K.-B., Jang, J., Park, W.: Wearable sign language translation system using strain sensors. Sens. Actuat. A: Phys. **331**, 113010 (2021)
4. Badhe, P.C., Kulkarni, V.: Indian sign language translator using gesture recognition algorithm. In: 2015 IEEE International Conference on Computer Graphics, Vision and Information Security (CGVIS), pp. 195–200. IEEE (2015)
5. Sinha, K., Miranda, A.O., Mishra, S.: Real-time sign language translator. In: Cognitive Informatics and Soft Computing, pp. 477–489. Springer, Singapore (2022)
6. Putra, Z.H.: Developing sign language translator by using OpenCV. J. Aksara Komputer Terapan **11** (2022)
7. Azizan, M.A., Zulkiflee, I., Norhashim, N.: Development of sign language translator for speech impairment person. In: Human-Centered Technology for a Better Tomorrow, pp. 115–131. Springer, Singapore (2022)
8. Cassim, M.R., Parry, J., Pantanowitz, A., Rubin, D.M.: Design and construction of a cost-effective, portable sign language to speech translator. Inform. Med. Unlock. **30**, 100927 (2022)
9. Srikanteswara, R., Niveditha, C.B., Sindhu Sai, A., Reddy, R.K., Akshayanjali, S.A.: Sign language interpreter. In: Computer Networks, Big Data and IoT, pp. 83–92 (2022)
10. Aezinia, F., Wang, Y., Bahreyni, B.: Touchless capacitive sensor for hand gesture detection. In: 2011 IEEE SENSORS, pp. 546–549. IEEE (2011)
11. Si, Y., Chen, S., Li, M., Li, S., Pei, Y., Guo, X.: Flexible strain sensors for wearable hand gesture recognition: from devices to systems. Adv. Intell. Syst. **4**(2), 2100046 (2022)
12. Telluri, P., Manam, S., Somarouthu, S., Oli, J.M., Ramesh, C.: Low-cost flex powered gesture detection system and its applications. In: 2020 Second International Conference on Inventive Research in Computing Applications (ICIRCA), pp. 1128–1131. IEEE (2020)

13. Jani, A.B., Kotak, N.A., Roy, A.K.: Sensor based hand gesture recognition system for English alphabets used in sign language of deaf-mute people. In: 2018 IEEE SENSORS, pp. 1–4. IEEE (2018)
14. Vaidya, O., Gandhe, S., Sharma, A., Bhate, A., Bhosale, V., Mahale, R.: Design and development of hand gesture-based communication device for deaf and mute people. In: 2020 IEEE Bombay Section Signature Conference (IBSSC), pp. 102–106. IEEE (2020)
15. Jeyasheeli, P.G., Indumathi, N.: Deep learning based indian sign language words identification system (2021)
16. Chu, X., Liu, J., Shimamoto, S.: A sensor-based hand gesture recognition system for Japanese sign language. In: 2021 IEEE 3rd Global Conference on Life Sciences and Technologies (LifeTech), pp. 311–312. IEEE (2021)
17. Jeyasheeli, P.G., Indumathi, N.: Sentence generation for Indian sign language using NLP. Webology **18**(SI01), 196–210 (2021)
18. Zhou, Z., Tam, V.W.L., Lam, E.Y.: A portable sign language collection and translation platform with smart watches using a blstm-based multi-feature framework. Micromachines **13**(2), 333 (2022)
19. Gaitonde, G., Bharathan, K.S.: SAHARA: communication system for deaf and mute (2021)
20. Sumathi, K., Ramkumar, L., Abhinesh, S.B., Vishwesh, S..: Notifying glove for dumb people. In: 2022 International Conference on Communication, Computing and Internet of Things (IC3IoT), pp. 1–5. IEEE (2022)
21. Krishnan, A., Vijay, A., Balaji, M., Sreeja, B.S.: Gesture recognizer and communicator using flex sensors and accelerometer with logistic regression. In: 2020 3rd International Conference on Intelligent Sustainable Systems (ICISS), pp. 391–394. IEEE (2020)
22. Atitallah, B.B., et al.: Comparative study of digital filters for a smart glove functionalized with nanocomposite strain sensor. In: 2021 18th International Multi-Conference on Systems, Signals & Devices (SSD), pp. 1366–1371. IEEE (2021)
23. Nascimento, L.M.S.D., Bonfati, L.V., Freitas, M.L.B., Mendes Junior, J.J.A., Siqueira, H.V., Stevan Jr., S.L.: Sensors and systems for physical rehabilitation and health monitoring—a review. Sensors **20**(15), 4063 (2020)
24. Chen, Y.-L., Hwang, W.-J., Tai, T.-M., Cheng, P.-S.: Sensor-based hand gesture detection and recognition by key intervals. Appl. Sci. **12**(15), 7410 (2022)
25. Speech Impairment, by Lealani Aflague, ED443 (2008). https://www.slideshare.net/lealani/speech-impairment-ed443-presentation
26. Bishop, D.V.M., Leonard, L.B.: Speech and Language Impairments in Children: Causes, Characteristics, Intervention and Outcome. Psychology Press (2000)

Implementation of Scrum Model for Virtual Learning in Mini Project Course During COVID-19

Gunjan Behl[1], Deepali Shahane[2](✉), and Kirti Mahajan[3]

[1] Vijay Patil School of Management, Navi Mumbai, India
mailto.gunjan83@gmail.com

[2] Dr. Vishwanath Karad MITWPU, Pune, Maharashtra, India
shahanedeepali@gmail.com

[3] Bharati Vidyapeeth Deemed to be University, IMED, Pune, India
kirti.mahajan@bharatividyapeeth.edu.in

Abstract. Since the outbreak of Coronavirus, the perspective of students, teachers, and parents towards education has changed significantly. The educational institutes across different continents have been forced to shift to online mode in COVID-19. All over the world, approximately 1.2 billion students' offline classrooms had been transformed into online classrooms using the virtual learning environment. The whole education model had changed dramatically and had introduced a new concept of virtual learning, whereby digital platforms had become the need of an hour for teaching and learning of today's education model. In a research study, it had been suggested that virtual learning has changed the scenario with increased accumulation of information, and an ever-lasting changing mode of operation. With this sudden shift away from the offline mode of the classroom, the education system has adopted a virtual learning platform. So, in the current situation, it had been observed that the lack of coordination among BCA students for developing their projects because no face-to-face communication was there between **students** and teachers, so as result it leads to poor quality projects and inadequate learning.

Keywords: Covid-19 · Scrum Model · Virtual learning · Agile approach · Waterfall model · Hybrid approach

1 Introduction

The normal lifestyle of people had been hampered across the globe in this pandemic and to overcome, these problems only higher education institutes and even schools had also shifted to online education with virtual education platforms. The alternative solution to ordinary face-to-face classes was virtual education in this pandemic situation. BCA projects are student-driven and guided by mentors. In this covid pandemic situation, due to lack of student-mentor interaction, it had affected the development of a student's project with poor quality and poor developing skills.

S. Singh et al. (Eds.): AIDSP 2023, CCIS 2440, pp. 57–68, 2025.
https://doi.org/10.1007/978-3-031-88762-8_6

In most cases, mentors guide the students in the project development to follow the traditional waterfall model. It requires lots of effort and time to develop projects using the waterfall model. To overcome the problem of poor quality, and time management, a case is considered where a comparative study of the Agile-Scrum model had been conducted with the Waterfall model for developing the BCA projects. For the development of BCA projects, two groups of students were formed where the waterfall model was followed by Group A and the agile scrum model was followed by Group B. This research focused on the use of the Scrum model during BCA project development and its outcomes.

In this Covid-19 pandemic situation, with a virtual learning environment, the development of the projects for BCA students was inspected by a team of project mentors and under their mentorship, the students completed the whole lifecycle of the project. The majority of the students faced network issues as they were residing in remote areas, therefore they were restricted with educational facilities. The geographical location of the students, network facility, resources available, and other factors was handled in the few meetings with minimal communication between students and project guides. The project quality gets hampered due to these above-mentioned factors. Therefore, this paper focuses on identifying the shortcomings of the existing process and highlights the outcomes of the agile approach used in the development of the project. Based on the above scenario, a study of an Agile and Waterfall model had been suggested and compared. This approach improved student-mentor collaboration, student monitoring, and the evaluation process.

2 Literature Review

Previous Work

Various notions of the waterfall model have been offered in the literature.

Adel Alshamrani [1] defines the waterfall model as very simple to implement as it should follow in a linear manner. It is easy to understand. The project is executed with phase-wise sign-off meetings.

S. Balaji [2] describes a waterfall model that follows the sequential manner approach for software development. In this model, after the completion of the requirement phase, we can start the design phase. Further changes in requirements will not be considered. Every phase is completed in a specified time, after that it moves to the next phase.

Agile is repetitive and evolving at every step, this approach has been used in the process instead of in-depth planning in the starting. It has very nimble and changing phases; these phases work out in parallel also [3].

Like software development, the teaching process can also make use of agile methods by considering the people involved, their varied interests, and the restricted access to the resources available in the system, to satisfy the needs of the current scenario. There are many more limitations concerning schedule, the rigidity of the deliverables, and many unknown factors that can be changing the circumstances. In this pandemic era, it has proven the point that the situation is beyond the control of stakeholders of the education system. So due to this, adapting to the new and changing teaching-learning process is a need for an hour.

Various notions of Scrum have been offered in the literature.

Pasquale Salza [3] shows the implementation of agile methodologies in education increase the energy level of the students to work together, in a goal-oriented, task-oriented, and effective way, this reflects their increasing interest in studying. This improves the energy level to work together, in a focused, and effective way. In his research, he has formalized agile methodologies in the educational context. Agile methodologies transform teaching and learning from knowledge transfer to knowledge generation via rich interaction and experience. The result of agile methodology is that mentors become facilitators, guides, and inspiration and students become self-directed learners.

Erika Lozada-Martine [4] defines the project activities are divided into Scrum software development agile methodology. It all depends on the log history in which all activities are stored. Team members determine and analyze the log history and looking at it they prioritize the activities which work should be done in the next iteration. In this technique, each activity is assigned to a team member. Every iteration can be of a maximum of 4 weeks and the working product will be given to the user for review.

Guillermo Rodriguez [5] states that the time, scope, and facilities limitations within academic contexts hamper students' experiential use of professional Agile Methods. In the context of the above, to enhance students' vulnerability to Scrum in teaching and learning, researchers developed Virtual Scrum with the help of educational tools such as a web browser, charts, blackboards, document viewers, and a calendar that simulates a Scrum-based team room. Researchers tested Virtual Scrum with a group of 45 students running a capstone project with and without Virtual Scrum support, where students' feedback showed that Virtual Scrum is a workable and useful tool in a Scrum team room to implement the different elements and to perform activities throughout the Scrum process. Students found that Scrum had enabled them to conduct the realistic collaboration practice that they gained from their different past experiences and was very valuable to their learning.

Larissa Freitas Santana [6, 7] represents in his paper the Scrum methodology which helps to manage the software project and pedagogical practices, wherein in this paper, researchers presented an investigation with the students of scientific initiation of a university from Brazil's northeast. The purpose of the study is to analyze the attainment of the students with the Scrum practices, as well the acumen of the learning of the methodology, which shows the positive and the negative aspects of students' involvement. In closing, the Scrum methodology defines that it helps students' involvement, and they easily manage projects even when the team has limited or no experience fostering learning, adding value and satisfaction to the student.

Rebecca Pope-Ruark [8] reveals in this paper researchers reveal that when the scrum was not introduced to the classroom, it had been found that for the completion of the projects, students consume more time, have less involvement, are less goal-oriented, and are dependent on supervised learning but after introducing Scrum to the classroom, it has added a new facet to group projects where the students learn collaborative learning, self-directing, self-learning and promote interactive learning. This is the new transformation for new classroom pedagogy.

Minjie Hu [9] represented that Scrum is designed for the teaching process in second-year software process courses for web development tasks. The researcher had focused on the following objectives:

1. How students can quickly and effectively handle the requirement change during the development of the project? 2. How to evaluate the extent of work done by the individual teachers while using a Scrum-oriented project-based course?

The focus was to build up students' technical development skills rather than project management methods. Hence, the objective was to develop students' abilities where they can enhance their technical development skills by using Scrum as a software project management method. Scrum focuses on accommodating the changes in software development during all sprints. The students were able to handle the changes.

Implementing the agile methodology in education is the key to this ever-changing era of education. Teaching must be flexible, and we cannot stick to our years' implemented methods. This process needs to be modified according to changing, distinct and various learning needs depending upon the student's interests, style, and capacity. Agile and flexible learning processes need to be adopted by students as well to cope with changing requirements of the market, technological advancements, and the need for research. A properly defined approach, with a prioritized teaching process and applying suitable strategies based on the needs of students is key for a good teaching activity. Students need to acquire a similar approach while learning as well. The teaching and learning process go together and cannot be processed alone [6, 7].

3 Going Agile with Scrum

Scrum focuses on robust, synchronic workflows [5, 10] It breaks projects into compact, continual periods. Scrum focuses on getting more work done speedily. By reviewing the regular sprints in the Scrum approach, we can recover time and effort for the development process. During development, the changes in scrum are agreeable at an early and late stage during development. Scrum focuses on getting more work done speedily.

What Makes Waterfall Unique?
In the waterfall model, the project is divided into small modules which follow linear and sequential phases. The output of one module works as the input for the other module. Changes can only be done at the requirement phase. There is no choice of making changes at later stages. So, it consumes a lot of time and effort. Hence, the projects which can be completed continuously without any backtracking changes are best suitable for the waterfall model.

Difference Between Scrum and Waterfall Model
Waterfall keeps the customers and stakeholders on hold and includes them by the time of the completion of the project. It may take extra time as reviewing is done at the end, that is completion only, if found inappropriate then Process is back to level 1. Work is broken down into two phases and the team works intently. The required linear documentation is done at the initial stage. During the requirement phase, the proper documentation takes place. The waterfall model works well with smaller projects [11].

In each of Scrum's approaches, its customers and stakeholders are included. The regular spin saves lots of time and money in the development process. Work is break-down into teams as individual responsibilities. The feedback is collected from product owners and stakeholders in the scrum approach. A Full stop customer is kept in the loop and constantly takes his place throughout the process of development. For handling difficult and complex projects, the institution's scrum method is more suitable [11].

4 Research Methodology

A case study approach was used for this research paper. Here, we had a BCA student single group for study purposes. The selected BCA group was studied intensively with minute details with the pattern of study, understanding level, quality of the project, and student monitoring and evaluation. The factors that affect the quality of the project such as the mindset of the students, level of understanding, the complexity of the project, and learning eagerness need to be considered. Through this case study method, the behaviour pattern of the students was studied. This research aims to find out a better approach to complete the project, focusing on improved quality and within the stipulated time along with fewer efforts. To complete the project, digital tools such as conferencing tools like zoom or Ms. Teams, open-source project management tools, google drive, and classrooms were used. BCA students were categorized into two groups: Group A and Group B.

5 Project Lifecycle

The BCA batch was divided into two groups for developing the projects. Two groups of students were formed: Group A follows the waterfall model and Group B follows the scrum approach for the development of their projects. While forming the groups, no specific criteria were followed and randomly students were picked up and asked to form a group.

6 Project Allotment Details

- Total no. of students: 60
- Total mentors:5
- Group size allowed max 6 students/group
- Total groups 10 to12 groups
- Groups per mentors:2 / 3 groups

7 Project Life Cycle Followed by Group A

Group A followed the sequential and linear approach which is the traditional way of software development. Here the project was planned in such a way that a detailed layout was drafted for the project. With this approach to development, it became easy to rectify the defects of one phase before moving to the next stage.

During the requirement phase, the students discussed requirements, took guidance from the mentors, and noted down the requirements. Students sought to understand and know what the mentors' expectations from the software are. After checking the feasibility of the requirements on different factors such as technical, economic, operational, etc., the requirements were frozen.

Documentation was done by the team throughout the development process. It kept a record of time and quality and checked if the process was followed sincerely. After the development of the project, the integration of different modules was done and, in the end, testing was carried out. It checked the final product was developed according to SRS prepared at the beginning.

The mentors reviewed the documents, highlighted the issues, and asked for a resubmission of the document. This process was repeated until a document was accepted. As a waterfall or linear approach was used, the modifications at later stages could not be accommodated and the tasks were not completed efficiently in the stipulated time (Fig. 1 and Table 1).

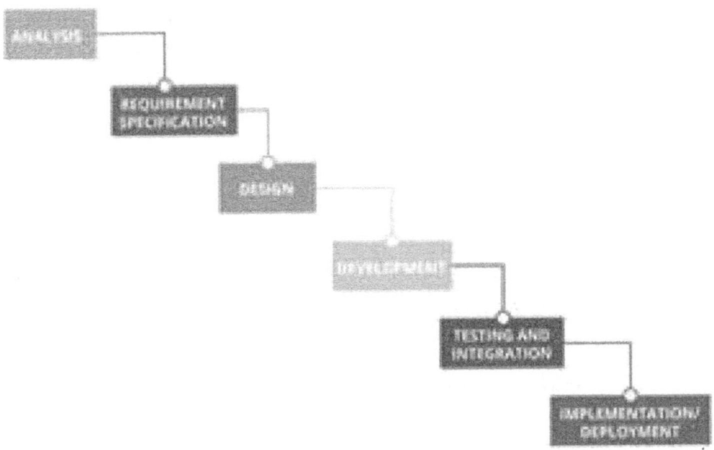

Fig. 1. The proposed model for group A

Table 1. Project deliverables

Deliverable	Duration	Marks
Software Requirement Specification (SRS)	4 weeks	15
Software design and development Method	7- 8 weeks	25
Technical Documentation	1 week	10
Final Report Submission	1week	10
Project Evaluation	1 week	40

8 Project Life Cycle Followed by Group B

To reduce the project complexity and understanding level of the BCA students, the proposed model of the scrum with agile methodology was suggested as it divides the tasks into smaller sprints(cycle). These are the roles that appear in the process while adapting the scrum approach to the educative context.

Project Owner: Subject teacher of software project management has performed this role. The owner put forth different project topics on which the project can be developed by the students. Topics given were student-centric topics that were open and can be modified to achieve the learning objectives of the subject. Students' interests and current trends, and applicability in business were considered. The project owner decides what to be built in the project.

Scrum Master: The mentor who acts as a Scrum master helps the team to perform best. The scrum master conducts meetings, during each sprint, and the documentation is maintained by an active team member who acts as a leader for that team [12].

Team Leader: He/she is the link between fellow team members and the mentor i.e., scrum master. The leader connects fellow team members and mentors, schedules the meeting, and gives important notifications to both parties.

Scrum Team: A team of students in the project forms a scrum team A scrum team works on the project. The team develops the project and delivers a complete product at the end. Responsibilities to be handled and the progression of the project is open to all members. The team members work in coordination, and they are self-managed.

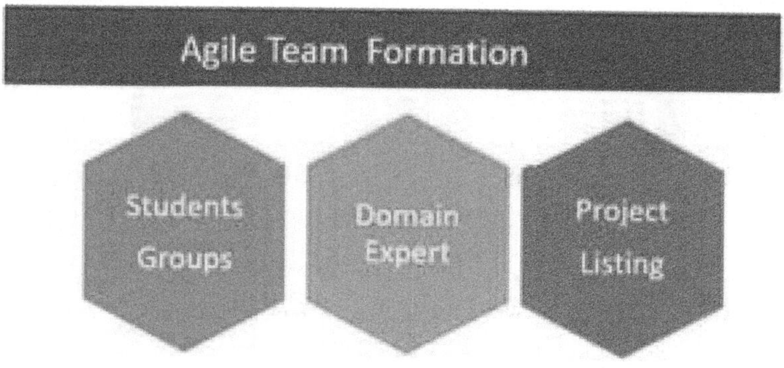

Fig. 2. Key Elements of Agile Team Formation for Team B

The Steps of the Project Development Life Cycle Are

The subject teacher gave topics, and the scrum master gave this proposal to the students for project development. The mentor acted as a scrum master. The project started with a proposal given by a mentor, a topic that students should develop. With simple planning, the process became easy and nimble. It became very easy to manage.

At the start of the sprint planning meeting was conducted, to define team formation, project mission, and activities planning. Scrum master conducted the first meeting which was called sprint 0. In this meeting, the project was explained, and different stages /steps were also discussed.

A sprint is a basic unit of work. The project tasks are considered time-boxed events with a maximum duration and are designed to allow critical transparency and inspection. Thus, the sprints are a set of activities, coherently organized to achieve the project mission, with a time duration of 2 weeks or less. A sprint is a basic unit of work.

The total project duration including the final evaluation of the project was 15 weeks each semester. The duration of project development was approximately 9 weeks. On average 3 sprints were conducted. The groups submitted deliverables during that time and documentation was presented. For better coordination, the roles within the team were allotted in rotation so that each member of the team was aware of all the activities, and the interest of all the team members was also maintained.

The concept of pair programming has also been applied to some extent so that one student was acting as a developer and another member became a reviewer/ tester in the process. It turned into a better-quality project.

Every sprint was conducted within 3 weeks duration. Sprint 1 was started by focusing on user stories that cover the user requirements followed by planning, execution, presenting the work to the customer, and then implementation. The work completed during this time was evaluated and retrospection was carried out.

Sprints 2 and 3 were started with a review of the product backlog and the rest of the activities were like the previous sprint. The final documentation was completed, and the report was delivered for evaluation of the project by examiners in 3 weeks' time (Fig. 3 and Table 2).

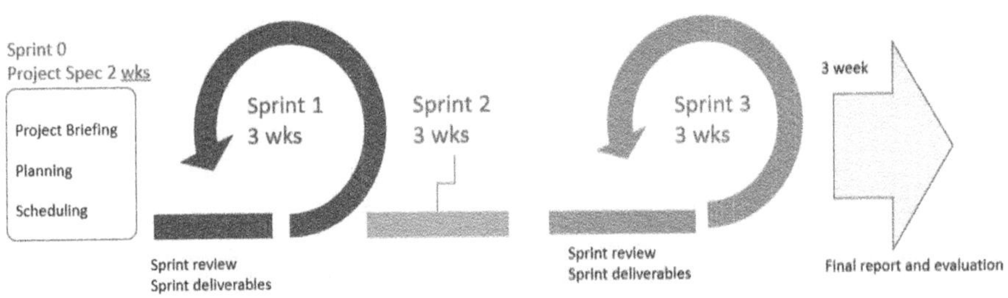

Fig. 3. Agile-based project execution plan for Team B

Table 2. Project Deliverables Using The Scrum Model

Deliverable	Duration	Marks
Sprint 0 Project Specification (PS) Identify Stakeholders User Stories Workshop High-Level Sprint Planning	2 weeks	15
Sprint 1 Deliverable Definition User Stories Planning Execution Demo Implementation Retrospection/Evaluation	3 weeks	15
Sprint 2 Product Backlog Review Sprint Planning Execution Demo Implementation Retrospection/Evaluation	3 weeks	15
Sprint 3 Planning Execution Demo Implementation Retrospection/Evaluation	3 weeks	15
Project Report	1 week	10
Project Evaluation	2 weeks	30

9 Findings and Suggestions

A comparison of the waterfall method and Scrum method has been done to check the efficacy of the Scrum approach based on students' quality and virtual learning. The impact of using the scrum model in the comparative study has been studied. It is found that this improves the student-mentor interaction through Scrum meetings.

1. Increase in the understanding of the project.
2. Peer Learning build-up the confidence level of the students.
3. Reduced the complexity of project development.
4. Improve and enhance the quality of the project.
5. Improvements in the student monitoring and assessment process.
6. Enhance the skills development of the student to meet their industry expectations.

7. Students could be assessed on different criteria as they were playing different roles throughout the project.
8. Students could apply different skills learned during the program such as soft skills which include presentation, communication, coordination, leadership, and many more.
9. They could apply project management skills and apply different concepts to reality.
10. Students residing in different locations having many infrastructural and resource issues applied different problem-solving techniques (Table 3).

Table 3. Evaluation Criteria For Project Success Quotient

Project Success Factors - Evaluation Criteria Mapping	Conventional	Agile
Increase in the understanding of the project	2	
Peer Learning builds up the confidence level of the students	1	3
Reduce the complexity of the project development	4	2
Increased the quality of the project	2	2
Improved the student monitoring and assessment process		4
Improved the development skills of the students that help them to meet the industry expectations	2	
Students could be assessed on different criteria as they were playing different roles throughout the project	1	
They could apply project management skills and apply different concepts to reality	3	
Miscellaneous aspects of the project	1	3

It has been observed that some students residing in remote areas are facing a lot of problems with student-mentor interaction and were unable to meet the sprint deadline assigned weekly. Therefore, to resolve such problems it has been suggested that a hybrid model of the scrum approach and waterfall model should be adopted for BCA projects. The proposed hybrid model helps in replacing the post-design phases of the Waterfall model with the scrum model. The hybrid approach will help the students to better understand, commanding, and developing of the project (Table 4).

Table 4. Rating For Project Success Quotient

	Group A Students	Group B Students	Mentor	Program Owner
Project Success Quotient	3	5	5	4
Scale: Low 1, Below 2, Avg 3, High 4, Very High 5				

The above table depicts the Project success quotient as perceived by different stakeholders of the project. It states that team members from Group B and mentors are more satisfied than Group A (Table 5).

Table 5. Soft Skill Development Matrix

Soft Skill Development Matrix	Traditional	Agile
Communication	High	Very High
Coordination	Average	High
Leadership	High	Average
Product Knowledge	Average	High
Distributed Team Working	Low	High
Problem Solving Skill Tracking	Low	High

Soft skill Development rating as perceived by students from the groups is summarized in the above table. The Agile Development method helps to improve the overall soft skills as compared to the traditional method of the students.

10 Conclusion

The researcher suggested a model for the BCA projects in this COVID-19 pandemic situation. Through this model, the student's involvement has been increased which is in the form of peer learning (where the students help each other to solve their problems of each other). Through Scrum meetings, this improves the student-mentor interaction through Scrum meetings. This will enhance the time management and task division throughout the project. The suggested model increases the understanding of the project and comparatively, it is better than the Waterfall model. The output of using this model has reduced the complexity of the project development. This will enhance the skills development of the BCA students to meet their industry expectations.

The use of the Scrum method for BCA software development projects has given exposure to the students, at the educational level and prepared them practically, for applying this agile approach in the professional software development projects at the industry level. Lockdown and travel restrictions have benefitted students to practice agile in real distributed teamwork. Thus, making them industry-ready. This development approach bridges the gap between educational and professional platforms.

The study is limited to only a group of students of the BCA program and further study needs to be conducted in the future to validate the findings considering other programs with various subjects and confirm the scalability of the framework.

Therefore, the paper stated that there was an enhancement in student engagement during the Scrum process.

Acknowledgement. The study was conducted in the Bharati Vidyapeeth's Educational institutions in India.

References

1. Alshamrani, A., Bahattab, A.: A comparison between three SDLC models waterfall model, spiral model, and incremental/iterative model. IJCSI Int. J. Comput. Sci. Issues **12**(1), 1694–0784 (2015)
2. Balaji, S.M., Murugaiyan, S.: Int. J. Inf. Technol. Bus. Manage. **2**(1) (2012)
3. Salza, P., Musmarra, P., &ferrucci, F.: (2018)
4. Lozada-Martinez, E., Naranjo, J.E., Garcia, C.A., Soria, D.M., Toscano, O.R.: (2019)
5. Rodriguez, G., Lvaro, A., Soria, M., Campo: (2015)
6. Santana, L.F., Santos, L.F.C.D., Silva, T.S.C., Villar, V.B., Rocha, F.G., Gonçalves, V.: Scrum as a platform to manage students in projects of technological development and scientific initiation: a study case realized at UNIT/SE. J. Inf. Syst. Eng. Manage. **2**(2), 2468–4376 (2017)
7. Pope-Ruark, R.: (2011)
8. Hu, M.: A case study of using scrum in teaching software process. In: The 9th Annual Conference of Computing and Information Technology Research and Education New Zealand (CITRENZ2018) and ITx (2018)
9. Pope-Ruark, R.: Know thy audience: Helping students engage a threshold concept using audience-based pedagogy. Int. J. Sch. Teach. Learn. **5**(1) (2011)
10. Shahane, D., Jamsandekar, D.P.: Study of critical success factors for adaption of agile methodology. IJCET **4**(4), 539–546 (2013)
11. Shahane, D.: Empirical study of agile methods in software development in Indian context. Doctoral dissertation, Bharati Vidyapeeth (2014)
12. Shahane, D., Jamsandekar, P., Shahane, D.: Factors influencing the agile methods in practice - literature survey & review. In: International Conference on Computing for Sustainable Global Development (INDIACom), pp. 556–560. IEEExplorer (2014)

Exploring Land Distribution Patterns and Network Structures in Indian Tribal Communities

Shanmuk Srinivas Amiripalli[1], Praveen Gupta[3], Rakesh Nayak[2], and Umashankar Ghugar[4](✉)

[1] Department of Computer Science and Engineering, GST, GITAM University, Visakhapatnam, Andhra Pradesh 530045, India

[2] Department of Computer Science and Engineering, OP Jindal University, Raigarh, Chhattisgarh, India
rakesh.nayak@opju.ac.in

[3] School of Engineering and Technology, Jaipur National University, Jaipur, India

[4] Department of CSE, University Institute of Engineering (UIE), Chandigarh University, Mohali, Punjab 140413, India
ughugar@gmail.com

Abstract. This research examines the socio-economic challenges faced by tribes in India, focusing on land distribution patterns and network structures. It highlights the importance of land for tribal communities and the impact of globalisation and liberalisation policies on issues such as poverty, health, education, and exploitation. The study proposes a methodology that analyses land distribution patterns within tribal families and utilises graph theory models to study diverse networks representing tribes in different areas. By understanding the behaviour and characteristics of these groups, the research aims to extract valuable insights for effective problem-solving. It also explores the implementation of government schemes in tribal areas while considering the preservation of government lands. The findings emphasise the significance of comprehending land ownership patterns and network structures in formulating suitable strategies for tribal development. The study contributes to a better understanding of the challenges faced by tribes. It underscores the roles of leadership, traditional conflict resolution methods, and government support in enhancing the socio-economic conditions of tribal communities.

Keywords: Network Science · Tribe Network · social behavior of tribes · land distribution · socio-economic culture

1 Introduction

Indigenous tribes in India have a rich cultural heritage and occupy a unique place in the country's social fabric. These tribes, often called Adivasis, inhabit various regions of the country, including remote and inaccessible areas. They are characterised by their distinct languages, customs, and belief systems, which set them apart from the mainstream Indian population. The tribes in India have a long history and have inhabited the

S. Singh et al. (Eds.): AIDSP 2023, CCIS 2440, pp. 69–76, 2025.
https://doi.org/10.1007/978-3-031-88762-8_7

land for generations, maintaining a harmonious relationship with nature and the environment [1]. They have developed intricate knowledge systems and practices deeply rooted in their traditional way of life [2]. Despite their diversity, tribes share commonalities, such as a solid connection to their land, a collective identity, and a communal way of living. Tribal communities have faced numerous challenges throughout history, including marginalisation, discrimination, and exploitation. Their isolation and geographic remoteness have often made it difficult to access basic amenities, quality healthcare, education, and economic opportunities [3]. As a result, poverty and socio-economic disparities persist within these communities. Globalisation and liberalisation policies have also significantly impacted tribal communities in India. While these policies have brought economic growth and development to certain parts of the country, they have also posed challenges to the traditional ways of life and cultural practices of tribes. The rapid pace of modernisation and the encroachment of external influences have disrupted their social structures and threatened their cultural identity [4].

Recognising the importance of preserving and uplifting tribal communities, the Indian government has implemented various schemes and policies for their welfare and development. These initiatives encompass education, healthcare, livelihood opportunities, and land rights. However, there is still a need for a deeper understanding of the specific challenges faced by tribes and the formulation of targeted strategies to address their unique needs [5]. This research aims to shed light on the socio-economic challenges faced by tribes in India, with a particular focus on understanding their land distribution patterns and network structures. By delving into these aspects, the study seeks to gain valuable insights into the behaviour and characteristics of tribal communities. It aims to contribute to developing practical solutions that respect their cultural heritage and empower them to improve their socio-economic conditions. By exploring the complexities of tribal life, the research endeavours to highlight the importance of recognising and preserving the diverse cultural identities of tribes in India. It underscores the need for inclusive policies, effective governance, and sustainable development approaches that consider the specific needs and aspirations of tribal communities. Ultimately, this research seeks to contribute to a better understanding of the challenges faced by tribes and pave the way for their holistic growth and well-being.

The paper has been organised in the following order: Sect. 1 describes the introduction part, Sect. 2 describes the literature review, Sect. 3 describes the proposed methodology, and Sect. 4 describes the result and discussion. Finally, Sect. 5 describes the conclusion.

2 Literature Review

Land is an indispensable natural resource for human survival, and the right to land is protected by the Indian Constitution as part of the right to life [6]. It serves as a vital source of income. It plays a significant role in social status, identity, security, and self-reliance for most people, especially indigenous communities whose entire way of life is intricately connected to their ancestral lands. However, in tribal regions, there are instances where land is forcefully acquired for "public use" and development, which raises concerns about the recognition and protection of tribal land rights [7]. The collective ownership of land and other resources is a distinguishing characteristic of tribal communities, and the increasing demand for land and the recognition of private property ownership by the state have presented challenges for tribals. Disputes over land ownership have become more prevalent among tribe members, different tribal groups, and between tribals and non-tribals [8]. Existing institutional mechanisms for resolving these disputes need to be updated and aligned with the evolving circumstances while also considering and enhancing the role of traditional conflict resolution methods. The difficulties tribals face when their collective land ownership and community norms of succession are not acknowledged during land acquisition or the implementation of forest laws necessitate empathetic attention and supportive measures [9].

Recognising the criticality of land rights for tribals, efforts have been made to address their vulnerabilities by formally recognising their land claims. Historically, tribals lacked formal suits for their cultivated land, leaving them vulnerable to eviction. Legal recognition of land rights protects them from potential exploitation by moneylenders and landlords and helps change public perception towards tribes. Strengthened administrative procedures and directives to safeguard tribal welfare have curtailed government officials' unauthorised collection of fees from villages, such as forest guards or police constables, improving conditions [10].

Tribal communities have a long-standing history of successfully managing their natural and communal resources. They possess a wealth of traditional knowledge systems that have allowed them to utilise resources while sustainably maintaining harmony with their surroundings. Their stewardship and resource management capabilities are evident in their deep understanding and respect for their environment. We must recognise that introducing new structures and resources may not align well with tribal cultures and may result in a diminished sense of ownership. Traditional community resources are best maintained when individuals rely on their cultural knowledge systems honed through generations living in harmony with nature. For successful integration and management of new resources, it is crucial to understand and respect tribal culture and community structures and engage them in the decision-making process from the early stages. This requires an inclusive approach that values their wisdom, seeks their input, and ensures their active participation [11].

3 Proposed Methodology

The research consists of multiple phases focused on analysing land distribution patterns, network structures, and their implications for tribal communities in India. In the first phase, the study examines how land is distributed within a tribal family in a specific tribal region, emphasising the significance of land for their survival and livelihood. This analysis sheds light on the socio-economic and cultural background of the tribe, highlighting the specific challenges they face regarding land ownership [12].

Moving into the second phase, the research extends its scope to encompass three different networks: tribes, non-tribes in non-tribal areas, and non-tribes living in tribal areas. These networks are represented using graph theory models, enabling a comprehensive understanding of the behaviour and interactions within each group. The analysis uncovers valuable insights into these networks' structural properties and characteristics by applying standard network science parameters, allowing for a deeper understanding of the dynamics at play.

The research extracts useful information from the graph analysis results in the subsequent phase. This information serves as a basis for addressing the challenges faced by tribal communities and developing improved strategies for their socio-economic development. The research also examines the implementation of government schemes in tribal areas, aiming to identify practical approaches that can enhance the well-being of tribal communities without encroaching on government lands. Overall, this research underscores the importance of comprehending land ownership patterns and network structures among different groups of people. By understanding these dynamics and utilising the insights gained, it becomes possible to devise tailored solutions to the problems faced by tribes, ultimately contributing to their overall socio-economic progress and well-being (Fig. 1).

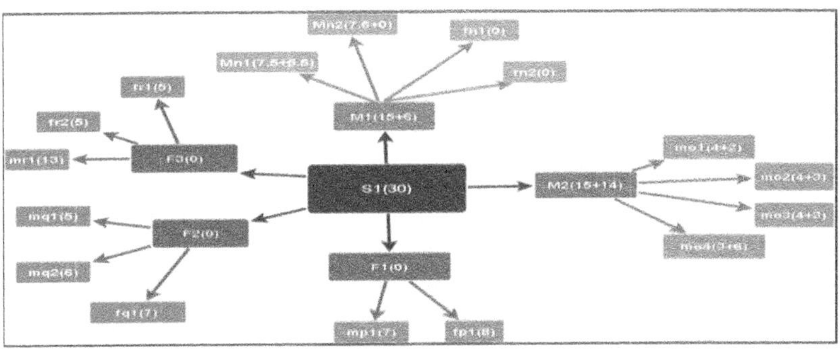

Fig. 1. Sample Tribe living in tribe area and their land acquisition.

In the given family tree, the eldest person, S1, has two sons (M1 and M2) and three daughters (F1, F2, and F3). S1 initially cultivated thirty acres of land. In tribal families, it is typical for land shares not to be given to female children. Therefore, in this case, the entire thirty acres of land was distributed equally between the two male children, resulting in each receiving fifteen acres. Tribal communities rely on agriculture as their primary means of sustenance, inheriting farming practices from their ancestors. Due to their isolation and dependence on agriculture, tribes often require additional land to support more prominent families. As a result, they may occupy nearby government lands and convert hills into cultivable areas. Observations from the family tree reveal that families with more male children tend to occupy more land for their future use. For example, M1, with two male children, occupies six acres of government land, while M2 occupies fourteen acres with four male children. Further down the tree, M2's sons, MO1, MO2, MO3, and MO4, occupy 2, 3, 3, and 6 acres of land, respectively.

Additionally, certain family members have migrated to urban areas for employment, resulting in their official land shares being fixed and no further land expansion. The total land occupied by S1's family, including both official and unofficial claims, is approximately 114.5 acres. The analysis shows that land distribution within the family is directly proportional to the number of male children. Families with more male children tend to occupy more government lands for survival and sustenance. This information highlights the relationship between family size, precisely the number of male children, and land acquisition in tribal communities. It demonstrates the importance of understanding these dynamics when considering land distribution patterns and their socio-economic implications within such communities (Fig. 2).

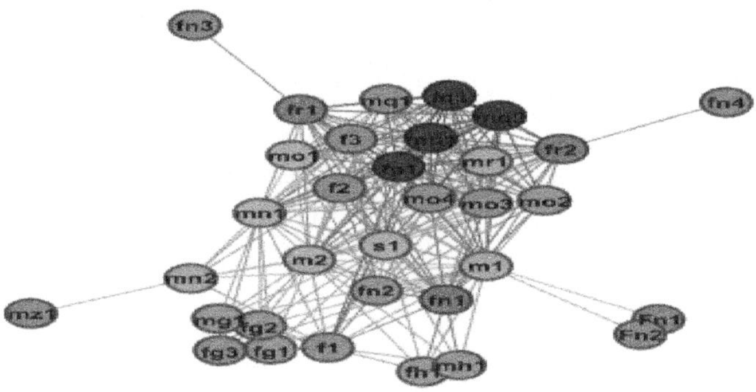

Fig. 2. Sample tribe family Graph network representation

Proposed Algorithm

1. *Begin the study by selecting tribes and non-tribes for analysis.*
2. *Collect comprehensive information about each family member, including their relationships and details regarding the land they possess.*
3. *Organise the gathered data into tree and graph theory models for further examination and analysis.*
4. *Utilise standard network science parameters to analyse the resulting graphs and derive insights from them.*
5. *Extract valuable information from the graph analysis to better understand the behavioural patterns within different groups of people.*
6. *Apply the extracted information to address the specific challenges faced by tribal communities and aid in developing more effective government strategies and interventions.*
7. *Conclude the study, contributing to the research on tribal communities and their socio-economic development.*
8. *End the research process.*

4 Results and Discussions

The table summarises the comparison between a tribe network and a non-tribe network based on various parameters and metrics. The tribe network consists of 37 nodes, indicating a more significant number of individuals within the tribe than the non-tribe network, which has 14 nodes. Regarding edges or connections between nodes, the tribe network has 236 edges, while the non-tribe network has 31 advantages. This indicates a higher level of interconnectedness within the tribe network (Table 1).

Table 1. Comparison of tribe and non-tribe characteristics

Parameters	Tribe	Non Tribe
Nodes	37	14
Edges	236	31
Average Degree	16	5.74
Modularity	0.172	0.104
Avg Cluster coefficient	0.797	0.765
Avg Path length	1.7	1.807
Betweenness centrality (max) %	31.3	42.01
Betweenness centrality (min) %	3.09	8.74
Authority(max) %	9.8	17.02
Authority(min) %	3.47	8.67
Closeness centrality (max) %	25.01	26
Closeness centrality (min) %	3.09	8.47
Cluster coefficient(max) %	18.6	38.5
Cluster coefficient(min) %	3.7	12.59
degree(max) %	15.75	34.03
degree(min) %	3.09	8.7
Eccentricity(max) %	73.02	42.22
Eccentricity(min) %	9.27	26
Eigenvector centrality(max) %	9.27	16.7
Eigenvector centrality(min) %	3.03	8.73
Harmonic closeness centrality(max) %	18.4	25.6
Harmonic closeness centrality(min) %	3.09	8.33
Hubs(max) %	9.27	16.77
Hubs(min) %	3.03	8.6

The average degree, representing the average number of connections per node, is 16 in the tribe network, while in the non-tribe network, it is 5.74. This suggests that individuals within the tribe network tend to have more connections on average compared to the non-tribe network. Modularity, which measures the division of the network into distinct communities, is 0.172 for the tribe network and 0.104 for the non-tribe network. This indicates a more robust community structure within the tribe network. The average cluster coefficient, which measures the tendency of nodes to form clusters or groups, is 0.797 in the tribe network and 0.765 in the non-tribe network. This suggests a high level of connectivity within immediate neighbourhoods in both networks. The average path length, representing the shortest path between any two nodes, is 1.7 in the tribe network and 1.807 in the non-tribe network. This indicates a relatively efficient flow of information within both networks. Additionally, the table provides information on various centrality measures, including betweenness centrality, authority, closeness centrality, and eigenvector centrality. These measures assess the importance and influence of nodes within the networks. In summary, the table highlights the structural and connectivity differences between the tribe network and the non-tribe network. The tribe network exhibits more nodes, higher interconnectedness, a more robust community structure, and a greater average degree. These findings provide insights into the network dynamics and social relationships within tribal and non-tribal communities.

5 Conclusion

The research findings highlight the importance of understanding land ownership patterns and network structures in tribal development. The analysis reveals that land holds immense social, economic, and cultural value within tribal communities. However, inadequate land rights recognition and protection have resulted in disputes and challenges. The tribe network demonstrates a more robust network structure with higher interconnectedness and a more substantial community presence than the non-tribe network. Both networks exhibit high connectivity within immediate neighbourhoods and efficient information flow. It is crucial to implement inclusive policies, effective governance, and sustainable development approaches that respect and preserve the cultural identities of tribes while addressing their specific needs and aspirations. Traditional conflict resolution methods, leadership roles, and government support play pivotal roles in overcoming challenges faced by tribal communities. By integrating these insights, policymakers can formulate strategies that promote the holistic development and well-being of tribal populations.

Acknowledgements. I am grateful to all my co-authors who helped during the writing of this article. Also, thanks to my affiliations for providing adequate infrastructure and kind support.

References

1. Batabyal, A., Mohapatra, P.K.: Land rights and livelihoods of tribals in India: a study of displacement and rehabilitation processes. Int. J. Sociol. Anthropol. **7**(1), 1–12 (2015)
2. Upadhayay, V.: Understanding tribals: a study on the role of forests in tribal life. SSRN Electron. J. (2017)
3. Sinha, A., Sinha, S.: Land rights and indigenous people in India: a critical analysis of policies and practices. Land Use Policy **84**, 194–204 (2019)
4. Ghosh, A.: Tribal land alienation and livelihood sustainability in India: a study on forest-dependent communities. Indian J. Agric. Econ. **73**(3), 356–370 (2018)
5. Das, D., Ghosh, P.: Dispossession of tribal lands: an analysis of forest land acquisition in India. Econ. Pol. Wkly **53**(29), 63–71 (2018)

6. Kujur, A.: Land alienation and livelihood crisis among the tribal communities of Jharkhand, India. In: Economic Anthropology and Development, pp. 75–89. Springer, Singapore (2017)
7. Bhattacharya, S.: Land alienation, dispossession, and livelihood vulnerability: a case study of tribal communities in West Bengal, India. J. Land Rural Stud. **4**(1), 27–46 (2016)
8. Fernandes, W.: Tribal land alienation in India: a historical perspective. Indian J. Hist. Sci. **50**(4), 535–545 (2015)
9. Mohanty, M., Gupta, A.: Socio-economic determinants of tribal land alienation: a case study from Odisha, India. J. Rural. Stud. **34**, 52–59 (2014)
10. Pathak, D., Choudhury, H.: Land alienation and livelihood security of tribal women in Odisha India. Indian J. Gend. Stud. **20**(3), 379–396 (2013)
11. Mahapatra, N.D.: Land alienation and tribal livelihoods: a study of the Dongria Kondh tribe in Odisha India. Anthropol. Today **28**(4), 6–10 (2012)
12. Das, D.: Land alienation among the tribal communities of Odisha India. Contributions to Indian Sociology **45**(1–2), 105–128 (2011)

Machine Learning Algorithms Design Pertaining to Cancer Dependent Categorization on Information of RNA Sequencing

Pravin V. Shinde[1(✉)], Rajesh Deshmukh[1], and Preeti S. Patil[2]

[1] Kalinga University, Raipur, India
psshinde92@gmail.com

[2] D.Y. Patil College of Engineering, Akurdi, Pune, India

Abstract. Due to delay in proper treatment and diagnosis, Cancer move toward one of major character that is accountable for fatalities globally. This happened due to uncontrolled and abnormal cell growth in human body, that can get transferred to different parts of human flesh. Sequence of Ribonucleic acid (RNA) checks variation that takes place in cells and assist in analyzing transcriptome in pattern of gene expression inside RNA fluids. In an early stages of cancer schemes of Machine learning contributes in cancerous cell prediction, if concerned information is made easily accessible. Main aim of this research is in building prototypes and group various kind of cancer. Hence, we deployed many machine learning schemes such as support vector machine (SVM), random forest (RF), multilayer perceptron (MLP), & k-nearest neighbors (KNN) for classifying samples as per their labeling. Out of Genotype-Tissue Expression (GTEX) & cancer genome Atlas (TCGA), for this thesis datasets were accumulated. On independent dataset (GTEX) machine learning techniques were tested and trained on TCGA data. Data representation thus received by applying stacked denoising autoencoders areutilized in training and testing prototypes. These models were not having very high performance; but, comparatively to others MLP exhibited better performance. Best characteristics that were chosen by applying SelectKBest, were utilized too for performance comparison. Outcome showed that K-nearest neighbor classifier provided good results, having precision of 85.12% whereas verified with self-sufficient information, with 98.4%. as accuracy of training.

Keywords: RNA · Cancer · KNN · SVM · Classification and ANN

1 Introduction

Abnormal cells develop and blowout of regulated systematic growth in many places inside a cell tissues, causing disease accumulation (cancerous). Normal cells develop, grow, and divide in the human body. In the case of a tumour, celluloid particles were substituted as it become worn-out or old [2]. Malignant tumours that have metastasized (dispersed) in other bodily portions are known to be metastatic cancer. A process by virtue of which cancerous celluloid disperse to different substrate of flesh came to be

S. Singh et al. (Eds.): AIDSP 2023, CCIS 2440, pp. 77–86, 2025.
https://doi.org/10.1007/978-3-031-88762-8_8

known as metastasis. In most cases, it is caused by a gene alteration, such as a DNA mutation [3]. IHME reports that cancer had become world's most and highest serious issues, affecting millions of lives. IHME's Fig. 1 depicts that Global Burden of Disease (GBD), displays total yearly cancer deaths across all ages and genders [4]. Cancer is taken accountable for second most reason for worldwide casualties. Around 9.6 million fatalities and 18.1 million new cases were projected worldwide in 2018. Worldwide Cancer Observatory (GCO) remains an information interchange portal which displays global cancer information. From International Agency for Research on Cancer (IARC) [5], Fig. 2 depicts the projected number of new cancer cases in 2018. Records showed that greater than 100 kinds of cancer that consist of nomenclature based on tissues or organs they form. As part of the world's biggest genetic data collection, The Cancer Genome Atlas (TCGA), Center for Cancer Genomic, National Cancers Institute from 2006 controls TCGA. Assisting in recognising the relevance of cancer genetics and changing therapeutic methods. We employed RNA sequencing data in this study. RNA-Seq is revolutionizing transcriptome research by capturing entire cells gene's readouts. Human's genome constitute entirely of DNA that instructs body in making cell. DNA should be capable in reading and transcripting or copying ribonucleic acids to carry out these instructions (RNA).

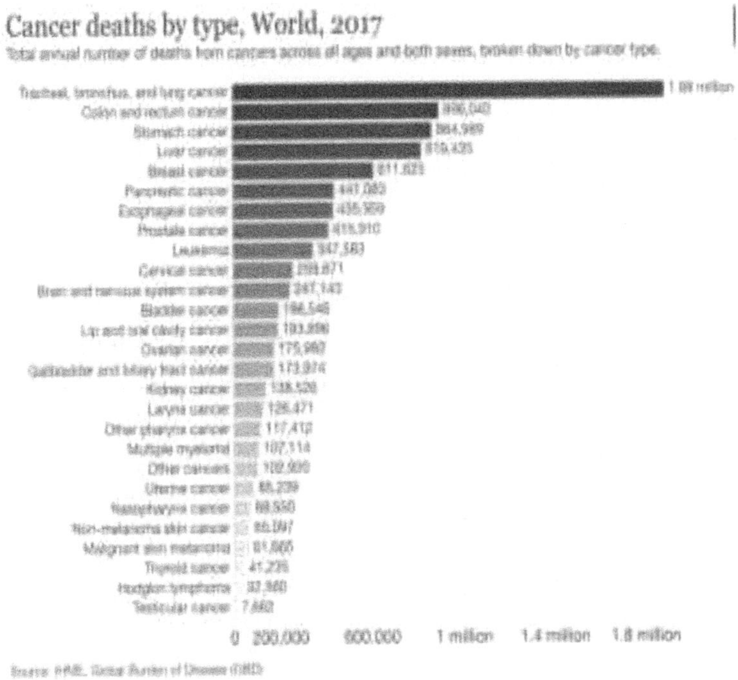

Fig. 1. Categorization of Cancerous deaths

To quantify expression distribution throughout the transcriptome, RNA-Seq is an accurate and sensitive method and helps scientist in receiving clear view of previous unchecked variation taking place state of disease. Researchers can assess gene status in cells and tissues using RNA sequencing data [7].

In today's environment, machine learning has numerous practical uses. These were found to be utilised with recognition of images, voice recognition for robotics, traffic pattern studies, weather forecasting, and other fields. Various method is provided by Machine learning, tools and techniques, which could assist in resolving prognostics and diagnostic issues in different health sectors.

Such techniques are utilized in answering fundamental difficulties with genetically produced medicines and in explaining how differences are present with DNA might impact disease risk [8]. So as to know deeply, mechanisms which is core in expression of gene expression a wide variety of machine learning tools had be evolved. Analysis result depicted, gene identification can be carried out by machine learning approach, its characteristics features, having Tran- scripted in initiation location by applying huge and complicated information of genomics [9]. Kochan et al. employed an RNA-seq data classification technique to choose genes [9]. We can readily forecast the genesis and spread of cancer using gathered data and relevant knowledge. For recognising patterns and relationships across complicated genomic information, machine learning has become a popular and frequently utilised technology.

Many research were dependent on application of algorithm of machine learning in micro arraying expression of gene information to group kind of cancer [10]. It has helped in forecasting gene expression in a cell. Strong machines learning algorithm had contributed in feasibility in pinpointing cancer stem before therapy.

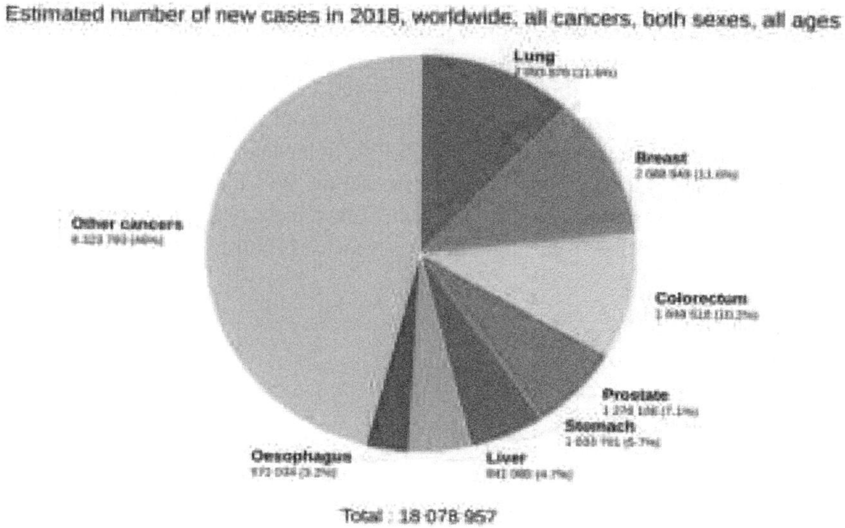

Fig. 2. New cases of all cancer

2 Literature Review

Sequencing of RNA could help scientists researchers with knowing classification of tumorous cells with its progress through closely observing transcriptome and gene expression variation with tumurous RNA-seq [11]. To get micro observation about transcriptome of cell, capabilities of higher throughput sequencing techniques is applied [12].

The most widely used approach for examining the transcriptome is RNA-seq (NGS). Classification based on information of gene expression leads to significant portion latest medicine for treatment personalization and diagnosis of disease. A helpful approach for analysing transcriptome data, RNA-seq has been around for decades [13]. Different RNA-seq applications are being investigated and fresh research projects are being carried out. This section will cover some prior experiments and research.

For categorising RNA-seq data, several approaches have been offered. The use of quintile renovated quadratic discriminants analysis pertaining to RNA-seq information has recently gained popularity [14].

According to the model, counts are somewhat negative binomial but are not independent for RNA-seq data. For the likelihood ratio test, the initial step was to filter out low- expression genes (LRT). Top m genes were classified using LRT statistics sorted by them. Their study compared the performance of qtQDA with other models like support vector machines (SVM), k-nearest neighbours (KNN), machine learning classifiers, logistic regression (LR), and others using freely accessible information sets (haplotype map (HapMap), prostrate cancer, and data cervical cancer,). It was computed by randomly splitting the data set into two pieces. Also, qtQDA did well with HapMap and cervical cancer information. This becomes computationally quicker than previous techniques and does not need parameter adjustment [14].

Researchers discovered tiny gene's fraction and categorised cancer classification information, species classification, etc. With and without power transformation, they used continuous and discrete classifiers like PLDA. In the end, they compared classifier performance to that of sparse models for gene extraction. With bulk and single-cell RNA-seq data, researchers created classifiers for detecting cancer types. A KNN smoothing was used to eliminate noise from single-cell RNA-seq. NN outperforms all other models tested.

3 System Model

Experimental workflow is below enlisted (see Fig. 3).

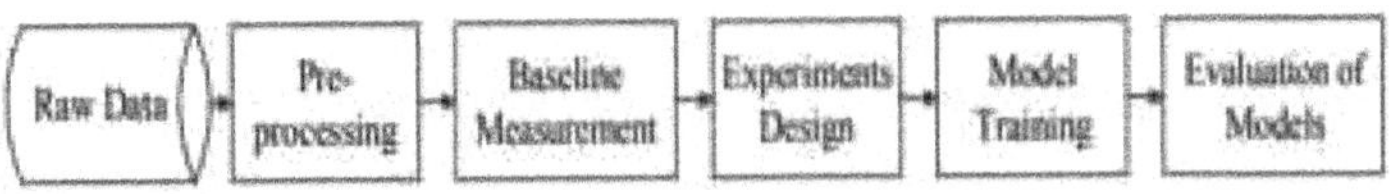

Fig. 3. Workflow overview

A. **Data selection.**
   Analysis project [5–7] from TCGA PanCancer provided raw data set for this research and taken down from synapse.org [8] that consist of cancer patient samples of about 10,471 from 33 kinds of cancer. Characteristics consisted of 20,531 genes; every gene constituted of dimension on levels of RNA expression levels. Entire measurements in raw information set is accumulated from Illumina HiSeq system [9–12].
B. **TCGA Barcode and Class Labelling.**
   Main aim of this study to in grouping of 33 various kinds of cancer, even prior to construction of grouping models, extra labelling of classes are required. Labels are produced depending on barcode of TCGA [13]. A unique barcode is provided to every sample in information set. Data regarding sample source site and type of cancer is gathered in barcode. We were capable in labelling individual sample with correct cancer group, by linking every barcode to respective kind of cancer [14–16].
C. **Pre-processing of Information.**
   With a reliable rate of forecasting in developing model of classification, an information pre-processing step is required. Primarily, raw data identified to make sure that there were no duplicate or missing details. Secondly, any column or row having all values as zero values were eliminated out of information set [17]. Like, around 212 attributes identified having all zero values were deleted . Post accomplishment of these two steps, reduction of attributes were done from 20,531 to 20,319 however total amount of samples retained were same. Later, many information pre-processing techniques were taken into consideration for training preparation of these models [18–20].

4 Selection of Features

There were many problems that can influence outcome of training, while analysing raw data,. Primarily, amount of samples was relatively lower compared to attributes count, hence choosing of feature or a feature decreasing techicque could be applicable [21].

Therefore, variance threshold and tree classifier [22] schemes were taken in selecting features. Features based on their importance scores of a subset is selected in tree classifier method. Features with difference having less than threshold is filtered out in variance threshold method . Therefore, data set of filter consist of very high variance features only. To observe which method yields a better outcome result, both data selection methods were applied in proceeding experimental setups [23].

Normalization: Relates to standard deviation of features widely distributed across all over raw data. Hence, two normalization schemes were deployed. Common feature scaling method is the min-max scaling. It scales every individual feature independently to align them in same range as provided [24].

Standardization scaling is another approach. In this method standardization of features is accomplished by eliminating mean of every data point and unit variance is scaled out of every individual features [25].

5 Experimental Evaluation

The TCGA dataset was divided into validation and training datasets [26, 27]. With training dataset various machine learning models were skilfully qualified and different hyper parameters too were aligned accordingly by applying cross- validation. Subsequent correcting hyperparameters, we received a model's best hyperparameter It was used to fit the whole TCGA dataset. A separate set of information (GTEX) is utilised in assessing model. Representation of information through marginalised stacked denoising autoencoders (mSDA) is used to illustrate outcomes coming of various machine learning techniques [28].

F-score and Chi-Squared tests was used in picking characteristics from the TCGA and GTEX datasets. From 38019 features, only 832 were chosen [29]. The reduced characteristics were used to compare models. To tweak hyperparameters, TCGA's information were applied in training various model. It uses GTEX data to evaluate model behaviour. With respect to complexity of time, chosen models were quicker with decreased characteristics [30].

Support Vector Machine (SVM)
Across all time behavior concerning support vector machine was depicted with Table 1. SVM was not able to accurately forecast entirely GI test cases. Only 42 samples out of 706 were accurately categorised, while all TCGA samples were correctly classified. This model's GTEX accuracy was 76.3% while its precision was 94%

Table 1. Result of SVM Using SelectKbest

Model	SVM
Best hyperparameter	C: 0.01, Kernel: linear
Best training accuracy	0.994
Test set accuracy	0.975
Accuracy on independent data (GTEX)	0.763

Random Forest (RF)
Table 2 provides entire behavior of models. The implementation was done using sklearn. We used TCGA to train and test the model, and it was 97.5% accurate! Model 38 also performed well and properly predicted the samples. GTEX data predictions were 97.5% accurate. Only a few GI samples were successfully identified, and this model did poorly on GTEX.

K Nearest Neighbor (KNN)
We had applied python's sklearn library during systems implementations. Behaviour all across regarding KNN was depicted with Table three. We had implemented GTEX representing as testing information that was entirely concealed for prototype schemes.

Table 2. Results of RF Using SelectKbest

Model	Random Forest
Best hyperparameter	Criterian = entropy, max depth = 12, split = 8
Best training accuracy	0.997
Test set accuracy	0.975
Accuracy on independent data (GTEX)	0.70

Resulting F1 score was about 83% and that of accuracy was 80%. Highest precision is attained through such models applied in chosen characteristics feature with GTEX We can see some test cases getting miscategorised pertaining to PCPG and COAD (Table 3).

Table 3. Result of KNN Using SelectKbest

Model	KNN
Best hyperparameter	N_neighbors = 10, weight = distance
Best training accuracy	0.968
Test set accuracy	0.956
Accuracy on independent data (GTEX)	0.80

Mulitilayer Perceptron (MLP)

Table 4 depicts MLP's total performance. On the TCGA and GTEX datasets, we reached 96% accuracy. In comparison to KNN, this model was faster to compute. Computationally, this model was quicker using the specified characteristics. We still have misclassified samples inside G.

Table 4. Outcome Data Representation Applied MLP

Model	Multilayer Perceptron
Best hyperparameter	Hidden_layer_size = 416, activation = tanh, alpha = 0.001
Best training accuracy	0.975
Test set accuracy	0.960
Accuracy on independent data (GTEX)	0.80

Figure 4 shows F1 score and accuracy of several model using independent information.

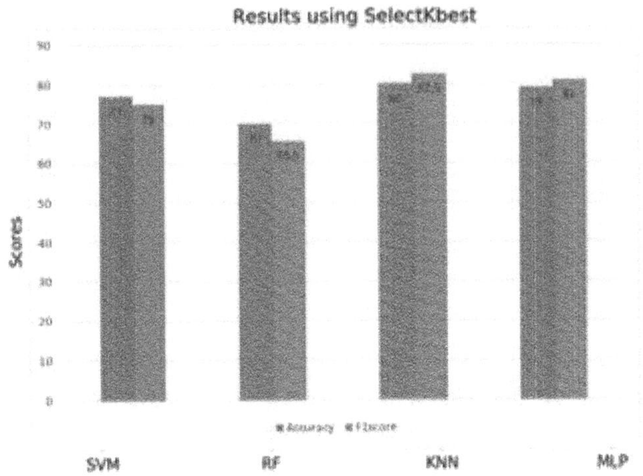

Fig. 4. F1 score and Accuracy for different models using autonomous data

6 Conclusion

To categorise distinct cancer classifications, machine learning was applied in this study. Cleanup of GTEX and TCGA training data. By using these datasets, the thesis sought to forecast the muscle origin of cancer. The data were encoded using a stacked denoising autoencoder (SDA). The SelectKBest algorithm from sklearn library was used to choose features. From 38019 to 832 characteristics were chosen. To train these algorithms, preprocessed data was fed into k-nearest neighbour, random forest, support vector machines (SVM) and multilayer perceptron. Model was built using the best hyperparameters from cross validation. Without and with representation of information, having specified characteristics, best model were used in several studies. While MLP performed quite efficiently with information displays employing stacked denoising autoencoder, precision and recall were both 66%. In Comparison with remaining prototypes, KNN performed best. Information post combining along with dividing label yielded precision of 95%, accuracy 85.1%, recall 85%. More data means better models. Choosing the right hyperparameters may improve accuracy and performance. When employing certain characteristics, MLP outperformed other methods. Preciseness and recall were 95% and 80%, respectively.

Difficulty in collecting datasets from many sources, models behaviour varies with having 2 varying information set. We were able in applying various pre-processing schemes priour to implementation of machines learning algorithm. To find the fastest method among those with comparable performance, we used time complexity. So we may choose an algorithm that balances space consumption and performance, we can analyse other models in the future. A dataset that offers valuable information for the researcher may help improve results and quality assessment in future studies. Deep learning may also be used to detect patterns in large datasets, potentially improving accuracy. It is possible to acquire additional data from diverse sources and use sophisticated machine learning algorithms to tackle the issue of sub clusters inside a cluster.

References

1. Saxena, S., et. al.: Machine learning based sensitivity analysis for the applications in the prediction and detection of cancer disease. In: 2020 IEEE International Conference on Distributed Computing, VLSI, Electrical Circuits and Robotics (DIS - COVER), pp. 1–4 (2020)
2. Kumar, A., et. al.: Machine learning based approaches for cancer prediction: a survey. SSRN Electron. J. (2020)
3. https://www.cancer.gov/about- cancer/understanding/what is cancer (2020). Accessed 15 Feb 2020
4. https://ourworldindata.org/cancer (2020). Accessed 16 Feb 2020
5. https://gco.iarc.fr/today/home (2020). Accessed 17 Feb 2020
6. https://www.cancer.gov/about nci/organization/ccg/research/structuralgenomics/tc ga (2020). Accessed 19 Feb 2020
7. https://www.genome.gov/about - genomics/fact sheets/Transcriptome - Fact – Sheet (2019). Accessed 19 Feb 2019
8. Khera, A.V., et al.: Genome - wide polygenic scores for common diseases identify individuals with risk equivalent to monogenic mutations. Nat. Genet. **50**(9), 1219–1224 (2018)
9. Lai, K., et al.: Artificial intelligence and machine learning in bioinformatics. Encycl. Bioinf. Comput. Biol. ABC Bioinform. **55**, 272 (2018)
10. Koçhan, N., et al.: A new local covariance matrix estimation for the classification of gene expression profiles in RNA seq data. bioRxiv, 766402 (2020)
11. van Ijzendoorn, D.G., et al.: Machine learning analysis of gene expression data reveals novel diagnostic and prognostic biomarkers and identifies therapeutic targets for soft tissue sarcomas. PLoS Comput. Biol. **15**(2), e1006826 (2020)
12. Maag, J.L., et al.: Novel aberrations uncovered in Barrett's esophagus and esophageal adenocarcinoma using whole transcriptome sequencing. Mol. Cancer Res. **15**(11), 1558–1569 (2017)
13. Chen, S., et al.: High - throughput sequencing of the transcriptome and chromatin accessibility in the same cell. Nat. Biotechnol. **37**(12), 1452–1457 (2020)
14. Zappia, L., et al.: Exploring the single - cell RNA - seq analysis landscape with the scRNA tools database. PLoS Comput. Biol. **14**(6), e1006245 (2018)
15. Koçhan, N., et al.: QtQDA: quantile transformed quadratic discriminant analysis for high - dimensional RNA - seq data. PeerJ **7**, e8260 (2019)
16. Mirza, B., et al.: Machine learning and integrative analysis of biomedical big data. Genes **10**(2), 87 (2019)
17. Alquicira-Hernandez, J., et al.: ScPred accurate supervised method for cell - type classification from single cell RNA-seq data. Genome Biol. **20**(1), 1–17 (2019)
18. Cascianelli, S., et al.: Machine learning for RNA sequencing - based intrinsic subtyping of breast cancer. Sci. Rep. **10**(1), 1–13 (2020)
19. Al - Obeidat, F., et al.: (CDRGI) Cancer detection through relevant genes identification. Neural Comput. Appl. 1–8 (2021)
20. Qi, R., et al.: Clustering and classification methods for single cell RNA - sequencing data. Briefings Bioinform. **21**(4), 1196–1208 (2020)
21. Wu, J., et al.: Breast cancer type classification using machine learning. J. Personalized Med. **11**(2), 61 (2021)
22. Vrahatis, A.G., et al.: Recent machine learning approaches for single - cell RNA - seq data analysis. In: Advanced Computational Intelligence in Healthcare – 7, pp. 65 – 79. Springer, Berlin, Heidelberg (2020)
23. Liñares-Blanco, J., et al.: Machine learning analysis of TCGA cancer data. PeerJ Comput. Sci. **7**, e584 (2021)

24. Alimadadi, A., et al.: Machine learning - based classification and diagnosis of clinical cardiomyopathies. Physiol. Genomics **52**(9), 391–400 (2020)
25. Charbuty, B., et al.: Classification based on decision tree algorithm for machine learning. J. Appl. Sci. Technol. Trends **2**(1), 20–28 (2021)
26. Wu, D., et al.: Machine learning and deep learning in genetics and genomics. In: Machine Learning in Dentistry, pp. 163–181. Springer, Cham (2021)
27. Mueller, F.B., et al.: AI (artificial intelligence) and hypertension research. Curr. Hypertens. Rep. **22**(9), 1–7 (2020)
28. Thakur, T., et al.: Gene expression - assisted cancer prediction techniques. J. Healthc. Eng. **2021**(1), 4242646 (2021)
29. Zhu, W., et al.: The application of deep learning in cancer prognosis prediction. Cancers **12**(3), 603 (2020)
30. Namkung, J., et al.: Machine learning methods for microbiome studies. J. Microbiol. **58**(3), 206–216 (2020)
31. Gartner, J.J., et al.: A machine learning model for ranking candidate HLA class I neoantigens based on known neoepitopes from multiple human tumor types. Nat. Carcinoma **2**(5), 563–574 (2021)
32. Vibert, J., et al.: Identification of tissue of origin also guided therapeutic applications in carcinomas of unknown primary using deep learning also RNA sequencing (transcuptomics). J. Mol. Diagn. **23**(10), 1380–1392 (2021)
33. Mahin, K.F., et al.: Panclassif: improving pan carcinoma classification of single cell RNA - seq gene expression data using machine learning. Genomics **114**(2), 110264 (2022)
34. Marx, V.: Machine learning, practically speaking. Nat. Methods **16**(6), 463–467 (2019)
35. Lopez-Rincon, A., et al.: Machine learning-based ensemble recursive feature selection of circulating miRNAs for carcinoma tumor classification. Carcinomas **12**(7), 1785 (2020)

Estimating Surface Vertical Displacement of Joshimath Using InSAR Data Processing Through Cloud Based OpenSAR Lab Environment

S. Sivaranjani and M. Geetha Priya(✉)

CIIRC - Jyothy Institute of Technology, Bengaluru 560082, India
geetha.sri82@gmail.com

Abstract. Areas vulnerable to landslides require an accurate estimate of surface displacement for prediction and monitoring purposes. Prediction of the degree and intensity of surface displacement assist in the protection of human and constructional designs. Deformation monitoring is one of the most advanced applications of the space-borne Interferometric Synthetic Aperture Radar (InSAR) approach. Because of its capacity to monitor large-scale deformation with millimeter accuracy, small baseline subset (SBAS) InSAR has been diversely used in geodetic fields. The current research study examines recent subsidence in Joshimath town of Chamoli district, Uttarakhand, utilizing Sentinel-1, Synthetic Aperture Radar (SAR) data for December 2022 and early January 2023 with a temporal baseline of 12 days. The data is analyzed in the OpenSAR lab Jupyter Notebook using the HyP3 and MintPy tools designed to run in specific conda environments. Processed data show that the surface vertical displacement gradually increased from 3.2 cm to 5 cm in several regions near Joshimath during the last week of December 2022 and the first week of January 2023. The average velocity reached roughly 32 cm/yr, indicating that certain zones are very dynamic. The northward and eastward sites (residential and commercial areas) of Joshimath have experienced severe subsidence, while the southward site has seen tolerable subsidence. Early identification of these surface deformations remains a difficult process, necessitating the development of an economical method for continually monitoring a large region. Avoiding unplanned construction activities and maintaining the underlying channel will overcome subsurface instability and prevent subsidence in the area.

Keywords: SBAS · HyP3 · MintPY · OpenSAR lab · Joshimath

1 Introduction

The contemporary design of technological progress and the ensuing population upsurge in the delicate ecosystems of mountain ranges has resulted in increased stress on the environment in the form of land use/land cover changes [1]. The depositional paleogeology of high alpine-altitude locations is under threat due to increasing temperatures

S. Singh et al. (Eds.): AIDSP 2023, CCIS 2440, pp. 87–95, 2025.
https://doi.org/10.1007/978-3-031-88762-8_9

[2]. Due to the perennial low-temperature environment, cold alpine and high-altitude locations are characterized by fragile ecological habitats with periglacial geomorphology such as glaciers, permafrost, rock glaciers, and subsurface ice [3]. Joshimath exists beneath the topography of the Himalayan orogenic belt, with a highly sheared landscape prone to erosion and high relief generating slope instability in that location [4]. Land subsidence is caused by a mixture of natural sediment compaction, groundwater extraction, geothermal fluids, coal, and other materials extracted by mining, and underground building. Land subsidence as a result of subterranean engineering work has generated various geological risks and hampered the sustainable development of urbanization [5]. A significant number of major land subsidence locations throughout the world are associated with the increased use of groundwater [6]. The state where Joshimath is located witnessed natural calamities in the years 2009, 2010, 2012, 2013, 2017, 2021, and 2023 [7–11]. InSAR (Synthetic Aperture Radar Interferometry) technology can be used to monitor large-scale ground deformation [12, 13]. Using InSAR data, researchers predicted ground subsidence in Joshimath [14, 15]. Among the many InSAR approaches, SBAS-InSAR is a new D-InSAR time series analysis approach developed by Berardino et al. that can minimize the most atmospheric influence [16]. In the proposed study, time series vertical displacement and average velocity have been estimated with cloud-based OpenSAR lab conda environment using Sentinel 1 SAR data with HyP3 products in MintPy of Jupyter Notebook.

2 Study Area

Joshimath (30° 33’ 05”N and 79° 34’ 30”E), also known as Jyotirmath, is a town in Chamoli District, Uttarakhand, located on the national highway (NH 7) flanked by Rishikesh and Badrinath [17]. The region’s slope ranges between 60 and 70°. The location map of the study area is depicted below in Fig. 1. It is an important transit point for travelers visiting Hindu pilgrimage sites. The town has a total area of 4676.63 square kilometers and a population of around 18,000–20,000 people (3,898 households) as per population census of India 2011. The elevation of the surface ranges from 1890 m to 2045 m above mean sea level in the Garhwal Himalayas. Joshimath has a Tundra environment with unpredictable meteorological conditions and experiences recurrent high-intensity rainfall [18]. The average annual temperature in the town is 5.97 °C, with 76.8 mm of precipitation and 132 wet days. The region’s ecological equilibrium is being harmed by rampant building, which is vulnerable to earthquakes and landslides. Joshimath, according to the 1976 Mishra Committee study, is built on a sand and stone deposit, not on the main rock, but on an ancient landslip[19]. The river currents of the Alaknanda and Dhauliganga in the north and south are also playing a role in bringing landslides to Vishnuprayag in Joshimath.

3 Data Used and Methodology

The current study utilized Sentinel-1 Single Look Complex (SLC) data of 5x20 m resolution in the Interferometric Wide (IW) swath operational mode with VV + VH polarization, accessed through Alaska Satellite Facility vertex (https://search.asf.alaska.edu/). Global Copernicus DEM GLO-30 of medium resolution which has a global coverage at 30m spatial resolution was utilized to convert SAR data to a higher-level output. The European Centre for Medium-Range Weather Forecasts (ECMWF) ERA5 climate reanalysis data (https://cds.climate.copernicus.eu/) is utilized to rectify tropospheric propagation delays. Table 1 lists the datasets used in the current study. The work methodology [20] adopted in the current study is represented in Fig. 2.

Table 1. Description of datasets used for the study

Data	Sentinel-1 SLC
Flight Path /Row	63/488
Orbit Direction	Descending
Date of Acquisition	December 03, 2022, December 15, 2022, December 27, 2022, January 08, 2023
Temporal Resolution (days)	12
Resolution (m)	Medium
DEM	GLO–30
Vertical Datum	EGM2008
Projection	UTM Zone-WGS84
ERA5 climate reanalysis pressure data	ERA5_N20_N40_E70_E90_20221203_01.grb ERA5_N20_N40_E70_E90_20221215_01.grb ERA5_N20_N40_E70_E90_20221227_01.grb ERA5_N20_N40_E70_E90_20230108_01.grb

The surface vertical displacement for Joshimath has been estimated using Sentinel-1 SAR SLC data from December 2022 to the first week of January 2023. SBAS images with a swath width of 250 km were used in the Interferometric Wide (IW) swath mode to construct interferograms using ASF HyP3 (On-Demand process). These unwrapped stacks of interferograms were inverted and corrected utilizing multithreaded processing in a cloud-based lab environment with MintPy in the OpenSAR lab Jupyter Notebook in a specialized conda environment (InSAR analysis).

During the On-Demand process in ASF Vertex, the incidence angle, water mask, and set MintPy options should be enabled. Block 1 amends unwrapping errors for the raw phase, and Block 2 amends noises from multiple sources to produce a time series of vertical surface displacement. The time gap between the master and slave images of an interferogram is used to calculate displacement. From this time series displacement, average velocity is estimated which corresponds to the entire study period.

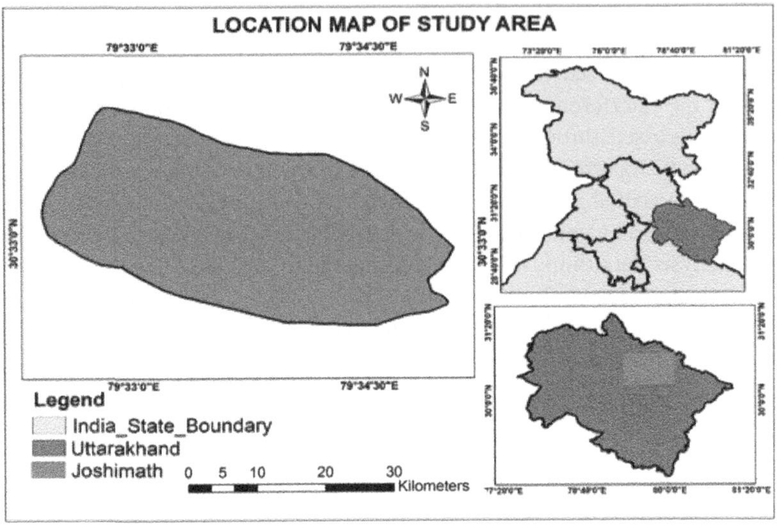

Fig. 1. Location map of the study area - Joshimath, Uttarakhand.

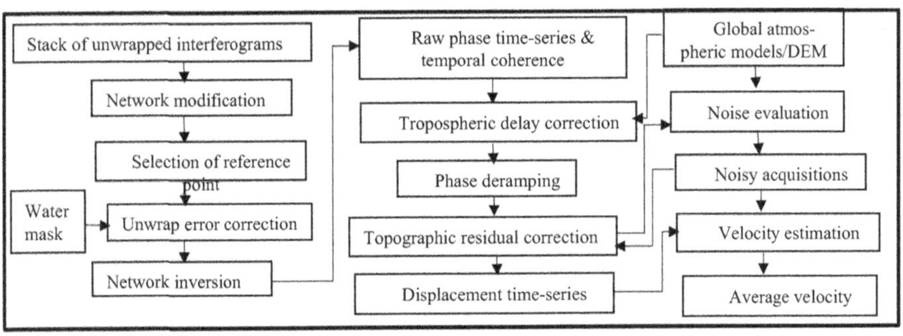

Fig. 2. Flowchart for work methodology

4 Results

Using the InSAR technique discussed in the methodology section, the surface vertical displacement of Joshimath town along with the average velocity for the time series has been obtained. From the estimated vertical displacement (Table 2 and Fig. 3) and average velocity results (Fig. 6), the interpretations and analysis are made on the Joshimath subsidence event.

For the entire time series analysis, the image acquired on 03 December 2022 has been considered as master image whereas the image acquired on 15 December 2022, 27 December 2022, and 08 January 2022 were considered slave images as per the requirement for InSAR processing. With a 12-day temporal baseline, the vertical displacement in Joshimath town is estimated from 0.78 cm to 4.18 cm with a mean displacement of 0.52 cm for the period 03 December to 15 December 2022. Higher displacement is observed in the places (buildings) in the northwest of Joshimath town. (Fig. 3a). For the

Table 2. Time series Vertical Displacement of Joshimath

Time Series	Temporal baseline (days)	Vertical Displacement (cm)
03–12–2022 to 15–12–2022	12	0.78–4.18
03–12–2022 to 27–12–2022	24	1.56–4.96
03–12–2022 to 08–01–2023	36	3.26–3.87

period 03 December to 15 December 2022 with a temporal baseline of 24 days, the vertical displacement is estimated from 1.56 cm to 4.96 cm in areas near the residential places (Fig. 3b). Mean displacement is observed to be 0.27 cm. Sample points of severe vertical displacement in buildings and residential areas of Joshimath during the study period are shown in Fig. 4. Severe displacement is observed in northeast locations indicating the subsidence of the region. On the southward side of the region, the displacement is not severe. Figure 5 displays actual field photos of houses above threatened demolition sites due to the subsidence of the land.

With 36 days temporal baseline, the vertical displacement is observed to be around 3.26 cm to 3.87 cm with a mean displacement of 0.76 cm from 27 December 2022 to 08 January 2023. The severity of the vertical displacement was reduced during this period (Fig. 3c). The estimated average velocity ranges from 32.8 cm/yr to 36.1 cm/yr. In the Joshimath town area with buildings, the higher velocity is observed to be around 25 cm/yr to 34 cm/yr and is represented in Fig. 6. Areas that underwent subsidence are observed to possess higher velocity compared to the other regions of the town. Due to the scattered rocks, and boulders from paleo landslide debris, the subsidence could have initiated. Loose soil with low bearing capacity and cohesive value causes weathering of the gneissic rocks in the town.

There could be an uplift in the previous times after which the subsidence occurred and got severe at the end of December 2022 to early January 2023 causing damage to people and property. The radiation temperature of the surfaces has risen in the Kedarnath-Joshimath region according to the literature. Surface water runoff has also increased as glaciers uphill are melting thereby causing more water discharge which will result in instability of the area in the future. For a disaster management system, the following technical factors would have been beneficial:

(a) The slopes have become extremely unstable in the previous few decades due to increased building, hydroelectric projects, and the enlargement of the National Highways.
(b) Due to poor drainage, inadequate sewage management, and increased waste seeping into the soil and loosening it from within, Joshimath's drainage planning has been disrupted, making it one of the most important elements that need to be investigated and redeveloped.
(c) Therefore, the government ought to impose limitations on large-scale building projects, including blasting or digging to remove rocks for road repairs and other works.

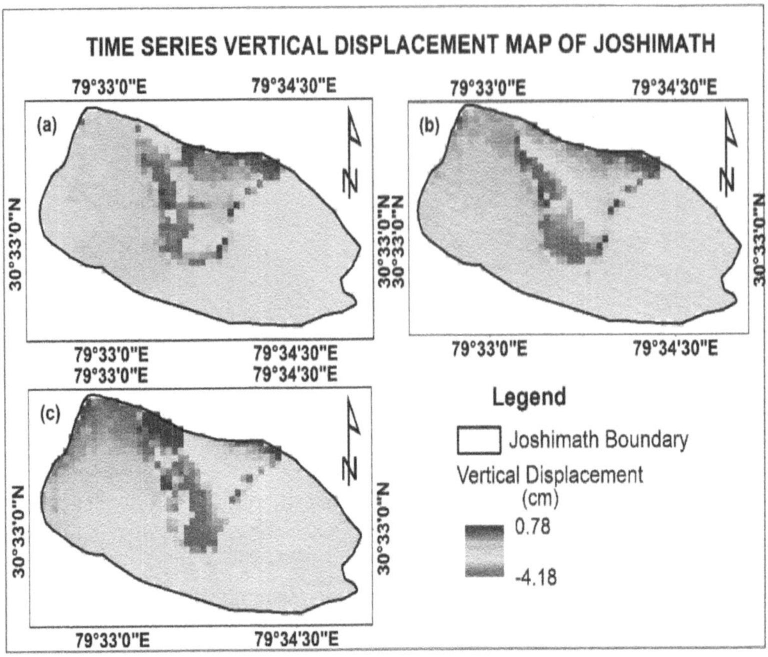

Fig. 3. Time series vertical displacement map of Joshimath (a) December 03–15,2022 (b) December 03–27,2022 (c) December 03, 2022–January 8, 2023

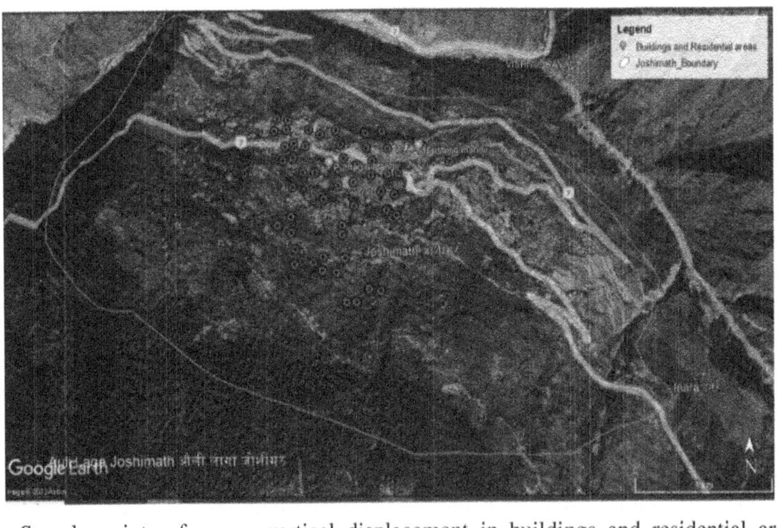

Fig. 4. Sample points of severe vertical displacement in buildings and residential areas of Joshimath

(d) Replanting the area will aid in retaining soil capacity, lessening the severity of the disaster in the future, especially in the city's vulnerable places for subsidence.

Fig. 5. Field photographs of the damaged buildings located in Joshimath town (a) Tilted house with severe cracks (b) Collapsed Temple with cracks (c) Boulders from paleo landslide debris seen after the building collapse (d) Badminton court with severe cracks. **(Source: PTI).**

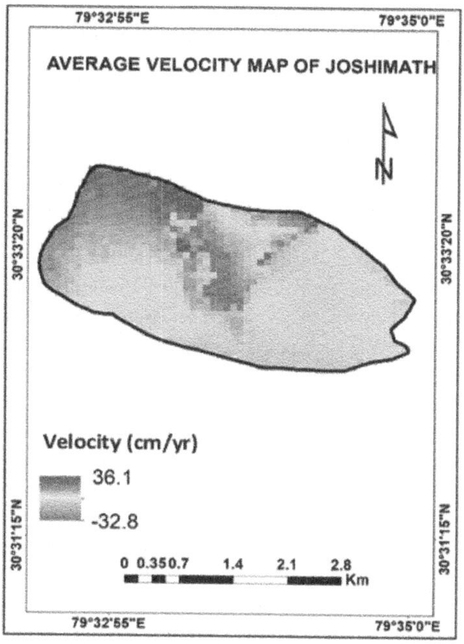

Fig. 6. Average velocity map of Joshimath for the time series from December 03, 2022, to January 8, 2023.

(e) As the town falls in the weak tectonic plate, regular monitoring of the location with better weather forecasting technology and extensive coverage is essential to protect the town from impending disasters.

5 Conclusion

Vertical displacement of Joshimath town has been estimated for the period December 2022 to early January 2023 using SBAS InSAR technique with HyP3, MintPy embedded in OpenSAR lab Jupyter Notebook available online free with cloud data storage. In this time series study, the subsidence of Joshimath is found to be severe at the places situated in the north of the town with a magnitude of 5 cm vertical displacement with a mean displacement of 2.7 cm. This severe displacement resulted in the subsidence of the town and caused more damage to the people and property. Average velocity ranges to 38 cm/yr in the places of subsidence point compared to other areas of the town. Time series analysis of displacement in a region could be done rapidly using the HyP3 products in MintPy and some potential subsidence disaster areas can also be discovered to prevent the reoccurrence. Understanding the factors such as the growing influence of changing climate and anthropogenic pressure will help to quantify the potential risks of such disasters in the near future. The government should regulate and impose standard building codes to reduce soil instability caused by unplanned construction activities.

Acknowledgment. The authors gratefully acknowledge the support rendered by PTICL and Dr.Krishna Venkatesh, Director, CIIRC, Jyothy Institute of Technology, Bengaluru. We also acknowledge the use of the ASF DAAC HyP3 On-Demand process in the OpenSAR lab environment through Jupyter Notebook.

References

1. Paudel, B., et al.: Review of studies on land use and land cover change in Nepal. J. Mt. Sci. **13**, 643–660 (2016). https://doi.org/10.1007/s11629-015-3604-9
2. Hansson, A., Anshelm, J., Fridahl, M., Haikola, S.: Boundary work and interpretations in the ipcc review process of the role of bioenergy with carbon capture and storage (BECCS) in limiting global warming to 1.5°C. Front. Clim. **3**, 1–14 (2021). https://doi.org/10.3389/fclim.2021.643224
3. Du, Q., et al.: SBAS-InSAR-based analysis of surface deformation in the eastern tianshan mountains. China. Front. Earth Sci. **9**, 1–16 (2021). https://doi.org/10.3389/feart.2021.729454
4. Rautela, P.: Ground subsidence: a silent disaster in Himalaya. Disaster Prev. Manag. An Int. J. **14**, 395–406 (2005). https://doi.org/10.1108/09653560510605054
5. Galloway, D.L., Bawden, G.W., Leake, S.A., Honegger, D.G.: Land subsidence hazards. L. Subsid. Hazards. **1164**, 74 (2008)
6. Krishnan, P.V.S., Kim, D.: Subsidence due to groundwater withdrawal in kathmandu basin detected by time-series PS-InSAR analysis. Korean J. Remote Sens. **34**, 703–708 (2018)
7. Sharma, S., Deen, S.: Impact of kedarnath flood tradegy on ecological and socio-economic setup: a geographical study. Shodh Sanchar Bull. **10**, 128–133 (2020)
8. National Institute of Disaster Management New Delhi, I.: Uttarakhand Disaster (2013)
9. National Disaster Management Authority, Ministry of Home Affairs, G. of I.: Detailed Report: Uttarakhand Disaster on 7 Feb 2021. pp. 1–60 (2022)

10. Disaster : Nature Fury : a Preimenarly Report on Uttarakhand Disaster
11. Kumar, A.: Demystifying a Himalayan tragedy: study of 2013 Uttarakhand disaster. J. Indian Res. **1**, 106–116 (2013)
12. Suresh, D., Yarrakula, K.: Subsidence monitoring techniques in coal mining: Indian scenario. Indian J. Geo-Marine Sci. **47**, 1918–1933 (2018)
13. Suresh, D., Kiran, Y.: ce pt us cr t. Geocarto Int. 0, 000 (2018). https://doi.org/10.1080/10106049.2018.1544289
14. Awasthi, S., Jain, K.: Analyzing the Land Subsidence activity in the Joshimath Region of Indian Himalayas Using Persistent Scatterer Interferometric Synthetic Aperture Radar (PSInSAR) (2023)
15. Garg, S., Karanam, V., Motagh, M.: Monitoring and Understanding Land Subsidence in Joshimath : An InSAR and Ground-based Study, p. 15976 (2023)
16. Wu, Q., Jia, C., Chen, S., Li, H.: SBAS-InSAR based deformation detection of urban land, created from mega-scale mountain excavating and valley filling in the Loess Plateau: the case study of Yan'an city. Remote Sens. **11**(14), 1673 (2019). https://doi.org/10.3390/rs11141673
17. Bhattacharya, A., Jugran, D.K.: Photo-int. Remote Sens. **8**, 8–10 (1982)
18. Agarwal, S., et al.: Identifying potential hotspots of land use / land cover change in the last 3 decades , Uttarakhand , NW Himalaya Department of Geology , HNB Garhwal University , Srinagar , India Department of Geology , University of Delhi , Delhi , India
19. Chandi Prasad Bhat: Bhatt Impact of natural disasters on environment HARC5513BHIP.pdf,https://lib.icimod.org/record/5290/files/BhattImpactofnaturaldisastersonenvironmentHARC5513BHIP.pdf (1993)
20. Yunjun, Z., Fattahi, H., Amelung, F.: Small baseline InSAR time series analysis: unwrapping error correction and noise reduction. Comput. Geosci.Geosci. **133**, 104331 (2019). https://doi.org/10.1016/j.cageo.2019.104331

An IoT-Based Smart Parking Management System

R. Priyadarshini(✉), Soham Kumar, Sudharshanan Balaji, B. Mukesh Kumar, and Gaurav Pendharkar

Vellore Institute of Technology, Chennai, Tamil nadu 600127, India
priyadarshini.r@vit.ac.in

Abstract. Traffic congestion and pollution have become frequent issues in cities due to the limited availability of free spaces and the increasing number of cars. This work aims to develop a parking system application to optimize the use of parking spaces in busy areas. The paper examines the technical aspects of the system, including hardware and software components, as well as the user interface and user experience. It uses the Arduino Uno as the controller, proximity and IR sensors, and a servo motor as the actuator. The real-time data logs collected by the sensors are transmitted to a Redis DB. Analytics are performed and displayed on a website to provide insights into parking usage statistics and revenue. A parking occupancy map is shown to the driver to indicate which parking areas are free or occupied. A pre-trained YOLOv3 model integrated with Raspberry Pi is used for real-time vehicle detection, and license plate numbers are extracted using optical character recognition technology. The proposed system has been able to reduce the waiting time for drivers to park their cars, ensure security, and provide analytical insights.

Keywords: smart parking · Arduino · IoT · object detection

1 Introduction

The global population is rapidly increasing, and the need for efficient urban planning and infrastructure comes with it. One of the biggest challenges urban areas face today is the issue of parking, as more people than ever own vehicles. This has led to a growing demand for smart parking systems that can help cities manage their parking resources more effectively [2].

A smart parking system is a technology-driven solution that improves the efficiency of parking management. The study's primary objective is to provide an effective solution to automate the parking process and increase its efficiency, ensuring drivers have a seamless parking experience. Rehman et al. [1] developed a smart parking system to reduce parking search time, fuel consumption, and CO2 emissions. The system was found to save up to 23.6% in fuel consumption and 9% in CO2 emissions according to the results. By minimizing the time and effort required to find a parking spot, the project can enhance the customer experience and optimize parking utilization.

S. Singh et al. (Eds.): AIDSP 2023, CCIS 2440, pp. 96–109, 2025.
https://doi.org/10.1007/978-3-031-88762-8_10

The Internet of Things (IoT) refers to the network of physical objects, devices, vehicles, and buildings embedded with sensors and software that allow them to connect and exchange data [3]. IoT has been widely adopted in various industries due to its ability to improve efficiency and productivity.

The proposed study uses advanced technologies to provide real-time information about available parking spaces to drivers. This improves convenience for drivers and reduces traffic congestion [4]. It also helps in optimizing parking utilization and enables new business models for parking operators. It can contribute to the growth of smart cities and improve the quality of life in urban areas. By leveraging data analytics, it can provide insights using which informed decisions can be made about parking infrastructure, creating a more sustainable and efficient parking system.

2 Related Work

Rehman, M. U., Shah, M. A., et al, 2018 propose an presents a decision support system that predicts car parking space availability using sensor data. [5] It utilizes a deep LSTM network to predict both overall and location-specific parking availability, as well as dayand hour-wise occupancy. However, the system has limitations, as it does not consider weather conditions or social events, and only accounts for parking lots, not roadside parking or traffic congestion. Future research aims to address these limitations. Wang, H., et al, 2020 suggest a parking reservation policy to address the needs of both service providers and users. [6] This policy is designed to minimize traffic congestion caused by drivers searching for parking spots. The RSPS system hardware consists of a sensor network, central server, and mobile device with Zigbee sensors, Bluetooth, and Wi-Fi communication. The software architecture includes the iRev as the central location for hosting applications and functions, with sensor network, web server, database, and real-time sensing data stream.The reservation-based system guarantees available parking spots to drivers who make reservations and allows them to choose the most convenient spot within their budget Pham, T. N., Tsai, M. F., et al, 2015 propose a system to assist users in finding a free parking space with the least cost by considering the distance and the number of free spaces in each car park. [7] The system uses RFID technology to calculate the percentage of free parking spaces in each car park. The proposed algorithm involves two processes: reservation and entering. In the reservation process, the user sends a request message to the system, which finds the car park with the least cost based on the function $F(\alpha, \beta)$ and sends the user a response message with the car park address and directions. In the entering process, the user is authorized to enter the car park via an ID or RFID card, and the system sends a response message to notify successful parking. The system updates the neighboring tables of neighboring nodes whenever there is a change in the value in the Counter.The proposed algorithm helps increase the chance of successful parking and decrease user waiting time.

Geng, Y., et al, 2013 proposed a"smart parking" system that allocates parking spots to drivers based on technology for detecting parking space availability and driver location. [8] MILP problems are used to ensure an efficient and fair allocation strategy. The research work focuses on selecting proper decision intervals and using pricing control to enhance fairness. Future works involve large scale depolyment which can be done on collaboration.

Vakula, D., et al, 2017 develop a low-cost smart parking system for smart cities using IoT technology. [9] The proposed smart parking system uses Arduino and NodeMCU ESP-32 to connect to the internet and send sensor data to the ThingSpeak cloud, utilizing ultrasonic and infrared sensors to detect occupancy and automate barrier control. It provides real-time information about parking availability and has the potential to reduce traffic congestion. The system can predict the number of expected vehicles and help with resource planning, making it a cost-effective and efficient solution for smart cities.The proposed IoT-based vehicle parking system can be highly useful for real-time data analytics and optimization of parking zones. Mainetti, L., Palano, L., etal, 2018 describes a Smart Parking System that utilizes IoT technologies such as WSN, RFID, NFC, and mobile to monitor outdoor parking lots. [10] The system consists of hybrid UHF RFID and IEEE 802.15.4-based WSN devices, a central server with advanced database management, and mobile apps for both traffic cops and drivers. It ensures parking spot availability and fee payments, with the potential for future work to allow users to increase paid fees.

3 Proposed Architecture

The proposed IoT architecture aims to automate various components of manual parking systems. It is pictorially represented in Fig. 1. The Arduino Uno is connected to two sensors, namely the Web Camera and an IR sensor, using a USB connection and electric signals, respectively. The sensors are positioned in close proximity to the boom barrier to monitor the entry and exit of a car. An electric impulse is used as the control signal to open the boom barrier using the actuator, a servomotor. When a car enters the parking lot, its image is stored locally and then saved to the Redis database using the Web Socket protocol. Finally, the web application interacts with the Redis database using the HTTP protocol. The overall parking system provides three major services, particularly fee

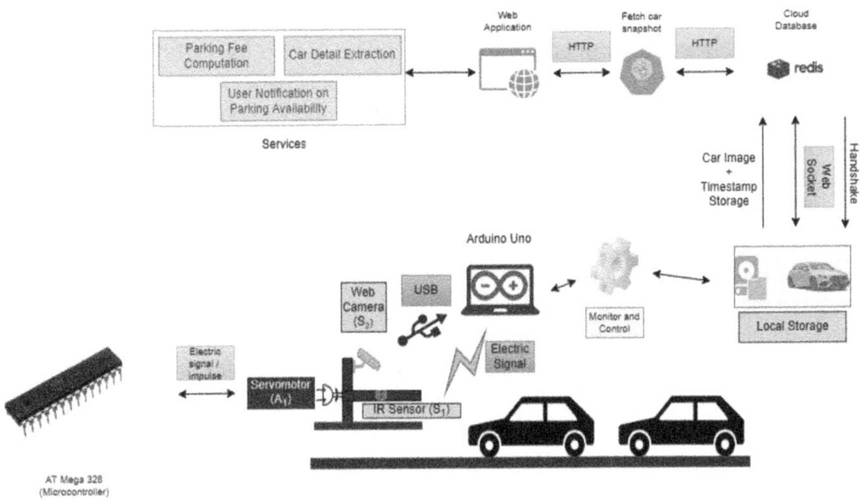

Fig. 1. IoT architecture for Smart Parking System

computation, car detail extraction, and user notifications for parking availability. Each of the components is explained in detail in the forthcoming sections.

4 Proposed Methodology

4.1 Hardware Components

The sensors are a camera module in Fig. 2e for detecting a vehicle and collecting its picture and license plate number, and the IR sensors in Fig. 2d for each gate if a vehicle is coming towards the gate. The actuators are two servomotors in Fig. 2c (one for the

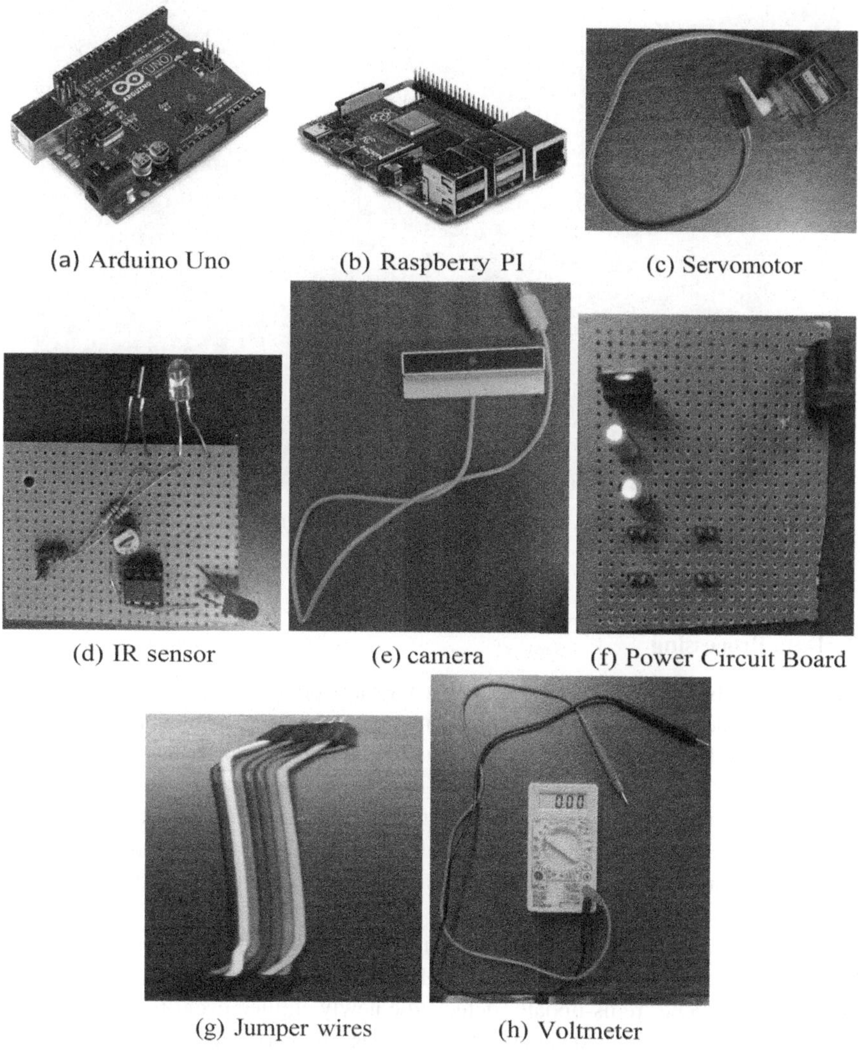

(a) Arduino Uno (b) Raspberry PI (c) Servomotor

(d) IR sensor (e) camera (f) Power Circuit Board

(g) Jumper wires (h) Voltmeter

Fig. 2. Individual hardware components

entry gate and one for the exit). The actuators are controlled by Arduino Uno's Micro-Controller Atmega 328 in Fig. 2a which instructs the server motors to open or close the gates based on the information collected from the sensors as per the algorithm shown below. The Arduino Uno communicates with Raspberry Pi in Fig. 2b by giving a signal, which further gives a POST request via HTTP to the Redis database. The counter variable keeps track of the number of free parking slots available. The overall assembly of the hardware is shown in Fig. 3 and the individual components are shown in Fig. 2.

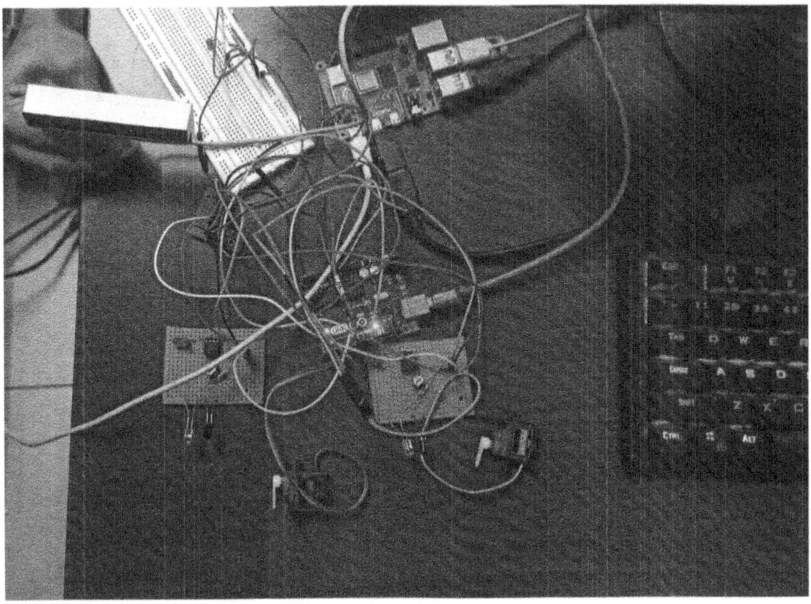

Fig. 3. Overall Hardware Connection

4.2 Data Processing

A snapshot of the incoming car is taken when a car is detected near the entry gate. The license plate is detected and cropped and the number is extracted using object recognition technology (OCR). The snapshot, the license plate number as well as the entry timestamp is collected and sent to the backend server. The backend server then handles the POST request and invokes a controller called createEvent which stores the received data in a Redis Database. While saving it in the database, an available free parking slot will be allotted to the entered car, and the allotted parking slot's id is removed from a Redis Set which maintains available parking slot ids.

This updation in the Redis database triggers a socket event called 'redisupdate'. Meanwhile, the frontend which is connected to the backend via sockets, is listening for events with the name 'redis-update' detects the newly emitted event from the backend and fetches the newly updated data from the database, then triggers a re-render operation in the Parking Slot Availability Dashboard.

4.3 Monitoring and Controlling

Monitoring and controlling refers to the ability to gather and analyze data from connected devices in order to make informed decisions and automate processes as shown in Algorithm 1. [12] It involves the use of sensors, actuators, and software to monitor and control various aspects of the environment. This allows for realtime adjustments and optimizations to improve efficiency and performance. In Algorithm 1, the sensor outputs are fetched from the respective pins on the Arduino represented by ard_i and *total parking spaces* denotes the maximum number of parking slots in the parking lot. *Pass counter to rasp*() is the function to propagate the current value of the counter to the Raspberry PI. Furthermore, *display parking lot full message*() and *display parking lot empty message*() are functions used the displaying the output for the full and empty parking lot in the console respectively.

4.4 Services

The smart parking system focuses on intelligent tolling which implies that the system is capable of identifying whether the customer has paid the parking fee or not and then opening the boom barrier for the exit. Hence, the proposed system requires the ability to extract the license plate from the car for payment verification while exiting the parking lot.

Algorithm 1 Monitoring and Controlling

```
ir₁ ← ard₇
ir₂ ← ard₄
servo₁ ← ard₁₀servo₂
← ard₁₁
output₁ to rasp ← ard₂output₂
to rasp ← ard₆
counter ← total_parking_spaces
while true do
    if ir₁ = 1 and camera = car and counter < total_parking_spaces then
        servo₁ ← 90clk
        Decrement counter
        output₁ to rasp() ← high signalservo₁
        ← 90aclk
    end if
    if ir₂ = 1 then
        servo₂ ← 90clk
        Increment counter
        output₂ to rasp() ← high signalservo₂
        ← 90aclk
    end if
    if counter = 0 then
        display_parking_lot_full_message()
        wait_for_car_to_leave()
    end if
    if counter = total parking spaces then
        display_parking_lot_empty_message()
        wait_for_car_to_enter()
    end if
end while
```

Car Detail Extraction. When a car enters the parking lot, the IR sensor detects that there is an object in close proximity to the boom barrier. The object needs to be a car otherwise the boom barrier shall not open. Hence, the YOLOv3 [11] object detection model is invoked to validate whether the object is a car or not. If the object is a car, then its license plate is extracted and cropped. Finally, the license plate content is converted to text using Optical Character Recognition (OCR) [13]. The image of the car, its license plate in image and text form, and some information about the customer along with the entry timestamp. On exit from the parking, the computed fee along with the exit timestamp is stored in the database. Figure 4 shows the process followed for car detail extraction and Fig. 5 depicts the whole process followed in details manner with the data being stored at each stage.

User Notifications for Parking Availability. A web application is developed to show the available parking spaces like it is available in the case of flight bookings and cinema bookings. Any user can get an idea of the number of available parking spaces to plan accordingly and also ensure less congestion at the parking lot entrance. Since the smart parking system uses a web application sending an automated email and SMS to the customer as alerts as shown in Fig. 6.

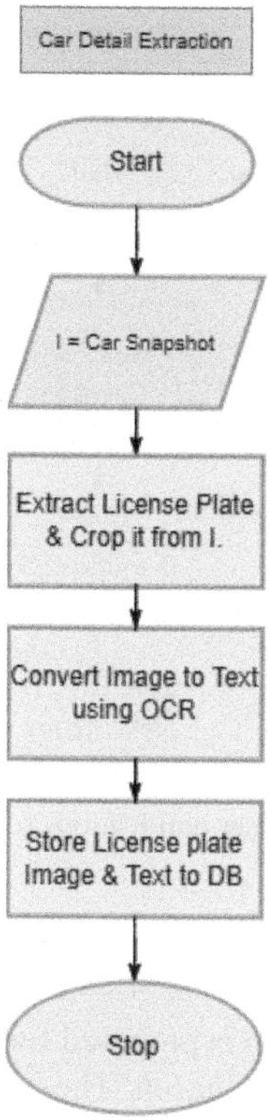

Fig. 4. The full process of car details extraction using object detection

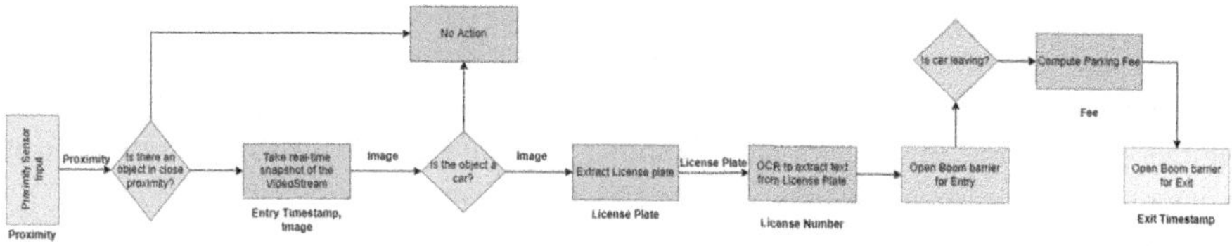

Fig. 5. Comprehensive Process diagram for car details extraction

Parking Fee Computation. Prior to the time of exit from the location, the customer can log in to the web application and make a payment at the payment gateway. The payment will be verified at the boom barrier and accordingly, the boom barrier will open. The verification process for the system involves a query to the database to check for a recent successful payment from the customer based on the license plate number as the primary key. Hence, the toll collection will not require any kind of manual intervention as shown in Fig. 7.

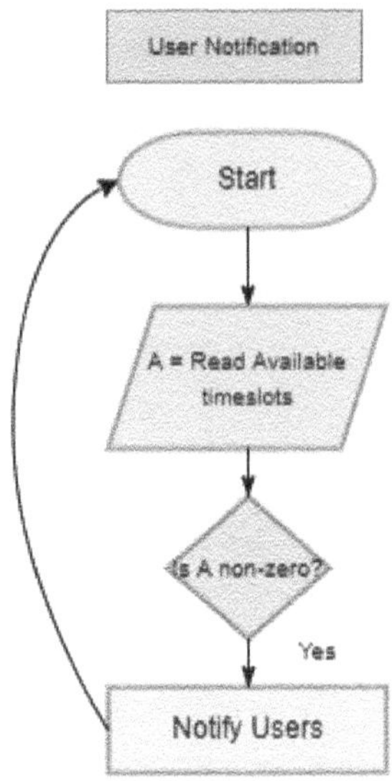

Fig. 6. The full process of user notifications for parking space availability

5 Result and Discussion

In the paper, a smart parking system is proposed for intelligent tolling eliminating any manual intervention at the tolls in parking lots. The system has the hardware arrangement, the object detection models, and the web application. The object detection model used is the YOLOv3 model pre-trained on the COCO dataset. An example of its detection is shown in Fig. 8 in which the model achieves a confidence of 84% and is able to extract the license plate. YOLOv3 is used for object detection because it is lighter for the hardware resources compared to YOLOv7 [14] and YOLOv6 [15].

The hardware is tested on an image of a car and on a real-time car in the university campus parking lot. Figures 9a and 9b shows the entry and exit to the parking lot by using a sample car image displayed from a mobile phone.

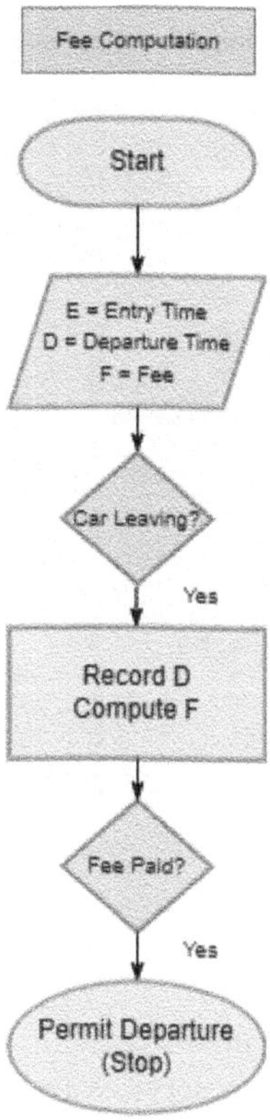

Fig. 7. The full process of fee computation and departure of the car

Vehicle is detected with an IoU score of 84 % and license plate is also detected. YOLOv3 achieved good accuracy in vehicle detection due to its speed, multi-scale detection, and anchor boxes. Its real-time processing, ability to handle objects of varying sizes, and predefined anchor boxes contribute to its effectiveness in accurately locating and classifying objects in images and videos.

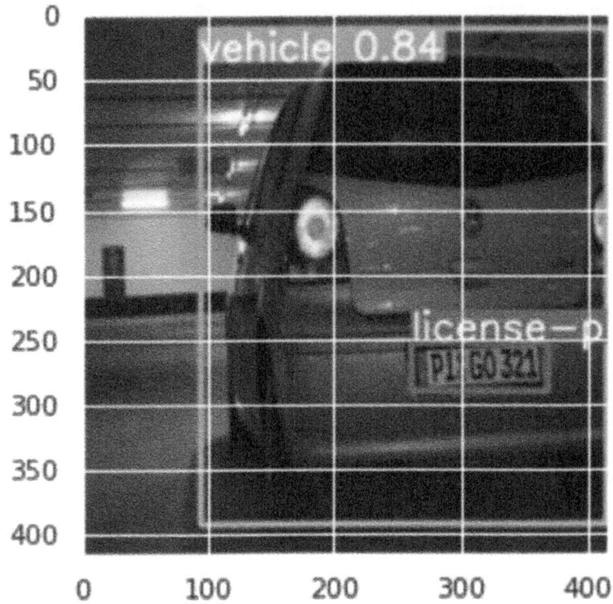

Fig. 8. Predictions of YOLOv3 for Car detection and License Plate detection

The whole setup is integrated with a website to show the available parking spaces. Figures 10a and 10b shows the different services available for the smart parking system in the web application. Whenever a car is parking the website refreshes to show the parking vacancy map where parked lots get dark and a counter for available number of slots decreases. In the real time logs the user name, registered mail id, the car manufacturer, plate number and time and date of entry is stored. Thus Redis is highly effective for storing real-time logs due to its blazing-fast in-memory data storage and retrieval capabilities. Its lowlatency response times ensure that log entries can be added and retrieved quickly, making it ideal for time-sensitive applications. It supports data structures like lists and streams, enabling efficient log management and retrieval for monitoring and analysis purposes.

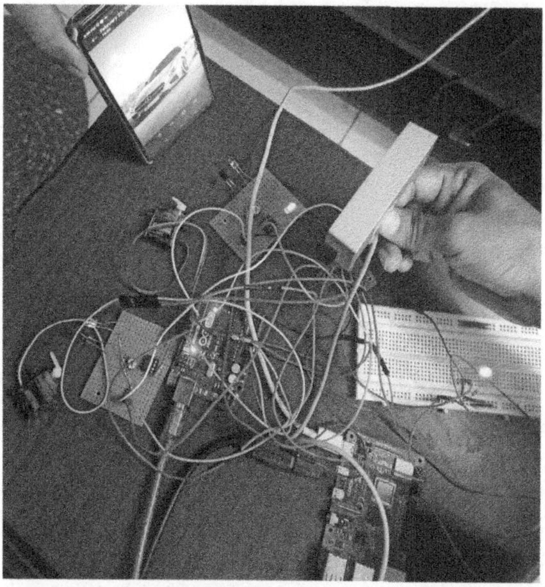

(a) Entry to the parking lot

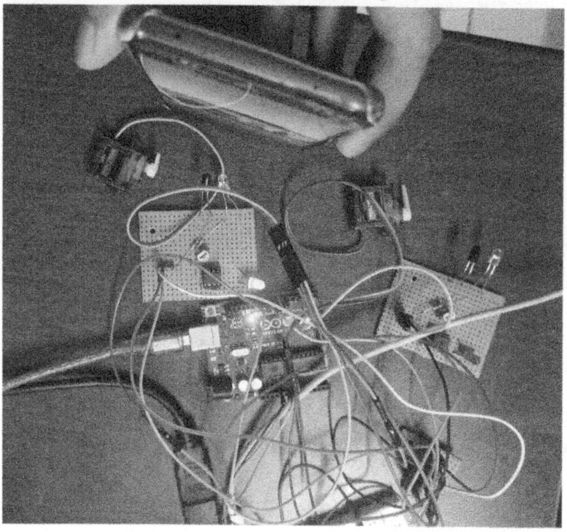

(b) Exit from the parking lot

Fig. 9. Experimenting with the smart parking system using a sample car image displayed on the mobile

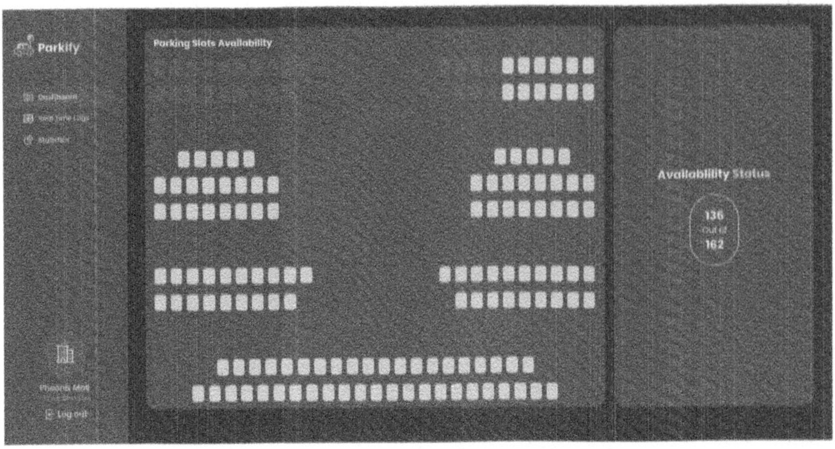

(a) Availability at the parking lot

(b) Sample for data logging

Fig. 10. The different services provided by the web application

6 Conclusion

The proposed smart parking system will help by providing a more efficient and convenient parking experience for drivers, while also optimizing the use of parking spaces and reducing traffic congestion. This system uses sensors and cameras to detect the availability of parking spots, reducing the time and frustration associated with finding a parking spot. Additionally, the intelligent tolling feature can automatically charge drivers for their parking, reducing the need for manual collection and processing of payments. This not only improves the overall efficiency of the parking system but also generates revenue for the operator.

7 Future Work

The proposed smart parking system is implemented on a small scale but can be expanded to large-scale implementation in the real world. Moreover, more software components like analytics dashboards with time series-based visualizations can be developed. The system has a lot of space in the real world, especially in various cities across India.

References

1. Rehman, M.U., Shah, M.A., Khan, M., Ahmad, S.A.: VANET based smart car parking system to minimize searching time, fuel consumption and CO 2 emission. In: 2018 24th International Conference on Automation and Computing (ICAC), pp. 1–6. IEEE (2018)
2. Su, K., Li, J., Fu, H.: Smart city and the applications. In: 2011 International Conference on Electronics, Communications and Control (ICECC), pp. 1028–1031. IEEE (2011)
3. Madakam, S., Lake, V., Lake, V., Lake, V.: Internet of Things (IoT): a literature review. J. Comput. Commun. **3**(05), 164 (2015)
4. Ujjwal, J., Bandyopadhyaya, V., Bandyopadhyaya, R.: Identifying key determinants for parking management to reduce road traffic congestion for congested cities-a structural equation modelling approach. Adv. Transp. Stud. **54** (2021)
5. Ali, G., et al.: IoT based smart parking system using deep long short memory network. Electronics **9**(10), 1696 (2020)
6. Wang, H., He, W.: A reservation-based smart parking system. In 2011 IEEE Conference on Computer Communications Workshops (INFOCOM WKSHPS), pp. 690–695. IEEE (2011)
7. Pham, T.N., Tsai, M.F., Nguyen, D.B., Dow, C.R., Deng, D.J.: A cloud-based smart-parking system based on Internet-of-Things technologies. IEEE Access **3**, 1581–1591 (2015)
8. Geng, Y., Cassandras, C.G.: New "smart parking" system based on resource allocation and reservations. IEEE Trans. Intell. Transp. Syst. **14**(3), 1129–1139 (2013)
9. Vakula, D., Kolli, Y.K.: Low cost smart parking system for smart cities. In: 2017 International Conference on Intelligent Sustainable Systems (ICISS), pp. 280–284. IEEE (2017)
10. Mainetti, L., Palano, L., Patrono, L., Stefanizzi, M. L., Vergallo, R.: Integration of RFID and WSN technologies in a Smart Parking System. In: 2014 22nd International Conference on Software, Telecommunications and Computer Networks (SoftCOM), pp. 104–110. IEEE (2014)
11. Redmon, J., Farhadi, A.: Yolov3: An incremental improvement (2018). arXiv preprint arXiv:1804.02767
12. Pavithra, D., Balakrishnan, R.: IoT based monitoring and control system for home automation. In: 2015 Global Conference on Communication Technologies (GCCT), pp. 169–173. IEEE (2015)
13. Qadri, M.T., Asif, M.: Automatic number plate recognition system for vehicle identification using optical character recognition. In: 2009 International Conference on Education Technology and Computer, pp. 335–338. IEEE (2009)
14. Wang, C.Y., Bochkovskiy, A., Liao, H.Y.M.: YOLOv7: trainable bag-of-freebies sets new state-of-the-art for real-time object detectors (2022). arXiv preprint arXiv:2207.02696
15. Li, C., et al.: YOLOv6: a single-stage object detection framework for industrial applications (2022). arXiv preprint arXiv:2209.02976

Study and Analysis of Substitution Cipher

Sarita Singh[1(✉)], Y. D. S. Arya[1], Shailesh Saxena[2], and Laxmi Shanker Maurya[3]

[1] Department of CSE, Invertis University, Bareilly, India
er.sarita2014@gmail.com
[2] SRMS College of Engineering and Technology, Bareilly, India
[3] SRMS College of Engineering and Technology and Research, Bareilly, India

Abstract. Data is shared via different medium over the insecure network. We are aware of the value of data in our lives. Numerous researchers have developed a variety of mechanisms to ensure data security; see here this study sheds light on papers that are specifically constructed with substitution ciphers and analyses them in terms of several metrics, including encryption and decryption times as well as CPU and memory usage. For secure communication via unsecured channels, like the internet, ciphers are necessary. Because internet communication is not secure and can be intercepted by hostile individuals, cyphers are required. Ciphers offers a number of advantages, including as confidentiality, speed, efficiency, and adaptability.

Keywords: Substitution Cipher · Security metcrics · Cipher · Data · Security

1 Introduction

The internet is the most powerful tool for communication and information exchange. No one is entirely responsible to manage the internet since it is a network of networks that is controlled by a variety of people but how to deliver a communication surreptitiously across an unsafe network is one of the basic and trivial problem in communication.

As a result, there are greater risks to information security. Three out of four (73%) employees who worked from home during the COVID-19 epidemic have not yet received any special cybersecurity awareness guidance or training from their employer, according to a recent Kaspersky report [6], How COVID-19 impacted the way people work. Furthermore, one in four (27%) workers reported having received phishing emails pertaining to COVID-19.

According to a poll of a select group of employees conducted globally, companies may not be adequately educating their staff on how to protect themselves from cyberattacks. Devices can become infected, and company data may be compromised, if malicious information is unintentionally downloaded from assaults like phishing emails. Additionally, many workers have increased their usage of "shadow IT," or internet services not authorised by their IT departments, for work-related purposes. Examples include video conferencing (70%), instant messaging (60%) and file storage services (53%). Andrey Dankevich, senior product marketing manager at Kaspersky, said, "When

S. Singh et al. (Eds.): AIDSP 2023, CCIS 2440, pp. 110–117, 2025.
https://doi.org/10.1007/978-3-031-88762-8_11

everything needs to change so dramatically, it is difficult to maintain 'business as usual'." IT and cybersecurity teams are under pressure to make it possible for employees to continue working safely as they adjust to the new reality of working from home. Cyber-incidents will only make this problem more harder, therefore it's critical to be on guard and ensure that working remotely is secure." [9]

Over 27 million Indian citizens, according to research from the 2021 Norton Cyber Safety Insights Report, have been victims of identity theft in the previous year. According to the survey, more than 59% of Indian adults encountered cybercrimes in the previous year, and those affected by these crimes spent a total of 1.3 billion hours attempting to fix the problems. [10]

Here we have defined the concept of cryptology which is the combination of Cryptography and Cryptanalysis. Cryptography means the art of making the cipher whereas Cryptanalysis means the art of breaking the cipher (Fig. 1).

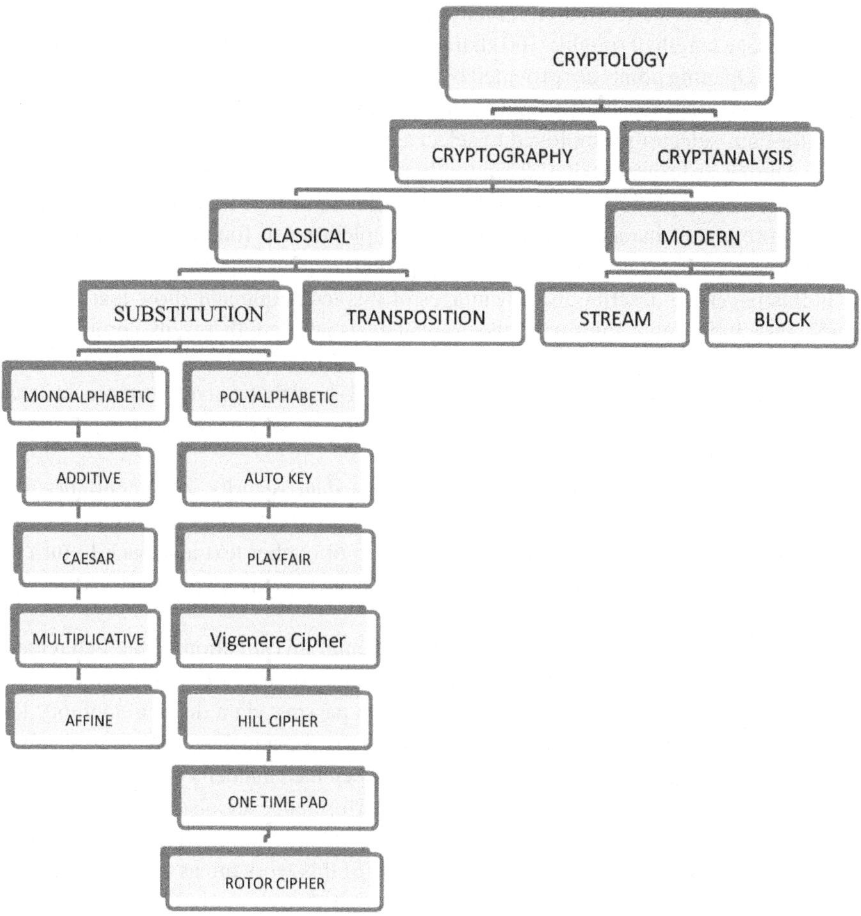

Fig. 1. Hierarchy of ciphers

In the above figure the detailed description of Cryptology and the types of Cryptography.

We have seen different Problem which is related to security of Data past several years and understand those problems and we have decided to analysis of existing Traditional Ciphers So in this Paper we have provide the different objectives

1. To provide the idea of Cryptology.
2. To explain the Strength or Weakness of Traditional Ciphers
3. To provide equivalent calculation power and battery usage of the Ciphers.

2 Related Work

In this paper the author described a algorithm along with some of the outcomes of a Pascal implementation. The procedure can also be used to break other straightforward cyphers in a generalised form. [1]

This paper describes a method for automatically solving simple substitution cyphers with a message length of roughly 160 characters and a space character suppressed during encryption. Opening points are provided by patterns found in the crypt. After appropriate filtering, words are fitted at these spots using a dictionary system. Following that, a tuple generator-cum-selector is employed to select a relevant tuple of words for decoding the cypher. During test runs, 70 of 100 randomly picked messages were solved to the point where manual completion was possible. When the algorithm successfully recognized 70% or more of the character places, manual completion was found to be rather simple. The algorithm took a long time to get to this point. [2]

In this paper we describe the advantages of this technique and show that it gives a simple, fast, and elegant solution to the cryptanalysis problem that is also promising for more sophisticated block ciphers. [3]

We describe the advantages of this technique and show that it gives a simple, fast, and elegant solution to the cryptanalysis problem that is also promising for more sophisticated block ciphers. [4]

In today's communication world, maintaining data security, confidentiality, and integrity is a major concern. In a communication network, cryptography is a technique that helps to achieve this goal. It uses an algorithm or cypher text approach to turn the information into cypher text. One such technique is the Playfair Cypher, which encrypts pairs of letters (bigrams or diagrams). [4]

In this paper the author presented an experimental investigation of the behavior of an HMM- based substitution cipher attack in order to quantify the security enhancement when mistakes occur while the cipher text is passing via a discrete memory less symmetric channel. The results show that a bound on the attack's performance can be directly connected to the mutual information between the channel's inputs and outputs. Rough linear estimates of the attack's high-end performance are also presented. [5]

In this study, the authors provide a step-by-step method for analyzing and breaking classical ciphers using CT2 tools. The basic goals of this work are as follows: (1) introduce historians and non-computer scientists to classical encryption, (2) introduce CT2, allowing people to break cipher on their own, and (3) describe our future ambitions for

CT2 in terms of (automated) cryptanalysis of classical ciphers. This study does not go into detail about the analysis methods employed, however it does provide references. [6]

In this research, the author's proposed a new symmetric substitution cypher methodology that takes advantage of the drawbacks of the symmetric substitution methods that are currently known to us by adopting a mechanism based on vowels, English language consonants, and numeric digits. Additionally, some fresh angles on encryption and decoding have been investigated. [7]

We present a unified cipher generative adversarial network (UC-GAN) that can conduct cipher text-to-plaintext translations across different domains (ciphers) using a single DL model in this research. The suggested approach, in particular, is based on unified unsupervised DL for the study of classical substitution ciphers. The simulation results have shown that the proposed strategy is feasible and performs well. Furthermore, we compared our experimental results to the findings of conditional GAN, which is cipher mapping between unpaired cipher text and plaintext in a single domain, and Cipher GAN, which is cipher mapping between unpaired cipher text and plaintext in a single domain. By learning only data without prior knowledge of three substitution ciphers, the proposed model achieved more than 97% accuracy. [8]

3 Material and Methods

In this section, we have described various materials used to accomplish our research work

3.1 Tools Used

All the eight classification algorithms used to build classifiers are implemented using following libraries of

Python: psutil library. – for measure encryption/decryption time, memory usage, and CPU usage

Scikit learn/sklearn – for Algorithm implementation.

Pandas – for dataset-related operations.

Matplotlib – for plotting.

Google Colaboratory – a free cloud service of Google was used to write and execute code in Python.

Steps of Research Methodology:

Step1: Problem Identification.

Step2: Literature

Review: Step 3: Create

Dataset

Step4: Implementation

Step5: Performance Measurement

Step6: Experimental Design

Step7: Testing and Data

Collection: Step8: Data Analysis

4 Results and Discussions

The main focus of our research is in this part. The experiment's results have been explained in this section, along with the pertinent discussions that were necessary to elaborate on the key findings. Table 1 and Table 2 provide the summary of this section:

Table 1. Monoalphabetic Cipher's

	Ciphers	Text Length (Plain text/Cipher Text)	Time Taken Encryption	Time Taken Decryption Time	Memory Used	CPU Usage
1	Ceaser Cipher	110000	0.0480976104	0.0480976	0.8626785278320	5.0%
			7363281	104736328	312 GB	
			s	1 s		
2	Additive cipher	110000	0.0870242118 8354492 s	0.0924661 159515380 9 s	0.7885437011718 75 GB	16.2%
3	Multiplicative Cipher	110000	0.0409481525 4211426 s	0.0995037 555694580 1 s	0.8011970520019 531 GB	21.9%
4	Affine Cipher	110000	0.0462865829 46777344 s	0.0474247 932434082 s	0.8031082153320 312 GB	6.5%

Here we are describing the graphical Representation of Mono alphabetic cipher in terms of Encryption time, Decryption Time, CPU Usage and Memory Used (Fig. 2).

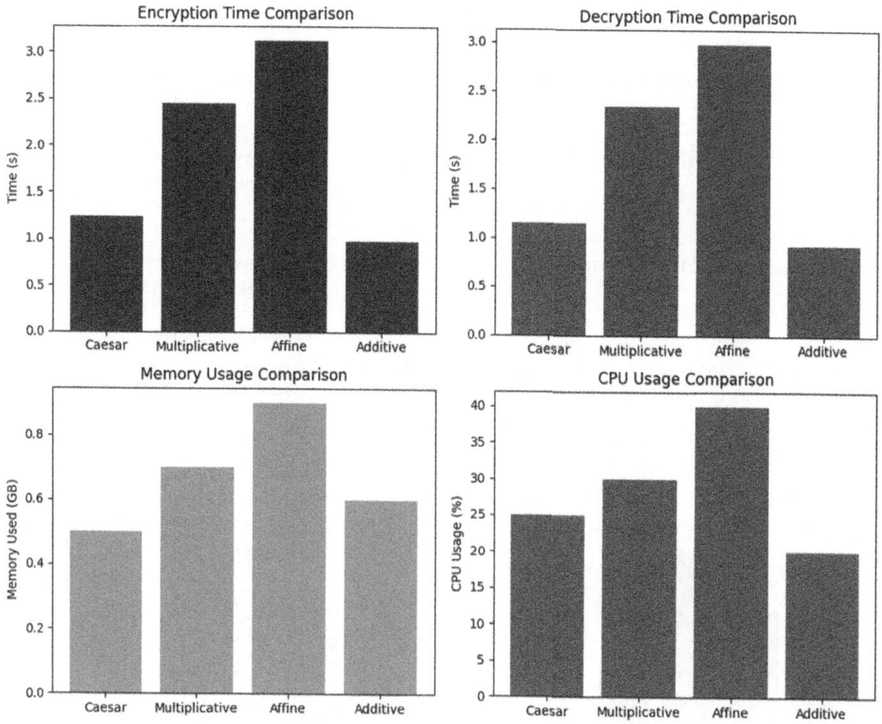

Fig. 2. Comparison Analysis of Monoalphabetic Ciphers

Table 2. Polyalphabetic Cipher's

	Ciphers	Text Length (Plain text)	Text Length (Cipher Text)	Time Taken Encryption	Time Taken Decryptio n Time	Memory Used	CPU Usage
1	Playfair Cipher	110000	100000	0.05552029609680176 s	0.050245046615600586 s	0.7938957214355469 GB	2.0%
2	Auto key cipher	110000	100000	0.03934049606323242 s	0.03582882881164551 s	0.8201103210449219 GB	16.2%
3	Hill Cipher	110000	100002	0.19393014907836914 s	0.09950375556945801 s	0.8490219116210938 GB	3.0%

(*continued*)

Table 2. (*continued*)

	Ciphers	Text Length (Plain text)	Text Length (Cipher Text)	Time Taken Encryption	Time Taken Decryption Time	Memory Used	CPU Usage
4	Vigenere Cipher	110000	110000	0.0403447151184082 s	0.03305530548095703 s	0.8679046630859375 GB	2.0%
5	One time Pad	110000	110000	0.025246858596801758 s	0.0289680957794189 45 s	0.8687591552734375 GB	3.0%
6	Rotor Cipher	110000	110000	0.020116090774536133 s	0.01873159408569336 s	0.8630905151367188 GB	3.0%

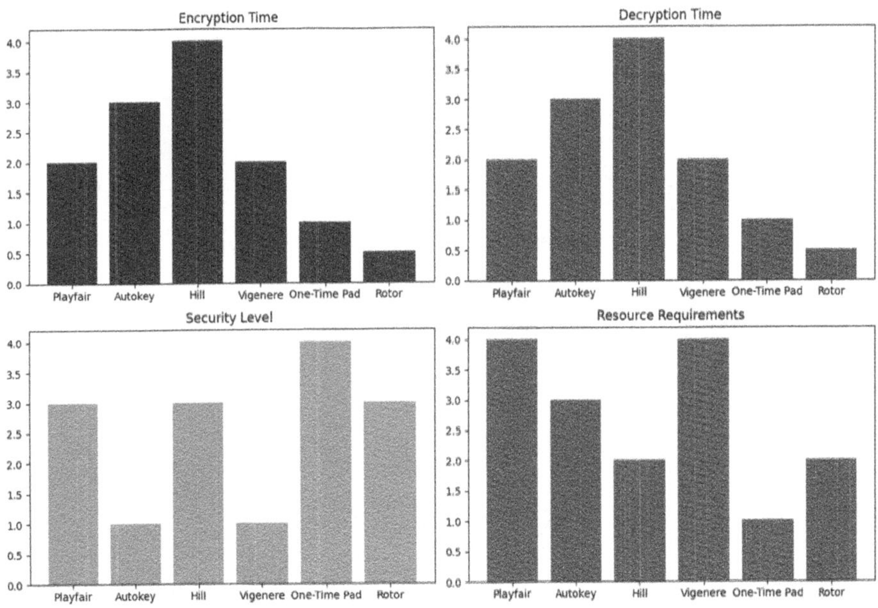

Fig. 3. Comparison Analysis of different Ciphers

5 Conclusion and Future Enhancements

In conclusion, we have used and examined a number of traditional ciphers, such as Playfair, Autokey, Hill, Vigenère, One-Time Pad, and a condensed Rotor Cipher. You've looked at elements including encryption/decryption time, security, key length, and resource needs through comparative analysis. Bar charts and other graphic representations were employed to graphically portray these comparisons. Our work demonstrates a thorough understanding of cryptography principles and how they are used.

Here are a few directions you may think about for future developments and research in the realm of cryptography: Explore more sophisticated encryption techniques like the AES (Advanced Encryption Standard) and RSA (Rivest-Shamir-Adleman) algorithms, which are frequently utilised in contemporary applications. Higher levels of security are offered by these algorithms.

References

1. Jakobsen, T.: A fast method for cryptanalysis of substitution ciphers. Cryptologia **19**(3), 265–274 (1995)
2. Ramesh, R.S., Athithan, G., Thiruvengadam, K.: An automated approach to solve simple substitution ciphers. Cryptologia **17**(2), 202–218 (1993)
3. Forsyth, W.S., Safavi-Naini, R.: Automated cryptanalysis of substitution ciphers. Cryptologia **17**(4), 407–418 (1993)
4. Omran, S.S., Al-Khalid, A.S., Al-Saady, D.M.: Using genetic algorithm to break a mono-alphabetic substitution cipher. In: 2010 IEEE Conference on Open Systems (ICOS 2010), pp. 63–67. IEEE (2010)
5. Sharma, N., Meghwal, H., Mehta, M., Kumar, T.: A review on playfair substitution cipher and frequency analysis attack on playfair. In: 2018 2nd International Conference on Trends in Electronics and Informatics (ICOEI), pp. 1–9. IEEE (2018)
6. Gross, N.L., Harrison, W.K.: An analysis of an HMM-based attack on the substitution cipher with error-prone ciphertext. In: 2014 IEEE International Conference on Communications (ICC), pp. 749–754. IEEE (2014)
7. Luthra, J., Pal, S.K.: A hybrid firefly algorithm using genetic operators for the cryptanalysis of a monoalphabetic substitution cipher. In: 2011 World Congress on Information and Communication Technologies, pp. 202–206. IEEE (2011)
8. Kopal, N.: Solving classical ciphers with CrypTool 2. In: HistoCrypt, pp. 149–010 (2018)
9. https://usa.kaspersky.com/about/press-releases/2020_kaspersky-research-finds-73-of-employees-have-not-received-remote-working-cybersecurity-guidance
10. https://indianexpress.com/article/technology/tech-news-technology/norton-survey-reveals-59-indians-have-dealt-with-cybercrime-in-past-12-months-7280071/

Simulations and Real-Time Control Using Time Abstraction

R. Srinivasan(✉) and M. Kavitha

Department of Computer Science and Engineering, Vel Tech Rangarajan Dr Sagunthala R&D Institute of Science and Technology, Chennai, India
srinivasanrajkumar28@gmail.com, Kavitha@veltech.edu.in

Abstract. This paper determines how simulated time can be implemented in python and how time can be regarded as a general level such that, it is independent of a specific time context using simulated or real-time. Simulated time is explored through literature search where it was discovered that there exist multiple classifications of simulation models based on the representation of time: Static, Continuous, and Discrete. A discrete event simulation model is then implemented which follows a process-based approach. Time in the simulation model is generalized with the introduction of an abstracted superclass called Time base. Time base introduces general time expressions that allow an entity to execute on the time base of the environment without specifically knowing which time context it is running on. The implementation was successful in changeably executing the simulation model in either time context, that is, simulated time and real-time. This paper confirms that generalized time concepts and simulation models can be applied with python but its effectiveness depends on the time precision requirements of the real-time systems.

Keywords: PID · LQR · Kalman Filter · ROS-Noetic · Gazebo

1 Introduction

Simulating the behaviour of a new product, production line, facility expansion, or other proposed or existing economic activity is a flexible technique. By running simulations and analysing the results, we may learn how a current system functions and what might happen if it were modified, or we can predict how a suggested new system would act. Instead of using the actual genuine system, we simulate it and perform tests on it. We do this because testing the model using trials is more efficient, affordable, and secure.For the design and enhancement of electrical systems, simulation tools have been widely used. Simulation tools have evolved in lockstep with the advancement of computing technologies since the mid-twentieth century. As a result, simulation tools have experienced tremendous performance improvements and consistent cost reductions. The Fourth Industrial Revolution is claimed to have a dramatic role in transforming the way we live, with the potential of shaping a better future for all. Industry 4.0 facilities include equipment with wireless connection and sensors that are connected to a system

S. Singh et al. (Eds.): AIDSP 2023, CCIS 2440, pp. 118–125, 2025.
https://doi.org/10.1007/978-3-031-88762-8_12

that can see the complete production line and make decisions on its own, despite the fact that the terms "Industry 4.0" and "fourth industrial revolution" are commonly used interchangeably. Internet of Things, Cloud Computing, Artificial Intelligence, Machine Learning, Cognitive Computing, Cyber-Physical Systems, Smart factories, and Dark factories are examples of industry 4.0 concepts. The digital twin is claimed to have an equal state to its corresponding physical twin and changing the digital twin should result in a change in the physical twin. The digital twin may be used to simulate future actions to guarantee well-chosen actions for its physical part. That is, the digital twin needs to be able to act on both real-time and simulated time bases. Thus, a general solution for real and simulated time could be used. A general time concept was introduced which allows entities of a system to act on a general time base and not be designed around a specific time concept. The main aim of this paper is to create a simulation framework for all the real- time systems to predict future actions and make the system time-consuming, efficient, and flexible.

2 Literature survey

A software tool and algorithm were presented by Fernando G. Gonzalez et al. [1] which can accept collections of models rather than one model. The modeling effort saved by using the approach presented here is approximately 50% of that of developing control models for a large-scale distributed system with seven levels of control hierarchy. Using control models from simulations is permissible. It is almost a requirement when designing a controller for large-scale distributed systems. As a result of this method, the modelling effort is reduced, the communication between controllers is automatically modelled, dynamic simulation is resolved, and the process of transferring system state from the actual system to the simulation model is simplified. According to Giovanni Lugaresi, et.al., [2] increased accessibility of data and flexibility of resources has opened up the possibility of developing methods that use real-time inputs in planning and control. There are many approaches in the literature for simulating, evaluating, and making decisions based on optimization criteria. The Industry 4.0 revolution identifies simulation as one of its pillars. In contrast, there is no widely accepted methodology or approach for dealing with real-time decision-making. In the coming years, simulation approaches can play a significant role in industry. An overview of the current research context and a brief summary of existing approaches are provided in this position paper, as well as considerations for implementing real-time simulation. Katja H. Sirvio, et. al., [3] describe the development of an accelerated co-simulation and testing platform for long-term power system simulations. Active network management functions for microgrids and smart grids will be developed and tested on the platform. Distribution-network-connected distributed energy resources can be used for a variety of technical flexibility services, for example, by conducting long-term simulations. Presented in this paper are the potential and issues associated with the developed method for accelerated real-time simulations. According to Li Xu et al. [4], many kinds of engineering software and digitalized equipment are often employed throughout the lifespan of industrial goods. As a result, enormous volumes of data are produced. However, due to their hysteresis and isolation from one another, these data are used inefficiently. In the past, systems have been verified, validated, and optimised using simulations based on theoretical and static models, but during system operation, the simulation application is not given any thought. In his paper, Luis Ibarra, et. al., [5], he examines a few current simulation technology uses. In an SG environment where quick development and innovation are essential, simulation tools are more than simply technological assets. Simulators could make it easier to work together and reach a technical/political consensus over existing electric grid standards. Due to the fact that this work is created from the standpoint of introducing a novice to the simulation world, the suggested literature may be found throughout the offered references. Furthermore, RT simulation should be shown as a workable method that may be applied to satisfy the technological demands of the present. R Ramya et al. [6] investigation examines the properties of the aforementioned multimachine system. It is crucial to analyse the characteristics of the many machines that make up a power system in real time. The nine-bus, three-unit system of the Western System Coordinating Council (WSCC) is modelled transiently using MATLAB/Simulink. A real-time interface is used to run the model in the digital signal processor (DSP) of the dSPACE hardware. The performance of the model is assessed in respect to various variations in mechanical torque and load. We investigate the resilience of the model to changes in network and system parameters as well as its transient behaviour in the event of a large network fault. The circumstances are simulated using the DS1104 and digital signal processor. Real-time digital simulation is briefly described by Rick Kuffel et al. [7] and examples of how it is used to the design and operation of modern, complex power systems are provided. The invention, testing, and use of protection and control methods used throughout the industry will be the main focus of this paper. The goal of real-time digital simulation was to bring together the adaptability and precision of digital simulation techniques with the in-the-moment response of analogue simulators. Once accomplished, a whole new level of accuracy and adaptability would be added to the method for creating, putting into use, and testing power system control and protection systems. The assessment of available tools by Robert Grepl et al. [8] considers usability, sample rate capabilities, needed software, and cost considerations. Most methods call for the writing of C code using Real-Time Workshop. However, Simulink's usual mode, which is especially advised for educational applications, may be used directly with the Real-Time Toolbox, for instance. In various example studies, the article describes how undergraduate students may work with real-time instructional models without having any prior experience of real-time systems. The list of strategies that are supported by the hardware and software tools includes code creation for embedded targets, high-performance hardware for rapid prototyping or hardware-in-the-loop simulation, and inexpensive general-purpose I/O cards with/without C code generation. Victor Zhidchenko, et al. in their study from [9] Dynamic and kinematic crane models are created to solve this problem. A digital twin created by the models can be utilised to facilitate crane operation. Crane dynamics, as opposed to kinematics, have a significant role in movement prediction since the crane may move swiftly. How- ever, the crane model should be simplified in order to be calculated faster than in real-time. Wenhe Yang, among others [10] Digital twins are accurate, functional, and communicative digital copies of real-world objects or subjects. As Internet of Things nodes, they enable networking and subsequently the automation of intricate value-added chains. Digital Twins can be tested and brought to life via the use of simulation techniques. At first, these Experientable Digital Twins (EDTs) only speak to one another online. The networks of interconnected EDTs that are created replicate numerous application scenarios and offer fresh starting points for complete simulation-based systems engineering. The EDTs are getting more and more specific with each application. Consequently, a whole digital representation of the relevant physical assets and their behaviours is gradually built up.

3 Proposed Work

3.1 General Architecture

An implication of this transparent view on concepts of time is that we introduce a general concept of time base. A time base may be specified to be a real-time time- base (RTTime base) representing a real-time concept, or a simulated-time time base (STTime base) representing a simulated-time concept. Furthermore, a time base may be bound (Bound Time base) to any other time base through a Time Transformation, which converts time from one time base to another. The probably most common way is to have a linear relationship between time bases.

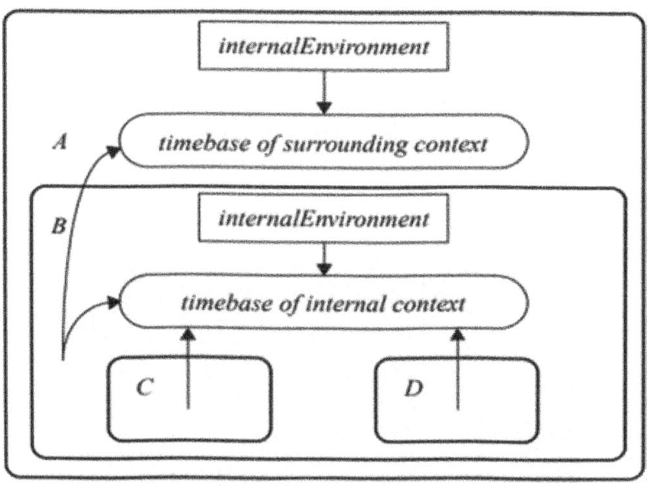

Fig. 1. Illustrates the hierarchically organized time concepts.

By connecting the time base, T1, of an activity, A, to e.g., a RTTime base, T2, we can let the time of T1 flow, for instance, twice the speed or half the speed of T2, i.e., without affecting the time variant expressions of A. As timing relates to hierarchical modelling in the way that different hierarchical levels may define and act against different time concepts. Since an activity may be present at two different levels, it may actually act towards both the time concept defined at the upper level, and towards the time concept defined locally as a part of its own locally defined environment. This modelling exemplifies this situation, where activity A has a time base internally defined. Then activity B may act against this time base. Furthermore, B defines an internal time base, accessible from B itself and from internal activities C and D. If B would not define its internal time base, then B, C and D, would adjust themselves to the time base of A, as illustrated in the Fig. 1

3.2 Proposed Architecture

Step 1: Make four module dependencies for this main program, named as environment, time base, machine, entity, global.

Step 2: Environment class defines environment functions such as setTimebase, getTimeBase, insert and remove entity which executes for both the sim- ulated and realtime timebase. Mainloop defines the timebase and future upcoming task and waiting function for all the waiting task in the row. It also insert and remove event for queuelock and futureEvents.

Step 3: The main timebase class in timebase module contains both simulated and realtime which is inherited to main timebase class, for all fuunc- tions like getTime, wait, WaitUntil, start functions.

Step 4: In machine module, Machine module contains main tasks in array with timepoint and life of machineTask for their execution time, timepoint and print the flow of task run.

Step 5: In entity module, machine class contains machineTask which evaluates the task whether it is in running state, waiting or in stopped state.

Step 6: Main file defines the timebase and it checks whether it is realtime or simulatedTime. Then machineTask in that environment calculates the time difference between start-time and end-time in environment.run() module. Figure 2 shows execute of all tasks and Fig. 3 shows the life cycles.

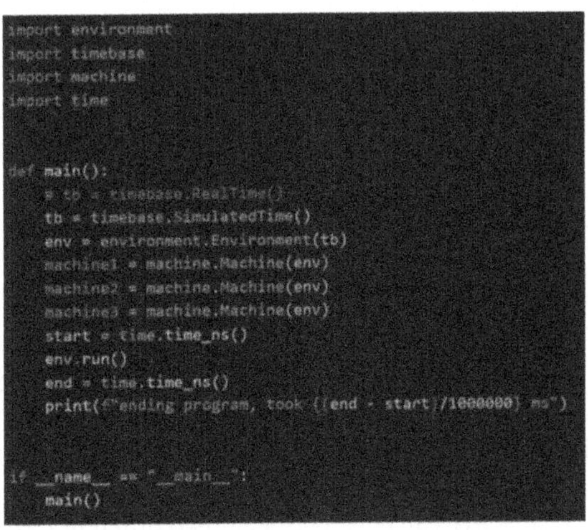

Fig. 2. Main program to execute all tasks

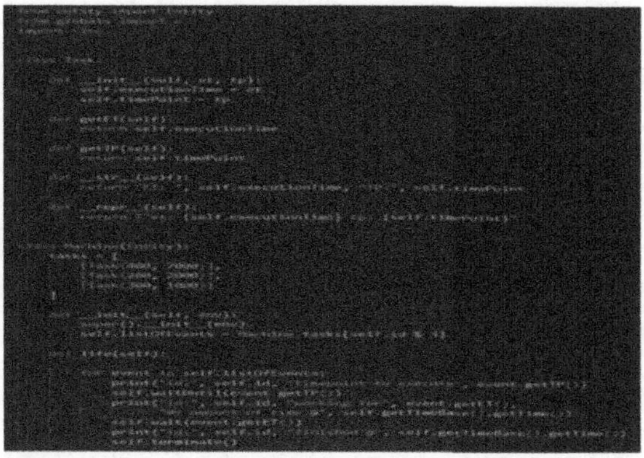

Fig. 3. A class for maintaining machines and its life cycle

4 Results and Discussions

In a simulation paper, the simulation is ultimately driven by the input data. The gathering of input data, analysis of the input data, and application of the analysis of the input data in the simulation model are all steps in this process. The input data may come from historical records or the simulation paper role of real-time data collection. Finding the theoretical distribution that best describes the input data is a step in the analysis. The theoretical distributions must be written into the simulation software code in order for the input data to be used in the model. Our input takes the form of jobs, each of which specifies an execution time point and an entity-starting time point. Figure 4 describes the simulated output.

```
id: 0 timepoint to execute 1000
id: 1 timepoint to execute 2000
id: 2 timepoint to execute 3000
started all entities
id: 0 released @ 1030
id: 0 working for 400 ms amount of time @ 1046
id: 0 finished @ 1489
id: 1 released @ 2018
id: 1 working for 200 ms amount of time @ 2033
id: 1 finished @ 2267
id: 2 released @ 3031
id: 2 working for 300 ms amount of time @ 3047
id: 2 finished @ 3394
no entities running, terminating
ending program, took 60007.4159 ms
```

Fig. 4. Sample output for the simulated

The output could be generated by combining with a discrete event model in such a way that no com- ponent of the model is specifically modelled to fit a certain time context. This allows the program to switch between the different time contexts by only changing the Timebase used by the environment, thus achieving a general solution. Figure 5 shows real time output and Fig. 6 describes the Simulated Output.

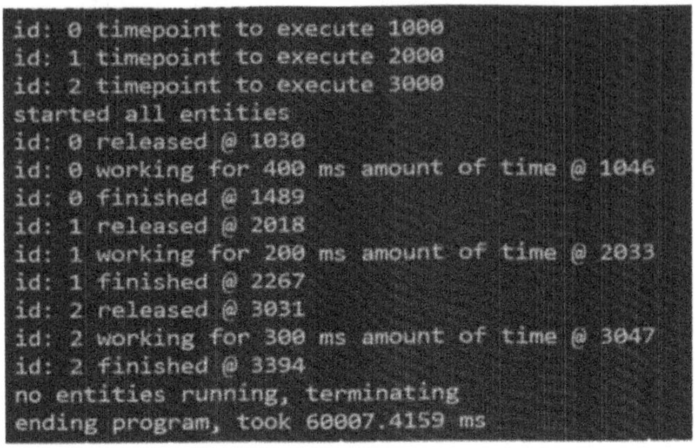

Fig. 5. Realtime Output

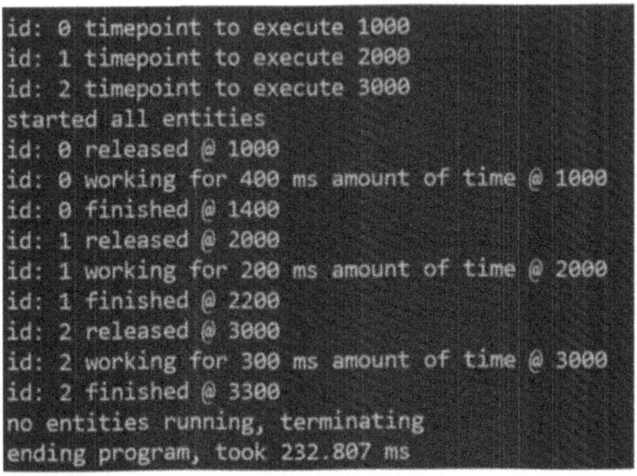

Fig. 6. Simulated Output

5 Conclusion

This paper concludes with two-time aspects, that is, simulated time can be implemented in python and time can be considered on an abstracted level, separate from specific time notions. Simulated time is simply a variable but is highly affected by the model using it and the way that model advances time. Model classifications were found which led to deeper research into discrete event simulation. The main way of advancing time in a discrete event simulation is through a method called next-event time advancement. The model makes use of a sorted event list where the events are sorted in ascending order based on what point in time they are to execute. The current time of the system is based on the current event N. When done executing the time is advanced to the point in time of N + 1. With the introduction of Timebase, one can implement time at an abstracted level where entities interact with the component independent of time context. Simulated timebase and real-Timebase are subclasses of Timebase and implement their methods to their specific design. The implementation integrated the concepts obtained from the literature search to construct a discrete event simulation model that can run in a simulated or real-time environment. This work could be expanded to include more concepts of timebases. This intro- duces a Bound Timebase and a simulated-real-Timebase. Bound Timebase would allow a system to run real time on different speeds. The purpose of this concept is to achieve some of the benefits of simulated time, that it executes faster than real time, while still keeping attributes of real time. The actual time provided by the Time-base is dependent on and bound to another Timebase through a time transformation that defines the relation between the two timebases. Simulated-real-Timebase would allow a system to move between events with discrete amounts but run on real time during events. These time concepts could also be expanded to support hierarchical systems in python to support internal and external environments. Here different combinations of timebases could exist within this structure such as an internal simulated time with external real time or vice versa.

References

1. Gonzalez, F.G.: Real-time simulation and control of large scale distributed discrete event systems. Procedia Comput. Sci. **16**, 177-186 (2013). ISSN 1877-0509
2. Lugaresi, G., Matta, A.: Real-time simulation in manufacturing systems: challenges and research directions (2018). https://doi.org/10.1109/WSC.2018.8632542
3. Sirvio¨, K., et al.: Accelerated real-time simulations for testing a reactive power flow controller in long-term case studies. J. Electr. Comput. Eng. 1–17 (2020). https://doi.org/10.1155/2020/8265373
4. Da Li, X., Xu, E.L., Li, L.: Industry 4.0: state of the art and future trends. Int. J. Prod. Res. **56**, 1–22 (2018). https://doi.org/10.1080/00207543.2018.1444806
5. Ibarra, L., Rosales, A., Ponce, P., Molina, A., Ayyanar, R.: Overview of real-time simulation as a supporting effort to smart-grid attainment. Energies **10**, 817 (2017). https://doi.org/10.3390/en10060817
6. Ramya, R., Selvi, K., Murali, K.: Real-time simulation and performance analysis of multimachine power systems using dSPACE simulator. SIMULA- TION. **92**, 63–75 (2016). https://doi.org/10.1177/0037549715620014
7. Kuffel, R., Forsyth, P., Peters, C.: The role and importance of real time digital simulation in the development and testing of power system control and protection equipment. IFAC-PapersOnLine **49**, 178–182 (2016). https://doi.org/10.1016/j.ifacol.2016.10.739
8. Grepl, R.: Real-time control prototyping in MATLAB/Simulink: review of tools for research and education in mechatronics. In: 2011 IEEE, ICM 2011 – Proceedings, pp. 881-886. (2011). https://doi.org/10.1109/ICMECH.2011.5971238
9. Zhidchenko, V., Malysheva, I., Handroos, H., Kovartsev, A.N.: Faster than real-time simulation of mobile crane dynamics using digital twin concept. J. Phys. **1096**, 012071 (2018). https://doi.org/10.1088/1742-6596/1096/1/012071
10. Yang, W., Tan, Y., Yoshida, K., Takakuwa, S.: Digital twin-driven simulation for a cyber-physical system in industry 4.0. (2017) https://doi.org/10.2507/daaam.scibook.2017.18

Comprehensive Way for Modeling of Femur Bone Fracture from CT Images

Aashish Joshi[1,2(✉)] and Kailash Karande[1]

[1] Department of E&TC Engineering, S.K.N Sinhgad College of Engineering, Korti, Pandharpur Dist-Solapur, Maharashtra 413304, India
ashu.joshi05@gmail.com

[2] Department of E&TC Engineering, Karmayogi Institute of Technology, Shelve, Pandharpur Dist-Solapur, Maharashtra 413304, India

Abstract. In previous few years development of modeling and methodology of surgical computation simplified the management of surgical operations for surgeons. It also reduces the risk associated. Home Accidents and osteoporosis are the two most common causes of fractures associated with femur bone. In old age patients, osteoporosis is responsible for bone weakness and fatigueness. Such Patients will benefit from early rehabilitation if they receive proper treatment during surgery. This paper depicts a work of issues of concern. With Biomedical CT image modelling of fractures associated with femur bone for surgical planning by modeling a 3D model by using CT image data. The purpose of this paper was to propose the work for bio-clinical image preprocessing in order to detect the significant ROI by providing an relevant technique. 3D model is useful to applications like 3D printing and another applications to ease surgery of the doctor.

Keywords: Femur Bone. · Image preprocessing · 3D CT image modeling · Biomedical Image analysis · proximal femoral fractures

1 Introduction

In old age patients Fractures occurs due to osteoporosis typically in proximal femur. Osteoporosis is a disease where Bone becomes so fragile and delicate such that even a slightest wrench can result in a fracture. This issue persist and grows more severe with age goes on increasing. Road Accident is one prominent cause for variety of fractures but fracture associated with femur is found to be major focusing. Surgical interventions are used in the treatment of hip fractures. If parameter analysis and image modelling are remembered for the procedure with ideal plate and screw placements and different dispositions will reduce the impact and likelihood of disappointment.

The advantages of femur bone fracture modelling can be seen in a variety of areas, such as medical diagnosis, surgical planning, education, and research. Here are some specific advantages:

Accurate Diagnosis: 3D models of femur bone fractures can aid in the accurate diagnosis and classification of fractures. They can better understand the extent and severity of the fracture by visualising it in 3D, which aids in treatment decisions.

S. Singh et al. (Eds.): AIDSP 2023, CCIS 2440, pp. 126–135, 2025.
https://doi.org/10.1007/978-3-031-88762-8_13

Surgical Planning: 3D models can be used by surgeons to plan surgical interventions for femur fractures. They can use the models to analyse the fracture pattern, plan the best approach, determine the need for hardware placement (such as screws or plates), and simulate bone fragment reduction or alignment.

Communication is Improved: 3D models improve communication between medical professionals and patients. Patients can better understand their condition, treatment options, and potential outcomes if they can visualise the fracture in a tangible and understandable format.

Educational Tools: Femur bone fracture models are useful educational tools for medical students, residents, and surgeons in training. They offer hands-on learning about fracture types, patterns, and treatment approaches. To improve surgical skills, interactive models can be used for simulation-based training.

Research and Development: 3D models of femoral bone fractures aid in fracture management and implant design. These models can be used by researchers to investigate fracture biomechanics, compare different treatment approaches, and test novel implant designs or surgical approaches.

Patient-Specific Treatment: Surgeons can better understand the unique characteristics of a patient's femur fracture using personalised 3D models. This allows for tailored treatment planning, allowing for optimal fracture alignment, reduction, and fixation, which can lead to better patient outcomes.

Preoperative Rehearsal: Surgeons can perform virtual preoperative rehearsals using 3D models. This entails simulating the surgical procedure on the model in order to become acquainted with the steps, anticipate problems, and improve surgical precision. It has the potential to shorten operating times and improve surgical outcomes.

2 Preliminaries of Modeling of Femur Bone

The technology advancement makes clinical tasks easier and saved a lot of time in medical applications requirement for operation. The surgeons and Biomedical Engineers are taking efforts to automate the surgeries to easiness and reduction in risk associated in surgeries. Fractures are most difficult kind of surgery with rehabilitation associated. In such critical issues 3D model will gives the clear cut idea to replace or provide appropriate implant for the mishap occur.

To make a 3D model of a femur bone fracture, you would typically do the following:

1. **Obtain reference information**: Obtain medical images of the fractured femur bone, such as CT scans or MRI scans. These images contain detailed information about the fracture and will be used to create the 3D model.
2. **Image segmentation**: In the acquired medical images, use specialized medical imaging software or tools to separate the fractured femur bone from the surrounding tissues. This procedure entails manually or automatically outlining the bone structure in order to distinguish it from other anatomical structures.

3. **Image Processing:** Once the bone has been segmented, use image processing techniques to improve the visualization of the fracture. This may entail adjusting contrast, removing artefacts, and smoothing the image to achieve the desired effect.
4. **3D Conversion:** To convert the segmented 2D images into a 3D representation, use medical imaging software. Stacking the segmented images allows you to reconstruct the fractured femur bone in three dimensions.
5. **Fracture reconstruction**: Use the 3D model to identify the specific characteristics and details of the femur fracture. This could include examining the fracture line, displacement, and any other relevant factors. The bone fragments can be aligned and assembled using software tools or by manually manipulating the 3D model.
6. **Surface reconstruction**: Create a smooth surface mesh for the fractured femur bone after reconstructing the fracture. The mesh should accurately represent the bone's shape and contours, including fractured areas. This step is typically carried out with the aid of specialized equipment.
7. **Visualization and refinement**: Once the 3D model has been created, you can view it from various angles to assess the fracture and its severity. Make additional adjustments or refinements to the 3D model as needed to ensure it accurately represents the actual fracture.

It should be noted that creating a 3D model of a femur bone fracture usually necessitates knowledge of medical imaging, segmentation techniques, and 3D modelling software. It is frequently carried out by medical professionals, radiologists, or specialized technicians who are equipped with the necessary knowledge and tools to work with medical images.

For a number of uses, including anatomical research and implant modelling, a three-dimensional (3D) bone prototype is required. Gauging and modelling load-bearing joints present a significant challenge for researchers and clinical specialists [1]. (Francois Girinon et al.2020). Let's review some articles on femur bone modelling. Figure 1 shows a picture of a femur bone and where it is in the human body for your convenience.

S.J Lee et al. [2] developed a 3-D FEM CAD model for femur bone using mathematical feature extraction obtained from CT image datasets. Two models were created based on this femoral dislocation simulation. The first is a non-cemented model (NC), in which the femur supports the hip screw, and the second model is a concrete-cemented model (CA), which includes a concrete mantle to encompass the hip screw. ITKSNAP open source software was used by K.C. Nithin Kumar et al. [3] to create three-dimensional CAD models for such proximal Femur had been created using CT data. For preprocessing, the created model was imported into Hyperworks. This modeling's main objective is to forecast alternative suitable materials for structural analysis. For the 3D displaying of the bone, V Sindhu et al. [4] used AutoCAD and SOLIDWORKS as CAD virtual products. SOLIDWORKS was used to import the femur drawings for 3D rendering. Four restrictions and the expelled cut and hurling procedures were used to create the 3D model in SOLIDWORKS. A goal of such a experiment is to locate surgical supplies and act as a virtual roadmap for meticulous placement. Shuqin Wu et al. [5] conducted an experiment using VC6.0 to create the FNF empty nail interior obsession activity programming and assessment framework based on specialists' exploration and Finite Element

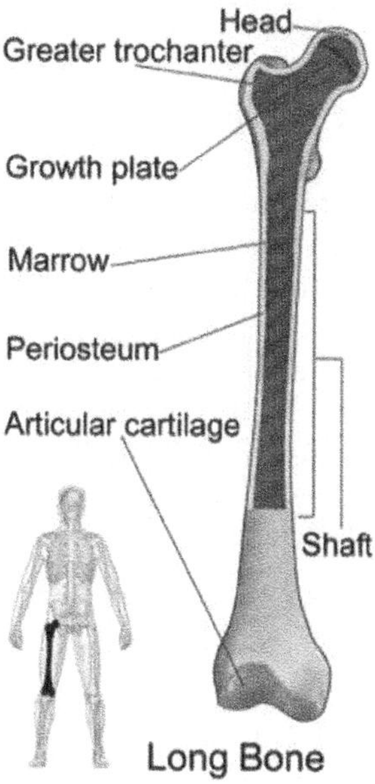

Fig. 1. Femur Bone Parts and Their Place in the Human Skeleton Blausen. Com staff (2014)

Analysis (FEA) data. The program computes ideal activity boundaries to provide an in-depth assessment of the surgical procedure. The activity writing computer program are dependent on the limited component investigation information, which was determined based on the activity needs. The main development includes the patient's individual information (load, age, etc.), the femur neck and fracture data measure, the image set improving details distinctive, deciding operation limits, deciding operation mobility, and many more. Radiographic images with self-alignment to account for low-portion preoperative planning were investigated by Kibeom Youn et al. [6]. Using parallel calibration and shape optimization processing, a 3D femur bone was created. In the author's recently developed technique mathematical structural modeling, 3d surface model is performed with the generation of structural coordinates in parallel with radiological data sets, so that measured properties can be adjusted as needed. In contrast to other reproduction techniques, the proposed method uses conventional radiograph frameworks and is simple to incorporate into current clinical practice. Dong Zheng et al. [7] used Computed Tomography (CT) images of an unbroken right femur of a 60-year-old medically stable Chinese female to create a femur bone model using Mimics software, with a cut thickness of 0.5mm and a size of 512512. The discarded femur tissue was placed on a volume veil and considered in a 3D model using an insertion calculation built into Mimics. The developed femur model was imported into Geomagic10.0 in a twofold stereolithography arrangement to create the strong model. After the surface was smoothed, the 3D Femur model was imported into Solidworks. Tanvir Faisal et al. [8] used the medical imaging software Mimics to obtain a 3D calculation of the femur in order to identify stresses at the femur neck region and recommend the prospect of crack formation by taking account for the 3D complex geometry of the femur. Mimics' interpolation computation feature was used to convert the 2D images into 3D models. The bone is identified using a dark edge subset of 226 to 2221 HU in the 3D model, which is a grey scaled estimate based on CT data. According to Alex C.T Choh et al. [9], One 3D FEM model was created using a CT dataset for a female's right femur. To divide datasets, multiple help has been used. The goal of this study was to determine the position of the lag screw in various intertrochanteric hip fractures. The fracture angle for osteoporotic and healthy bone is found to be different in experiments. Fang Chen et al. [10] built a femur bone model from a Chinese dataset to represent the IM nail. To create 3D femur replicas, the CT data were saved in the Digital Imaging and Communication in Medicine (DICOM) format and fed into the modelling program Mimics. The steps of Mimics' veil creation, district developing, and remeshing were used to create this 3D femur model.

The Femur bone outline is shown below, along with the software and image type that were used, as shown in Table 1.

Table 1. Literature Review of Different Femoral Bone Modelling Software

Sr.No	Author	Type of input image	Software Used
1	S.J Lee et al.	CT	NA
2	K.C. Nithinkumar et al.	CT	ITKSNAP
3	V.Sindhu et al.	CT	SOLIDWORKS and AutoCAD
4	Shuqin Wu et al.	CT	VC 6.0
5	Kibeom Youn et al.	CT	Mimics
6	Dong Zheng et al.	CT	Mimics
7	Tanvir Faisal et al.	CT	Mimics
8	Alex C.T Choh et al.	CT	Mimics
9	Fang Chen et al.	CT (DICOM format)	Mimics

The pie chart in Figure 2 provides an overview of the influential software for the papers that are alluded to.

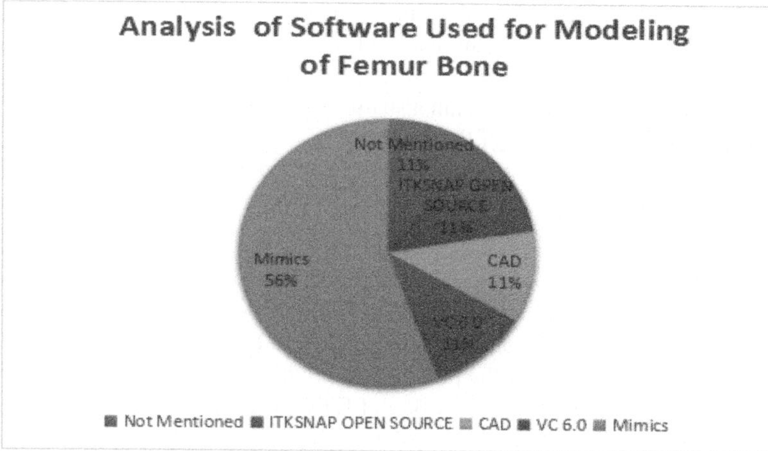

Fig. 2. Software for femur bone analysis Modeling

3 Modeling of Femur Bone

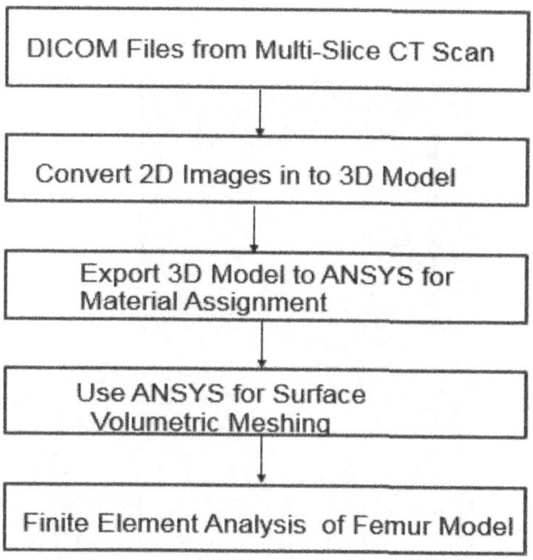

Fig. 3. Workflow for Femur Bone Modeling

A popular piece of free, open-source software known as 3D Slicer is used in imaging studies related to medicine and other fields. A growing international user and developer community backs 3D Slicer. 3D model of femur bone has been developed by using 3D slicer software. The modality used was CT (Compute Tomography) dataset samples collected from Diagnostic center. The medical file format used were dcm i.e. DICOM (Fig. 3).

Steps for Femur Bone Model

1. Bringing the DICOM images into 3D Slicer is step one.

2. To display the 3D view, enable the volume rendering option.
3. Choose the bone type, which eliminates all the undesirable elements like skin and muscles.
4. Modify the frequency scale to obtain a sharp 3D model of the component.
5. Cut out the necessary portion of the volume.
6. Edit the portion that was cropped for a better 3D model.
7. To get a finished part, activate the model effect option.
8. Save this component as an STL/IGS file.

Software imported DICOM files. The required changes have been made for progressing the work. The above steps were used to create bone model. The starting view of the of the files for visualization is as shown below in Fig. 4.

Select the type of bone after the preset selection is complete. Because we used CT scan images in this work, a preset selection of CT bones was chosen, and as shown in Fig. 4, only bones are visible in three-dimensional view, getting rid of all extraneous material. Shift is another option for adjusting By removing the unnecessary components, the appropriate bone can be chosen. The crop volume module is then used to perform cropping. Cropping is used here to get the image's correct femur bone. Immediately after cropping, a new volume is produced in which additional work to complete the model is performed. Figure 4 displays the new volume that was cropped.

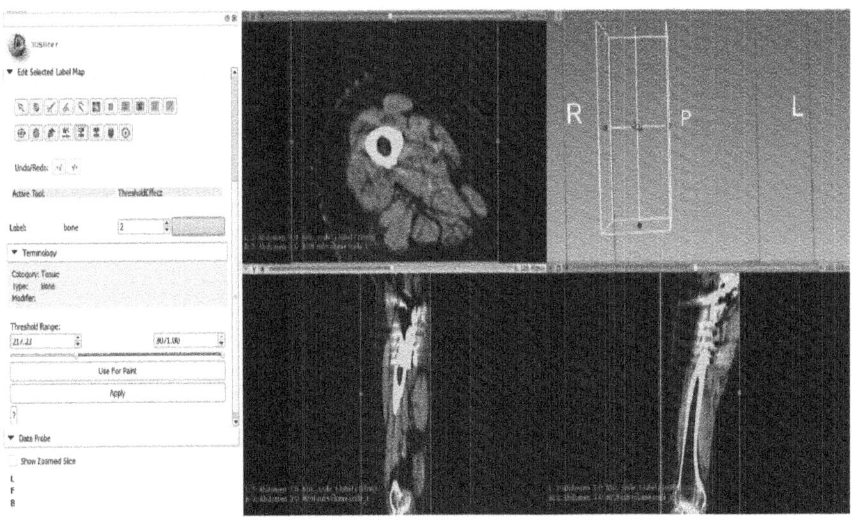

Fig. 4. Four different view at a time in 3D Slicer

Once the new volume is completed, the editing module will assist in making additional changes. The threshold frequency for bone is changed in the editing module to only bring to light the bone part. The CAD model of the right femur bone is obtained after applying this model effect. As illustrated in Fig. 5(a). The obtained CAD model has numerous difficulties, anomalies, and discontinuities since it is already going for a prosthesis with implantations. The paint effect is used in the editor module to obtain the best model with no errors. Figure 6 depicts the best following the application of the paint effects, a 3D CAD model of the right femur is shown (b).

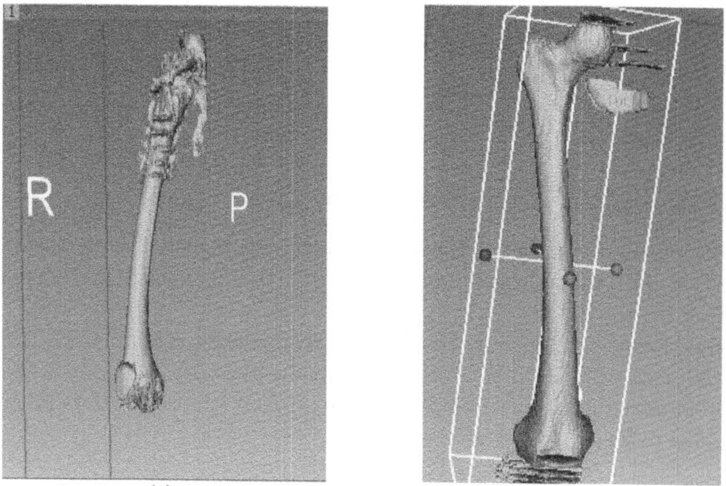

(a) (b)

Fig. 5. (A) 3D model Effect (b) Femoral bone 3D CAD model

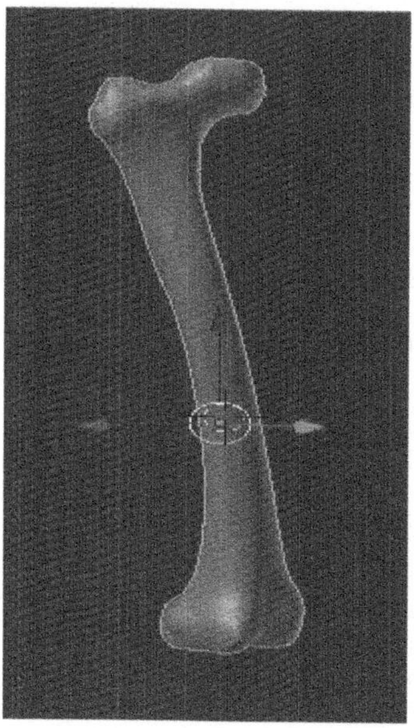

Fig. 6. 3D Model of Femur Bone by Manual Finishing

Material Assignment Problems

1. The femur bone is a complexly structural components made up of compact and cancellous bone.
2. Compared to cancellous bone, which is an isotropic material, compact bone is an anisotropic material.
3. As a result, it is challenging to assign material properties along each direction of the bone model.

4 Conclusion

In the view of work done completed following observations are made:

- Meshing of femur bone is required for fast and good geometrical representation
- 3D slicer is a good nonproprietary software used to create the human femur's CAD model
- This same model created is utilized for application areas, including the 3D printing of a thighbone to be used as a prototype, conducting various analyses on a femur bone, conducting an impanation research on a femur bone, as well as other clinical application to facilitate doctors' treatments.

References

1. Girinon, F., Gajny, L., Ebrahimi, S., Dagneaux, L., Rouch, P., Skalli, W.: Quasi-automated reconstruction of the femur from bi-planar X-rays. Comput. Methods Biomech. Biomed. Eng. Imaging Vis. **8**(5), 529–537 (2020)
2. Lee, S.J., Kim, B.J., Kwon, S.Y., Shin, J.W., Kim, J.K., Tack, G.R.: Viability of bone cement augmented compression hip screw system for the treatment of intertrochanteric fracture: a biomechanical analysis. In: 2001 Proceedings of the 23rd Annual EMBS International Conference, Istanbul (2001)
3. Nithin Kumar, K.C., Griya, N., Shaikh, A., Chaudhry, V., Chavadaki, S.: Structural analysis of femur bone to predict the suitable alternative material. In: Materials Today: Proceedings Published by Elsevier Ltd., p. 5, 4 Dec (2019)
4. Sindhu, V., Soundarapandian, S.: Three-dimensional modelling of femur bone using various scanning systems for modelling of knee implant and virtual aid of surgical planning. Measurement **141**, 190–208 (2019)
5. Wu, S., Pan, H, Pei, B.: Research of biomechanics parameters optimization of femur neck fracture fixation. In: 3rd International Conference on Biomedical Engineering and Informatics (BMEI 2010), Yantai (2010)
6. Youn, K., Park, M.S., Lee, J.: Iterative approach for 3D reconstruction of the femur from un-calibrated 2D radiographic images. Med. Eng. Phys. Elsevier Ltd. **50**, 89–95 (2017)
7. Zheng, D., Wang, L., Gong, H.: Finite element analysis of functional graded proximal femoral nail antirotation for alleviating the complications of intertrochanteric fracture treatment. In: 2013 6th International Conference on Biomedical Engineering and Informatics (BMEI), Published by IEE Explore, Hangzhou (2013)
8. Faisal, T.R., Luo, Y.: Stress variations owing to single-stance load and sideways fall result in fracture at proximal femur. In: 2015 IEEE 12th International Symposium on Biomedical Imaging (ISBI), Brooklyn (2015)

9. Choh, A.C., Chong, D.Y.:Biomechanical analysis of the placement of fixation lag screw in different intertrochanteric hip fracture angles. Biosurf. Biotribol. IET Southwest Jiaotong Univ. **5**(2), 52–56 (2019)
10. Chen, F., et al.: Clustering of morphological features for identifying femur cavity subtypes with difficulties of intramedullary nail implantation. IEEE J. Biomed. Health Inform. **22**(4), 1209–1217 (2018)

'Useful' Information Measure for Complex Fuzzy Set with Divergence Measure and Computation of Hybrid Ambiguity

Mahima Poonia[1], Rakesh Kumar Bajaj[2(✉)], and Varun Shukla[3]

[1] ABES Engineering College, Ghaziabad 201009, Uttar Pradesh, India
[2] Jaypee University of Information Technology, Waknaghat, Solan 173234, Himachal Pradesh, India
rakesh.bajaj@juitsolan.in
[3] Pranveer Singh Institute of Technology, Kanpur 209305, U.P, India

Abstract. In the complex real-life uncertain situations, the role of complex fuzzy sets become significant for obtaining desired solutions for utility based various uncertain problems. In the present article, we have studied the concept of a 'useful' fuzzy information measure for the complex plane of a fuzzy set which expands the area of solving problems related to uncertainties in complex situations. Later on, various properties related to the notion of total ambiguity and directed divergence have also been explained in detail to understand the novel concept in this direction.

Keywords: Fuzzy Set · Complex Fuzzy Set · 'useful' Measures · Hybrid Ambiguity and Divergence

1 Introduction

Various problems caused due to uncertainties have been first dealt with the help of the Fuzzy set [1] (FS) theory of Zadeh which has a direct impact on medicine [2], decision-making [3], and engineering sciences [4]. The proposed theory leads the trail for many researchers to extend the concept and propose several innovative studies to deal with the problem caused due to imprecision or lack of information. Atanassov introduced a new concept of Intuitionistic fuzzy set [5] (IFS) which added a non-membership function to the existing theory of FS. Similarly, various new concepts (Hesitant fuzzy set [6], picture fuzzy set [7], Pythagorean fuzzy set [8], and many more) have been proposed and exist in the literature which plays a significant role in dealing with the vague data available on the real plane. It has been noticed that many researchers have used these new concepts and proposed various information measures and methodologies to deal with problems of decision-making. Also, various similarity measures on picture fuzzy sets have been applied in many decision making problems [9–12]. Further, the picture fuzzy extensions have been utilized in various fields [13–16].

From the available data in the literature, it has been noticed that divergence [17], entropy [18], similarity [19], and distance [20] measures play a vital tool in analyzing

S. Singh et al. (Eds.): AIDSP 2023, CCIS 2440, pp. 136–147, 2025.
https://doi.org/10.1007/978-3-031-88762-8_14

uncertainties associated with existing theories. The degree of discrimination and similarity between the two sets have been calculated by using distance and similarity measures. Similarly, Luca and Termini [18] introduced a nonprobability global measure to obtain the degree of fuzziness known as the entropy measure. The authors have also presented some basic properties (I-IV) given below which an entropy measure must satisfy to be a valid measure.

"Consider P and Q to be two FSs. Then, $E : FS(X) \to [0,1]$ is entropy if it satisfies the following conditions:

I. *(Sharpness)- In the case of crisp set* $E(P) = 0$;
II. *(Maximality) -* $\mu_P(x) = 0.5 | x \in X$, *then,* $E(P) = 1$;
III. *(Resolution) -* $E(P) \leq E(Q) | P$ *is shapened version of* Q, *then*
   i. If $\mu_P(x) \leq \mu_Q(x) | \mu_Q \leq 0.5$ *and*
   ii. If $\mu_Q(x) \leq \mu_P(x) | \mu_P \leq 0.5$
IV. *(Symmetry) –* $E(P) = E(\overline{P})$ *where P complement is* $\overline{P}$."

Further, Luca and Termini [10] combined the concept of entropy with Shannon's entropy and presented the measure given below:

$$E(P) = -\sum\nolimits_{i=1}^{n} [\mu_P(x_i)\log(\mu_P(x_i)) + (1 - \mu_P(x_i))\log(\mu_P(x_i))].$$

Similarly, Hooda and Bajaj [21] 'useful' measure in the case of fuzzy set and presented various properties (integrated ambiguity, directed divergence and constrained optimization) to increase the understanding of the concept.

According to the references of the studies presented by various authors, it has been concluded that the entropy measure plays a significant role in obtaining solutions to a wide range of uncertain problems, but this measure ignores the database depending on the time or phase component. This drawback has been first noticed by Ramot who introduced the theory of Complex fuzzy sets [22] (CFS) which extended the FS theory to the complex plane of unit disc and added information related to the phase component.

This study has been further extended for the case of Complex intuitionistic fuzzy set [23] (CIFS), Cohesive fuzzy set [24] (CFS) and Complex neutrosophic set [25] (CNS). Various properties related to the information measures have been presented on the complex plane to obtain the solution of various complex problems.

In the current work, we have introduced the concept of a 'useful' fuzzy measure on a complex plane to widen its area of solving problems with real-life applications. Later, various theorems and properties related to ambiguity have been discussed in detail for a better understanding of the concept.

The rest of the article has been arranged in the following manner: Sect. 2 consists of the basic preliminaries. In Sect. 3, the 'useful' information measure on the complex plane has been proposed in detail. In Sect. 4, useful complex fuzzy measure of ambiguity has been proposed. Section 4 proposes the useful complex fuzzy directed divergence measure. Finally, manuscript ends with the Sect. 5 which includes the conclusion and scope for future work.

2 Preliminaries

In this section, we have presented some basic definitions for the initial understanding of the concept.

Definition 1. Complex Fuzzy Set (CFS) [22].*"Consider a set of ordered pair* $P = \{(u, \mu_u(u))|u \in \mathcal{U}\}$ *in the universe of discourse* $\mathcal{U}$ *s.t.* $\theta_P : \mathcal{U} \to \{h : h \in H, |h| \le 1\}$, *where* $\theta_P(u) = \alpha_P(u)e^{i\tau_P(u)}, i = \sqrt{-1}, \alpha_P(u) \in [0,1] \& \tau_P(u) \in [0,2\pi]$."

Definition 2. Shannon's Entropy (SE) [26]. "Consider the probability $\mathbb{P} = (\mathbb{P}_1, \mathbb{P}_2, \ldots, \mathbb{P}_n)$ with relation set $\mathbb{X} = \mathbb{X}_1, \mathbb{X}_2, \ldots, \mathbb{X}_n$. Then, in this experiment, the information has been given by:

$$H(\mathbb{P}) = -\sum\nolimits_{i=1}^{n} \mathbb{P}_i \log \mathbb{P}_i.$$

Definition 3. Ambiguity Measure (ABM) and generalized 'useful' measure [21]. "Consider the subset P of set X such that P characterized by the membership function $\mu_P : X \to [0,1]$. Then, this must satisfy the following condition:

$$\mu_P(x) = \begin{cases} 0 \; or \; 1, \; if \; x \in P \; or \; x \notin P \; and \; there \; is \; nor \; ambiguity \\ 0.5, \; if \; there \; exist \; maximum \; probability \; whether \; x \in P \; or \; x \notin P. \end{cases}"$$

Definition 4. Complex Fuzzy Entropy (Type A)[27]. *"Let us assume two complex fuzzy sets P and Q maps in the interval* $[0, 1](CFS(\mathcal{U}) \to [0, 1])$ *is known as an entropy measure (A-type) if it satisfies the following axioms:*

I. $H(P) = minimum, \; iff \; ||\mu_P(u)|| = 0, 1, \; \forall P \in U.$
II. $H(P) = maximum, \; if \; ||\mu_P(u)|| = \frac{1}{2}, \; \forall P \in U.$
III. $H(P) \le H(Q) \; if \begin{cases} |\mu_P(u)| \le |\mu_Q(u)| \; when \; |\mu_Q(u)| \le 0.5, \\ |\mu_P(u)| \le |\mu_P(u)| \; when \; |\mu_Q(u)| \ge 0.5. \end{cases}$
IV. $H(P) = H^c(P)$."

Definition 5. Complex Fuzzy Entropy (Type B)[27]. *"Let us assume two complex fuzzy sets P and Q maps in the interval* $[0, 1](CFS(\mathcal{U}) \to [0, 1])$ *is known as an entropy measure (A-type) if it satisfies the following axioms:*

I. $H(P) = minimum, \; iff \; \alpha_P(u) = 0 \; or \; 1, \; \& \beta_p(u) = 0, \; \forall P \in \mathcal{U}.$
II. $H(P) = maximum, \; if \; \alpha_P(u) = 0.5 \; and \; \beta_p(u) = 2\pi \; \forall P \in \mathcal{U}.$
III. $H(P) \le H(Q) \; if \begin{cases} \alpha_P(u) \le \alpha_Q(u), \beta_p(u) \ge \beta_Q(u) \; when \; \alpha_Q(u) \le \frac{\beta_Q(u)}{2\pi}, \\ \alpha_P(u) \ge \alpha_Q(u), \beta_p(u) \le \beta_Q(u) \; when \; \alpha_Q(u) \ge \frac{\beta_Q(u)}{2\pi}. \end{cases}$
IV. $H(P) = H^c(P)$."

3 'Useful' Complex Fuzzy Information Measure

In this section, we have introduced a utility measure having both quantity and quality measures commonly known as 'useful' complex fuzzy information measures in detail for better clarity of the concept.

Definition *6.* Let us consider the probability $\mathbb{P} = \{(\mathbb{P}_1, \mathbb{P}_2, \ldots, \mathbb{P}_n), \mathbb{P}_i \epsilon (0,1] \forall i = 1,2,3,\ldots,n\}$ and utility $\mathbb{U} = \{\mathbb{U}_1, \mathbb{U}_2, \ldots, \mathbb{U}_n\} | \mathbb{U}_i > 0, \forall i = 1,2,3,\ldots,n$ in universe ($\mathcal{U} = \{u_1, u_2, \ldots, u_n\}, \forall i = 1,2,3,\ldots,n$). Then, the fuzzy set ($\mathcal{A}$) defined in universe ($\mathcal{U}$) with the degree of membership, the function has been denoted by $\mu_{\mathcal{A}}(u_i)$ such that the. $(\mu_{\mathcal{A}}(u_1), \mu_{\mathcal{A}}(u_2), \ldots, \mu_{\mathcal{A}}(u_n) \forall i = 1,2,3,\ldots,n)$ lies in interval [0, 1], where these ambiguities $\mu_{\mathcal{A}}(u_i)$ cannot be assumed as probabilities as the sum of these functions cannot satisfy the condition of probability. Now, extending the concept of Luca and Termini [18] and Pal and Bezdek [28] from the real plane to the complex plane and proposing the following 'useful' measures:

$$H_1(\mathcal{A}; \mathbb{P}; \mathbb{U}) = -\frac{\sum_{i=1}^{n} \mathbb{U}_i \mathbb{P}_i \left[|\mu_{\mathcal{A}}(u_i)| log |\mu_{\mathcal{A}}(u_i)| + (1 - |\mu_{\mathcal{A}}(u_i)|) log (1 - |\mu_{\mathcal{A}}(u_i)|) \right]}{\sum_{i=1}^{n} \mathbb{U}_i \mathbb{P}_i}, \tag{1}$$

$$H_2(\mathcal{A}; \mathbb{P}; \mathbb{U}) = \frac{\sum_{i=1}^{n} \mathbb{U}_i \mathbb{P}_i [|\mu_{\mathcal{A}}(u_i)||1 - \mu_{\mathcal{A}}(u_i)|]}{\sum_{i=1}^{n} \mathbb{U}_i \mathbb{P}_i}, \tag{2}$$

$$H_3(\mathcal{A}; \mathbb{P}; \mathbb{U}) = -\frac{\sum_{i=1}^{n} \mathbb{U}_i \mathbb{P}_i \left[\alpha_{\mathcal{A}}(u_i) log(\alpha_{\mathcal{A}}(u_i)) + (1 - \alpha_{\mathcal{A}}(u_i)) log(1 - \alpha_{\mathcal{A}}(u_i)) + \frac{\beta_{\mathcal{A}}(u_i)}{2\pi} log\left(\frac{\beta_{\mathcal{A}}(u_i)}{2\pi}\right) + \left(1 - \frac{\beta_{\mathcal{A}}(u_i)}{2\pi}\right) log\left(1 - \frac{\beta_{\mathcal{A}}(u_i)}{2\pi}\right) \right]}{\sum_{i=1}^{n} \mathbb{U}_i \mathbb{P}_i}, \tag{3}$$

$$H_4(\mathcal{A}; \mathbb{P}; \mathbb{U}) = \frac{\sum_{i=1}^{n} u_i p_i \left[\alpha_{\mathcal{A}}(u_i)(1 - \alpha_{\mathcal{A}}(u_i)) + \frac{\beta_{\mathcal{A}}(u_i)}{2\pi} \log\left(1 - \frac{\beta_{\mathcal{A}}(u_i)}{2\pi}\right) \right]}{\sum_{i=1}^{n} u_i p_i}. \tag{4}$$

Hence, the above-presented equations are known as 'useful' complex fuzzy information measures.

Theorem 1. The equation given by (1) and (2) of measures given in definition 6 have been validated with the help of axioms given in Definition 4.

Proof: The axioms of definition have been proved below:

I. When $H_1(\mathcal{A}; \mathbb{P}; \mathbb{U}) = minimum$,

Then,

$$\sum_{i=1}^{n} \mathbb{U}_i \mathbb{P}_i \left[|\mu_{\mathcal{A}}(u_i)| \log |\mu_{\mathcal{A}}(u_i)| + (1 - |\mu_{\mathcal{A}}(u_i)|) \log (1 - |\mu_{\mathcal{A}}(u_i)|) \right] = 0,$$

$$\Rightarrow |\mu_{\mathcal{A}}(u_i)| \log |\mu_{\mathcal{A}}(u_i)| + (1 - |\mu_{\mathcal{A}}(u_i)|) \log (1 - |\mu_{\mathcal{A}}(u_i)|) = 0,$$

That is $|\mu_{\mathcal{A}}(u_i)| = 0 or 1, \forall i = 1,2,\ldots n$.
Next, when $H_2(\mathcal{A}; \mathbb{P}; \mathbb{U}) = 0$,
Then, $\sum_{i=1}^{n} \mathbb{U}_i \mathbb{P}_i [|\mu_{\mathcal{A}}(u_i)||1 - \mu_{\mathcal{A}}(u_i)|] = 0$,
$\Rightarrow [|\mu_{\mathcal{A}}(u_i)||1 - \mu_{\mathcal{A}}(u_i)|] = 0$,
This also proves that either $|\mu_{\mathcal{A}}(u_i)| = 0 \ or \ 1, \forall i = 1,2,\ldots n$.

II. To prove $H_1(\mathcal{A}; \mathbb{P}; \mathbb{U}) = maximum$.

Let us differentiate the Eq. (1) w.r.t. $|\mu_{\mathcal{A}}(u_i)|$.

$$\frac{\partial H_1(\mathcal{A};\mathbb{P};\mathbb{U})}{\partial|\mu_{\mathcal{A}}(u_i)|}=\sum_{i=1}^{n}u_ip_i\log\frac{1-|\mu_{\mathcal{A}}(u_i)|}{|\mu_{\mathcal{A}}(u_i)|}$$

For maximum cases,

$$\frac{\partial H_1(\mathcal{A};\mathbb{P};\mathbb{U})}{\partial|\mu_{\mathcal{A}}(u_i)|}=0,$$

$\Rightarrow |\mu_{\mathcal{A}}(u_i)|=0.5$. It means that the measure has the maximum value at this point. Similarly, the second measure given by Eq. (2) has been differentiated w.r.t. $|\mu_{\mathcal{A}}(u_i)|$.

$$\frac{\partial H_2(\mathcal{A};\mathbb{P};\mathbb{U})}{\partial|\mu_{\mathcal{A}}(u_i)|}=\sum_{i=1}^{n}u_ip_i(\,|1-2\mu_{\mathcal{A}}(u_i)|),$$

$\Rightarrow |\mu_{\mathcal{A}}(u_i)|=0.5$. It means that the measure has the maximum value at this point.

III. Now, we can say that the measures are decreasing in the interval (0.5,1] and increasing in the interval [0,0.5).

Case 1: $|\mu_{\mathcal{A}}(u_i)|\leq|\mu_{\mathcal{B}}(u_i)|\ when\ |\mu_{\mathcal{B}}(u_i)|\leq 0.5$.
Then, $H_1(\mathcal{A};\mathbb{P};\mathbb{U})\leq H_1(\mathcal{B};\mathbb{P};\mathbb{U})$.
Case 2: $|\mu_{\mathcal{A}}(u_i)|\geq|\mu_{\mathcal{B}}(u_i)|\ when\ |\mu_{\mathcal{B}}(u_i)|\geq 0.5$.
This again, gives the same result which is $H_1(\mathcal{A};\mathbb{P};\mathbb{U})\leq H_1(\mathcal{B};\mathbb{P};\mathbb{U})$.
Similarly, we can prove it for $H_2(\mathcal{A};\mathbb{P};\mathbb{U})$ measure.

IV. The complement of the two measures $H_1(\mathcal{A};\mathbb{P};\mathbb{U})$ & $H_2(\mathcal{A};\mathbb{P};\mathbb{U})$ can easily be obtained by replacing $\mu_{\mathcal{A}}(u_i)=1-\mu_{\mathcal{A}}(u_i)$ or $1-\mu_{\mathcal{A}}(u_i)=\mu_{\mathcal{A}}(u_i)$.

It is visible that.

$$H_1(\mathcal{A};\mathbb{P};\mathbb{U})=H_1^c(\mathcal{A};\mathbb{P};\mathbb{U});$$

$$H_2(\mathcal{A};\mathbb{P};\mathbb{U})=H_2^c(\mathcal{A};\mathbb{P};\mathbb{U}).$$

Hence, all the axioms have been satisfied by the two proposed measures. This validates the 'useful' entropy measures $H_1(\mathcal{A};\mathbb{P};\mathbb{U})$ & $H_2(\mathcal{A};\mathbb{P};\mathbb{U})$.

Theorem 2. The equations given by (3) and (4) of Definition 6 have been validated with the help of axioms given in Definition 5.

Proof: The proof of Theorem 2 is the same as Theorem 1.
Hence, proved.

4 'Useful' Complex Fuzzy Measure of Ambiguity

In this section, the integrated ambiguity in the case of 'useful' complex fuzzy measure has been discussed in detail.

Luca and Termini [18] initiated the concept of combining the uncertainties of probability and fuzzy theories and proposed the concept of 'useful' fuzzy measures which have both quantity and quality measures. The basic underline of the studies already exists in the literature has been given below to increase the understanding of the concept.

(1). Shannon's theory first calculated the average of uncertainties present in the experiment given in Definition 2.

(2). Luca and Termini [18] proposed the quality measure commonly known as fuzzy entropy given as:

$$H(\mathbb{A}) = -\sum_{i=1}^{n}\left[\mu_{\mathbb{A}}(u_i)\log\mu_{\mathbb{A}}(u_i) + (1-\mu_{\mathbb{A}}(u_i))\log(1-\mu_{\mathbb{A}}(u_i))\right].$$

(3). The quantity average of the complete set ambiguity calculated by Hooda and Bajaj [21] has been given by:

$$S(\mathbb{A},\mathbb{P}) = -\sum_{i=1}^{n} u_i\left[\mu_{\mathbb{A}}(u_i)\log\mu_{\mathbb{A}}(u_i) + (1-\mu_{\mathbb{A}}(u_i))\log(1-\mu_{\mathbb{A}}(u_i))\right].$$

The total fuzzy information has been obtained by

$$H(\mathbb{A},\mathbb{P}) = H(\mathbb{P}) + S(\mathbb{A},\mathbb{P}).$$

Thus, the author obtained it as 'useful' measure and named it as 'useful' total ambiguity measure given as:

$$H_T(\mathbb{A},\mathbb{P},\mathbb{U}) = H(\mathbb{P},\mathbb{U}) + H(\mathbb{A},\mathbb{P},\mathbb{U})$$

where,

$$H(\mathbb{P},\mathbb{U}) = -\frac{\sum_{i=1}^{n} u_i\mathbb{P}_i\log\mathbb{P}_i}{\sum_{i=1}^{n} u_i\mathbb{P}_i} \quad and$$

$$H(\mathbb{A},\mathbb{P},\mathbb{U}) = \frac{\sum_{i=1}^{n} u_i\mathbb{P}_i\mu_{\mathbb{A}}(u_i)\log\mu_{\mathbb{A}}(u_i) + (1-\mu_{\mathbb{A}}(u_i))\log(1-\mu_{\mathbb{A}}(u_i))}{\sum_{i=1}^{n} u_i\mathbb{P}_i}.$$

Next, this information has been extended from a complex plane to a real plane which can help in tracking the pattern if required. Secondly, the total information can calculate the randomness of copious amounts of data. Recently, P.K. Singh [29] used the Shannon entropy on the complex fuzzy lattice with crisp properties and tracked the period using the set on Shannon entropy. This also lacks the quality factor which we have calculated in the present manuscript.

1) Now, we have considered the average measure from Hooda and Bajaj [21] given by:

$$H(\mathbb{P},\mathbb{U}) = -\frac{\sum_{i=1}^{n} u_i\mathbb{P}_i\log\mathbb{P}_i}{\sum_{i=1}^{n} u_i\mathbb{P}_i} \tag{5}$$

2) Then, the amount of uncertainty or ambiguity calculated with the help of the proposed 'useful' complex fuzzy entropy measure given as:

$$H_b(\mathcal{A},\mathbb{P},\mathbb{U}) = -\frac{\sum_{i=1}^{n}\mathbb{U}_i\mathbb{P}_i H_{bi}(\mathcal{A})}{\sum_{i=1}^{n}\mathbb{U}_i\mathbb{P}_i}, \tag{6}$$

where $b = 1, 2, 3$ & 4.

That is when $b = 1$,

$$H_{1i}(\mathcal{A}) = |\mu_{\mathcal{A}}(u_i)|\log|\mu_{\mathcal{A}}(u_i)| + (1 - |\mu_{\mathcal{A}}(u_i)|)\log(1 - |\mu_{\mathcal{A}}(u_i)|).$$

When $b = 2$,

$$H_{2i}(\mathcal{A}) = |\mu_{\mathcal{A}}(u_i)||1 - \mu_{\mathcal{A}}(u_i)|.$$

When $b = 3$,

$$\begin{aligned} H_{3i}(\mathcal{A}) &= \alpha_{\mathcal{A}}(u_i)\log(\alpha_{\mathcal{A}}(u_i)) + (1 - \alpha_{\mathcal{A}}(u_i))\log(1 - \alpha_{\mathcal{A}}(u_i)) \\ &+ \frac{\beta_{\mathcal{A}}(u_i)}{2\pi}\log\left(\frac{\beta_{\mathcal{A}}(u_i)}{2\pi}\right) + \left(1 - \frac{\beta_{\mathcal{A}}(u_i)}{2\pi}\right)\log\left(1 - \frac{\beta_{\mathcal{A}}(u_i)}{2\pi}\right). \end{aligned}$$

When $b = 4$,

$$H_{4i}(\mathcal{A}) = \alpha_{\mathcal{A}}(u_i)(1 - \alpha_{\mathcal{A}}(u_i)) + \frac{\beta_{\mathcal{A}}(u_i)}{2\pi} log\left(1 - \frac{\beta_{\mathcal{A}}(u_i)}{2\pi}\right)$$

3) Thus, the total ambiguity for proposed 'useful' complex fuzzy entropy measures has been given by:

$$H_{bt}(\mathcal{A}, \mathbb{P}, \mathbb{U}) = H(\mathbb{P}, \mathbb{U}) + H_b(\mathcal{A}, \mathbb{P}, \mathbb{U}), \tag{7}$$

where $b = 1, 2, 3, 4$.

5 'Useful' Complex Fuzzy Directed Divergence Measure

In this section of the manuscript, we have explained the directed divergence measure of two complex fuzzy sets.

Let $\mathcal{A}$ and $\mathcal{B}$ be two complex fuzzy sets on points $u_1, u_2, \ldots, u_n$. Then, the complex-valued membership functions have been given by $\mu_{\mathcal{A}}(u_1), \mu_{\mathcal{A}}(u_2), \mu_{\mathcal{A}}(u_3), \ldots, \mu_{\mathcal{A}}(u_n)$ and $\mu_{\mathcal{B}}(u_1), \mu_{\mathcal{B}}(u_2), \mu_{\mathcal{B}}(u_3), \ldots, \mu_{\mathcal{B}}(u_n)$. Then, the Bhandari and Pal divergence concept has been extended to a complex plane and the following equation has been obtained.

$$I_1(\mathcal{A}, \mathcal{B}) = \sum\nolimits_{i=1}^{n}\left[|\mu_{\mathcal{A}}(u_i)|\log\frac{|\mu_{\mathcal{A}}(u_i)|}{|\mu_{\mathcal{B}}(u_i)|} + (1 - |\mu_{\mathcal{A}}(u_i)|)\log\frac{|1 - \mu_{\mathcal{A}}(u_i)|}{|1 - \mu_{\mathcal{B}}(u_i)|}\right], \tag{8}$$

$$\begin{aligned} I_3(\mathcal{A}, \mathcal{B}) = \sum\nolimits_{i=1}^{n}\Big[&\alpha_{\mathcal{A}}(u_i)\log\left(\frac{\alpha_{\mathcal{A}}(u_i)}{\alpha_{\mathcal{B}}(u_i)}\right) + (1 - \alpha_{\mathcal{A}}(u_i))\log\left(\frac{1-\alpha_{\mathcal{A}}(u_i)}{1-\alpha_{\mathcal{B}}(u_i)}\right) \\ &+ \frac{\beta_{\mathcal{A}}(u_i)}{2\pi}\log\left(\frac{\beta_{\mathcal{A}}(u_i)}{\beta_{\mathcal{B}}(u_i)}\right) + \left(1 - \frac{\beta_{\mathcal{A}}(u_i)}{2\pi}\right)\log\left(\frac{1-\frac{\beta_{\mathcal{A}}(u_i)}{2\pi}}{1-\frac{\beta_{\mathcal{B}}(u_i)}{2\pi}}\right)\Big] \end{aligned} \tag{9}$$

where it must satisfy the condition that is $\begin{cases} if\,|\mu_{\mathcal{A}}(u_i)| \neq |\mu_{\mathcal{B}}(u_i)|, I(\mathcal{A}, \mathcal{B}) \neq 0 \\ if\,|\mu_{\mathcal{A}}(u_i)| = |\mu_{\mathcal{B}}(u_i)|, I(\mathcal{A}, \mathcal{B}) = 0 \end{cases}$

and $\begin{cases} \alpha_{\mathcal{A}}(u_i) \neq \alpha_{\mathcal{B}}(u_i), \beta_{\mathcal{A}}(u_i) \neq \beta_{\mathcal{B}}(u_i) \neq 2\pi, I(\mathcal{A}, \mathcal{B}) \neq 0 \\ \alpha_{\mathcal{A}}(u_i) = \alpha_{\mathcal{B}}(u_i), \beta_{\mathcal{A}}(u_i) = \beta_{\mathcal{B}}(u_i) = 2\pi, I(\mathcal{A}, \mathcal{B}) = 0 \end{cases}.$

Similarly, we have extended the concept to our proposed measure and introduced the following equation for obtaining a 'useful' complex fuzzy entropy measure for directed divergence for $\mathcal{A}$ and $\mathcal{B}$ sets.

$$I_1(\mathcal{A},\mathcal{B},\mathbb{P},\mathbb{U})=\frac{\sum_{i=1}^{n}\mathbb{U}_i\mathbb{P}_i\left[|\mu_{\mathcal{A}}(u_i)|\log\frac{|\mu_{\mathcal{A}}(u_i)|}{|\mu_{\mathcal{B}}(u_i)|}+(1-|\mu_{\mathcal{A}}(u_i)|)\log\left(\frac{1-|\mu_{\mathcal{A}}(u_i)|}{1-|\mu_{\mathcal{B}}(u_i)|}\right)\right]}{\sum_{i=1}^{n}\mathbb{U}_i\mathbb{P}_i} \tag{10}$$

$$I_3(\mathcal{A},\mathcal{B},\mathbb{P},\mathbb{U})=\frac{\sum_{i=1}^{n}\mathbb{U}_i\mathbb{P}_i\left[\begin{array}{c}\alpha_{\mathcal{A}}(u_i)\log\left(\frac{\alpha_{\mathcal{A}}(u_i)}{\alpha_{\mathcal{B}}(u_i)}\right)+(1-\alpha_{\mathcal{A}}(u_i))\log\left(\frac{1-\alpha_{\mathcal{A}}(u_i)}{1-\alpha_{\mathcal{B}}(u_i)}\right)+\frac{\beta_{\mathcal{A}}(u_i)}{2\pi}\log\left(\frac{\beta_{\mathcal{A}}(u_i)}{\beta_{\mathcal{B}}(u_i)}\right)+\\ \left(1-\frac{\beta_{\mathcal{A}}(u_i)}{2\pi}\right)\log\left(\frac{1-\frac{\beta_{\mathcal{A}}(u_i)}{2\pi}}{1-\frac{\beta_{\mathcal{B}}(u_i)}{2\pi}}\right)\end{array}\right]}{\sum_{i=1}^{n}\mathbb{U}_i\mathbb{P}_i} \tag{11}$$

where $\mathbb{U}_i>0\forall i$.

Next, a 'Useful' complex fuzzy symmetric divergence measure has been given by

$$J_b(\mathcal{A},\mathcal{B},\mathbb{P},\mathbb{U})=I_b(\mathcal{A},\mathcal{B},\mathbb{P},\mathbb{U})+I_b(\mathcal{B},\mathcal{A},\mathbb{P},\mathbb{U}) \tag{12}$$

where $b=1,2,3,4$.

Theorem 3. Through this theorem, we validate the proposed divergence measure $I_b(\mathcal{A},\mathcal{B},\mathbb{P},\mathbb{U})\ \geq\ 0\ \ (b=1,3)$, where if $|\mu_{\mathcal{A}}(u_i)|\ =\ |\mu_{\mathcal{B}}(u_i)|$ or $\alpha_{\mathcal{A}}(u_i)\ =\ \alpha_{\mathcal{B}}(u_i)$, $\beta_{\mathcal{A}}(u_i)=\beta_{\mathcal{B}}(u_i)=2\pi\forall i=1,2,..n$.

Proof: Suppose

$$\sum_{i=1}^{n}|\mu_{\mathcal{A}}(u_i)|=\ell,\sum_{i=1}^{n}|\mu_{\mathcal{B}}(u_i)|=m,\sum_{i=1}^{n}\mathbb{U}_i\mathbb{P}_i=n.$$

Next,

$$\sum_{i=1}^{n}\mathbb{U}_i\mathbb{P}_i\left[|\mu_{\mathcal{A}}(u_i)|\log\frac{|\mu_{\mathcal{A}}(u_i)|}{|\mu_{\mathcal{B}}(u_i)|}\right]\geq n\ell\log\frac{\ell}{m}. \tag{13}$$

Similarly, we have

$$\sum_{i=1}^{n}\mathbb{U}_i\mathbb{P}_i\left[(1-|\mu_{\mathcal{A}}(u_i)|)\log\left(\frac{1-|\mu_{\mathcal{A}}(u_i)|}{1-|\mu_{\mathcal{B}}(u_i)|}\right)\right]\geq n(n-\ell)\log\frac{n-\ell}{n-m} \tag{14}$$

From Eqs. (13) and (14), we have

$$\sum_{i=1}^{n}\mathbb{U}_i\mathbb{P}_i\left[|\mu_{\mathcal{A}}(u_i)|\log\frac{|\mu_{\mathcal{A}}(u_i)|}{|\mu_{\mathcal{B}}(u_i)|}+(1-|\mu_{\mathcal{A}}(u_i)|)\log\left(\frac{1-|\mu_{\mathcal{A}}(u_i)|}{1-|\mu_{\mathcal{B}}(u_i)|}\right)\right]\geq n\left(\ell\log\frac{\ell}{m}+(n-\ell)\log\frac{n-\ell}{n-m}\right) \tag{15}$$

$$f(\ell)=\ell\log\frac{\ell}{m}+(n-\ell)\log\frac{n-\ell}{n-m},$$

Differentiating the above equation w.r.t ℓ. We get,

$$f'(\ell) = \log\frac{\ell}{m} - \log\frac{n-\ell}{n-m};$$

$$f''(\ell) = \frac{1}{\ell} + \frac{1}{n-\ell} > 0.$$

Hence, this proves that $f''(\ell) > 0$ is a convex function with a minimum value. Thus, when $\mathcal{A} = \mathcal{B}$, then $f(\ell) = 0$. This implies that function only exists when $f(\ell) > 0$ as when $\ell = m$ it vanishes. This is already known that $\sum_{i=1}^{n} \mathbb{U}_i\mathbb{P}_i > 0$ therefore, $I_1(\mathcal{A}, \mathcal{B}, \mathbb{P}, \mathbb{U}) \geq 0$ and vanishes when $\mathcal{A} = \mathcal{B}$.

This validates our divergence measure $I_1(\mathcal{A}, \mathcal{B}, \mathbb{P}, \mathbb{U})$.

Now we will prove the theorem for $I_3(\mathcal{A}, \mathcal{B}, \mathbb{P}, \mathbb{U})$.

Suppose

$$\alpha_{\mathcal{A}}(u_i) = r, \alpha_{\mathcal{B}}(u_i) = s, \beta_{\mathcal{A}}(u_i) = w, \beta_{\mathcal{B}}(u_i) = z, \sum_{i=1}^{n} \mathbb{U}_i\mathbb{P}_i = n.$$

Next,

$$\sum_{i=1}^{n} \mathbb{U}_i\mathbb{P}_i\left[\alpha_{\mathcal{A}}(u_i)\log\left(\frac{\alpha_{\mathcal{A}}(u_i)}{\alpha_{\mathcal{B}}(u_i)}\right)\right] \geq nr\log\frac{r}{s}, \tag{16}$$

$$\sum_{i=1}^{n} \mathbb{U}_i\mathbb{P}_i\left[(1-\alpha_{\mathcal{A}}(u_i))\log\left(\frac{1-\alpha_{\mathcal{A}}(u_i)}{1-\alpha_{\mathcal{B}}(u_i)}\right)\right] \geq n(n-r)\log\frac{n-r}{n-s}, \tag{17}$$

$$\sum_{i=1}^{n} \mathbb{U}_i\mathbb{P}_i\left[\frac{\beta_{\mathcal{A}}(u_i)}{2\pi}\log\left(\frac{\beta_{\mathcal{A}}(u_i)}{\beta_{\mathcal{B}}(u_i)}\right)+\right] \geq n\frac{w}{2\pi}\log\frac{w}{z}, \tag{18}$$

$$\sum_{i=1}^{n} \mathbb{U}_i\mathbb{P}_i\left[\left(1-\frac{\beta_{\mathcal{A}}(u_i)}{2\pi}\right)\log\frac{1-\frac{\beta_{\mathcal{A}}(u_i)}{2\pi}}{1-\frac{\beta_{\mathcal{B}}(u_i)}{2\pi}}\right] \geq n(n-\frac{w}{2\pi})\log\frac{n-\frac{w}{2\pi}}{n-\frac{z}{2\pi}}. \tag{19}$$

From Eqs. (16) and (17) the amplitude part of Eq. (11) has been obtained in the form of the assumed value.

$$\sum_{i=1}^{n} \mathbb{U}_i\mathbb{P}_i \left[\alpha_{\mathcal{A}}(u_i)\log\left(\frac{\alpha_{\mathcal{A}}(u_i)}{\alpha_{\mathcal{B}}(u_i)}\right) + (1-\alpha_{\mathcal{A}}(u_i))\log\left(\frac{1-\alpha_{\mathcal{A}}(u_i)}{1-\alpha_{\mathcal{B}}(u_i)}\right)\right] \geq n\left[r\log\frac{r}{s} + (1-r)\log\frac{1-r}{1-s}\right].$$

Let us consider,

$$f(r) = r\log\frac{r}{s} + (n-r)\log\frac{n-r}{n-s},$$

Next,

$$f'(r) = \log\frac{r}{s} + \log\frac{n-r}{n-s},$$

$$f''(r) = \frac{1}{s} + \frac{1}{n-s} > 0 \tag{20}$$

From Eqs. (18) and (19) the amplitude part of the Eq. (11) has been obtained in the form of the assumed value.

$$\sum_{i=1}^{n} \mathbb{U}_i \mathbb{P}_i \left[\frac{\beta_{\mathcal{A}}(u_i)}{2\pi} \log\left(\frac{\beta_{\mathcal{A}}(u_i)}{\beta_{\mathcal{B}}(u_i)}\right) + \left(1 - \frac{\beta_{\mathcal{A}}(u_i)}{2\pi}\right) \log\left(\frac{1 - \frac{\beta_{\mathcal{A}}(u_i)}{2\pi}}{1 - \frac{\beta_{\mathcal{B}}(u_i)}{2\pi}}\right)\right]$$
$$\geq n\left(\frac{w}{2\pi}\log\frac{w}{z} + (n - \frac{w}{2\pi})\log\frac{n - \frac{w}{2\pi}}{n - \frac{z}{2\pi}}\right)$$

Take $\theta_1 = \frac{w}{2\pi}$ and $\theta_2 = \frac{z}{2\pi}$,
Then,

$$\sum_{i=1}^{n} \mathbb{U}_i \mathbb{P}_i \left[\frac{\beta_{\mathcal{A}}(u_i)}{2\pi} \log\left(\frac{\beta_{\mathcal{A}}(u_i)}{\beta_{\mathcal{B}}(u_i)}\right) + \left(1 - \frac{\beta_{\mathcal{A}}(u_i)}{2\pi}\right) \log\left(\frac{1 - \frac{\beta_{\mathcal{A}}(u_i)}{2\pi}}{1 - \frac{\beta_{\mathcal{B}}(u_i)}{2\pi}}\right)\right] \geq n\left(\theta_1 \log\frac{\theta_1}{\theta_2} + (n-\theta_1)\log\frac{n-\theta_1}{n-\theta_2}\right).$$

Let us consider,

$$f(\theta_1) = \theta_1 \log\frac{\theta_1}{\theta_2} + (n-\theta_1)\log\frac{n-\theta_1}{n-\theta_2},$$

Next,

$$f'(\theta_1) = \log\frac{\theta_1}{\theta_2} + \log\frac{n-\theta_1}{n-\theta_2}$$

$$f''(\theta_1) = \frac{1}{\theta_1} + \frac{1}{n-\theta_1} > 0 \tag{21}$$

From Eqs. (20) and (21), it has been obtained that our functions have a minimum value, and these kinds of functions are known as convex functions.

Hence, similarly, this validates our divergence measure $I_3(\mathcal{A}, \mathcal{B}, \mathbb{P}, \mathbb{U})$.

6 Conclusion and Scope for Future Work

In the present article, four new 'useful' measures have been proposed and presented in detail for a better understanding of the concept. The 'useful' fuzzy measures have the key benefit of obtaining information related to the quality of the uncertainties present in the real world which has been ignored by various authors in the literature. The four uncertainties cannot only capture the quantity and quality of incomplete data available to the reviewer. This also adds information related to the phase term which helps to track the period or cycle of the experiment. Later, in the article properties related to hybrid ambiguity and directed divergence have also been studied which increases the basic information of the proposed theory. All information adds the basis of the problem caused by the fuzziness of the experiments and helps the researchers to obtain solutions to the given problems.

In future, it lays the foundation of the 'useful' Complex Neutrosophic fuzzy entropy which again adds the two important independent membership functions which increase the understanding of various current issues created due to lack of availability of imprecise data.

References

1. Zadeh, L.A., Klir, G.J., Yuan, B.: Fuzzy sets, fuzzy logic, and fuzzy systems: selected papers. World Sci. (6) (1996)
2. Adlassnig, K.P.: Fuzzy set theory in medical diagnosis. IEEE Trans. Syst. Man Cybern. **16**(2), 260–265 (1986)
3. Hong, D.H., Choi, C.H.: Multicriteria fuzzy decision-making problems based on vague set theory. Fuzzy Sets Syst. **114**(1), 103–113 (2000)
4. Kabir, S., Papadopoulos, Y.: A review of applications of fuzzy sets to safety and reliability engineering. Int. J. Approximate Reasoning **100**, 29–55 (2018)
5. Li, D.F., Wang, Y.C., Liu, S., Shan, F.: Fractional programming methodology for multi-attribute group decision-making using IFS. Appl. Soft Comput. **9**(1), 219–225 (2009)
6. Torra, V.: Hesitant fuzzy sets. Int. J. Intell. Syst. **25**(6), 529–539 (2010)
7. Cuong, B.C., Kreinovich, V.: Picture fuzzy sets. J. Comput. Sci. Cybern. **30**(4), 409–420 (2014)
8. Peng, X., Selvachandran, G.: Pythagorean fuzzy set: state of the art and future directions. Artif. Intell. Rev. **52**, 1873–1927 (2019)
9. Singh, A., Dhumras, H., Bajaj, R.K.: On green supplier selection problem utilizing modified topsis with r-norm picture fuzzy discriminant measure. In: 2022 5th International Conference on Multimedia, Signal Processing and Communication Technologies (IMPACT), pp. 1–5, Aligarh, India (2022)
10. Aggarwal, S., Dhumras, H., Bajaj, R.K.: On banking site selection decision making problem utilizing similarity measures of picture fuzzy soft sets. In: 2022 5th International Conference on Multimedia, Signal Processing and Communication Technologies (IMPACT), pp. 1–5, Aligarh, India (2022)
11. Sharma, E., Dhumras, H., Bajaj, R.K.: On banking site selection problem utilizing novel picture fuzzy discriminant measure. In: 2023 IEEE International Students' Conference on Electrical, Electronics and Computer Science (SCEECS), pp. 1–5, Bhopal, India (2023)
12. Tiwari, A.A., Dhumras, H., Bajaj, R.K.: On parametric picture fuzzy information measure in pattern recognition problem. In: 2023 IEEE International Students' Conference on Electrical, Electronics and Computer Science (SCEECS), pp. 1–5, Bhopal, India 2023
13. Muskan, D.H., Shukla, V., Bajaj, R.K.: On medical diagnosis problem utilizing parametric neutrosophic discriminant measure. In: 2023 IEEE International Students' Conference on Electrical, Electronics and Computer Science (SCEECS), pp. 1–5, Bhopal, India (2023)
14. Bansal, P., Dhumras, H., Bajaj, R.K.: On t-spherical fuzzy hypersoft sets and their aggregation operators with application in soft computing. In: 2022 5th International Conference on Multimedia, Signal Processing and Communication Technologies (IMPACT), pp. 1–6, Aligarh, India 2022
15. Dhumras, H., Bajaj, R.K.: On some new similarity measures for picture fuzzy hypersoft sets with application in medical diagnosis. Emergent Converging Technol. Biomed. Syst. **1040**, 119–130 (2023)
16. Dhumras, H., Bajaj, R.K., Shukla, V.: On utilizing modified TOPSIS with R-norm q-rung picture fuzzy information measure green supplier selection. Int. j. inf. tecnol. **15**, 2819–2825 (2023)
17. Montes, I., Pal, N.R., Janiš, V., Montes, S.: Divergence measures for intuitionistic fuzzy sets. IEEE Trans. Fuzzy Syst. **23**(2), 444–456 (2014)
18. De Luca, A., Termini, S.: A definition of a non-probabilistic entropy in the setting of fuzzy sets theory. Inf. Control. **20**, 301–312 (1972)
19. Lee-Kwang, H., Song, Y.S., Lee, K.M.: Similarity measure between fuzzy sets and between elements. Fuzzy Sets Syst. **62**(3), 291–293 (1994)

20. Xie, X.L., Beni, G.: A validity measure for fuzzy clustering. IEEE Trans. Pattern Anal. Mach. Intell. **13**(08), 841–847 (1991)
21. Hooda, D.S., Bajaj, R.K.: 'Useful' fuzzy measures of information, integrated ambiguity and directed divergence. Int. J. Gen. Syst. **39**(6), 647–658 (2010)
22. Ramot, D., Milo, R., Friedman, M., Kandel, A.: Complex fuzzy sets. IEEE Trans. Fuzzy Syst. **10**(2), 171–186 (2002)
23. Garg, H., Rani, D.: Some results on information measures for complex intuitionistic fuzzy sets. Int. J. Intell. Syst. **34**(10), 2319–2363 (2019)
24. Xue, X., et al.: On cohesive fuzzy sets, operations and properties with applications in electromagnetic signals and solar activities. Symmetry. **15**(3), 595 (2023)
25. Ali, M., Smarandache, F.: Complex neutrosophic set. Neural Comput. Appl. **28**, 1817–1834 (2017)
26. Lin, J.: Divergence measures based on the shannon entropy. IEEE Trans. Inf. Theory **37**(1), 145–151 (1991)
27. Bi, L., Zeng, Z., Hu, B., Dai, S.: Two classes of entropy measures for complex fuzzy sets. Mathematics **7**(1), 96 (2019)
28. Pal, N.R., Bezdek, J.C.: Measuring fuzzy uncertainty. IEEE Trans. Fuzzy Syst. **2**(2), 107–118 (1994)
29. Singh, P.K.: Crisply generated complex fuzzy concepts analysis using Shannon entropy. Neural. Process. Lett. **54**(6), 5643–5667 (2022)

Empirical Comparison of Machine Learning Algorithms to Identify Duplicate Questions on Various Social Media Forums

Karthik Sridhar(✉) and Pranita Mahajan

SIES GST, Navi Mumbai 400706, India
karthiksridhar8879@gmail.com

Abstract. Multiple questions with the same intent can cause a lot of wastage of time to the active readers because they will have to spend a lot of time trying to find information to their questions and the best possible answer to that question, this also creates confusion to the reader since there are multiple versions of the same answer. This paper provides a solution to this problem by identifying the duplicate questions on social media. The paper is an empirical comparison of a random model, Logistic Regression, Linear Support Vector Machine (SVM) and XGBoost Machine Learning (ML) algorithms. The experiment shows improved accuracy using Matthew's Correlation Score (MCC) metric with the XGBoost algorithm.

Keywords: Duplicate Questions · Regression · Support Vector Machine · XGBoost · Accuracy · Mathew's Correlation Coefficient (MCC)

1 Introduction

In Social media forums, most questions that are left unanswered have duplicates.The regularly added questions have to be checked with the pre-existing questions to find the similarity, and if found out to be duplicated, they must be discarded or linked to older questions. Multiple questions with the same intent can cause much wastage of time to the active readers because they will have to spend much time finding information to their questions and the best possible answer to that question. A duplicate question creates confusion for the reader since there are multiple versions of the same answer. The duplicate question may cause the user to miss a good response, and the writers are unlikely to answer the duplicate question again. Hence, identifying the duplicate questions is very important. Identifying and addressing the problem of duplicate questions reduces redundancy and helps deliver the best responses to the asked query and, hence, improves the user's overall experience.

There is a lot of research being done related to duplicate questions, how to restrict it and how to reduce it. Most of them are related to statistical techniques but only using the sample techniques to solve our problem leads to lower accuracy, recall and precision scores. Hence, here we have come up with advanced machine learning algorithms to solve the problem by identifying the duplicate pair of questions.

S. Singh et al. (Eds.): AIDSP 2023, CCIS 2440, pp. 148–160, 2025.
https://doi.org/10.1007/978-3-031-88762-8_15

This paper studies the state of the art on identifying duplicate questions and performs an empirical comparison of Machine Learning algorithms to classify questions on social media as duplicate or unique. The experiment involves pairing up duplicate questions to identify whether these question pairs already exist, that is they have already been asked by someone else. The implemented method calculates the similarity of user query with existing questions to classify if the question is duplicate, that is already answered. The various machine learning models like Logistic Regression, Linear SVM, XG Boost have been implemented to identify duplicate questions and the classifier performance is validated by using various key accuracy measures such as confusion matrix, recall, precision, and MCC score.

For example, we are comparing sentences like "How tall are you?" and "What is your height?". The two questions convey the same meaning and hence are duplicates of each other. Hence to find ML models like Logistic Regression, Linear SVM, XGBoost. Before applying the model we apply various Natural Language Processing (NLP) techniques such as stop word removal, stemming, lemmatization, and many more. Model predicts output by using the Tfidf Countvectorizer. In Sect. 2 critical review of literature related to duplicate QA identification is performed. Section 3 discusses the proposed framework along with the results achieved. Paper is concluded in Sect. 4 with future scope.

2 Literature Review

The state of art has contributed many researches regarding duplicate pairs regarding text featurization to find the best possible accuracy score. Sven Hartrumpf et. al. Explained a similarity join approach which aims to find duplicate strings having similarities higher than a given threshold [1]. Semantic Textual Similarity by Daniel Bar et al. [2] is computed using various models like log-linear regression where they are combining multiple text similarity measures of varying complexities. Duplicate records in the database are found using learnable string similarity measures where the training of the data has been done using SVM's is done by Mikhail Bilenko [3]. Duplicate detection in programming questions have been explained in detail wherein various techniques and measures have been stated by Wei Emma Zhang et al. [4].The detection of duplicates and doing cross validation and performance metrics has been done by Qingyu Chen et al. [5].The similarity between two text is derived using cosine similarity also the semantic meaning of the text is preserved using the cosine similarity has been explained by Faisal Rahutomo et al. [6]. Emmanoil K. Ikonomakis et al. have written about text classification, information extraction and summarization using automatic text classification techniques [7]. This is a survey paper based on the naïve Bayes algorithm, here the various advantages and disadvantages of the algorithm have been stated Pouria Kaviani et al. [8]. The advantage that SVM has over other machine learning models has been explained in this paper, here in this paper how SVM's can be computationally and theoretically analysed to achieve good performance is given by M.A. Hearst et al.[9]. This paper mainly focuses on the overview of the most important concepts of Logistic Regression where its working is shown in an imbalance dataset by Maher Mallouf [10]. The various advantages of Matthews correlation coefficient (MCC) over F1 score is given over here by Davide Chicco et al. [11]. The detection and categorization of various research papers making use of TF-IDF and LDA has been shown by Sang-Woon Kim and Joon-Min Gil [12]. How the semantic sensitive TF-IDF can be used to determine word relevance in documents is shown by Amir Jalilifard et al. [13]. A Case Study on Automatic Classification of Global Terrorist Attacks is done over here where they have made use of Stochastic Gradient Descent (SGD) Classification, done by Shadi Diab [14] (Table 1).

The state of art discusses various Machine Learning algorithms as shown in Table 2. The literature review shows scope of improvements in feature engineering to improve the precision. Section 3 is an experimental detail to overcome some of the challenges identified in Table 2.

3 Experiment and Results

This section explains our experiment and results in detail. We have performed exhaustive feature engineering to achieve precise results. This section is a detailed explanation of the dataset used in this paper along with the proposed methodology. We have used visualization to elaborate the results of feature engineering.

3.1 Dataset

This section details the dataset used for the experiment, we have used the Kaggle dataset [25]. The data contains 6 columns labelled as id, qid 1, qid 2, question 1, question 2 and is_duplicate furthermore, there are 404,290 rows showing with the class label as duplicate or not. Here each question is denoted by a unique id qid1 and qid2. These questions are stored in question1 and question2 columns. The question's present have special characters that need to be inspected before processing the data.

Here we are provided with is duplicate which is representing whether the question pairs are duplicate or not by 1 and 0 and we have to find the best function for it.

We have extracted 17 features out of our dataset such as minimum and maximum common word count, minimum common token count, maximum common token count, minimum common stop word count, maximum common stop words count, cleaned question1, cleaned question2, checking whether the first word of the questions are equal, checking whether the last word of the questions are equal, absolute length difference, mean length, longest common substring ratio, fuzz ratio, fuzzy partial ratio, token sort ratio, token set ratio by using various statistical operations. The feature engineering step, to synthesize relevant features, is explained in Sect. 3.2.iii.

3.2 Proposed Method

Here we have observed that each and every word is not that important to make sense of the entire context of the data. Only some of the words present in the data changes the context of the entire data. Such words which are important in our data are stored in the form of a token and they are further divided into 2 parts: words and stop_words.

Tokens Containing words having high significance.

Words Tokens which are not a part of stopwords.

Stop_wrods Tokens which are part of stopwords.

Table 1. .

Sr. No	Title/Author & year	Dataset	Methodology/Algorithm	Good points/Improvements/Comment/(future scope)
1	[1] Sven Hartrumpf et al (2010)	Not mentioned in the paper	SemDupl (Semantic Duplicate) have been used to detect obfuscated plagiarisms and semantic duplicates	How SemDupl is better than n-grams has been explained over here. Various other state of the art techniques like CErkenner, N-grams, ShallowChecker have been used
2	[2] Daniel Bar et al (2012)	MSR-Paraphrase [15] MSR-Video, Microsoft [16] SMTeuroparl: WMT2008 development dataset (Europarl section)	Explicit Semantic Analysis has been carried out to measure the degree of semantic relations in sentences. Various types of n-grams have been used to preserve the semantic relationship, also Pearson correlation coefficient has been employed over here	Here the UKP system has been used which is very robust, but it doesn't give high accuracy
3	[3] Mikhail Bilenko et al (2003)	Restaurant Dataset [17] Cora Dataset [18] Citeseer Dataset [19]	Support Vector Machine (SVM) has been used in novel- space based measure and learnable text similarity string edit distances to improve the duplicate detection accuracy	Support Vector Machines have been efficiently used over here for string and record similarity computation, here this model outperforms all the traditional models which helps in improving the model performance
4	[4] WEI EMMA ZHANG et al (2017)	Over 130k association pairs of known duplicate pairs of questions from - (stackoverflow quora)	Likelihood based techniques have been used over here to find the similarity for this many techniques like cosine similarity k-NN, Stochastic Gradient Descent (SGD), Random Forest, Support Vector Machines (SVM) Logistic Regression have been used	Various state of the art technology has been used to get the best accuracy from the model to find duplicate questions. Various machine learning algorithms have been used inside DupDetector which is a state of art technique for identifying the duplicates
5	[5] Qingyu Chen et al (2016)	The association number is in [20]	NB, DT and SVM are used over here along with many other algorithms specific to genomic data have been implemented	Supervised methods used over there show substantial promise. Quality results have been achieved while keeping it robust
7	[7] M. IKONOMAKIS et al (2005)	Not mentioned in the paper	Feature selection is done to bring the dimensionality down. Feature transform is carried out shorten the feature set size. Machine Learning Algorithms like NB, k-NN, SVM have been used which have shown the best accuracy	Here a lot of new features have been extracted. Here basic machine learning models have only been used, ensemble models could have been used for better accuracy

(continued)

Table 1. *(continued)*

Sr. No	Title/Author & year	Dataset	Methodology/Algorithm	Good points/Improvements/Comment/(future scope)
8	[8] Kaviani et al (2017)	Not mentioned in the paper	NB is a classification type of algorithm which is based on categorical set of variables and which is based on Bayes theorem	A survey on Naive Bayes Algorithm has been done over here where the advantages and disadvantages of the algorithm have been very nicely discussed
9	[9] M. A. Hearst et al (1998)	[21] [22]	SVM offers support vectors, support vectors are machine learning algorithms wherein various kernels can be directly input into space, various measures can be integrated into it like polynomial, RBF, radial, etc	Various ways in which SVM's can be used and modified is given over here. Making use of SVM's we can easily deal with non linear data by adding some additional parameters. Also we can integrate many other dimensionality reduction techniques with SVM's
10	[10] Maher Maalouf (2011)	Not mentioned in the paper	LR is a type of supervised leaning algorithm which is mainly used for classification where the probability of the targeted variable is predicted	In this paper they have shown us how Logistic Regression provides great means when modelling binary and multi class response variables
11	[11] Davide Chicco et al (2020)	[23]	Matthews correlation coefficient (MCC) is the measurement used for binary classification	In this paper they have shown how MCC gives much more stable results even in case of unbalanced dataset as compared to the F1 score
12	[12] Sang-Woon Kim et al (2019)	The data set has been taken from FGCS journal [24]	TF-IDF and LDA have been used over here to categorize similar research papers	A study has been does where many similar research papers are categorized, and how TF-IDF and LDA can be very useful for the classification task
13	[13] Amir Jalilifard et al (2021)	Not mentioned in the paper	Tf idf is an algorithm used for finding out the importance of a word in a sentence	In this paper how TF-IDF score is much better as compared to other classification supervised algorithms like SVM, NB, has been described
14	[14] Shadi Diab (2019)	Not mentioned in the paper	Stochastic gradient descent. Classification is used for fine tuning the hyper parameters for better classification	Here a case study has been done over here where it is shown how SGD is helpful in increasing the efficiency of LR,SVM and other machine learning models

We are using a pretrained model from spacy to train our data **(en_core_web_sm).** This acts as way better inputs for training our model as compared to any other inputs.

Hereafter we are converting our text data into a format which is understandable to our model which is into vector form. We will be understanding the data by doing data exploration, after analysing our data we will be doing data cleaning so that we can get best results from our model and avoid any kind of faulty results. Further we will be adding various kinds of features which will help us make more sense of the given data. Finally before inputting our data for training our model we will be converting our data into a form which is understandable to our models by converting it into vectors. At last we will be doing the result analysis of our model.

The various steps which are a part of our proposed model are as follows (Fig. 1):-

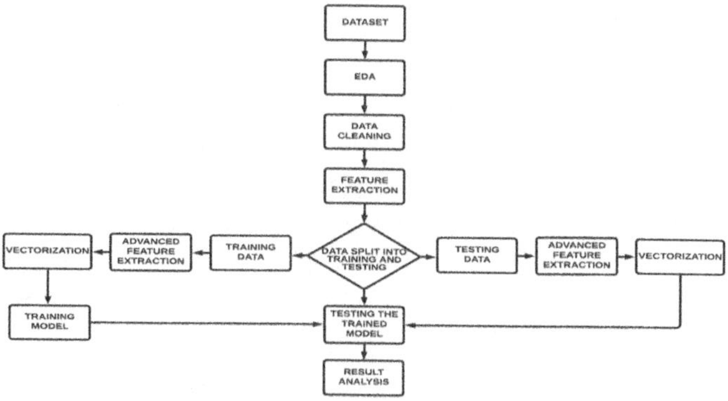

Fig. 1. Block Diagram of Proposed Model

I. Exploratory Data Analysis (EDA). This section details the pre-processing performed on dataset to extract and synthesize meaningful features to make sense of data in every possible manner.

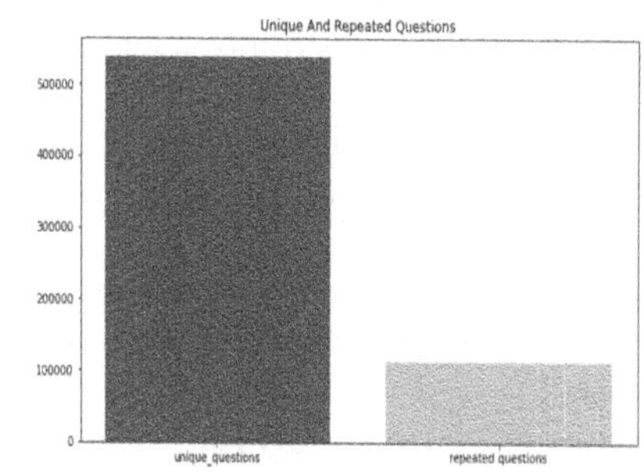

Fig. 2. Box Plot of Unique and Repeated Questions

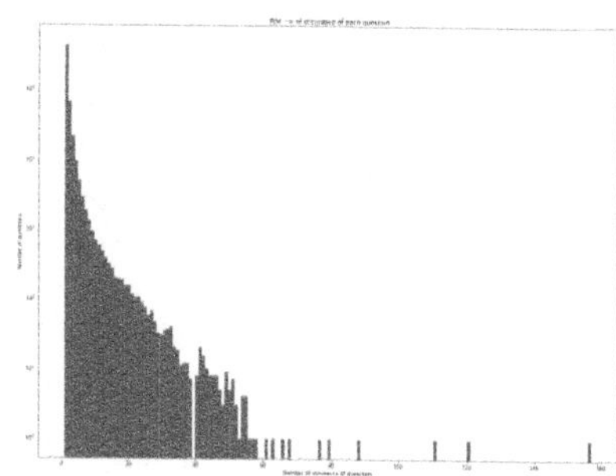

Fig. 3. PDF of Number of occurrence of each question

Figure 2 is a box plot of unique and repeated questions shows imbalance target class with unique questions as majority class. There are 537933 UNIQUE questions and 111780 REPEATED questions in our dataset. Figure 3 is a PDF plot representing the occurrence of each question. Plot shows very few questions are repeated more times and a lot of questions have been repeated very few times. The most number of times a single question has been repeated is 157.

Exploratory Data Analysis cites that the process of performing the initial exploration on the dataset so as to discover various patterns, and to spot any kind of anomalies, to test the hypothesis and to check the various assumptions made with the help of statistical and graphical based representations.

II. Data Cleaning. This section details the process to clean data by eliminating redundant information and retaining important data. Data cleaning starts with basic fillna function from Python to fill missing values and continues by removing html tags, stopwords and special characters.

III. Feature Extraction. Only two levels of headings should be numbered. Lower level headings remain unnumbered; they are formatted as run-in headings.

This section explains synthesized features from the existing dataset. Other than the semantic or syntactic similarity between the questions, there are many other factors which can contribute in identifying the relationship between question pairs. The proposed system synthesizes such features from the dataset by performing statistical analysis on question pairs. This section explains the feature generation process in detail. Following features are generated for depicting the relationship between Question pairs from the dataset precisely.

a. q1lenandq2len - The length of question 1 and 2 is calculated by using a simple len() function from the Numpy library of Python.
b. q1_wrods and q2_wrods = Number of words present in question 1 and 2 is calculated by token count.
c. common_word = Number of unique words which are common and present in both question 1 and question 2. This is an important feature as more no of common words implies higher similarity between questions.
d. longest_substring_ratio: It is the ratio of length the longest common substring to the minimum length of the token count of Q1 and Q2 longestsubstrratio = len(longest commonsubstr)/(min(len(q1 token), len(q2 token))
e. Along with above two features we have generated features such as fuzzratio, fuzzy partialratio, token sort ratio, token set ratio, by using the libraries from Python to find string similarity between question pairs.
f. word_share = This feature is calculated by, Number of to common unique words which are present in question 1 and question 2 divided by total number of words in question 1 + total number of words in question 2.
g. cwc_min and cwc_max: Few more features are generated to identify and visualize the separation between question pairs precisely. These features are generated as follows

    - The ratio of the minimum count is represented as (common word count)/(min(len (q1word), len(q2 word)) the max ratio is common word count)/(max(len (q1 word, len (q2word))
    - cscmin: The ratios of the minimum count is represented as (common stop count/(min (len (q1stop), len (q2stop))
    - cscmax: The ratios of the maximum count is represented as (common stop count/(max (len (q1stop), len (q2stop))
    - ctcmin: The ratios of the minimum token count is represented as common_token_count/(min (len(q1 token), len(q2token))
    - ctcmax: The ratios of the maximum token count is represented as common token count/(max(len (q1token), len (q2token))

h. last word equal and first word equal: To check whether the last word is same a the first word of the second question = int(q1token[-1] = = q2token[-1]).
i. Abslendiff: There is high probability of similar length of duplicate questions Duplicate questions. Hence the absolute length difference is calculated by:
   Abslendiff = abs(len (q1token) – len (q2token))
j. Meanlen: It is the average of the token length of both the questions (len (q1 token) + len (q2token))/2
k. Word total = It is the total number of words present in both of the question in totality.
l. Semantic Feature Generation: The process of feature generation is extended to understand the semantic relationship between question pairs. Hence semantic features are extracted using some advanced Natural Language Processing (NLP) techniques. The dataset is preprocessed by removing the stopwords and stemming using a snowball stemmer. The data is further cleaned to remove html tags and remove some of the special characters.

The result of above feature engineering is depicted in Figs. 4, 5, 6 and 7 On the completion of the feature extraction step the plot between predictor and target variable shows considerable separation.

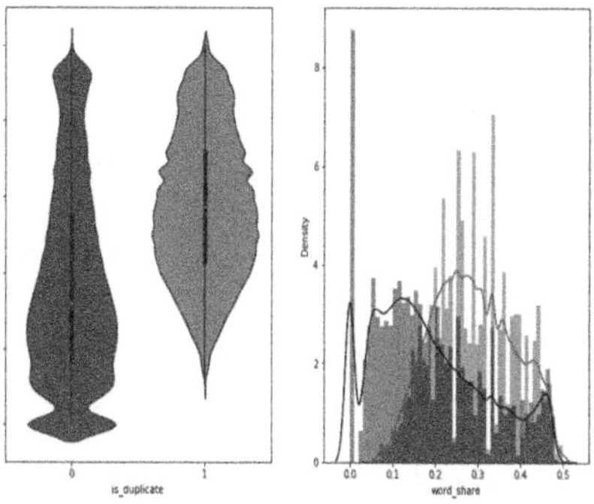

Fig. 4. word_share vs is_duplicate.

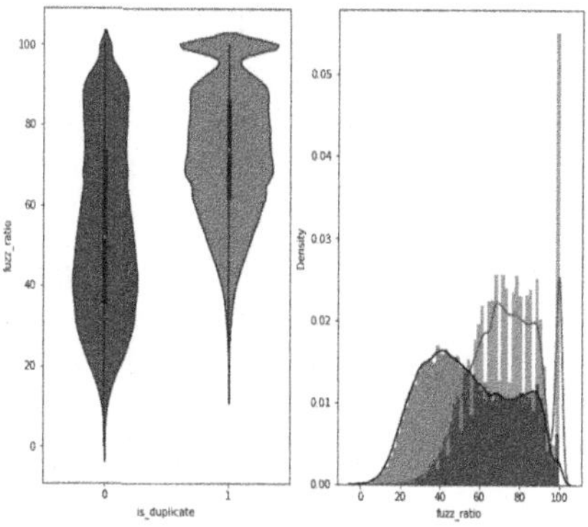

Fig. 5. fuzz_ratio vs is_duplicate.

word_ share, fuzz_ratio, longest_substring_ratio, token_ratio all these features have been engineered and have been plotted against is_duplicate because these features show the maximum spread. Larger the spread the more is the significance of the feature. Here we also use fuzz ratios like fuzz ratio, fuzz partial ration, token sort ration, token set ratio. Now we again plot violin and pair plots to see the separation (Fig. 8).

Some of the basic features such as common token count, common word count, common stop word count and token sort ratio have been plotted against each other using the plot. Here looking at the heat map and pdf we can determine the importance of the features. These features show considerably more separation as compared to that of the basic features without using nlp. We also plot a word cloud to find out the occurrence of each word (Fig. 9).

We have made use of the word cloud to represent the words which are present more number of times in our dataset. The larger the size of the word, the greater is the

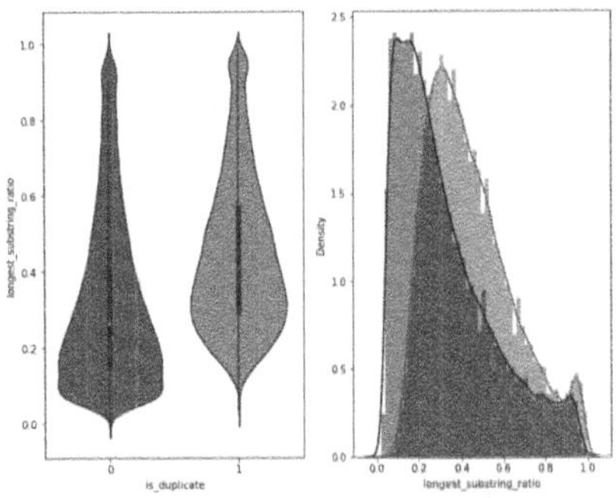

Fig. 6. longest_substring_ratio vs is_duplicate.

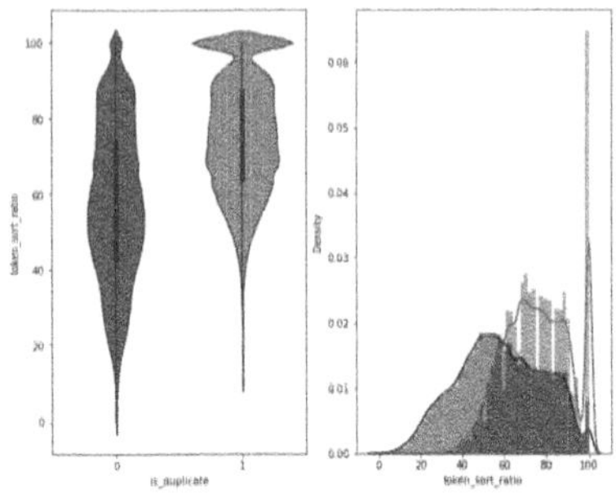

Fig. 7. token_sort_ratio vs is_duplicate.

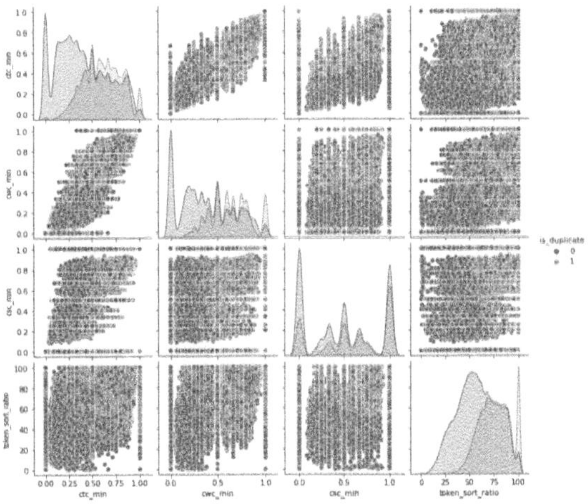

Fig. 8. Pair plot of important features against themselves

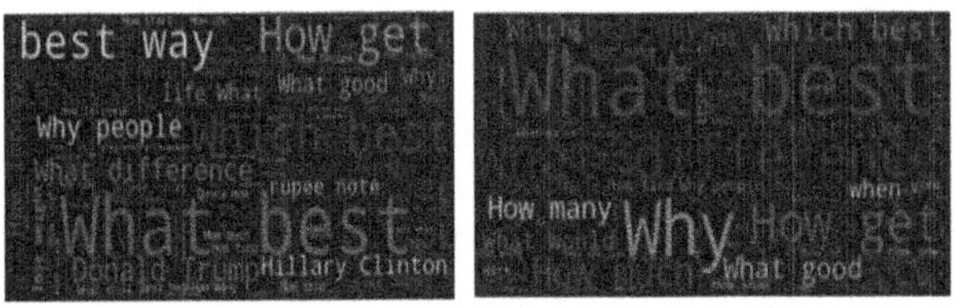

Fig. 9. Word Cloud of Question1 and Question 2

importance of it. Dimensionality reduction is the process where we try to reduce the dimension of the data, while retaining the original meaningful information.

Here we are performing dimensionality reduction and transforming the data into 2 dimensions and 3 dimensions. Here we are making use of the **Plotly** library (Figs. 10 and 11).

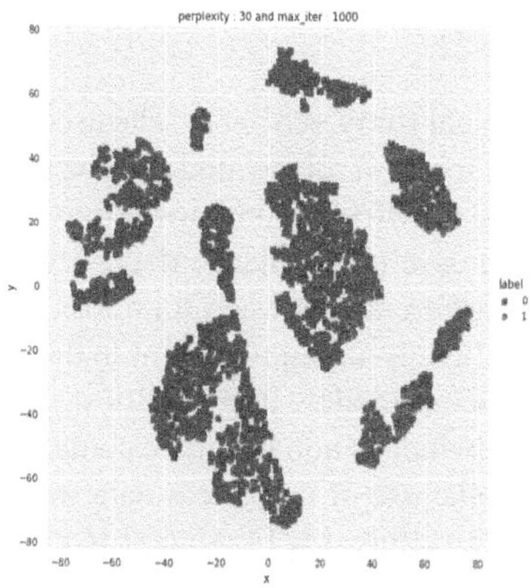

Fig. 10. TSNE 2D plot

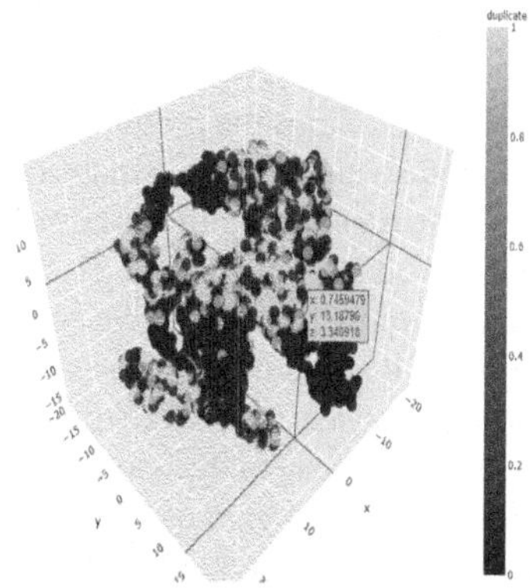

Fig. 11. TSNE 3D plot

Vectorization. To compare the question pairs semantically, we have used the tfidf model from the Spacy library of Python. The model builds a context vector of each question and calculates the similarity based on the simantics between question pairs by preserving the context. The TF-IDF score is calculated and is used to transform each question by making use of the weighted average. Here we use a already trained GLOVE model which comes free with "Spacy". https://spacy.io/usage/vectors-similarity [26]. It has been trained on the Wikipedia data.

Model Selection. Initially a random model (dumb model) is build to compare accuracy scores with it and get an estimation about the performance of proposed model.State of art shows poor performance of KNN and Naïve Bayes due to curse of demientionality [27]. The next models which are built are linear models. We built logistic regression and linear svm to observe the performance. Here we train our model using SGDClassifier and also calibrate our models. Figure 12 shows the block diagram of models used to build the proposed system. Result section explains the comparative analysis of the used models and shows how SGDClassifier outperforms the other models.

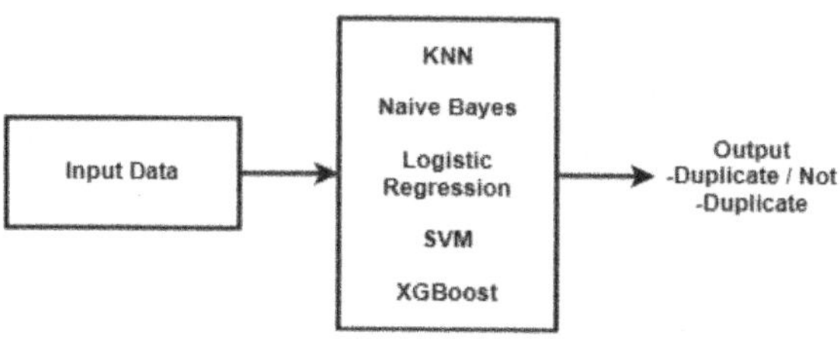

Fig. 12. Block Diagram of Selected Models

3.3 Result

This section explains the performance of our empirical study, see Table. The used performance metrics are confusion matrix, precision and recall of all the models. The model is also evaluated using Matthews correlation coefficient (MCC) as it removes bias towards correctly predicted labels and gives fair output in case of imbalance data. After comparing the confusion matrix we see that there is difference between train loss and test loss, we see that the model is overfit. We see that there is an improvement in the overall performance when we move from linear to non linear models like XGBoost. Because XGBoost makes use of decision tree data structure which is non linear we come to know that our data would give us better results if we make use of non linear data structures, because it proves beneficial in the case of overfitting (Figs. 13, 14, 15 and 16).

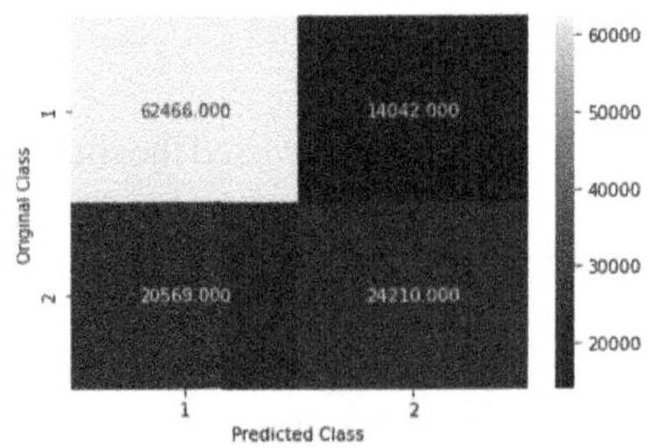

Fig. 13. Random Model

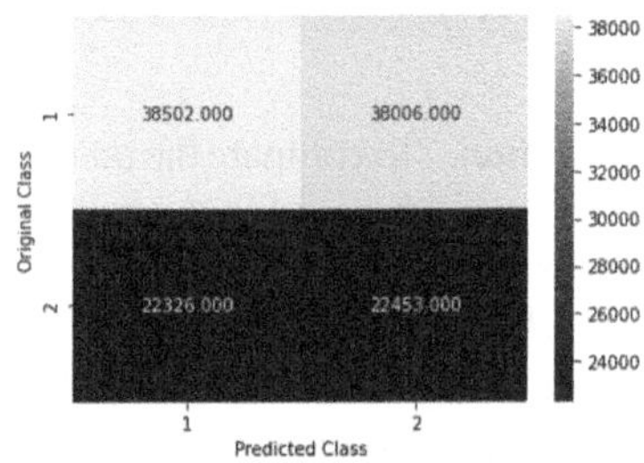

Fig. 14. Logistic Regression

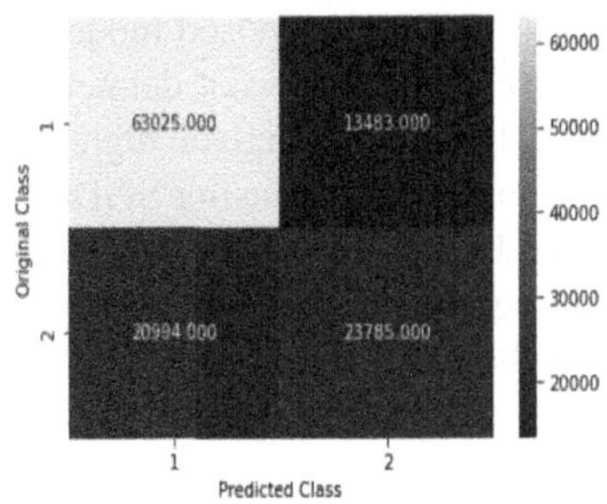

Fig. 15. SVM

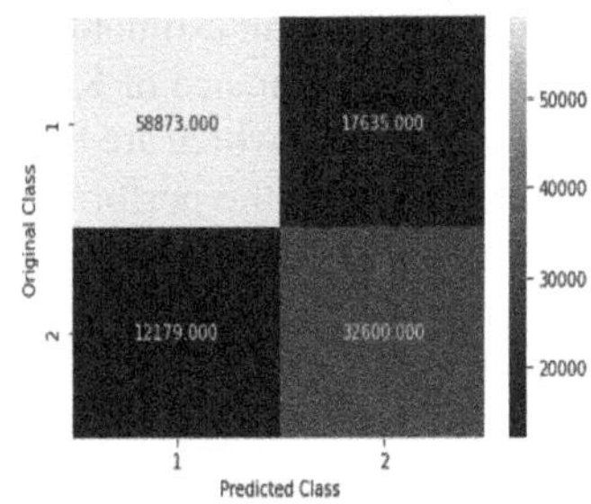

Fig. 16. XGBoost

$$\mathrm{MCC} = \frac{(TP \times TN) - (FP \times FN)}{\sqrt{(TP + FP)(TP + FN)(TN + FP)(TN + FN)}}$$

Table 2. Result Analysis

Method	Log Loss	Accuracy	Precision	Recall	F1 Score	MCC Score
Random Model	0.8836	0.5026	0.5014	0.3714	0.4267	0.0045
Logistic Regression	0.5097	0.7146	0.5407	0.6329	0.5832	0.3709
Linear SVM	0.5175	0.7157	0.5312	0.6382	0.5798	0.3713
XGBoost	0.4510	0.7542	0.7280	0.6489	0.6862	0.4874

4 Conclusion

The experimental comparison performed in this paper proves improved performance with XGboost over Linear, Logistic and SVM linear models. To capture the non-linearity in features this work can be extended by hyper parameter tuning and kernel tricks with cross validation.

References

1. Hartrumpf, S., vor der Bruck, T., Eichorn, C.: Semantic duplicate identification with parsing and machine learning, text, speech and dialogue. In: 13th International Conference, TSD 2010, pp. 6–10, Brno, Czech Republic, September (2010)
2. Bär, D., Biemann, C., Gurevych, I.: UKP: Computing semantic textual similarity by combining multiple content similarity measures. In: Conference: Proceedings of First Joint Conference on Lexical and Computational Semantics (*SEM), January (2012)
3. Bilenko, M., Mooney, R.J.: Adaptive duplicate detection using learnable string similarity measures. In: Ninth ACM SIGKDD International Conference on Knowledge Discovery and Data Mining (KDD-2003), pp. 39–48, Washington DC, August (2003)
4. Zhang, W.E., Sheng, Q.Z., Lau, J.H., Abebe, E., Ruan, W.: Duplicate detection in programming question answering communities. ACM Trans. Internet Technol. **18**(37), 1–21 (2017)
5. Chen, Q., Zobel, J., Zhang, X., Verspoor, K.: Supervised Learning for Detection of Duplicates in Genomic Sequence Databases. Published online 4 Aug (2016)
6. Rahutomo, F., Kitasuka, T., Aritsugi, M.: Semantic Cosine Similarity. In: The 7th International Student Conference on Advanced Science and Technology ICAST 2012, Seoul, South Korea, October (2012)
7. Ikonomakis, E.K., Kotsiantis, S., Tampakas, V.: Text classification using machine learning techniques. WSEAS Trans. Comput. **4**(8), 966–974 (2005)
8. Kaviani, P., Dhotre, S.: Short survey on naive bayes algorithm. Int. J. Adv. Res. Comput. Sci. Manage. **04**(11), 607–611 (2017)
9. Hearst, M.A., Dumais, S.T., Osuna, E., Platt, J., Scholkopf, B.: Support vector machines. IEEE Intell. Syst. Appl. **13**(4), 18–28 (1998)
10. Maalouf, M.: Logistic regression in data analysis: an overview. Int. J. Data Anal. Tech. Strat. **3**(3), 281–299 (2011)
11. Chicco, D., Tötsch, N., Jurman, G.: The advantages of the matthews correlation coefficient (MCC) over F1 score and accuracy in binary classification evaluation. BMC Genomics **21**(6), 1–13 (2020)

12. Kim, S.W., Gil, J.M.: Research paper classification systems based on TF-IDF and LDA schemes. Human-centric Comput. Inform. Sci. **9**(30), 1–21 (2019)
13. Jalilifard, A., Caridá, V.F., Mansano, A.F., Cristo, R.S., da Fonseca, F.P.C.: Semantic sensitive TF-IDF to determine word relevance in documents. In: Data Science Team - Digital Customer Service, Itaú Unibanco, São Paulo, Brazil Federal University of Minas Gerais, Brazil", 25 January
14. Diab, S.: Optimizing stochastic gradient descent in text classification based on fine-tuning hyper-parameters approach. A case study on automatic classification of global terrorist attacks. Int. J. Comput. Sci. Inform. Secur. **16**(12), 155–160 (2019)
15. https://www.microsoft.com/en-us/download/details.aspx?id=52398
16. https://paperswithcode.com/dataset/msr-vtt
17. https://www.cs.utexas.edu/users/ml/riddle/data/restaurant.tar.gz
18. https://www.cs.utexas.edu/users/ml/riddle/data/cora.tar.gz
19. https://linqs-data.soe.ucsc.edu/public/lbc/citeseer.tgz
20. https://bitbucket.org/biodbqual/duplicate_detection_repository/
21. http://www.ics.uci.edu/~mlearn/ML
22. http://research.microsoft.com/~jplatt/smo.html
23. https://archive.ics.uci.edu/ml/datasets/Mesothelioma%C3%A2%E2%82%AC%E2%84%A2s+disease+data+set+
24. https://www.sciencedirect.com/journal/future-generation-computer-systems
25. https://www.kaggle.com/c/quora-question-pairs#evaluation
26. https://spacy.io/usage/vectors-similarity
27. https://dataespresso.com/en/2017/10/24/comparison-between-naive-bayes-and-logistic-regression/
28. https://spacy.io/models/en [en_core_web_sm]
29. https://towardsdatascience.com/k-nearest-neighbors-and-the-curse-of-dimensionality-e39d10a6105d
30. https://plotly.com/python/

Intrusion Detection System Using Machine Learning

Shruti Wadhwa[1] and Monika Singh[2(✉)]

[1] Nidus, Chandigarh, India
[2] Chandigarh University, Mohali, India
monika.e11032@cumail.in

Abstract. Intrusion Detection System monitors the network resources in an effort to find unsuitable network activity. Attackers constantly create new vulnerabilities and attack strategies intended to weaken the defense. Obtaining user credentials that give access to the network and data is the goal of several attacks. For network security, a Network Intrusion Detection System is crucial since it enables to react to hostile activity. With the growth of the Internet, attackers, spammers and criminals have become a greater menace. Due to the frequency of such dangers, intrusion detection systems (IDS) have emerged as a crucial component of computer networks. Most of the previous IDS have focused on the classification of attacks as normal or intrusion. In this paper, a novel approach of creating intrusion detection system is created using different algorithms of machine learning. The proposed system can classify the type of attacks. We have used three types of classifier models SVM, SGD, and Adaptive Boost with a different set of features. The method is examined using knowledge discovery on the KDDcup99 dataset. The findings demonstrate that the suggested system has a low false alarm rate and a high accuracy of attack detection.

Keywords: IDS · KDDcup99 Dataset · SVM · SGD · Adaptive Boost

1 Introduction

The purpose of intrusion detection systems (IDS) is to detect malicious and suspicious behaviour in network traffic. It makes detection real-time on the network, which aids in maximising intrusion detection. IDS encompasses a variety of approaches aimed at detecting traffic in a variety of ways. IDS focuses on recognizing logging information, potential events, and reporting efforts. IDS is being used by the groups with the goal of identifying faults with security policies and documenting current security risks.

IDS is now a must for practically every enterprise. It aids in the recording of data related with observed actions, as well as the alerting of security administrators and the generation of reports. This technology continuously scans the network for unusual behaviour. The same convention-based assaults that the system is vulnerable can render network intrusion detection systems (NIDS) frameworks useless. Incorrect data and TCP/IP stack attacks might cause a NIDS to crash.

S. Singh et al. (Eds.): AIDSP 2023, CCIS 2440, pp. 161–170, 2025.
https://doi.org/10.1007/978-3-031-88762-8_16

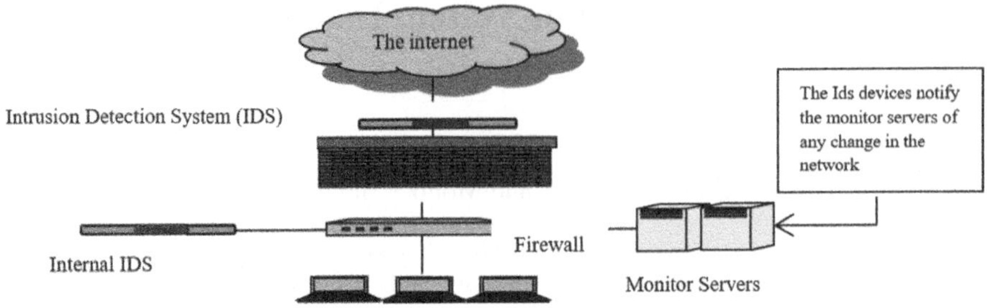

Fig. 1. Intrusion Detection Systems (IDS) [17]

The purpose of IDS is to examine the usefulness of control activities by detecting testimony of attacks. This kind of proceedings are frequently required to improve handling risks connected with widespread susceptibility like infected laptops, e-mail borne viruses, and human factors as shown in Fig. 1. To be functioning, however, an IDS solution requires expert staff and disciplined processes, as well as a machine that is thoughtfully configured and has tuned observing configuration.

2 Literature Survey

Srivastav and Krishna [1] proposed a layered structure that was utilised in conjunction with the neural network to create a successful framework for interruption detection. Using the KDD 1999 dataset for knowledge discovery and data mining, this framework has done a variety of things. The results show that the suggested framework has a high level of assault detection precision and a low rate of false alarms. **Bhavsar and Waghmare** [2] proposed an IDS based on SVM algorihm with a shorter training period. They performed data pre-processing in order to minimize the SVM training time. The NSLKDD'99 dataset was used in the experiments. They also looked at how SVM performance varied depending on the kernel function. The results suggest that if the data is appropriately pre-processed, the Gaussian kernel can shorten the learning time of SVM. Support vector machines and other machine learning algorithms such as traditional classification methods, hybrid algorithms, ensemble methods, and hybrid ensemble methods were compared by **Chauhan et al.** [3]. For training and testing, they used the terrorist attack dataset. Following the training of various models, these models were utilized to forecast which terrorist groups are behind terrorist attacks. The SVM method is the best, according to the evaluation findings.

Stevanovic and Pederson [4] presented a novel stream construct discovery framework that depends on machine learning for distinguishing botnet system activity. The outcomes demonstrate that keeping in mind the end goal to accomplish precise identification activity streams should be observed for just a constrained time period and number of bundles per stream. This shows a solid capability of utilizing the proposed approach inside of the online location structure. **Karim et al.** [5] presented a complete audit of the most recent best-in-class procedures for botnet discovery. **Livadas et al.** [6] focused on checking system activity, distinguishing abnormalities and digital assault movement designs, and, a posteriori, fighting the digital assaults and alleviating their belongings. **Singh et al.** [7] proposed a new scalable quasi-real-time intrusion detection model. This model is designed for detecting attacks from peer-to-peer botnet traffic on the Apache Hadoop environment using Hive and Mahout. **Feizollah et al.** [8] presented 100 examination works distributed somewhere between 2010 and 2014 with the point of view to highlight choices in versatile malware recognition. They classified accessible elements into four gatherings, in particular, static components, dynamic elements, cross-breed elements, and application metadata. **Karim et al.** [9] presented the investigation of mobile botnet attacks by investigating assault vectors and in this way display a very much characterized topical scientific classification. **Suad Mohammed Othman et al.** [10] introduced the Spark-Chi-SVM model. This author used ChiSqSelector which is used for selection of features on the Apache Spark (Big Data platform), & SVM classifier to build a model. Researcher uses KDD99 dataset to train and assess the model. **T.Saranya et al.** [11] investigates the ML algorithms used for IDS in various situations, as well as some experiments employing MLA and ML techniques such as LDA, CART, & Random Forest on the KDD'99 Cup data set. **Mukesh Kumar Yadav et al.** [12]. The analytical examinations of current IDS are the main purpose of this paper. Additionally, they look into useful data sources and techniques for building an effective IDS using single, hybrid, and ensemble algorithms.

3 Problem Formulation

3.1 Gap Analysis

According to the detailed study, it has been noticed that in the existing work, fewer machine learning algorithms are used and there is less comparison of algorithms done to find out which gives the accurate result. Also, a feature set with 41 features has been used which increases the time of processing of detection of intrusion. Secondly, Information gain and correlation are not considered when choosing features. As, feature selection in earlier systems is unable to give information about various infiltration kinds. Additionally, learning is very polynomial and does not produce accurate results.

3.2 Problem Statement

The purpose of the research is to create an IDS by finding the important features with the help of machine learning.

4 Proposed Methodology

The proposed method of designing IDS is divided into three modules viz. Module of Feature Selection, Module of Feature Extraction, and Regular Module.

4.1 Regular Module

In this module, the framework is prepared to utilize all the 41 features using the following steps:

1	KDD-99 dataset with 41 features.
2	Input the features & labels to be given to Linear SVM, SGD & Adaptive Boost models.
3	Perform the test on these models and calculate the precision, recall & accuracy.

4.2 Module of Feature Selection

Finding a subset of pertinent features to incorporate into a model is a process is featuring selection [13]. Dimensionality reduction and feature selection are not the same. Both approaches attempt to reduce the number of variables in a dataset, but dimensionality reduction reduces the number of variables by creating new combinations of attributes, whereas feature selection just includes and excludes attributes that are currently included in the data. Feature selection is valuable in and of itself. However, it largely acts as a filter, muting both the required and unnecessary aspects. Various feature selection techniques, including the Filter, Wrapper, and Hybrid methods, are classified in Fig. 2.

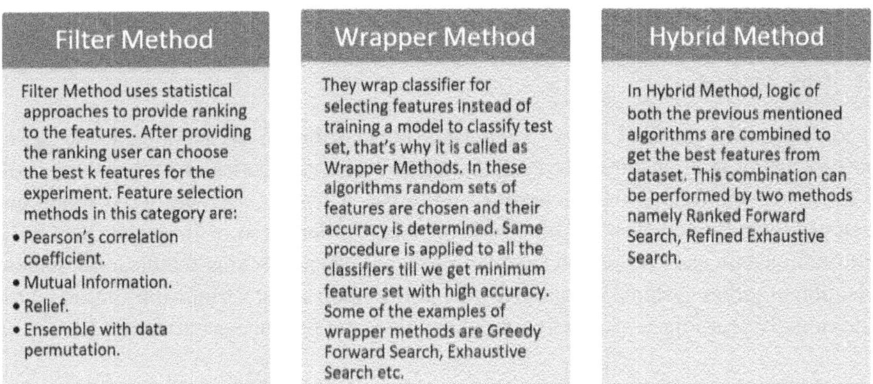

Fig. 2. Feature Selection Techniques

Instead of using all 41 features and testing with SVM, SGD, and Adaptive Boost, feature selection is done by using correlation to determine some of the best components. We adopt the correlation technique because it depicts feature similarity based on distribution while decreasing feature misbalancing in the form of distribution. The following steps occur during the feature selection phase:

1	KDD-99 dataset with 41 features.
2	Feature Selection by correlation method.
3	Input the feature & labels into Linear SVM, SGD & Adaptive Boost & make three models.
4	Perform the test on these models & calculate the precision, recall & accuracy.

4.3 Feature Extraction

One of the most challenging elements of doing a thorough data analysis is the enormous number of variables involved. A classification algorithm may overfit training examples and perform poorly on new samples because of the memory and processing power requirements of a large number of variables. The following operations are performed during the feature extraction phase are mentioned below:

1	KDD-99 dataset with 41 features.
2	Feature extraction with KPCA.
3	Input the feature & labels into Linear SVM, SGD & Adaptive Boost & make three models.
4	Perform the test on these models and calculate the precision, recall & accuracy.

4.4 Performance Evaluation

The following is a list of significant performance parameters used to examine the findings:

Precision:

$$\Pr ecision = TP/(TP + FP) \tag{1}$$

Recall:

$$\text{Recall} = TP/(TP + FN) \tag{2}$$

Accuracy:

$$\text{Accurancy} = (TN + TP)/(TN + TP + FN + FP) \tag{3}$$

TP, FP, FN, and TN, respectively, stands for the numbers of True Positives, False Positives, False Negatives, and True Negatives.

4.5 Algorithms for Classification

4.5.1 SVM Linear

SVM [13] is a supervised ML algorithm which can be used for the regression, as well as the classification. To classify data, classifier uses the idea of hyperplanes.

4.5.2 Adaptive Boost

Adaboost improves classification accuracy and can be applied with a variety of classifiers. Overfitting isn't a problem here. The bootstrap estimate is the average of the individual estimations and is used to calculate the variance of a statistic (parameter).

4.5.3 Stochastic Gradient Descent

For training a range of models, including (linear) support vector machines, logistic regression, and graphical models, stochastic descent is a prominent machine learning technique [14, 15].

4.5.4 Correlation Coefficient

A tool for assessing how closely two variables' movements are connected is the correlation coefficient. The correlation coefficient has a range of values from -1.0 to 1.0. An error has been made if the estimated correlation is larger than 1.0 or less than -1.0.

4.5.5 Kernel Principal Component Analysis (KPCA)

Huge volumes of high-dimensional data are a common difficulty in data mining. It's challenging to identify meaningful patterns in massive amounts of data. The answer is the 'KPCA' dimension reduction approach.

4.6 Dataset

The training and validation input vectors are taken from the KDD 99 data set. The Defence Advanced Research Project Agency (DARPA) intrusion detection evaluation programme served as the basis for its creation [15]. In order to collect raw TCP dump data, MIT Lincoln Lab, a participant in this study, built up a simulation of a typical LAN network [16]. They created a virtual LAN that was infected with several kinds of attacks and functioned as a typical setting. It was converted into connection records from the original data source. 41 characteristics were retrieved for each link, as shown in Fig. 3. Each link was classified as either secure or vulnerable to a certain attack. There are 39 different attacker kinds, and they can be divided into four major groups as shown in Fig. 4.

5 Results and Discussion

We carry out three different types of experiments. The frameworks are created based on the primary analysis and include all 41 elements. The second experiment used correlation to choose features rather than Linear SVM, SGD, and Adaptive Boost to use all 41 features. In the third experiment, instead of using all 41 features with SVM, Adaptive Boost & SGD and analyzing the results, features were extracted with KPCA in order to identify the best components.

5.1 Results with Normal Module

First, it was determined how well the three classification methods, Linear SVM, SGD, and adaptive Boost, performed. Without feature extraction or feature selection, Fig. 5 displays the performance of the Linear SVM, SGD, and Adaptive Boost algorithms for each of the 41 features in the KDD99 data set. SGD has a substantial result when 41 features are used. Linear SVM, like Adaptive Boost, produced encouraging results. SVM takes a long time to train and works best with a well pre-processed dataset. Our evaluation results show that AdaBoost with SVM has an accuracy of 84%, whereas SVM has an accuracy of 90.91%. SGD solves the problem of unbalancing by providing a 96% accuracy.

Number	Feature name	Data type	Description
1	Duration	Continuous	Length of the connection
2	protocol type	Nominal	Connection protocol
3	Service	Nominal	Destination service
4	flag	Nominal	Status flag of the connection
5	Srcbytes	Continuous	Bytes Sent from source to destination
6	Dst_bytes	Continuous	Bytes Sent from destination to source
7	Land	Nominal	1 If the connection is from/to the same host/port; 0 otherwise
8	Wrong fragment	Continuous	wrong fragment
9	urgent	Continuous	urgent packets
10	Hot	Continuous	hot indicators
11	Num failed logins	Continuous	failed login attempts
12	Logged in	Nominal	1 If successfully logged in; 0 otherwise
13	Numcompromised	Continuous	compromised conditions
14	Root_shell	Nominal	1 if root shell is obtained; 0 otherwise
15	Su_attempted	Nominal	1 if the "su root" command is attempted; 0 otherwise
16	Num_root	Continuous	root accesses
17	Num_file_creations	Continuous	file creation Operations
18	Num_shells	Continuous	shell prompts
19	Num_access_files	Continuous	operations on access control files
20	Num_outbound_cmds	Continuous	outbound commands in an FTP Session
21	Is_hot_login	Nominal	1 if the login belongs to the hotlist; 0 Otherwise
22	Is_guest_login	Nominal	1 if the login is a guest login; 0 otherwise
23	Count	Continuous	connections to the same host as the current connection in the past two seconds
24	Srv_count	Continuous	connections to the same service as the current connection in the past two seconds(same-service connections)
25	Serror_rate	Continuous	connections that have "SYN" errors(same-host connections)
26	Srv_serror_rate	Continuous	connections that have "SYN" errors(same-service connections)
27	Rerror_rate	Continuous	connections that have "REJ" errors(same-host connections)
28	Srv_rerror_rate	Continuous	connections that have "SYN" errors(same-service connections)
29	Same_srv_rate	Continuous	connections to the same service(same-service connections)
30	diff_srv_rate	Continuous	connections to different services
31	Srv_diff_host_rate	Continuous	connections to different hosts(same-service connections)
32	dst_host_count	Continuous	Count of connections having the same destination host
33	Dst_host_srv_count	Continuous	connections having the same destination host and using the same service
34	Dst_host_same_Srv_rate	Continuous	connections having the same destination host and using the same service
35	dst_host_diff_Srv_rate	Continuous	different services on the current host
36	Dst_host_same_src_Port_rate	Continuous	connections to the current host having the same Srcport
37	Dst_host_srv_diff_Host_rate	Continuous	connections to the same service coming from different host
38	Dst_host_serror_Rate	Continuous	connections to the current host that have an S0 error
39	Dst_host_srv_serror_rate	Continuous	connections to the current host and specified service that have an S0 error
40	Dst_host_rerror_rate	Continuous	connections to the current host that have an RST error
41	Dst_host_srv_Rerror_rate	Continuous	connections to the current host and specified service that have an RST error

Fig. 3. Feature set of KDD'99 Dataset (41 features)

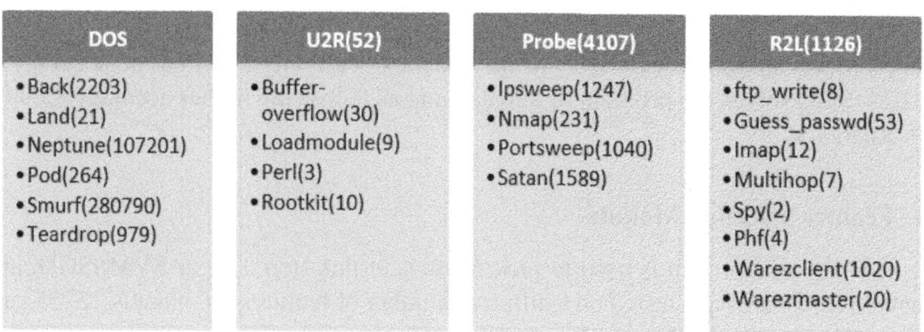

Fig. 4. Attacks Type in KDD'99 Dataset

5.2 Feature Extraction Module

KPCA is used to extract features in this step. Individually, Linear SVM, SGD, and Adaptive Boost are evaluated on 35, 25, and 15 features, respectively (Fig. 6).

SVM is a powerful classifier, therefore Adaboost with a strong classifier is not practical, according to Feizollah et al. [17]. Our evaluation results show that Adaboost

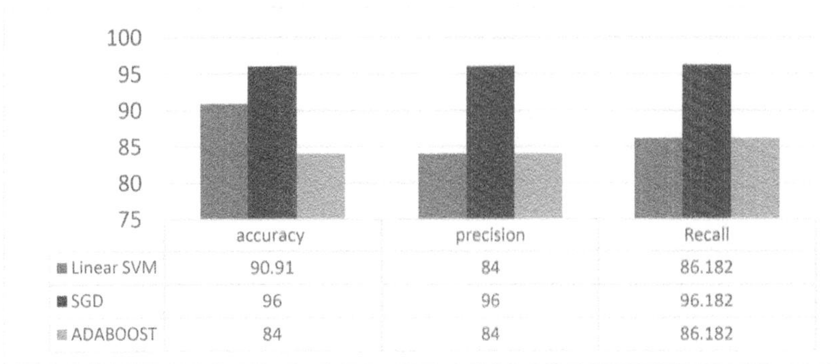

Fig. 5. Evaluation of Performance with All 41 Features

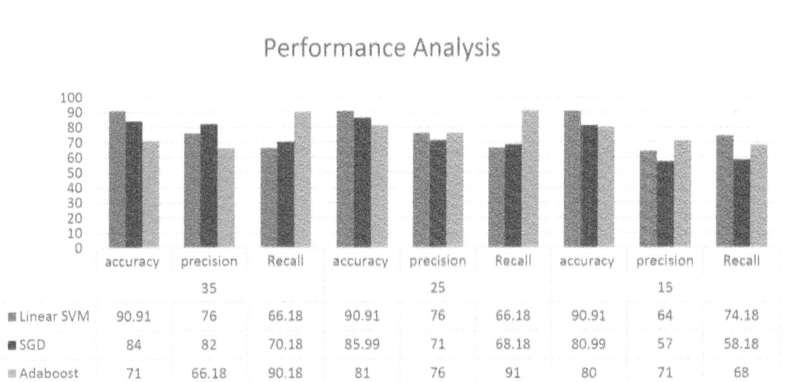

Fig. 6. Performance Evaluation of Feature Extraction Module

with SVM has an accuracy of 81%, while Adaboost with only SVM has an accuracy of 90.91%. SGD solves the problem of unbalancing as delivering higher accuracy 85.99% than Adaboost.

5.3 Feature Selection Module

The correlation approach is used to pick features at this step. Linear SVM, SGD, and Adaptive Boost are each tested on a different number of features set, namely 35, 25, and 15.

Finally, the results of feature selection using correlation revealed that Linear SVM and SGD have a greater accuracy rate than adaptive boost, as shown in Fig. 7, which is derived from individual comparative study of all three mentioned classification algorithms at different feature sets. SVM is a powerful classifier, therefore Adaboost with a strong classifier is not practical, according to Feizollah et al. [17]. Our evaluation results show that Adaboost with SVM has an accuracy of 81 percent, while Adaboost with only SVM has an accuracy of 90.91 percent. SGD solves the problem of unbalancing by providing 86 percent more accuracy than Adaboost with 35 characteristics.

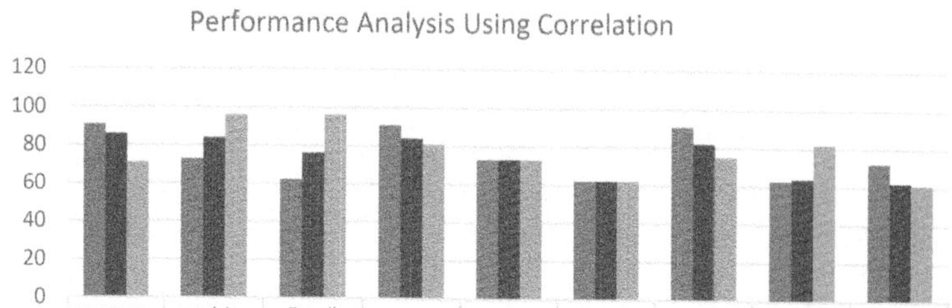

	accuracy	precision	Recall	accuracy	precision	Recall	accuracy	precision	Recall
		35			25			15	
Linear SVM	90.91	73	62.18	90.91	73	62.18	90.91	63	72.18
SGD	86	84	76.18	84	73	62.18	81.99	64	62.18
Adaboost	71	96	96.18	81	73	62.18	75	82	61.23

Fig. 7. Performance Evaluation for Feature Selection Module

6 Concluding Remarks and Future Work

With the growing use of the internet in everyday life, it can be stated that intrusion detection is a growing area of research in computer security and systems. Most intrusion detection systems rely on a single classifier calculation to categorize system activity as normal. This research focuses on forecasting an optimum classifier model for various types of attacks. We run three sorts of experiments: one is standard, the second is feature selection using correlation, and the third is feature extraction using KPCA, Linear SVM, SGD, and Adaptive boost, with the outcomes analyzed. These findings demonstrate that SGD outperforms alternative classifiers on average. There is still room for progress in terms of proposing systems that can identify all forms of assaults while also reducing the feature set through feature selection and feature extraction using various classifiers. For future work, will apply the machine learning algorithm on different dataset rather than KDD99 and will compare the results.

References

1. Giacinto, G., Roli, F., Didaci, L.: Fusion of multiple classifiers for intrusion detection in computer networks. Pattern Recognit. Lett. **24**(12), 1795–1803 (2003)
2. Bhavsar, Y.B., Waghmare, K.C.: Intrusion detection system using data mining technique: support vector machine. Int. J. Emerg. Technol. Adv. Eng. **3**, 581–586 (2013)
3. Chauhan, H., Kumar, V., Pundir, S., Pilli, E.: A comparative study of classification techniques for intrusion detection. In: International Symposium on Computational and Business Intelligence, pp. 1–6 (2013)
4. Shaheen, M.A., Ibrahim, H.E., Badr, S.M.: Adaptive layered approach using machine learning techniques with gain ratio for intrusion detection systems. Int. J. Comput. Appl. (0975–8887) **56**(7), 10–16 (2012)
5. Lin, S.W., Ying, K.C., Lee, C.Y., Lee, Z.J.: An intelligent algorithm with feature selection and decision rules applied to anomaly intrusion detection. Appl. Soft Comput. J. **12**(10), 3285–3290 (2012)

6. Karim, A., Salleh, R., Shiraz, M., Shah, S., Awan, I., Anuar, N.: Botnet detection techniques: review, future trends, and issues. Comput. Electron. **15**(11), 943–983 (2014)
7. Singh, K., Guntuku, S.C., Thakur, A., Hota, C.: Big data analytics framework for peer-to-peer botnet detection using random forests. Inf. Sci. **278**, 488–497 (2014)
8. Mukherjee, S., Sharma, N.: Intrusion detection using naive bayes classifier with feature reduction. Procedia Technol. **4**, 119–128 (2012)
9. Panda, M., Abraham, A., Patra, M.R.: A hybrid intelligent approach for networkintrusion detection. Procedia Eng. **30**(2011), 1–9 (2012)
10. Othman, S.M., Ba-Alwi, F.M., Alsohybe, N.T., et al.: Intrusion detection model using machine learning algorithm on big data environment. J. Big Data **5**, 34 (2018). https://doi.org/10.1186/s40537-018-0145-4
11. Saranya, T., Sridevi, S., Deisy, C., Chung, T.D., Khan, M.K.A.A.: Performance analysis of machine learning algorithms in intrusion delection system: a review. Procedia Comput. Sci. **171**, 1251–1260 (2020). https://doi.org/10.1016/j.procs.2020.04.133. ISSN 1877-0509
12. Yadav, M.K., Sharma, K.P.: Intrusion detection system using machine learning algorithms: a comparative study. In: 2021 2nd International Conference on Secure Cyber Computing and Communications (ICSCCC), pp. 415–420 (2021). https://doi.org/10.1109/ICSCCC51823.2021.9478086
13. Srivastav, N., Challa, R.K.: Novel intrusion detection system integrating layered framework with neural network. In: Proceedings. 2013 3rd IEEE International Advances Computing Conference IACC, vol. 35, no. 2, pp. 682–689 (2013)
14. Finkel, J.R., Kleeman, A., Manning, C.D.: Feature-based conditional random field parsing. In: Proceedings Annual Meeting of the ACL (2008)
15. Darken, C., Moody, J.: Fast adaptive k-means clustering: some empirical results In: Proceedings of International Joint Conference on Neural Networks (IJCNN). IEEE (1990)
16. Ranawana, R., Palade, V.: Multi-classifier systems: review and a roadmap for developers. Int. J. Hybrid Intell. Syst. **3**, 35–61 (2006)
17. Feizollah, A., Anuar, N.B., Salleh, R., Wahab, A.W.A.: A review on feature selection in mobile malware detection. Digit. Investig.Investig. **13**, 22–37 (2015)

An Efficient Framework for Multi-level Lung Cancer Prediction Using Support Vector Machine Classifier

Ashok K. Patil[1], Siddanagouda S. Patil[2(✉)], M. Prabhakar[1(✉)], and Vineet Kumar[3]

[1] School of Computer Science and Engineering, REVA University, Bangalore, India
prabhakar.m@reva.edu.in

[2] University of Agricultural Sciences, Bangalore, India
spatilsidasb@gmail.com

[3] Workplace if-A People Analytics Company, Bangalore, India

Abstract. This work mainly focuses on the need for collaborative learning framework that allows for early prediction of disease detection in patients. Principal component analysis comes out as a very effective approach for classifying the target classes. PCA successfully combines related qualities and creates a dispersed exhibition of its constituents. The number of principle components to be retained is determined by examining the screen plot. With a small amount of data, Support Vector Machines (SVM) beats other classification algorithms. The components obtained will be sent to the SVM which classifies the cancer based on Multi-Level and helps in prediction of malignancy of cancer, the early dangerous stage will urge clinical specialists to offer those patients extra attention. The results obtained through the proposed framework have achieved accurate results in terms of various performance parameters like accuracy. Precision, recall and F-measure and then confusion matrix is drawn to validate the proposed model and helps in detecting various stages of malignancy at an early stage.

Keywords: Multi-Level-Lung Cancer · PCA · SVM · Machine Learning · Prediction Model

1 Introduction

Malignant growth, often known as carcinoma, is one the main cause of death in the world as on today. Lung carcinoma (1.6 million) causes the majority of deaths among the various types of cancers that occur, such as breast, liver, stomach, and colorectal cancers [1]. Lung cancer accounts for 13% of all malignancies assessed in 2012, according to overall data. Lung Carcinoma tends to be more morphological in nature that spreads rapidly [8]. Small Cell and Non-Small Cell Lung Cancers (SCLC and NSCLC) are the two types of lung cancer [2]. Lung illness can be caused by both factors which are hereditary and non-hereditary [3]. Tobacco is the most common cause of lung cancer. It has a negative impact on the health of chain smokers and others who come into touch with them [4]. Nonsmokers' lungs can be impacted by this cancer as a result of air pollution, working in an environment where more chemical exposure, asthma due to cold [4]. There are some types of cancer which are caused inherited in nature.

S. Singh et al. (Eds.): AIDSP 2023, CCIS 2440, pp. 171–182, 2025.
https://doi.org/10.1007/978-3-031-88762-8_17

The various stages of lung cancer [27] are described below with their effect is described in Table 1:

Table 1. .

Stages	Explanation
Stage 1	The cancer is confined only in the lungs
Stage 2	The cancer is in the lungs and can have metastasized to the lymph nodes adjacent to the lungs
Stage 3	Cancer is localized in the lungs and has spread to the lymph nodes in the center of the chest
Stage 3A	The cancer has become extensive, however, is localized to one side of the lung
Stage 3B	The cancer has metastasized to the lymph nodes on both the sides of the lungs
Stage 4	The cancer has metastasized to both the lungs or has

Medical practitioners' primary care is required for quick survival. The type of lesion will be identified, which will suggest the need for additional treatment.

S.S. Ray, et al., [12] has developed an underlying architecture of microarray data referred as Granular Self Organizing Map(GSOM) [12]. This GSOM technique was mainly built by combining both SOM technique and the Fuzzy rough set, where the rough of fuzzy sets uses the decision classes and the fuzzy reflexive relation is used to control the uncertainties in the information class. The dependence of each gene (attribute) was computed by fuzzy rough sets by using the cluster of GSOM output. The performance of the developed GSOM technique was analyzed in light of Dunn-index, DB-index, fuzzy entropy and the β-index The conventional SOM technique includes three major concerns; difficult to identify the optimal map size, clustering results completely depends on initial weight vectors and also it needs large quantity of training data.

[11] Describes the current state of the art for lung cancer diagnosis using computer knowledge, including technical difficulties and validation. When compared to other methods, the CAD procedure is the gold standard in terms of speed and accuracy [12].

The majority of datasets in the world are accessible via web-based networking, and features relevant for tasks such as classification and prediction can be extracted from this massive amount of data [13]. In [14], the Random Forest Classifier is mainly used for classifying various types of features into target vector classification.

In [15], Different classifiers like of ANN, SVM and NB were combined with existing blocks of classifiers to develop a model for effective lung cancer patients in sequential manner.

Table 2. Various research works carried out based on Lung Cancer Identification

S.No.	Name	Year	Overview
1	Comparison of Lung Cancer Detection Algorithms [19]	2019	Applied Principal Component Analysis, K-Nearest Neighbors, Support Vector Machines, Naïve Bayes, Decision Trees, and Artificial Neural Networks machine learning methods before and after preprocessing to detect anomaly
2	A Deep Learning Based Approach to Lung Cancer Identification [16]	2018	A 3D neural network-based diagnosis with TensorFlow library which effectively works even with fewer samples
3	Lung Cancer Detection Using Image Processing and Machine Learning HealthCare [17]	2018	SVM classifier and image processing are collaborated on CT scan lung images to enable early detection
4	Small-Cell Lung Cancer Detection Using a Supervised Machine Learning Algorithm [18]	2017	A novel neural-network-based algorithm with entropy degradation method (EDM) to detect small-cell lung cancer (SCLC) from CT images
5	A Survey on Lung Cancer Detection using Image Data Analysis and Machine Learning [25]	2017	This study aims to highlight the significance of data analytics and machine learning
6	Lung Cancer Classification Using Deep Learned Features on Low Population Dataset [20]	2017	A deep autoencoder classification mechanism which works on low population and unstructured data
7	Deep Learning Application Trial to Lung Cancer Diagnosis for Medical Sensor Systems [22]	2016	Personal diagnosis is enabled using sensing data and deep learning techniques for feature selection
8	Convolutional Neural Networks for Lung Cancer Screening in Computed Tomography (CT) Scans [23]	2016	Special properties of CNN networks like spatial invariance, which extract features easily, are applied for lung cancer detection
9	Computer-Aided Diagnosis System Based on Machine Learning Techniques for Lung Cancer [21]	2012	Feature extraction using wavelet transform is applied on lung cancer dataset to achieve high accuracy
10	Predictive Analysis of Lung Cancer Recurrence [24]	2011	The predictive analysis of lung cancer recurrence based on non-small cell lung cancer carcinoma gene expression data using data mining and machine learning techniques

Table 3. Data Considered for the Features of Skewness

Features	Skew-Values
Loss of Weight	0.470156
Breath Shortness	0.480384
Allergy caused by Dust	−0.233203
Hazards at occupation	−0.270883
Risk at Genetic	−0.1285225
Chronic-Disease	−0.109587
Balanced_Diet	−0.053694
Air Pollution	0.195541
Alcohol use	0.023802
Genetic_Risk	−0.079225
Chronic_Lung_Disease	−0.209587
Passive-Smoker	0.472199
Smoking	0.407958
Coughing-of-Blood	0.262642
Chest Pain	0.189506
Fatigue	0.857201

2 Proposed Methodology of the Work

This work presents an efficient framework for Lung Cancer Prediction Using Support Vector Machine for Multi-Level Classification; Fig. 1 describes the flow of the work carried out.

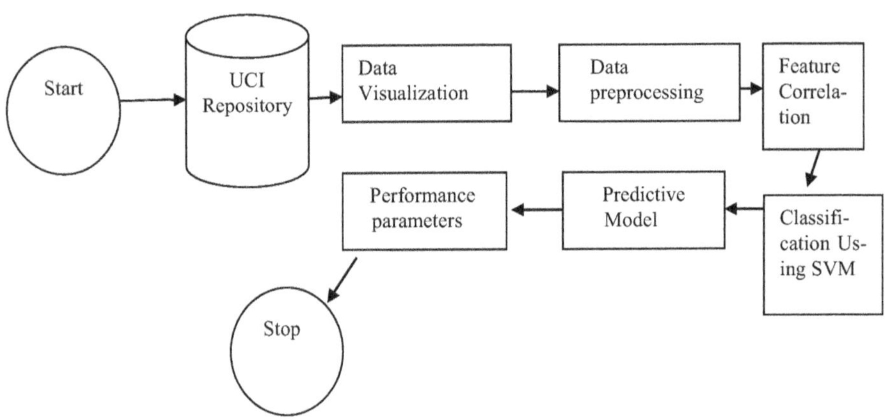

Fig. 1. Proposed flow diagram

2.1 Collection of Data

The Lung Cancer Dataset UCI repository provided the database for this study [26]. This data set contains around 610 occurrences with three multivariate target classes can be referred as pre-malignant (P, count-198), malignant (M, count-218), and benign (B, count-194) lung cancer tumours. Air Disease, Obesity, Blood Coughing, Fatigue, Smoking, Chest Pain, Blood Coughing, Fatigue, Chronic Lung Weight Loss, Shortness, Pollution, Alcohol Use, Dust Allergy, Occupational Hazards, Genetic Risk, Fingers, Frequent Cold, Dry Cough, Wheezing, Swallowing difficulty, Clubbing of and Snoring are among the various features listed in the considered dataset.

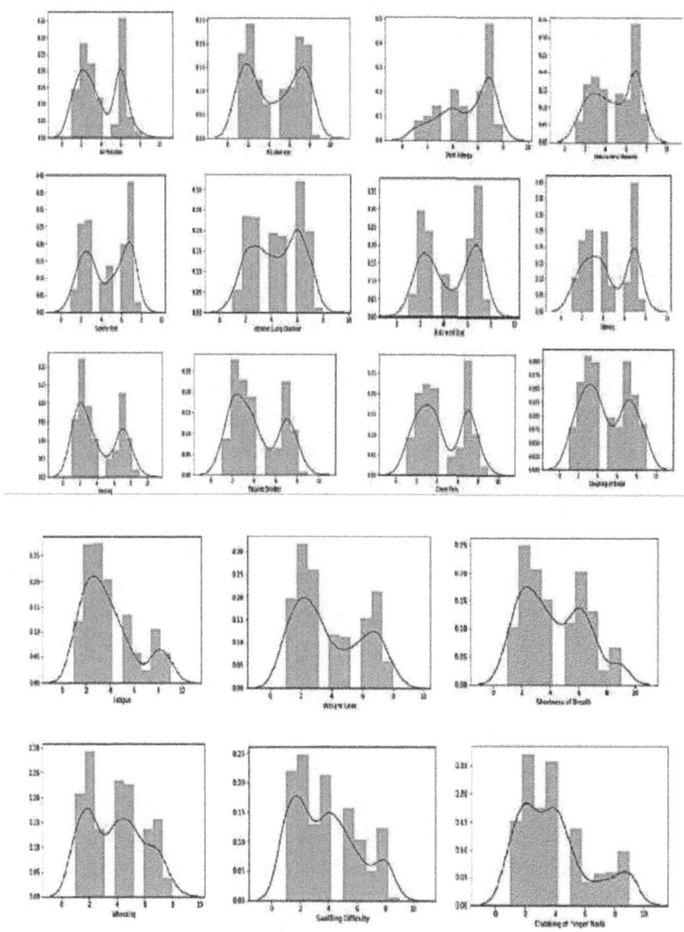

Fig. 2. Visualizing the data features

2.2 Data Visualization

The process of transforming data into abstract visuals that follow certain patterns or trends is known as visualization. It encourages analysts to consider a wide range of options at the application level. Python has a number of fantastic diagramming packages, including as pandas and Matplotlib that can generate various highlighted histograms, scatter plots and the curves density [28]. Figure 2 depicts the skewed features in the Table 2. These are the combination of both histogram sand the density cures.

2.3 Data Skewness

The asymmetry in statistics is represented by the skewness of the data. The curve will appear to be distorted to one side or the other [29]. The Comparison between resultant distribution and the normal distribution can be quantified to validate the discrepancy (Table 3).

- If the mean is > median, the data is "Positively skewed,"
- If the mean is < median, the data is "Negatively skewed."

2.4 Preprocessing of the Data

Data Pre-processing is a crucial prerequisite for any data analysis. It is normally a good idea to organize the data in such proper format, later this data is shown to the calculation done by the machine learning techniques. The use of data preparation techniques to improve the capabilities of classification systems is widely recognised [31]. This contains exercises such as allocating numerical attributes to targets, dealing with missing numerical and normalizing highlights [30].

To improve the effectiveness of the model we're considering, some minor details like age of the patient, id of the patient and gender are removed. There are now 3 targets, 21 characteristics, and 599 examples in the database. After lowering the dimensionality, Finally the csv file is generated to show the completeness of the preparation.

2.5 Dimensionality Reduction Using Principal Component Analysis (PCA)

The selection of features procedures are carried out by PCA in the context of reduced dimensionality. The unfocused characteristics can cause a significant drop in the prediction model's performance rate. Feature selection is the process of reducing the feature dimension for the dataset considered [38] by selecting the important features required for classification. These informative are used to break down the information in a straightforward manner for computations that do not require the processing of extraneous components. [37] Depicts PCA as a useful data visualization technique. PCA analyses the correlation of characteristics using linear algebra as a mathematical foundation [39]. The PCA feature space transformation is depicted in Fig. 3 that has two different target classes like Benign and Malignant and their samples are distributed across the axes. The early identification of Pre-malignant class that is shown alongside in the Fig. 5 can also benefit from this type of alteration. Some of the PCA's relevant marks.

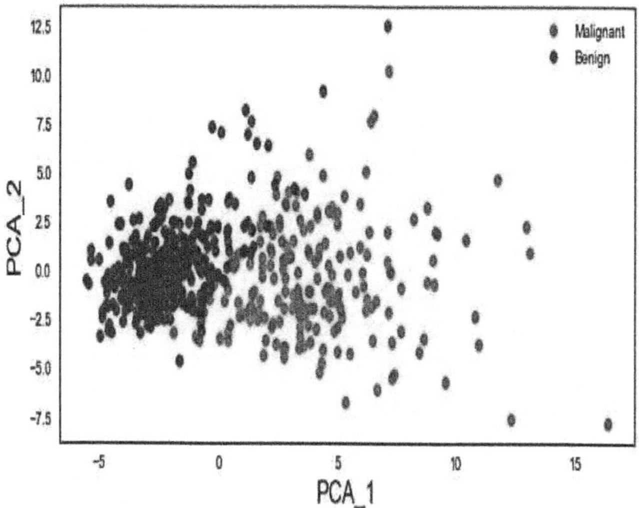

Fig. 3. Dimensionality reduction using PCA

In Fig. 3 describes the Dimensionality reduction using the various numbers of PCA components referred as PCA 1 and PCA 2 m that are mainly used to indicate the two different components to modify the space of features. Scree plot in Fig. 4 describes the importance of PCA, Scree plot is an analytical tool to determine the PCA function are working well or not based on the indicator[40].

Where in we can observe PCA 1 has the most diverse in natured when compared to PCA 2. In Fig. 4 at point 2 displaying an elbow curve, that indicated the production of PCA 1 and PCA 2 output required.

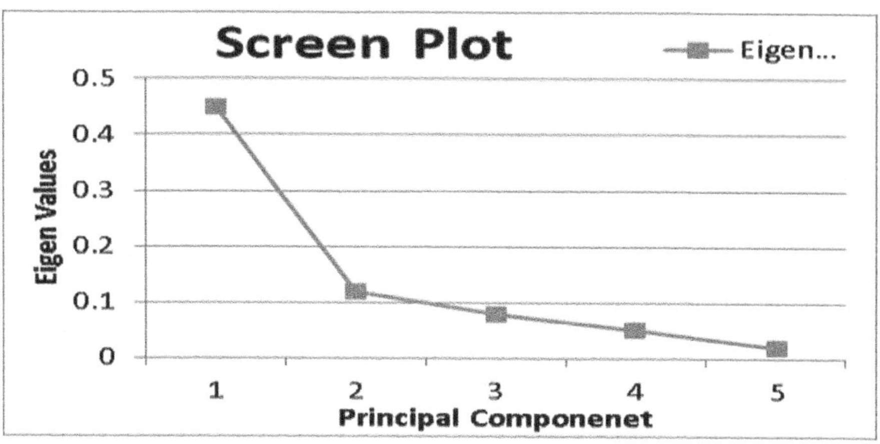

Fig. 4. Analytical Screen Plot

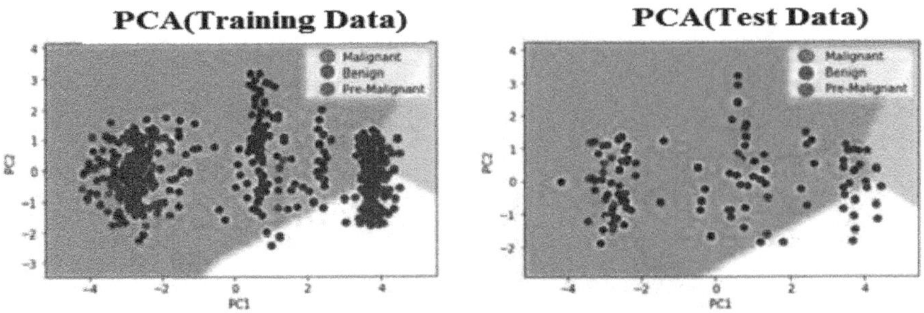

Fig. 5. The proposed PCA with 3 target class

2.6 Feature Correlation- for Selecting Features

Feature correlation is technique that describes how each feature is contributing towards the target class; during this we can get both negative and the positive correlation. In [42] the relevance of the features are summarized The values of Positive correlation imply that features are directly proportional to the class, whilst negative correlation values suggest that features are inversely proportional to the class. "Perfectly correlated" is a correlation value of one. In this feature correlation the correlation value of two different features of the malignant class referred as x and y. If such on one of those feature to be deleted.

2.7 Classification Using SVM

The logical grouping of data is known as classification. It can make decisions based on both unstructured and structured data. Most of the classification techniques has significant impact on health care systems. In [33] clearly explained how ANN is used over categorizing data based on neural network congestion with a fuzzy system to identify asthma. [35] Shows how signaling approaches and machine learning can be used to diagnose epilepsy. [6] Uses a categorization task to investigate the chances of Alzheimer's disease. [37] Investigates the use of data mining to combat malware and denial-of-service attacks. To maintain security, texts are trimmed and categorized in an automatic manner [40].

SVM Classifier is the most popular classification technique that has a beautiful way of processing nonlinear data. [32] Explains the SVM classification strategy in detail. SVM's essential tool, the hyper plane distributes the data points broadly. The hyper plane creates the decision boundary based on the support vectors that are closed to right and left hyper planes that employs linear SVC model.

In the health domain, SVM classifiers are widely used. [38] Used a GLCM (Gray level Co-occurrence Matrix) and an SVM (Support Vector Machine) to create 2-level DWT pictures. [40] Proposes a chip-based system with SVM flavor. The many SVM algorithms for accurately detecting breast cancer are explained.

3 Results and Discussion

The confusion matrix and classification report are used to acquire the results. The binary classification outputs are used to create a confusion matrix. This matrix can be used to determine various performance parameters like classification accuracy, true-positive-rate (TPR), error rate, false-positive rate (FTR), false negative rate (FNR) and true-negative rate (TNR). The main factor for identification systems is an efficient model with higher score. As a result of this the developer will strive to reduce the error rate. Equations 1 and 2. Represents the accuracy and error rate of the proposed model, respectively. 2 × 2 matrices are commonly used in cancer prediction. It is depicted in Fig. 6 as a 3 × 3 matrix, along with the report of the classification.

The various performance parameters like Accuracy, recall, precision and f1-Measure of this proposed classification method will be presented in the Classification Report. Table 4 lists the various parametric definitions considered in the work.

$$\text{Accuracy} = \frac{(TPR + TN)}{Total} = 0.89 \tag{1}$$

$$\text{Error Rate} = \frac{(FPR + FN)}{Total} = 0.25 \tag{2}$$

Table 4. Parametric definitions

S.No	Parameter	Meaning
1	True Positive (TP)	Correct positive predictions
2	False Positive (FP)	Incorrect positive predictions
3	True Negative (TN)	Correct negative predictions
4	False Negative (FN)	Incorrect negative predictions
5	Error Rate	Total number of incorrect predictions
6	Accuracy	Total number of correct predictions
7	Precision	The number of correct positive predictions divided by the total number of positive predictions
8	F1-score	Harmonic mean of precision and recall

```
confusion matrix
[[160   4   6]
 [  0 178  14]
 [  6  36 136]]
```

```
Confusion matrix plot of SVC
```

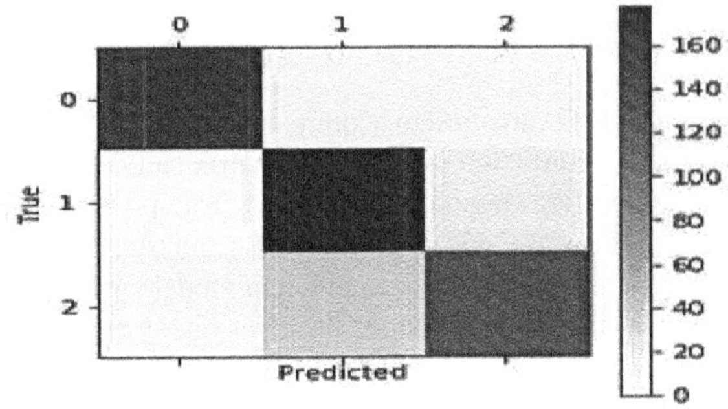


```
Classification report
             precision    recall  f1-score   support

          B       0.96      0.94      0.95       170
          M       0.82      0.93      0.87       192
          P       0.87      0.76      0.81       178

avg / total       0.88      0.88      0.88       540
```

Fig. 6. Overall performance of the proposed SVM model

4 Conclusion and Future Work

The importance of the pre-malignant stage in the early diagnosis of lung cancer is highlighted in this study. Both the innocent and real data is analyzed and cleaned. The proposed SVM-classifier's performance is seen as a clear indicator of high prediction accuracy and distinction. The information buried in the dataset might be properly labeled and identified using the confusion matrix. The propped technique performance is measured using the generated classification report using confusion matrix and the computations that go with it. In the 2*2 Matrix the error rate, accuracy of the proposed model is equal to 3*3 matrixes. This classification improves the performance parameters like f1-score, recall and the support are all listed in the Classification Report. The study paves the way for further improvement at the optimization level, ensuring higher performance over machine learning techniques.

References

1. Ferlay, J., Soerjomataram, I., Dikshit, R., Eser, S., Mathers, C., Rebelo, M.: Cancer incidence and mortality worldwide: sources, methods and major patterns in GLOBOCAN 2021. Int. J. Cancer **136**(5), 359–386 (2021)
2. Gadgeel, S.M., Ramalingam, S.S., Kalekerian, G.P.: Treatment of lung Cancer. adiol Clin. North Am., 961–74 (2020)

3. Ignatious, S., Joseph, R.: Computer aided lung cancer detection system. In: Proceedings of 2018 Global Conference on Communication Technologies (GCCT 2018), 7342723
4. American Cancer Society. https://www.cancer.org/cancer/types/lung-cancer/causes-risks-prevention/what-causes.html
5. Wang, W., Wu, S.: A study on lung cancer detection by image processing. In: International Conference on Communications, Circuits and Systems, Guilin, pp. 371–374 (2020)
6. Sarkar, A., Sadhu, S.B., Thakur, S., Sengupta, A.M., Chatterjee, J.: Multimodal characterization of radiologically detectable lung lesions. In: 38th Annual International Conference of the IEEE Engineering in Medicine and Biology Society (EMBC), Orlando, FL, pp. 1422–1425 (2020)
7. Greenberg, A.K., Yee, H., Rom, W.N.: Preneoplastic lesions of the lung. Resp. Res. **3**(1) (2002)
8. Klebe, S., Henderson, D.W.: Facts and fiction: premalignant lesions of lung tissues. Pathology **45**(3), 305–315m (2013)
9. Hirayama, K., et al.: Extraction of GGO regions from chest CT images using deep learning. In: 17th International Conference on Control, Automation and Systems (ICCAS), Jeju, 2017, pp. 351–355 (2017)
10. Yu, K.-H., et al.: Predicting non-small cell lung cancer prognosis by fully automated microscopic pathology image features. Nat. Commun. (2016)
11. El-Baz, A., et al.: Computer-aided diagnosis systems for lung cancer: challenges and methodologies. Int. J. Biomed. Imag., 1–46 (2013)
12. Bhavanishankar, K., Sudhamani, M.V.: Techniques for detection of solitary pulmonary nodules in human lung and their classifications -a survey. Int. J. Cybern. Inf. (IJCI) **4** (2015)
13. Shandilya, S., Chandankhede, C.: Survey on recent cancer classification systems for cancer diagnosis. In: International Conference on Wireless Communications, Signal Processing and Networking (WiSPNET), Chennai, pp. 2590–2594 (2017)
14. Sharma, M., Singh, R., Bhattacharya, M.: Classification of breast tumors as benign and malignant using textural feature descriptor. In: IEEE International Conference on Bioinformatics and Biomedicine (BIBM), Kansas City, MO, pp. 1110–1113 (2017)
15. Hosseinzadeh, F., KayvanJoo, A., Ebrahimi, M., Goliaei, B.: Prediction of lung tumor types based on protein attributes by machine learning algorithms. SpringerPlus, vol. 2, no. 1, p. 238 (2013)
16. Cengil, E., Cinar, A.: A deep learning based approach to lung cancer identification. In: International Conference on Artificial Intelligence and Data Processing (IDAP) (2018)
17. Rahane, W., Dalvi, H., Magar, Y., Kalane, A., Jondhale, S.: Lung cancer detection using image processing and machine learning healthcare. In: International Conference on Current Trends Towards Converging Technologies (ICCTCT), Coimbatore, pp. 1–5 (2018)
18. Wu, Q., Zhao, W.: Small-cell lung cancer detection using a supervised machine learning algorithm. In: International Symposium on Computer Science and Intelligent Controls(ISCSIC), Budapest, pp. 88–91 (2017)
19. Günaydin, Ö., Günay, M., Şengel, Ö.: Comparison of lung cancer detection algorithms. In: Scientific Meeting on Electrical-Electronics & Biomedical Engineering and Computer Science (EBBT), Istanbul, Turkey, pp. 1–4 (2019)
20. Salaken, S.M., Khosravi, A., Khatami, A., Nahavandi, S., Hosen, M.A.: Lung cancer classification using deep learned features on low population dataset. In: IEEE 30th Canadian Conference on Electrical and Computer Engineering (CCECE), Windsor, ON, pp 1–5 (2017)
21. Al-Absi, H.R.H., Samir, B.B., Shaban, K.B., Sulaiman, S.: Computer aided diagnosis system based on machine learning techniques for lung cancer. In: International Conference on Computer & Information Science (ICCIS), Kuala Lumpeu, pp. 295–300 (2017)
22. Shimizu, R., et al.: Deep learning application trial to lung cancer diagnosis for medical sensor systems. In: International SoC Design Conference (ISOCC), Jeju, pp. 191–192 (2016)

23. Rao, P., Pereira, N.A., Srinivasan, R.: Convolutional neural networks for lung cancer screening in computed tomography (CT) scans. In: 2nd International Conference on Contemporary Computing and Informatics (IC3I), Noida, pp. 489–493 (2016)
24. Srivastava,S., Rathi, M., Gupta, J.P.: ACC 2011, Part I, CCIS, vol. 190, pp. 260–269 (2011)
25. Mahale, A., et al.: A survey on lung cancer detection using image data analysis and machine learning. Int. J. Innov. Res. Comput. Commun. Eng. (2018)
26. UCI Lung Cancer Dataset. https://archive.ics.uci.edu/ml/datasets/Lung+Cancer
27. Stages and Survival Rate of Lung Cancer. https://www.epainassist.com/chest-pain/lungs/stages-and-survival-rate-of-lung-cancer
28. Introduction to Data Visualization in Python. https://towardsdatascience.com/introduction-to-data-visualization-in-python-89a54c97fbed
29. Machine LearningBasics: Data Skewness and handling. https://medium.com/mlrecipies/machine-learning-basics-data-skewness-and-handling-778ec7adb18c
30. What Steps should one take while doing Data Preprocessing?. https://hackernoon.com/what-steps-should-one-take-while-ng-data-preprocessing-502c993e1caa
31. Deshmukh, D.H., Ghorpade, T., Padiya, P.: Improving classification using preprocessing and machine learning algorithms on NSL-KDD dataset. In: International Conference on Communication, Information & Computing Technology (ICCICT), Mumbai, pp. 1–6 (2015)
32. Ghosh, A., Barman, S.: Prediction of prostate cancer cells based on principal component analysis technique. Procedia Technol. **10**, 37–44 (2013)
33. Ghosh, S., Mondal, S., Ghosh, B.: A comparative study of breast cancer detection based on SVM and MLP BPN classifier. In: 2014 First International Conference on Automation, Control, Energy and Systems (ACES), Hooghy, pp. 1–4 (2014)
34. Jhajharia, S., Varshney, H.K., Verma, S., Kumar, R.: A neural network based breast cancer prognosis model with PCA processed features. In: International Conference on Advances in Computing, Communications and Informatics (ICACCI), Jaipur, pp. 1896–1901 (2016)
35. Yadav, A., Jamir, I., Rajeshwari, R., Sohani, J.M.: Breast cancer prediction using SVM with PCA feature selection method. Int. J. Sci. Res. Comput. Sci., Eng. Inf. Technol. (2020)
36. Ilakkiya, G., Jayanthi, B.: Liver cancer classification using principal component analysis and Fuzzy neural network. Int. J. Eng. Res. Technol. (IJERT) **2**(10) (2013)
37. Song, F., Guo, Z., Mei, D.: Feature selection using principal component analysis. In: International Conference on System Science, Engineering Design and Manufacturing Informatization, Yichang, pp. 27–30 (2010)
38. Partridge, M., Jabri, M.: Robust principal component analysis, neural networks for signal processing X. In: Proceedings of the IEEE Signal Processing Society Workshop (Cat. No.00TH8501), Sydney, NSW, Australia, pp. 289–298 (2000)
39. Sehgal, S., Singh, H., Agarwal, M., Bhasker, V., Shantanu: Data analysis using principal component analysis. In: International Conference on Medical Imaging, m-Health and Emerging Communication Systems (MedCom), Greater Noida, pp. 45–48 (2014)
40. How to read PCA biplots and scree plots. https://blog.bioturing.com/2018/06/18/how-to-read-pca-biplots-and-scree-plots/

Optimized Neural Network for Prediction of Neurological Disorders

Swagat Karve, Siddheshwar Gangonda, Ganesh Birajadar(✉), Vaibhav Godase, Rahul Ghodake, and Altaf O. Mulani

SKN Sinhgad College of Engineering, Pandharpur, MH, India
gbirajadar@gmail.com

Abstract. There are an alarming number of people on the planet who are suffering from neurological problems. The EEG signal is a chaotic time series signal that changes fast depending on the patient's condition. The nature of the signals changes dramatically from normal to severe situations, with changes in amplitude and frequency. Early detection and prediction of these signals is a very difficult task. The target of this research is to forecast individual state signals. The work is carried out by using the Generalized Regression Neural Networks (GRNN) variation of the Radial basis Function Neural Network (RBFNN), but it requires careful selection of the spread factor. Selecting an exact spread factor is not an easy task, it necessitates time-consuming and painstaking experiments. The spread factor for GRNN is determined by searching the particles in the swarm. Implémentation of Particle Swarm optimization (PSO) in conjunction with GRNN dramatically improved GRNN's prediction accuracy for a variety of neurological illnesses.

Keywords: Electroencéphalogramme (EEG) · Particle Swarm Optimisation (PSO) · search space · radial basis function neural network · generalised régression neural network

1 Introduction

The Neurological illnesses are one of the most frequent nervous system ailments, affecting patients of different ages, races, ethnicities. There may be no observable reason for neurological problems in many cases. There are a variety of neurological diseases syndromes that are identified by a specific set of symptoms. [7]. Neurological illnesses are one of the most frequent nervous system ailments, affecting people of all ages, races, and ethnicities. There may be no observable reason for neurological problems in many cases. There are a variety of neurological diseases syndromes that are identified by a specific set of symptoms.Seizures, convulsions, and neurological diseases are all names used to describe generalised tonic-clonic seizures. Focal seizures happen once electrical activity is prohibited to a special location of the brain. Seizures can occasionally progress to generalised seizures, which involve the entire brain. Secondary generalisation is the term for this. There are three types of partial seizures:

S. Singh et al. (Eds.): AIDSP 2023, CCIS 2440, pp. 183–191, 2025.
https://doi.org/10.1007/978-3-031-88762-8_18

Seizures that do not disrupt awareness or memory are called simple seizures. Complex seizures alter awareness and recall of events that occurred before, during, and soon after the seizure, as well as behaviour. Sleep in slow waves—This is a unique opportunity. Syndrome of uncommon neurological illnesses [9]. When it does occur, it usually occurs in the middle of a child's life. It's also known as a'slow-sleep continuous spike-wave. There is no known aetiology for this syndrome. It generally occurs in children who have already been diagnosed with a neurological condition.

Brain death is entier and irréversible loss of cérébral function. Brain death is utilised as a requirement for légal death in several juridictions, but it is explained differently in each. The word "brain death" is being applied to a variety of scenarios in which some regions of the brain continue to function while others die. Although a prominent medical lexicon states that "brain death" and "cerebral death" are synonymous (death of the cerebrum) [11, 12]. This research aims to forecast future samples of epileptic patients in various phases of brain functionality.

A. **Generalized Régression Networks**

Generalized Régression Networks (GRNs) are versions of radial basis function networks [18, 19].Standard regression is its base, which is known by kernel regression. They're used to approximate functions or estimate continuous variables. Feed forward neural networks use iterative training and learning techniques, although they don't. The structure of network is similar to a Radial basis function (RBF) network, with exception of the layer number 2. Layer 2 is somewhat differed. In layer 2, dot product is calculated in between inputs and weights, which is normalised later. Generalised regression network including specific second layer is depicted in Fig. 1 below.

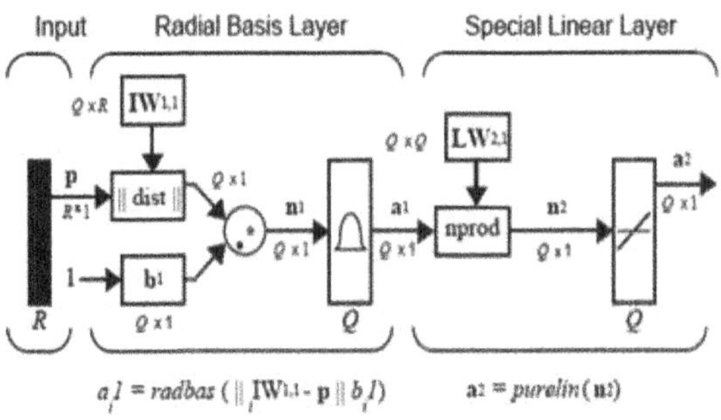

Fig. 1. Generalized Regression Network

With calculating transpose of weights of layer (P'), the bias b designed to column array 0.8326/SPREAD, also distance between the input and the weight of neuron set to half, the first layer creates same neurons set equal to set of inputs represented as P. The spécial or layer 2 computes dot product of weights of layer with input array (Stage 1 outcome) and normalizes the inputs by the sum of the components (as opposed to RBF). Weights of layer are fixed with target T, despite fact that this layer has the same amount of neurons as input/target/vectors. [20].

In GRNN, spread factor is particularly significant. The input vector encompasses a huge area when the spread is large, and neurons in first layer respond with considerable outcomes. As a result, for smaller value of spread, curve of RBF shows steepness. The spread factor is particularly significant in GRNN. The input vector is encompassed by a huge area when the spread is large, and layer 1 neurons respond with considerable outputs. As a result, when the spread is small, the radial basis function is quite steep. As a result, the slope smooths down as the spread expands, allowing most neurons to respond to an input vector. With design input vectors that are closer to the new input vector, the network employ average of weights of targetted vectors. More neurons will respond as the spread expands, allowing the network to function more smoothly. [13, 14, 15, 16].

One, two, or three parameters control the accuracy of the aforementioned networks:

1. Hidden units count or rbf,
2. The concealed unit's centres

The third factor is spread.

In first step of Learning, the RBF centers in the hidden layer are picked at random from some input dataset. A stochastic gradient approach is applied to identify the weights between the outcome and hidden layers. Most difficult part of this procedure is figuring out how many centers are needed to adequately cover the input vector space.

B. **Generalized Regression Network using PSO to SPREAD factor**

When tested on real-world applications and benchmark datasets, the Generalized Regression Network was utilised to discover appropriate value for the spread factor, resulting in accurate results [25, 26]. Finding SPREAD is a time-consuming and challenging task that, in some cases, can be infuriating. Typically, an SPREAD constant is selected from the range [0 1]. However, a two-digit precision can be estimated using the trial and error method. A balance between time complexity and acceptable mean squared error is required to experimentally identify an optimal SPREAD and the permitted mean squared error. PSO finds an SPREAD value by automatically altering the search space and converges early for the required mean squared error. Precision is no longer a problem for the PSO-optimized network. It can exhaust all double precision data types to get the SPREAD factor.

2 Methodology

Process of optimising GRNN is as explained below, and the EEG datasets utilised here are organised in following way:

Ptrain – array of inputs used to train the layer

Ttrain –Target array utilized as training array

Ptest – test array

Ttest – Target array utilized as test array

For each dataset, signals are stored as separate.mat extension. At the start of the programme file, the same is loaded.

1. Particle count was between 20 and 30. In the vast majority of cases, 20 was plenty.

2. Spread factor only matters in the GRNN network. The location of the spread-related particles is determined at random.
3. Next, the GRNN network is trained for 10–20 iterations using particle location value as spread, regardless of the goal of low MSE.
4. Outcome that corresponds to the best particle in the network with the greatest spread is kept.
5. The identical network, but with PSO's best spread factor, is now used.
6. The input vector and target vector are then used to train the network.
7. The MSE for the training samples is then determined.
8. The trained network has to undergo through test using different test samples. The MSE is then calculated by correlating the acquired data to the real target.
9. Error is plotted graphically and actual response of trainning and testing data is also plotted.
10. Mean Square Error is calculated for both datasets.

3 Database

At the CIIMS in Nagpur, MH, IN, Particular epileptic person with different mental condition is tested to collect dataset. The following situations were taken into consideration:

1. Normal
2. Generalized neurological disorder
3. Focal neurological disorder
4. Slow wave
5. Brain Death

EEG data collected specified over 16 EEG channels at a sampling frequency of 256 Hz. In exception of brain death case, signal was stored at rate of three thousand seconds with comprising aprroximate four lakh sample points, a total of two thousand five hundred sample points are stored across uni channel with average of ten seconds. The current anticipated value was determined using five previous sample points explained below:

$$w(t) = w(t-1)w(t-2)w(t-3)w(t-4)w(t-5) \tag{1}$$

Aforementioned numbers are calculated considering training and testing samples for a dataset ranging from sample 6 to 2000. As a result, the total number of samples collected was 1995. Fifteen hundred sample points got utilised in training, while the exceeding samples got utilised in testing. After that, processed array was normalised using highest number as divider. For training and testing, the data arrays were 5 by 1500 and 5 by 495. The figure below show the proposed outcome compared to the outcome of PSO-GRNN, as well as an error graph of PSO-GRNN in identification of various patient situations, including normal state, generalised neuron impairement, focal nervous abnormality, slow wave nervous abnormality, brain death.

4 Results

The result illustrates a PSO dépendent Generalized Regression neural network upon several nervous system based diseases states. The computed accuracy along with Mean Square Error of different states of nervous diseases has shown in tabular way. Suggested algorithm achieved an MSE of 0.003786 and an accuracy of 99.62 (Figs. 2, 3, 4, 5, 6, 7, 8, 9, 10 and 11).

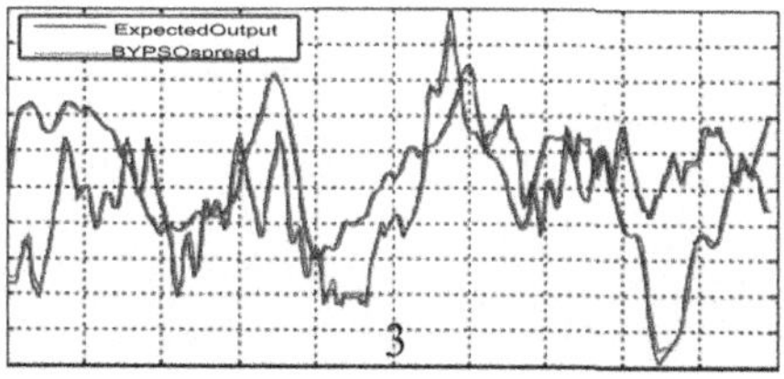

Fig. 2. Proposed Vs Stadard for person without disorder

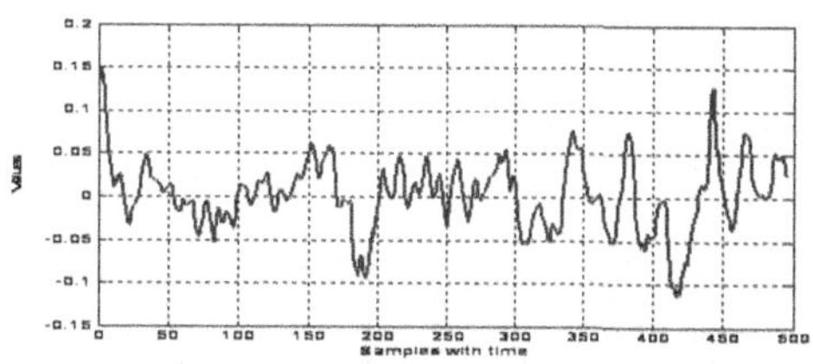

Fig. 3. Graphical presentation of Error: Standard with person without disorder

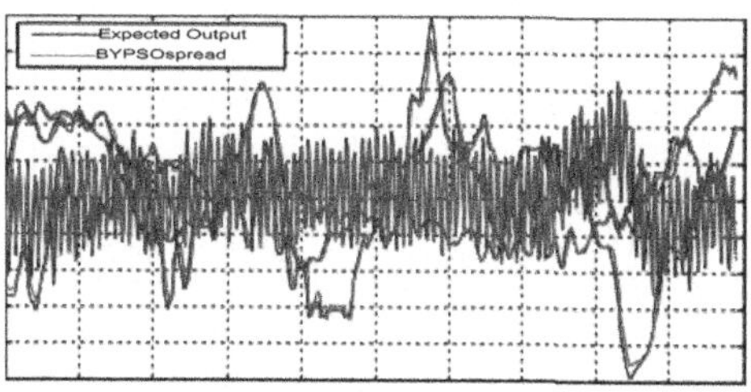

Fig. 4. Proposed Vs Stadard for person with disorder

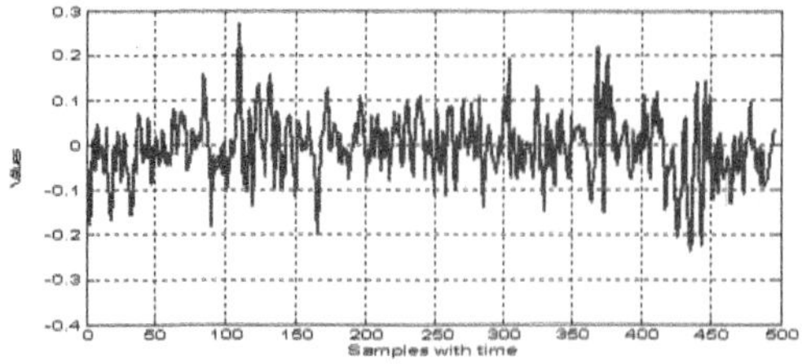

Fig. 5. Graphical presentation of Error: Standard with person with disorder

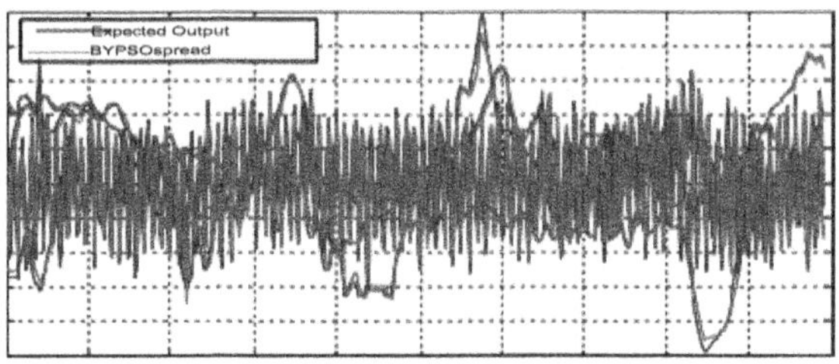

Fig. 6. Proposed Vs Stadard for person with focal neurological disorder

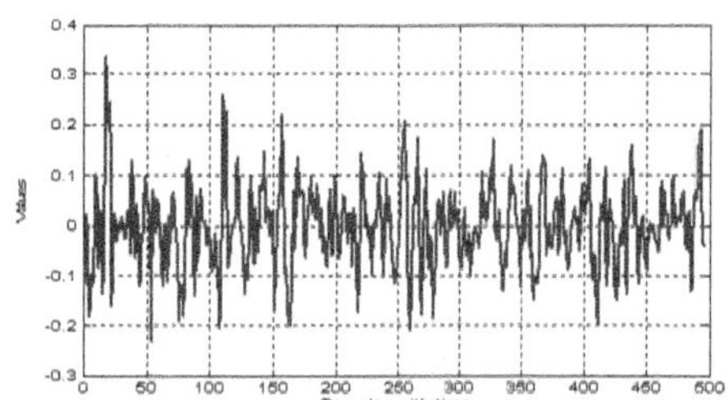

Fig. 7. Graphical presentation of Error: Standard with person with focal neurological disorder

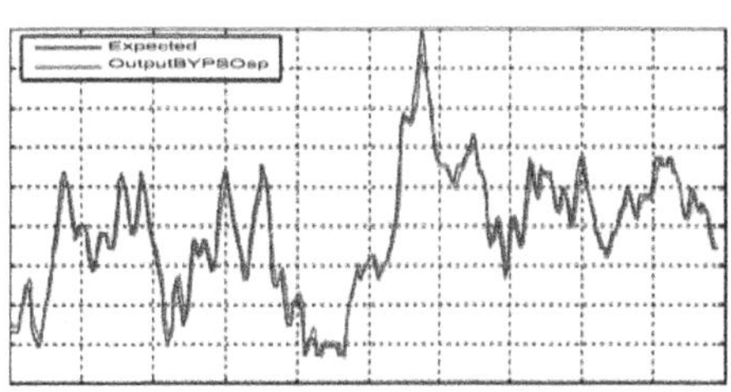

Fig. 8. Observed outcome with outcome obtained with PSO-GRNN showing slow wave disorder

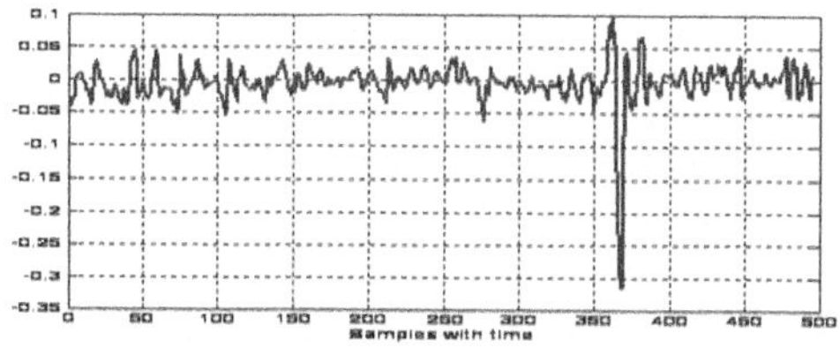

Fig. 9. Graphical presentation of Error: PSO-GRNN showing slow wave disorder

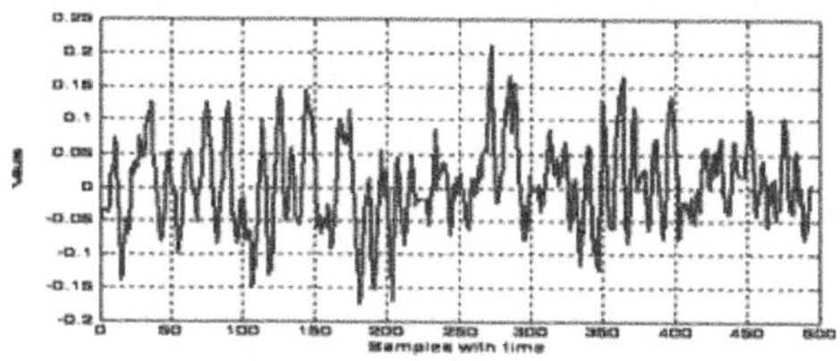

Fig. 10. Comparision of obtained outcome with PSO-G outcome

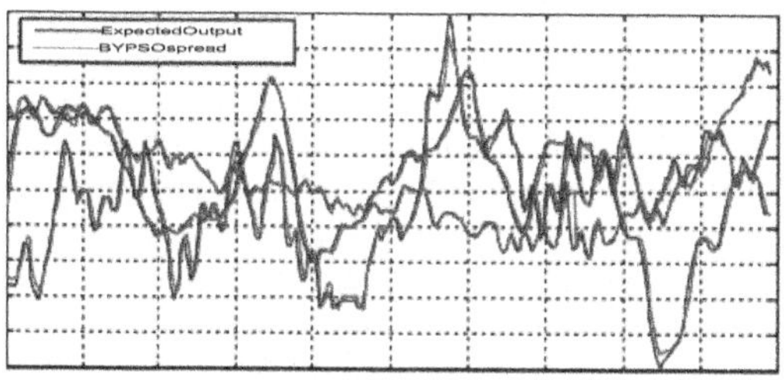

Fig. 11. Graphical presentation of Error with PSO-GRNN for Brain Death case involving prediction

Table 1. MSE, PSO related spread factor and accuracy cosideration of different nervous abnormalities.

Sr.	Condition	Mean Square Error	SF by PSO	Accuracy
1	Without Disorder	0.0016367	0.077536	99.8363
2	General Neurological disorder	0.0051768	0.096843	99.4823
3	Focal Neurological disorder	0.006942	0.068344	99.3058
4	Slow Wave	0.0010115	0.012214	99.8989
5	Brain Death	0.0041671	0.10883	99.5833

5 Conclusions

By tuning the parameters, such as MSE and spread factor, the suggested RBF + PSO based prediction technique demonstrates high prospective outcomes. In worst-case scenario, PSO can provide an estimate of network dimensions and epochs for preceding iteration. The network can converge quickly and effectively if the minimal and maximal restrictions for particle locations and speeds are suitably defined. When trained and testing samples are previously standardised, with a familiar parameter, PSO-based NN performs better (Maxima point in set). RBN implementations demonstrated way of obtaining similar accurate spread factor value. In small iterations, only PSO can find the best one (see Table 1).

References

1. Srinivasan, V., Eswaran, C., Sriraam, N.: Approximate entropy-based epileptic EEG detection using artificial neural networks. IEEE Trans. Inf. Technol. Biomed. **11**(3), 288–295 (2007)
2. Ebersole, J.S., Pedley, T.A.: Current Practice of Clinical Electroencephalography. third edition, Lippincott Williams and Wilkins (2003)
3. Alotaiby, T.N., Alshebeili, S.A., Alshawi, T., Ahmad, I., Fathi, E.: EEG seizure detection and prediction algorithms: a survey. EURASIP, J. Adv. Sig. Process., 1–21 (2014)
4. Samiee, K., Kovacs, P., Gabbouj, M.: Epileptic Seizure classification of EEG time-series using rational discrete short-time fourier transform. IEEE Trans. Biomed. Eng. **62**(2), 541–552 (2014)
5. Andrzejak, R.G., Lehnertz, K., Mormann, F., Rieke, C., David, P., Elger, C.E.: Indications of nonlinear deterministic and finite-dimensional structures in time series of brain electrical activity: Dependence on recording region and brain state. PHYS. Rev. E **6**, 1–8 (2001)
6. Lesca, G., et al.: Epileptic encephalopathies of the Landau-Kleffner and continuous spike and waves during slow-wave sleep types: genomic dissection makes the link with autism, vol. 9, pp. 1526–1538 (2012)
7. Ahammad, N., Fathima, T., Joseph, P.: Detection of epileptic seizure event and onset using EEG. BioMed. Res. Int., Hindavi **2014**, 1–7 (2014)
8. Qu, H., Gotham, J.: A patient-specific algorithm for the detection of seizure onset in long-term EEG monitoring: possible use as a warning device. IEEE Trans. Biomed. Eng. **44**(2), 115–122 (1997)
9. Nasehi, S., Pourghassem, H.: A novel epileptic seizure detection algorithm based on analysis of EEG and ECG signals using probabilistic neural network. Aust. J. Basic Appl. Sci. **5**(12), 308–315 (2011)
10. Tzallas, A.T., Tsipouras, M.G., Fotiadis, D.I.: Automatic seizure detection based on time-frequency analysis and artificial neural networks. Comput. Intell. Neurosci., NCBI (2007)
11. McGrogan, N.: Neural network detection of epileptic seizures in the electroencephalogram. [Ph.D. thesis] Oxford University, Oxford (1999)
12. Mgbe, C.O., Mom, J.M., Igwue, G.A.: Performance evaluation of generalized regression neural network path loss prediction model in macrocellular environment. J. Multi. Eng. Sci. Technol. **2**(2), 204–208 (2015)
13. Tomandl, D., Schober, A.: A modified GRNN with new efficient training algorithms as a Black box – tool for data analysis. Elsevier, Neural Netw. **14**, 1023–1034 (2001)
14. Specht, D.F.: A general regression neural network. IEEE Trans. Neural Netw. **2**(6), 568–576 (1991)

15. Hannan, S.A., Manza, R.R., Ramteke, R.J.: Generalized regression neural network and radial basis function for heart disease diagnosis. Int. J. Comput. Appl. **7**(13), 7–13 (2010)
16. Azuaje, F., Dubitzky, W., Lopes, P., Black, N., Adamsom, K.: Predicting coronary disease risk based on short-term RR interval measurements: a neural network approach. Artif. Intell. Med. **15**, 275–297 (1999)
17. David, P.A.: A radial basis neural network for the analysis of transportation data. thesis submitted at University of South Florida Scholar Commons (2004)
18. Martin, T.H., Howard, B.D., Mark, B.: Neural network design. Cengage Learn. (2008)
19. Shivanandan, S.N., Sumathi, S., Deepa, S.N.: Introduction to Neural Networks using Matlab 6.0. Tata McGraw Hill Education Private Ltd. (2010)
20. Eberhart, R., Kennedy, J.: A new optimizer using particle swarm theory. In: Sixth International Symposium on Micro Machine and Human Science, IEEE, pp. 39–43 (1995)
21. Kalyankar, P.A., Mulani, A.O., Thigale, S.P., Chavhan, P.G., Jadhav, M.M.: Scalable face image retrieval using AESC technique. J. Algebraic Statist. **13**(3), 173–176 (2022)
22. Mulani, A.O., Mane, P.B.: Fast and Efficient VLSI implementation of DWT for image compression. Turk. J. Comput. Math. Educ. (TURCOMAT) **12**(13) (2021)
23. Mulani, A.O., Mane, P.: Secure and area efficient implementation of digital image watermarking on reconfigurable platform. Int. J. Innov. Technol. Explor. Eng. (IJITEE) **8**(2), 1 (2018)
24. Birajadar, G., Bhyri, C.: Comprehensive survey on EEG analysis for detecting brain disorders. Mukt Shabd J. IX (VI), 2258–2262 (2020)
25. Birajadar, G., Bhyri, C.: Epilepsy Identification using EEG signal monitoring. Turk. J. Comput. Math. Educ. (TURCOMAT) **12**(2) (2021)
26. Kennedy, I., Eberhart, R.: Particle swarm optimization. In: International Conference on Neural Networks, Perth, Australia, IEEE, vol. 4, pp. 1942–1948 (1995)
27. Shi, Y., Eberhart, R.: Parameter selection in particle swarm optimization. In: Proceedings of the Seventh Annual Conference on Evolutionary Programming, pp. 591–601 (1998)
28. Eberhart, R.C., Shi, Y.: Comparing inertia weights and constriction factors in particle swarm optimization. In: Proceedings of the 2000 Congress on Evolutionary Computation, vol. 1, pp. 84–88 (2000)
29. Clerc, M., Kennedy, J.: The particle swarm - explosion, stability, and convergence in a multidimensional complex space. IEEE Trans. Evol. Comput. **6**, 58–73 (2002)
30. Kiranyaz, S., Pulkkinen, J., Yildirim, A., Gabbouj, M.: Multi- dimensional particle swarm optimization in dynamic environments. Expert Syst. Appl. **38**(3), 2212–2223 (2011)
31. Kshirsagar, P., Akojwar, S.: Hybrid heuristic optimization for benchmark datasets. Int. J. Comput. Appl. **146**(7), 0975–8887 (2016)
32. Kshirsagar, P., Akojwar, S.: Novel approach for classification and prediction of non linear chaotic databases. In: International Conference on Electrical, Electronics, and Optimization Techniques (2016). References should be listed at the end of the paperusing the following Suranaree Journal of Science and Technology patterns
33. Kondekar, R.P., Mulani, A.O.: Raspberry pi based voice operated robot. Int. J. Recent Eng. Res. Dev. **2**(12), 69–76 (2017)
34. Shinde, M.R.S., Mulani, A.O.: Analysisof biomedical image using wavelet transform. Int. J. Innov. Eng. Res. Technol. **2**(7), 1–7 (2015)
35. Mane, P.B., Mulani, A.O.: High speed area efficient FPGA implementation of AES algorithm. Int. J. Reconfigurable Embedded Syst. (IJRES) **7**(3), 157–165 (2018)

A Comparative Study of Various Techniques for Vehicle License Plate Detection and Recognition

G. Devasena and S. Ramalakshmi(✉)

Department of Computer Science and Engineering, Indian Institute of Information Technology, Tiruchirappalli, India
{devasenag,sramalakshmi}@iiitt.ac.in

Abstract. Technologies and services aimed at autonomous vehicles and Intelligent Transportation Systems (ITS) are transforming different aspects of everyday life. Automatic Vehicle License Plate Detection (LPD) and License Plate Recognition (LPR) systems for vehicles have various potential uses, discovering a stolen vehicle, monitoring the flow of traffic, automatic car parking systems, surveillance solicitations, toll systems, etc. This paper presents a comprehensive performance comparison of many Automatic Number Plate Recognition (ANPR) algorithms, including that incorporating computer vision, as well as a comprehensive analysis of current methodologies and advancements in ANPR. With the improvement of sophisticated machine learning algorithms, numerous techniques have been developed by researchers in the past decades for efficient LPD and LPR. Even with the greatest algorithms, ANPR system implementation may require multiple methods for optimum accuracy. However, the detection of License Plate (LP) is a more challenging task concerning the nature of the number plate, camera quality, non-standard formats, scene complexity, contrast problems, indoor/outdoor or day/night images, distortion tolerance, camera mount position, software tools may affect its effectiveness. This paper not only describes the numerous strategies used for LPD and LPR but also compares their performance and gives suggestions to help researchers choose the optimal technique for their work.

Keywords: Intelligent Transportation Systems · License Plate Detection · Automatic Number Plate Recognition · License Plate Recognition · License Plate · Machine Learning

LIST OF ABBREVIATIONS

LPD	License Plate Detection
LPR	License Plate Recognition
OCR	Optical Character Recognition
LP	License Plate
VIN	Vehicle Identification Number
ITS	Intelligent Transportation Systems
HED	Holistically Nested Edge Detection

S. Singh et al. (Eds.): AIDSP 2023, CCIS 2440, pp. 192–205, 2025.
https://doi.org/10.1007/978-3-031-88762-8_19

CNN Convolutional Neural networks
HOG Histogram of Oriented Gradients
SSD Single-Shot Detector
VGG Visual Geometry Group
LoG Laplacian of Gaussian
YOLO You Only Look Once
ANPR Automatic Number Plate Recognition

1 Introduction

Land vehicles are the preferred means of traveling to common people and transporting goods from one place to another. The purpose of using vehicles increases day by day. Hence, there is an urgent need to track, analyze and monitor the vehicles with the help of ITS [5,11] for electronic toll collection, highway surveillance, urban logistics, traffic management, etc. This system utilizes digital image processing methodologies to recognize vehicles in the images that are captured through high-end cameras fitted on the traffic signals and toll booths. In general, a vehicle can be identified by its Vehicle Identification Number (VIN) or by its LP number [16]. Each vehicle has a unique identification number called the VIN number or chassis number, which remains constant throughout the lifetime of the vehicle.

Figure 1 shows the vehicle identification system's block diagram. Any system that recognizes LP initially acquires vehicle images from CCTV cameras. After capturing the vehicle image, it should be subjected to pre-processing techniques [15,16] to decrease the unwanted noise present in the image.

The next step is to find the position of the LP in the image. The rules and regulations for the LPs vary from one country to another. LPs are typically rectangular in shape and some characteristics like color, dimension, font styles, and character size are common [3,22]. Based on these characteristics, the locality of the LP is identified and segmented. The cropped LP is subjected to morphological operations to increase the quality of the characters present in the image. In [15], Lubna developed a system capable of detecting LPs by retrieving all of its four edges.

The final step of the system is the character segmentation and recognition that rely on the accuracy of LP extraction [6,15]. Owing to zoom and camera distance, the segmented characters might vary in scale and thickness. There is a possibility that the characters will be destroyed, skewed, or distorted due to noise or other factors. After all the characters are segmented, they are recognized by using template matching. If any post-processing is required, then it can also be embedded in the basic architecture of the system. This work addresses real-time, flexible, and distinctive algorithms for LPD and LPR under varied weather and environmental circumstances, plate formats and languages, low plate clarities, high vehicle speeds, and numerous traffic situations in real-world applications.

The organization of the paper is as follows. Section 2 provides a summary of the LPD techniques, Sect. 3 summarizes the Character Recognition techniques and Sect. 4 describes a summary of LPR Techniques. Finally, the conclusion is given in the last section.

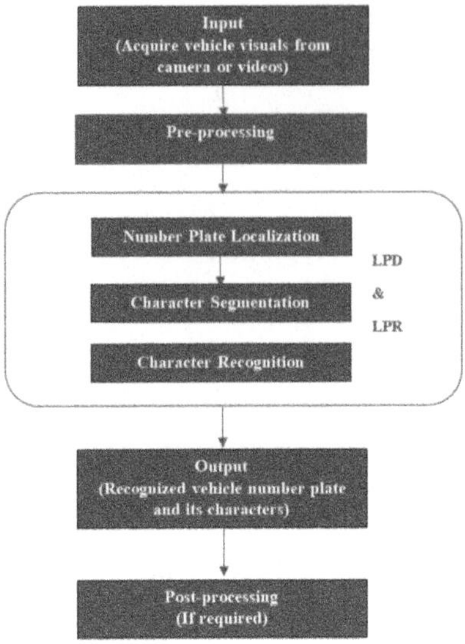

Fig. 1. LPD and LPR block diagram

2 LP Detection Techniques

Detecting the place of the vehicle in the image is a successful way in the LPD pipeline for detecting LPs, which ultimately reduces the search area and the amount of false positives [16]. It is crucial to detect the LP without finding the edges. In general, edges determine the boundaries between regions in an image. Several methods have been proposed for detecting the exact edges, including the use of filters, enhancing techniques, edge detection, neural networking algorithms, etc. [7,15]. The summarization of various LPD techniques is presented in Fig. 2 and reviewed in the next section, along with their benefits and limitations.

2.1 LP Retrieval Using Edge Detection

The LP is always a rectangle and has a set ratio of width to height. It all starts with resizing the images to a fixed ratio. Using a predetermined dataset, the authors evaluated and contrasted the results of a wide variety of bounding box identification methods [15,17], 8.

As evident from the research works [19,20, 8], the Sobel filter, a simple method, was implemented to successfully extract edges. The gradient approximation of the image is calculated with this function. As an outcome of the Sobel operator, each pixel of an image has either the normal to a vector or the gradient vector. After image acquisition, the image is applied with a Sobel filter to detect edges in the image. In comparison with other techniques, Sobel is quite simple to search for smooth edges. However, edge detection is inaccurate in a noisy environment which is evident from Fig. 3 (b) for the input image given in Fig. 3 (a).

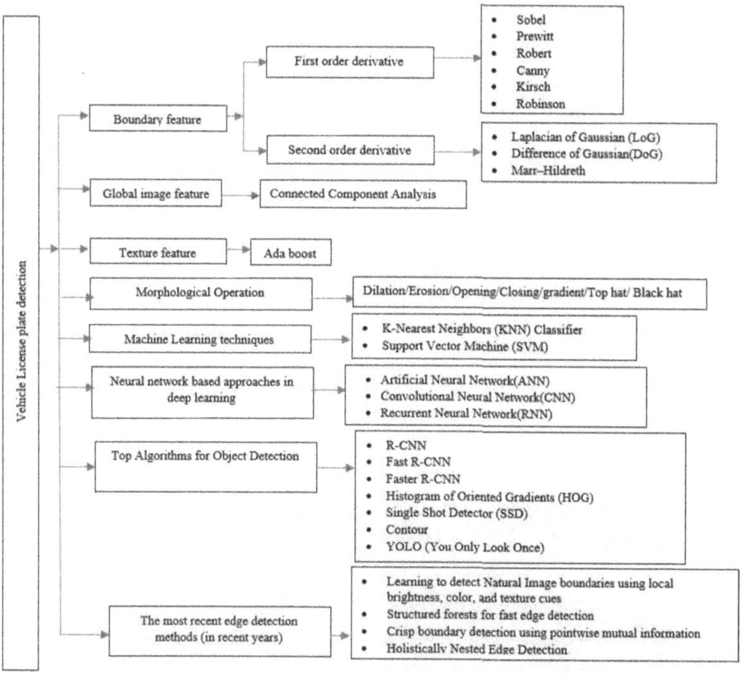

Fig. 2. Vehicle LP detection techniques

Fig. 3. Experimental result (a) Input (b) Sobel result (c) LOG result

Unlike the Sobel edge detector, the Laplacian edge detector employs a second-order derivative kernel both in vertical and horizontal directions, as mentioned in [15] and [16]. As the mask is intended to estimate a second derivative measurement on the image, it is extremely sensitive to noise. To remove the noise, the LoG was introduced. In LoG, the Laplace operator is applied to the input after it is smoothened with a Gaussian filter and the result of the same is shown in Fig. 3 (c). However, corner accuracy is comparatively lesser than the expected accuracy.

Next, the Canny method and the Marr-Hildreth algorithm are modern approaches for extracting edge features. The Marr–Hildreth algorithm is based on detecting edges of continuous curves by analyzing the brightness variation in real-time. The Marr-Hildreth approach is straightforward to use, however it relies on zero crossings. It works through the convolution of the image with the Laplacian of Gaussian function [21]. In contrast to

Marr- Hildreth, the Canny approach is preferred due to its edge accuracy, low errors, and noise reduction, as well as double thresholding. Canny detects a wide variety of edges in photos using multi-stage algorithms [16,21]. The canny method links the strong and weak edges as shown in Fig. 4 (a) and it is clear that the algorithm obtains adequate corner points in a wider area, with a maximum density in the intended edge area. In the end, it obtains the features of the image without modifying them in any way and is hence considered as one of the best algorithms. However, the main issues are computational complexity and long processing times.

In most approaches, the Canny edge detector is utilized to detect LPs [15,16]. The Canny, on the other side, has the following limitations [23]. Setting the lower and upper thresholding values needs experimentation and visual validation. Hysteresis thresholding values that are effective for one image may not even be effective for another. Moreover, several preprocessing steps are necessary for Canny edge detection and only local changes are taken into account [23].

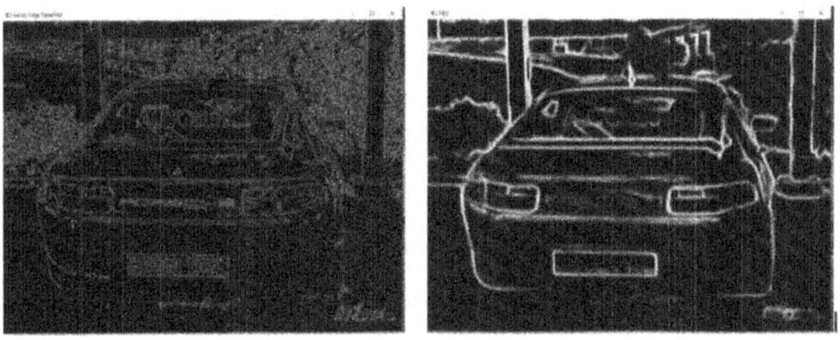

Fig. 4. (a) Canny result (b) HED result

End-to-end deep neural networks are being used in Holistically Nested Edge Detection (HED) to overcome the drawbacks of the Canny edge detector [23]. HED is a learning-based end-to-end feature extraction system that performs image-to-image forecasting with a trimmed VGG- like convolutional neural network [23]. HED utilizes the output of intermediate layers. The output of previous layers is referred to as side output, and the output of all five convolutional layers is merged to produce the final predictions. Evaluating the size of the feature maps produced at each layer tends to vary and it is useful in analyzing the image from multiple perspectives. Especially in comparison to other deep learning methods, HED is not only more accurate but also faster. In Fig. 4 (b), the HED algorithm is used to detect the boundaries of an object for the input image given in Fig. 4 (a). In comparison to other algorithms, HED can understand the image's content [23].

Many researchers suggested a morphology-based method for obtaining LPs from car images, which is followed by character segmentation and recognition. It takes two inputs: one is our original image, while the other is a structural element or kernel that helps determine the nature of operations such as morphological gradient [18], closing, erosion, tophat, opening, dilation, and blackhat. For identifying structural elements for morphological operations, characteristics such as LP width and LP height, individual character height, and distances are considered [1,4]. Figure 5 shows the output of several morphological operations.

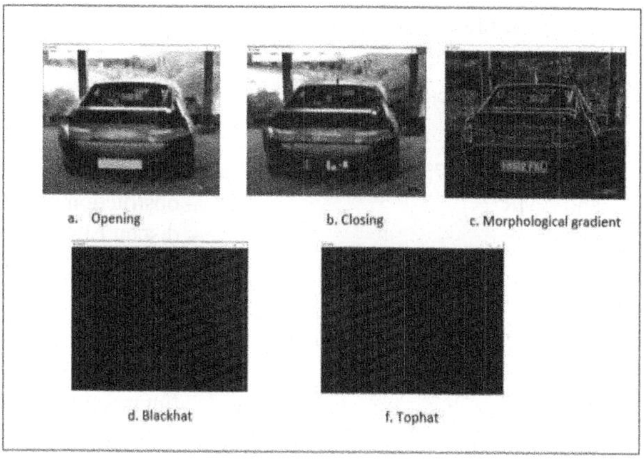

Fig. 5. Results of Morphological operations

2.2 *LP* Extraction Using Feature Learning

Among the most crucial and effective algorithms in the computer vision community is YOLO, which integrates several highly innovative computer vision concepts from the scientific community [6]. A region-based method is implemented by all existing algorithms for detecting an intended object in an image. YOLO differs greatly from region-based methods [15]. YOLO utilizes a single neural network [43] to determine bounding box coordinates and classifier Fig. 6 (c).

2.3 LP Extraction Using Texture Features

Factors that are used in the topmost algorithms like Fast RCNN, Faster RCNN, HOG, SSD, Contour, and YOLO are reviewed in several research studies. Navneet Dalal and Bill Triggs came up with the idea of Histograms of Oriented Gradients (HOG). Based on normalized histograms of gradient inclinations, this method is used. Because local gradients or edge directions can often be used to characterize an object's appearance and shape [15], even without knowing the exact gradients or edge directions, this technique, as defined by Datal and Triggs in [1], is very useful. These are carried out by dividing the image window into segments with the following measurements: 8×8, 16×16, and these blocks are further divided into smaller spatial regions named cells that have a similar dimension [14]. Tests have also shown that 8×8 blocks and 2×2 cells work best. The perfect detection process of HOG is shown in Fig. 6 (a). HOG captures pixel intensity gradients and information, and it also serves as a good texture descriptor. Due to HOG's large feature vectors, storage costs are high and feature vector comparisons are time-consuming and extremely expensive to compute (see Table 1).

Table 1. Summary of LPD techniques

Ref	Detection Methods	Dataset	Accuracy	Pros	Cons	Future Opportunities
[25]	Scale adaptive boosting algorithm	Caltech and AMR Database	97.72%	High accuracy	Lacks light, obstruction of distant LPs	Morphological operations could be used to enhance segmentation
[21]	AdaBoost and HOG features	Vehicle images dataset	89.60%	Beneficial for detecting LPs in complex environments	Dataset needs further improvement for better performance	This dataset will be used for AYPR
[28]	CO Artificial MET	Custom dataset	Detection rate 93.78%	Can detect LPs under different perspectives	No specific drawbacks mentioned	Improved method can be implemented
[29]	KNN Classifier	Custom dataset with video	93%	Estimated performance improvement	No specific drawbacks mentioned	Web application can be incorporated
[30]	SAI Edge Detector	Custom dataset	96.12%	Simplifies database processing	No specific drawbacks mentioned	A genetic technique can be implemented
[32]	SVM	Custom dataset with 100 images	90.56%	Improved performance	Brightness of images significantly affects accuracy	Enhanced brightness or feature extraction can improve accuracy
[33]	PWMI approach	Custom dataset with video	96.7%	Extremely good performance	No specific drawbacks mentioned	Can be further improved
[34]	AdaBoost based on 7-HOG features	Custom dataset (11,337 LP images)	98.7%	High accuracy	Complex numbers require specific training focus	Further enhancement for generalization
[38]	MLP Neural Network	Custom dataset	96.17%	High accuracy	Difficult to detect black LPs	Can be used with additional training techniques

Fig. 6. (a) HOG result (b) Contour result (c) YOLO result

3 Character Segmentation

Many optical character recognition systems initially segment the characters before classifying them. The efficiency of character segmentation is entirely dependent on the extraction of an LP from the image. Having identified the location of the plate region, the next step is to segment the characters present in the LP. Depending on the rectangular projection, the LP can be segmented [10, 12]. The isolated LP may have challenges including contrast issues, fluctuating lighting circumstances, or being orientated at different angles. Hence, pre-processing such as de-skewing, de-blurring, or other approaches may be required before segmenting the letters. The character segmentation methodologies are discussed in the following subsections.

3.1 Character Segmentation Using Projection Profiles

For character segmentation, pixel projection in both vertical and horizontal directions is employed [16, 23]. To extract individual characters, vertical projection is used in the binary LP, followed by horizontal projection to evaluate the start and endpoints of characters. In [33, 15], vertical projection is used for character extraction in addition to noise reduction analysis of the character sequence. After reviewing the techniques, it is determined that the approach that uses both horizontal and vertical pixel projections is the simplest and most often used. However, prior information of the character computation is essential. The projection values may be affected by the image quality and noise.

3.2 Character Segmentation Using Deep Learning

Neural networks have emerged as a new solution for character segmentation, which employs CNN for computer vision tasks [2,16]. The CNN is given a localized LP as input, and the bounding boxes of each character are created as output. Furthermore, numerous deep learning-based LP recognition pipelines avoid explicit character segmentation, resulting in fewer parameters and lower computing costs [7, 13].

4 LP Recognition Techniques

In ANPR systems, segmented recognition of characters is the last step. Variations in character size and thickness may occur in the segmented output [16] as a result of zooming in and out of the scene. The characters may be broken, slanted, or distorted as a result of noise. The numerous LPR approaches are described in Fig.7 and discussed in the next section, along with their advantages and disadvantages.

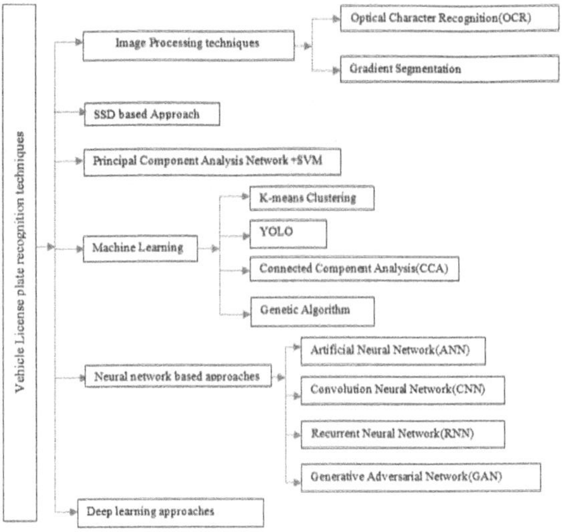

Fig. 7. Vehicle LP Recognition Techniques

4.1 Character Recognition Using Template Matching

Character recognition's simplest method is template matching. It is a cross-correlation technique in which the extracted character's similarity is compared or analyzed to a collection of template characters. Template Matching Technique is used to recognize segmented characters with variable degrees of success in [38].

4.2 Character Recognition Using Extracted Features

All pixels are involved in the template matching procedure, which increases the processing time. Character Recognition uses an additional approach to feature extraction since it decreases processing time and removes pixels that are not important [15]. For character recognition, SVM is preferred. In [31], and improved SVM-based technique is proposed for challenging and complex LPs. Vertical and horizontal binary character projection yields the feature map in [2, 31, 24].

4.3 Character Recognition Using Tesseract OCR

To segment and recognize the character [38], Pytesseract OCR is used. It is capable of extracting printed text from images directly as shown in Fig. 8. Though Tesseract OCR is quite powerful, it has some limitations. The handwritten text is not recognized well in some scenarios, as evident from Fig. 8 (Table 2).

Table 2. Summary of LPR techniques

Ref	Detection Methods	Dataset	Accuracy	Pros	Cons
[35]	OKM-CNN	MA English Stanford Dataset	98.1%	Performs in real-time	Needs more analysis on how the model works
[36]	CA, KNN	Arabic-Egyptian, English LP Dataset	96.47%	Identifies Egyptian and English LPs	Optimization is needed for better performance
[38]	Edge-based evaluation	English, Malaysian LPs	94.4%	Suitable for real-life applications	Font variability affects detection
[38]	MLP	Private Dataset	93%	Good performance	Lower detection rate; needs expansion
[39]	YOLOv3	Korean, English LP Dataset	Higher than 97%	Applicable across multiple countries	Needs improvements in character segmentation
[40]	CycleGAN-based CNN	CCPD, AOLP, Private Dataset	Not specified	Handles diverse conditions	Difficulties with highly distorted LPs
[41]	SSD-based approach	Chinese LP dataset (16 types)	99.1%	Works in real-time	Optimization required for accuracy consistency
[42]	Compressed English Character Recognition	CV Toolbox Image Set	98.61%	Effective sensing technique	Not specified
[43]	Fast YOLOv2 and YOLOv3	Caltech, English, Chinese LP Dataset	96.8%	Fast and efficient	Not specified
[44]	Gradient-based segmentation	English LP Dataset (78 images)	94.87%	Accurate segmentation	Not specified

(*continued*)

Table 2. (*continued*)

Ref	Detection Methods	Dataset	Accuracy	Pros	Cons
[45]	Image De-noising, Rectification	English AOLP-RP, VTLPs Dataset	93.08%	CNN-based approach	Not specified
[46]	Convolutional Neural Network (CNN)	Chinese LP Dataset (19,020 images)	82.5%	Good trade-off between accuracy and speed	Struggles with yellow-base LPs

Fig. 8. Tesseract OCT Output

4.4 Character Recognition Using ANPR

ANPR technique has been contributed to ITS and does not need any human interference [9,15]. ANPR method can identify and recognize vehicle-based characters on their LPs using recognition techniques [15,17]. To segment and identify characters, the ANPR algorithm is used after applying the LP extraction using morphological operations [8,4] as shown in Fig.9. The benefits of ANPR involve real-time monitoring and surveillance, as well as producing reliable results. However, if the scanning quality is bad, the OCR may produce unsatisfactory results [38]. Moreover, it does not work well for images with occlusion, distortion, and complex backgrounds.

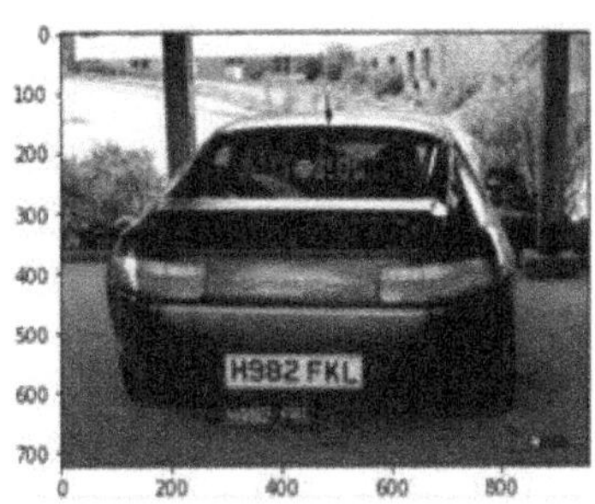

Fig. 9. ANPR result

5 Conclusion and Future Research

The survey of an automatic vehicle LP detection system is conducted and reviewed in this paper. The system is essential to the growth of a smart and intelligent transportation system. From the review of various papers, different methods and algorithms are discussed for LP detection, character segmentation, and character recognition under varying environments. Some of the algorithms are implemented and the corresponding visual outputs are presented in this paper.

There are still a number of obstacles in ANPR research. For instance, regardless of region, more robust algorithms for non-standardized formats are required. Recent developments in Deep Learning facilitate computer vision systems to explore a vast array of concepts, such as safe autonomous driving, object detection in unconstrained scenarios, automatic image analysis for a variety of applications, and so on. This system can be taught and evaluated using YOLO and other real-time object detectors. When RFID and ANPR are combined, they might make a hybrid system that can be used in the future for a range of intelligent transportation system applications.

References

1. Roy, C., Hossen, M.K., Nag, D.: License plate detection and character recognition system for commercial vehicles based on morphological approach and template matching. In: International Conference on Electrical Engineering and Information Communication Technology (ICEEICT), pp. 1–6 (2016)
2. Kosala, G., Harjoko, A., Hartati, S.: License plate detection based on convolutional neural network: support vector machine (CNN-SVM). In: Proceedings of the International Conference Video Image Process, pp. 1–5 (2017)
3. Israni, S., Jain, S.: Edge detection of license plate using sobel operator. In: Proceedings of the International Conference on Electrical, Electronics, and Optimization Techniques (ICEEOT), pp. 3561–3563 (2016)
4. Eswar, N., Gowri Shankar Reddy, D.: Morphological operation based vehicle number plate detection. Int. J. Eng. Res. **V9**(02), 428–433 (2020). https://doi.org/10.17577/ijertv9is020064
5. Sonavane, K., Soni, B., Majhi, U.: Survey on automatic number plate recognition (ANR). Int. J. Comput. Appl. **125**(6), 1–4 (2015). https://doi.org/10.5120/ijca2015905920
6. Patel, D., Shah, A.P.: Automatic number plate recognition system (ANPR): a survey. Int. J. Comput. Appl. **69**(9), 21–33 (2013). https://doi.org/10.5120/11871-7665
7. Ren, S., He, K., Girshick, R., Sun, J.: Faster R- CNN: towards real-time object detection with region proposal networks. IEEE Trans. Pattern Anal. Mach. Intell. **39**(6), 1137–1149 (2017)
8. Kocer, H.E., Cevik, K.K.: Artificial neural networks-based vehicle license plate recognition. Procedia Comput. Sci. **3**, 1033–1037 (2011)
9. Girshick, R.: Fast R-CNN. In: Proceedings of the IEEE International Conference on Computer Vision, pp. 1440–1448 (2015)
10. Liu, W., et al.: SSD: Single shot multibox detector. In: Proceedings of the European Conference Computer Vision, pp. 21–37 (2016)
11. Bhardwaj, S.M.: Review paper on automated number plate recognition techniques. Int. J. Emerg. Res. Manage. Technol. **6**(15), 2278–9359 (2015)
12. Ban, K.D., Yoon, Y., Yoon, H.S., Kim, J.: Number detection in natural image with boosting classifier, 525–526 (2012). https://doi.org/10.1109/URAI.2012.6463060

13. Sun, X., Wu, P., Hoi, S.C.H.: Face detection using deep learning: an improved faster RCNN approach. Neurocomputing **299**, 42–50 (2018)
14. Yaseen, N.O., Al-Ali, S.G.S., Sengur, A.: An efficient model for automatic number plate detection using HOG feature from new North Iraq vehicle images dataset. In: IISEC 2019 – Proceedings (2019)
15. Lubna, Mufti, N., Shah, S.A.A.: automatic number plate recognition: a detailed survey of relevant algorithms. Sensors (2021)
16. Parvin, S., Rozario, L.J., Islam, M.E.: Vehicle number plate detection and recognition techniques: a review. Adv. Sci., Technol.Eng. Syst. J. **6**(2), 423–438 (2021)
17. Bulan, O., Kozitsky, V., Ramesh, P., Shreve, M.: Segmentation and annotation-free license plate recognition with deep localization and failure identification. IEEE Trans. Intell. Transp. Syst. **18**(9), 2351–2363 (2017)
18. Wang, R., et al.: "License plate detection usinggradient information and cascade detectors. Optik– Int. J. Light Electron Opt. **125**(1), 186–190 (2014)
19. Malik, J., Belongie, S., Leung, T., Shi, J.: Contour and texture analysis for image segmentation. IJCV **43** (2001)
20. Dollár, P., Zitnick, C.L.: Structured forests for fast edge detection. In: CVF (2013)
21. Isola, P., Zoran, D., Krishnan, D., Adelson, E.H.: Crisp boundary detection using pointwise mutual information. In: ECCV 2014, Part III, LNCS 8691, pp. 799–814 (2014)
22. Martin, D.R., Fowlkes, C.C., Malik, J.: Learning to detect natural image boundaries using local brightness, color, and texture cues. IEEE Trans. Pattern Anal. Mach. Intell. **26**(1) (2004)
23. Xie, S., Tu, Z.: Holistically-nested edge detection. In: 2015 IEEE International Conference on Computer Vision, © IEEE (2015)
24. Lukac, M., Tanizawa, R., Kameyama, M.: Machine learning based adaptive contour detection using algorithm selection and image splitting. Interdisc. Inf. Sci. **18**(2), 123–134 (2012). https://doi.org/10.4036/iis.2012.123
25. Molina-Moreno, M., Gonzalez-Diaz, I., Diaz-De- Maria, F.: Efficient scale- adaptive license plate detection system. IEEE Trans. Intell. Transp. Syst. **20**(6), 2109–2121 (2019). https://doi.org/10.1109/TITS.2018.2859035
26. Yaseen, N.O., Al-Ali, S.G.S., Sengur, A.: An efficient model for automatic number plate detection using HOG feature from new north Iraq vehicle images dataset. In: 1st International Informatics and Software Engineering Conference: Innovative Technologies for Digital Transformation, IISEC 2019 – Proceedings (2019). https://doi.org/10.1109/UBMYK48245.2019.8965573
27. Yepez, J., Ko, S.B.: Improved license plate localisation algorithm based on morphological operations. IET Intel. Transport Syst. **12**(6), 542–549 (2018). https://doi.org/10.1049/iet-its.2017.0224
28. Babbar, S., Kesarwani, S., Dewan, N., Shangle, K., Patel, S.: A new approach for vehicle number plate detection. In: 2018 11th International Conference on Contemporary Computing, IC3 2018, 1–6 (2018). https://doi.org/10.1109/IC3.2018.8530600
29. Akshay lepcha, K.H.P., Adithya, S.R., Anuraag, G., Sathish, K.I.: A novel methodology for license plate detection using KNN classifier. J., F.O.R. Adv., A. Sci. **5**(3), 582–586. ISSN NO: 2394–8442
30. Tejas, K., Ashok Reddy, K., Pradeep Reddy, D., Rajesh Kumar, M.: Efficient licence plate detection by unique edge detection algorithm and smarter interpretation through IoT. ArXiv (2017)
31. Miyata, S., Oka, K.: Automated license plate detection using a support vector machine. In: 2016 14th International Conference on Control, Automation, Robotics and Vision, ICARCV 2016 (2016). https://doi.org/10.1109/ICARCV.2016.7838653
32. Chen, R., Luo, Y.: An improved license plate location method based on edge detection. Phys. Procedia **24**, 1350–1356 (2012). https://doi.org/10.1016/j.phpro.2012.02.201

33. Ban, K.D., Yoon, Y., Yoon, H.S., Kim, J.: Number detection in natural image with boosting classifier. In: 2012 9th International Conference on Ubiquitous Robots and Ambient Intelligence, URAI 2012, (Urai), pp. 525–526 (2012). https://doi.org/10.1109/URAI.2012.6463060
34. Wazalwar, E., Oruklu, J.S.: esign flow for robust license plate localization. In: IEEE International Conference on Electro Information Technology (2011). https://doi.org/10.1109/EIT.2011.5978590
35. Pustokhina, I.V., et al.: Automatic vehicle license plate recognition using optimal K-means with convolutional neural network for intelligent transportation systems. IEEE Access **8**, 92907–92917 (2020). https://doi.org/10.1109/ACCESS.2020.2993008
36. Yousif, B.B., Ata, M.M., Fawzy, N., Obaya, M.: Toward an optimized neutrosophic k-means with genetic algorithm for automatic vehicle license plate recognition (ONKM-AVLPR). IEEE Access **8**, 49285–49312 (2020). https://doi.org/10.1109/ACCESS.2020.2979185
37. Arafat, M.Y., Khairuddin, A.S.M., Paramesran, R.: Connected component analysis integrated edge based technique for automatic vehicular license plate recognition framework. IET Intel. Transport Syst. **14**(7), 712–723 (2020). https://doi.org/10.1049/iet-its.2019.0006
38. Rehman, S.U., Ahmad, M., Nawaz, A., Ali, T.: An efficient approach for vehicle number plate recognition in Pakistan. Open Artif. Intell. J. **6**(1), 12–21 (2020). https://doi.org/10.2174/1874061802006010012
39. Henry, C., Ahn, S.Y., Lee, S.W.: Multinational license plate recognition using generalized character sequence detection. IEEE Access **8**, 35185–35199 (2020). https://doi.org/10.1109/ACCESS.2020.2974973
40. Zhang, L., Wang, P., Li, H., Li, Z., Shen, C., Zhang, Y.: A robust attentional framework for license plate recognition in the wild. ArXiv, pp. 1–10 (2020). https://doi.org/10.1109/tits.2020.3000072
41. Yao, L., Zhao, Y., Fan, J., Liu, M., Jiang, J., Wan, Y.: Research and Application of license plate recognition technology based on deep learning. J. Phys.: Conf. Ser. **1237**(2) (2019)
42. Jokić, N.V.: License plate recognition with compressive sensing based feature extraction. ArXiv, pp. 1–4 (2019)
43. Laroca, R., Zanlorensi, L.A., Gonçalves, G.R., Todt, E., Schwartz, W.R., Menotti, D.: An efficient and layout- independent automatic license plate recognition system based on the yolo detector. ArXiv (2019)
44. Kumar, G., Barman, A., Pal, M.: License plate tracking using gradient based segmentation. In: IEEE Region 10 Annual International Conference, Proceedings/TENCON, pp. 1737–1740 (2019). https://doi.org/10.1109/TENCON.2019.8929688
45. Lee, Y., Lee, J., Ahn, H., Jeon, M.: SNIDER: single noisy image denoising and rectification for improving license plate recognition. In: Proceedings - 2019 International Conference on Computer Vision Workshop, ICCVW 2019, pp. 1017–1026 (2019). https://doi.org/10.1109/ICCVW.2019.00131
46. Zhu, L., Wang, S., Li, C., Yang, Z.: License plate recognition in urban road based on vehicle tracking and result integration. J. Intell. Syst. **29**(1), 1587–1597 (2020). https://doi.org/10.1515/jisys-2018-0446

ChatGPT: A Review of Language Models for Conversational AI

Rahul Kumar[1(✉)], Ashok Kumar[1], Saurabh Singh[2], and Penubaka Balaji[1]

[1] Department of Computer Science and Engineering, K L Deemed to be University, Vaddeswaram 522302, Andhra Pradesh, India
rahulkumar1680@gmail.com, pbalaji@kluniversity.in

[2] Computer Science and Engineering Department, Maharana Pratap Engineering College, Uttar Pradesh, Kanpur 209217, India

Abstract. Recent developments in machine learning and natural language processing (NLP) have considerably helped conversational AI systems. ChatGPT, which is based on the GPT-3.5 design and has attracted a lot of attention, is one particular no-table model. ChatGPT stands out for its exceptional ability to provide human-like responses in a conversational situation. This ground-breaking technology makes use of sophisticated artificial intelligence techniques to produce natural language responses in response to inputs or cues. Its uses are widespread and include content production, customer service, and natural language processing.

This article explores the background and technology of GPT, highlighting its generative pre-trained transformer model and its adaptability to a range of language-based tasks. It also looks at how ChatGPT uses this technology to serve as an advanced chatbot, capable of engaging in sophisticated conversations.

In this paper, we provide a comprehensive review of ChatGPT, exploring its architecture, training methodology, strengths, limitations, and applications. We also discuss the ethical considerations and potential challenges associated with deploying ChatGPT in real-world scenarios.

Keywords: ChatGPT · AI · Libraries · Natural Language · Chatbot · OpenAI · GPT · GPT-3.5 · Generative Pre-Trained Transformer

1 Introduction

1.1 Background and Significance of Conversational AI

Conversational AI is the branch of artificial intelligence that focuses on developing intelligent machines that can communicate with humans in natural language. Conversational AI is significant because it has the potential to transform human-computer interactions. Conversational AI systems can improve a number of elements of our life, including customer service, virtual assistants, language translation, information retrieval, and more by enabling robots to interpret and produce human-like language.

Several causes are responsible for the growth of conversational AI. First off, there is a growing need for automated systems that can comprehend and respond to user inquiries

S. Singh et al. (Eds.): AIDSP 2023, CCIS 2440, pp. 206–215, 2025.
https://doi.org/10.1007/978-3-031-88762-8_20

and requests due to the increasing dominance of digital communication channels like messaging applications and voice assistants. Second, improvements in machine learning, deep learning, and natural language processing (NLP) methods have made it.

1.2 Role of Language Models in Conversational AI Systems

Language models play a central role in conversational AI systems. A language model is a statistical model that learns the patterns and structures of human language by analyzing large amounts of text data. It enables machines to understand, generate, and respond to human language in a coherent and contextually appropriate manner.

In conversational AI systems, language models are used for several tasks. They are employed for natural language understanding (NLU), where they analyze user inputs and extract relevant information and intents. Language models help in entity recognition, sentiment analysis, and intent classification, allowing the system to comprehend user queries accurately.

Language models are also used for natural language generation (NLG), where they generate human-like responses based on the given context and user input. They can generate coherent and contextually appropriate responses by leveraging the knowledge acquired during training on large text corpora.

1.3 Motivation for Studying ChatGPT

To create more sophisticated and effective conversational AI systems, researchers are studying ChatGPT, a variation of OpenAI's GPT (Generative Pre-trained Transformer) model. ChatGPT demonstrates the capacity to hold open-ended discussions, comprehend a range of user inputs, and produce responses that are cogent and contextually appropriate.

The study of ChatGPT also enables researchers to gain insights into the broader field of natural language processing and artificial intelligence. It provides opportunities to explore the frontiers of language understanding and generation, and contribute to the development of more advanced conversational AI systems that can benefit various domains and applications.

1.4 Research Objectives: The Main Objectives of This Research Are

- To provide a comprehensive understanding of the development and capabilities of ChatGPT as a language model for conversational AI.
- To identify and analyze the limitations and challenges associated with ChatGPT and its application in conversational AI.
- To explore the various domains and applications where ChatGPT can be effectively utilized, such as customer service, virtual assistants, and education.
- To evaluate the performance and user feedback of ChatGPT in real-world scenarios.
- To discuss the implications of ChatGPT's advancements and potential future directions for research and development in conversational AI.
- To discuss with difference between search Engine and ChatGPT.

2 Architecture of ChatGPT

Building on OpenAI's GPT (Generative Pre-trained Transformer) architecture, ChatGPT is a novel language model. It makes use of the language generation capabilities of transformer-based models, which have shown remarkable success in identifying contextual dependencies and generating responses that are both coherent and appropriate for the given context.

A deep neural network with numerous layers of transformers is used to implement ChatGPT's architecture. These transformers are made specifically to handle sequential input, such text written in natural language, allowing the model to produce outputs that are both coherent and human-like. A sizable corpus of text data is used to train ChatGPT, allowing the model to discover complex correlations and patterns between words, phrases, and sentences. Because training is iterative, the model can continuously enhance its performance.

2.1 Overview of the GPT-3.5 Architecture

The GPT-3.5 architecture is comprised of a series of transformer encoder-decoder layers. The encoder is responsible for processing the input sequence, while the decoder is responsible for generating the output sequence. However, it is noteworthy that ChatGPT diverges from traditional sequence-to-sequence models in its utilization of the decoder. Rather than employing it for autoregressive language generation, ChatGPT leverages the decoder to retrieve and integrate prior context from the ongoing conversation.

2.2 Transformer-Based Models for Language Generation

The most popular neural network design for many natural language processing tasks, especially language production, has emerged as transformers. These models make use of processes for self-attention, allowing them to concentrate on various words in the input sequence while producing each word in the output sequence.

The primary advantage of transformer-based models lies in their capability to capture long-range dependencies within the input sequence, a critical aspect for language generation tasks. By attending to all the words in the input sequence, transformers can generate responses that are more contextually informed and coherent compared to conventional recurrent neural networks (RNNs).

2.3 Pre-training and Fine-Tuning Procedures

ChatGPT follows a two-step process: pre-training and fine-tuning.

1. Pre-training: During the pre-training phase, a language model like ChatGPT is trained on a large corpus of publicly available text from the internet. This corpus serves as the training data for the model to learn language patterns, grammar, facts, and some level of reasoning.

2. Fine-tuning: After pre-training, the model undergoes fine-tuning using a more specific dataset that is carefully generated and curated. This dataset includes demonstrations and comparisons to guide the model towards desired behavior. Human reviewers play a crucial role in providing feedback and evaluating model outputs to ensure the system aligns with the desired guidelines, values, and safety standards.

3 Training Methodology

3.1 Dataset Selection and Preparation

The training of ChatGPT involves the selection and preprocessing of datasets. The initial pre-training phase typically involves a large corpus of publicly available text data from the internet. This dataset can consist of books, articles, websites, and other sources of diverse language patterns and styles [11].

The preprocessing of the dataset involves tokenization, where the text is divided into smaller units called tokens. Tokens can be words, sub-words, or characters. Additionally, special tokens such as [CLS] (for classification) and [SEP] (for separating segments) may be inserted to assist in training. The tokenized dataset is then used to create input-output pairs for the language modeling objective.

3.2 Training Objectives

During pre-training, the primary objective is masked language modeling (MLM). The model is trained to predict the masked or hidden words within a sentence given the context of the surrounding words. This objective helps the model learn the statistical patterns and structures of language.

In the case of ChatGPT, the fine-tuning phase involves training the model on custom datasets created by OpenAI. These datasets include demonstrations of correct behavior and comparisons to rank different responses. The fine-tuning process helps align the model's behavior with desired conversational outcomes [3].

Reinforcement learning techniques can also be used to optimize the model's responses based on rewards or evaluations from human feedback.

4 Strengths of ChatGPT

1. Natural Language Understanding and Generation Capabilities: ChatGPT exhibits strong natural language understanding and generation capabilities. It can comprehend and generate human-like language, allowing for meaningful and contextually appropriate conversations with users.
2. Contextual Understanding and Response Coherence: ChatGPT leverages the transformer architecture and self-attention mechanisms to capture contextual dependencies. It can maintain a coherent understanding of the conversation history and generate responses that align with the given context, resulting in more coherent and relevant interactions [4].

3. Handling of Various Conversational Scenarios: ChatGPT demonstrates versatility in handling a wide range of conversational scenarios. It can engage in open-ended conversations, answer questions, provide explanations, and perform tasks like language translation, summarization, and more. Its flexibility makes it useful for different applications and domains [4].
4. Transfer Learning and Few-shot Learning Abilities: ChatGPT benefits from pretraining on large-scale datasets, which enables it to acquire general language understanding capabilities. It can then be fine-tuned on specific tasks or domains with relatively small amounts of task-specific data, making it adaptable to new conversational contexts with few-shot learning. This transfer learning capability reduces the need for extensive training data for each specific task.

5 Limitations of ChatGPT

1. Sensitivity to Input Phrasing and Prompt Engineering: ChatGPT can be sensitive to slight changes in input phrasing, resulting in different responses. Users may need to carefully craft their queries or prompts to get the desired output, which can be a limitation in terms of user experience and convenience [11].
2. Generating Plausible Yet Incorrect or Biased Responses: ChatGPT may generate responses that sound plausible but are factually incorrect or biased. Since it learns from large text corpora, it can inadvertently reproduce inaccuracies, propagate stereotypes, or exhibit biased behavior. Mitigating these issues requires ongoing efforts in data curation, fine-tuning, and bias detection and reduction techniques [11].
3. Lack of Common-Sense Reasoning and Factual Accuracy: While ChatGPT has impressive language generation capabilities, it may struggle with common sense reasoning and factual accuracy. It may generate responses that lack real-world knowledge or provide incorrect information, especially when faced with ambiguous or complex queries.
4. Handling Out-of-Domain or Ambiguous Queries: ChatGPT may not perform well when faced with queries outside its trained domain or when the user's intent is ambiguous. It may provide generic or unhelpful responses or ask clarifying questions repeatedly without providing a meaningful answer. Improving the system's ability to handle out-of-domain queries and better handle ambiguity remains an active area of research [11].

6 Applications of ChatGPT

1. Virtual Assistants and Customer Support Systems: ChatGPT can be used to develop virtual assistants and customer support systems that engage in natural language conversations with users. It can assist users in obtaining information, resolving queries, and providing personalized assistance, enhancing the user experience and reducing the burden on human customer support agents [4].
2. Content Generation and Creative Writing: ChatGPT can aid in content generation and creative writing tasks. It can generate text for articles, blog posts, product descriptions, and social media content. Writers and content creators can utilize ChatGPT to overcome writer's block, brainstorm ideas, or receive suggestions for improving their writing.

3. Educational Tools and Language Learning: ChatGPT can serve as an educational tool by providing explanations, answering questions, and facilitating language learning. It can assist students with homework, provide tutoring in various subjects, and offer language practice exercises. ChatGPT's conversational nature makes it engaging and interactive for learners.
4. Chatbot Development and User Engagement: ChatGPT can be leveraged for chatbot development and user engagement in various applications. It can power chatbots on websites, messaging platforms, or mobile apps, providing personalized and interactive experiences for users [9, 10]. ChatGPT's ability to understand and respond to user queries makes it useful for guiding users, collecting information, and assisting with tasks.
5. Impact of ChatGPT on different fields

- Academics: Academics: ChatGPT has been used as a learning tool for students and a research tool for academics, facilitating knowledge acquisition and idea generation. [5]
- Cybersecurity: ChatGPT has the potential to enhance cybersecurity measures by identifying vulnerabilities, suggesting improvements, and analyzing data to detect threats. However, it can also be a potential avenue for cyber-attacks if not properly secured [2].
- Healthcare: ChatGPT has been utilized in healthcare to provide information and assistance to patients, aid in diagnosis, and support medical research and decision-making. [2]
- Software Development: ChatGPT has found applications in software development by providing code suggestions, answering programming-related queries, and assisting developers in problem-solving and debugging. [2]
- Researchers and Scholars: ChatGPT has been valuable to researchers and scholars, offering a tool for exploring new ideas, generating content, and facilitating collaboration within the academic community. [2]

7 Ethical Considerations

1. Bias and Fairness in Language Generation: Ethical concerns arise when AI models like ChatGPT exhibit biases in their language generation. Biases present in training data can be reflected in generated responses, potentially perpetuating stereotypes or discrimination. It is crucial to address these biases by carefully curating datasets, employing diverse training data, and conducting ongoing bias detection and reduction efforts.
2. Responsible Deployment and Potential Misuse: The responsible deployment of conversational AI systems is essential to ensure their positive impact. Developers and organizations should consider the potential misuse of such systems, including the spread of misinformation, malicious manipulation, or unethical use. Clear guidelines, policies, and oversight mechanisms should be established to prevent abuse and mitigate harm. [12]
3. Ensuring User Privacy and Consent: Conversational AI systems process user interactions and may collect personal information. It is important to prioritize user privacy and obtain informed consent for data collection and usage. Transparent practices,

secure data storage, and compliance with privacy regulations help maintain user trust and protect their sensitive information.
4. Mitigating Harmful or Offensive Content: Language models like ChatGPT should be equipped with mechanisms to detect and mitigate the generation of harmful or offensive content. Offensive language, hate speech, or abusive behavior should be actively addressed to create a safe and inclusive user experience. Human reviewers, content moderation, and user feedback can help in identifying and addressing such issues.

8 Challenges and Future Directions

1. Improving Contextual Understanding and Response Coherence: Enhancing the contextual understanding and coherence of AI models like ChatGPT is a significant challenge. Future directions involve refining the ability of models to grasp nuanced contexts, maintain a coherent conversation flow, and generate responses that align with the given context. Advancements in training strategies and architectures can contribute to addressing this challenge [6].
2. Incorporating External Knowledge and Fact-Checking Mechanisms: Integrating external knowledge sources and fact-checking mechanisms into conversational AI systems is crucial for improving factual accuracy and reducing the generation of false or misleading information. Future directions involve developing methods to retrieve and incorporate reliable information from trusted sources to enhance the system's responses.
3. Addressing Biases and Promoting Inclusivity: Tackling biases and ensuring inclusivity is an ongoing challenge in conversational AI. Future directions involve advancing techniques for bias detection and reduction, promoting diversity in training data, and involving diverse perspectives during the model development process. Efforts should be made to create AI systems that respect and cater to the needs of users from diverse backgrounds (Table 1).
4. Developing Hybrid Models for Task-Oriented and Conversational AI: The integration of task-oriented capabilities with conversational AI is an emerging direction. Hybrid models that can switch between goal-oriented tasks and open-ended conversations would enable more seamless and efficient interactions. This direction involves developing models that can understand and execute specific commands or requests while still maintaining conversational abilities [12].

Table 1. Difference Between Google Search Engine and ChatGPT

Sno	Type	Google Search Engine	ChatGPT
1	Purpose	The primary purpose of the Google Search Engine is to deliver accurate and pertinent information to users based on their specific search queries	The main objective of ChatGPT is to address natural language questions and offer conversational support to users, aiming to provide them with helpful and interactive assistance
2	Functionality	The Google Search Engine scours the web, indexing a wide range of online content, including web pages, images, videos, and other relevant resources. This extensive indexing allows it to generate and present pertinent results to users based on their search queries	The key feature of ChatGPT is its use of artificial intelligence (AI) and natural language processing (NLP) to fully understand and address queries from users
3	User Interaction	A search query is entered into the Google Search Engine by typing relevant keywords or phrases. Following the user's query, the search engine generates a list of relevant results based on the information provided. User exploration and selection are then possible based on these results	ChatGPT engages in natural language conversation with users, facilitating real-time interaction by providing prompt and dynamic responses. Through its conversational capabilities, ChatGPT aims to simulate human-like communication, allowing users to engage in a fluid and interactive dialogue, receiving immediate feedback and assistance
4	Search Quality	The Google Search Engine employs intricate algorithms and a variety of ranking factors to ensure the delivery of the most pertinent results to users	ChatGPT harnesses advanced NLP (Natural Language Processing) and AI (Artificial Intelligence) technologies to effectively comprehend user questions and furnish relevant answers
5	User Experience	The Google Search Engine offers a text-based interface that allows users to input their search queries and retrieve information	ChatGPT provides users with a conversational interface that enables interaction and facilitates the exchange of information
6	Accuracy	The Google Search Engine endeavors to provide users with accurate and up-to-date information	ChatGPT also uses advanced AI and NLP technology to provide accurate and relevant answers to users

(*continued*)

Table 1. *(continued)*

Sno	Type	Google Search Engine	ChatGPT
7	Search Speed	The Google Search Engine is known for its fast and efficient search results. With millions of web pages indexed and an optimized infrastructure, it can generate search results within milliseconds	ChatGPT is designed to provide real-time responses to users. By employing advanced AI and NLP technology, ChatGPT can rapidly process and understand user queries, enabling it to generate prompt and timely responses
8	Personalization	The Google Search Engine offers personalized search results by taking into account various factors, including a user's search history, location, and other relevant information	ChatGPT is designed to provide personalized responses to users based on their past interactions and preferences
9	User Data	The Google Search Engine does collect and store user data, including search history and location data, to provide personalized search results and targeted advertisements	ChatGPT also collects and stores user data for personalized responses and to improve the accuracy of its AI and NLP technology
10	Privacy	Google Search Engine has privacy policies in place to safeguard user data and ensure its protection	ChatGPT primary function is to provide information and answer questions to the best of my knowledge and abilities
11	Data Sources	The Google Search Engine indeed searches and indexes a vast amount of online content to provide relevant search results. It scours the entire web, indexing content, audio,video and other types of online content	ChatGPT does not have direct access to a specific database or structured information. It is trained on a diverse range of internet text data to generate responses based on patterns and associations it has learned during training
12	Search Types	It gives search using web, also used image, audio and video also	ChatGPT is primarily designed to address natural language questions and provide conversational assistance to users
13	User Feedback	the Google Search Engine allows users to provide feedback on search results to help improve their accuracy and relevance	ChatGPT does not have a built-in feedback mechanism for users to directly provide feedback on its responses

9 Conclusion

ChatGPT, as a conversational AI system, possesses strengths in natural language understanding and generation, contextual coherence, versatility in handling various conversational scenarios, and transfer learning abilities. However, it also has limitations such as sensitivity to input phrasing, potential for generating incorrect or biased responses, lack of common-sense reasoning, and challenges in handling out-of-domain or ambiguous queries.

The future of conversational AI holds great potential for advancements and improvements. Addressing the limitations of current systems, such as bias reduction, fact-checking, and better handling of user queries, is crucial. Ethical considerations, responsible deployment practices, and user privacy protection will continue to play a significant role in shaping the future of conversational AI.

References

1. Agrawal, Ajay, Joshua Gans, and Avi Goldfarb.: ChatGPT and how AI disrupts Industries. Harvard Bus. Rev., 1–6 (2022)
2. Kalla, D., Smith, N., Samaah, F., Kuraku, S.: Study and analysis of chat GPT and its impact on different fields of study. Int. J. Innov. Sci. Res. Technol. **8**(3) (2023)
3. Diaz, M.: How to use ChatGPT: everything you need to know (2023)
4. Else, H.: Abstracts written by ChatGPT fool scientists. Nature **613**(7944), 423 (2023)
5. Lund, B.D., Wang, T.: Chatting about ChatGPT: how may AI and GPT impact academia and libraries? Libr. Hi Tech News **40**(3), 26–29 (2023)
6. Mandelaro, J.: How will AI chatbots like ChatGPT affect higher education (2023)
7. Mok, A., Zinkula, J.: ChatGPT may be coming for our jobs. Here are the 10 roles that AI is most likely to replace (2023)
8. Sachdev, S.: ChatGPT and its impact on society. The Times of India (2023)
9. Timothy, M.: The 3 best alternatives to ChatGPT. MUO (2023)
10. Vasylkiv, B.: Limitations and ethical considerations of using ChatGPT. Incora European software development company (2023)
11. Kumar, R., Mishra, A.K., Singh, D.K.: Packet loss avoidance in mobile Adhoc network by using trusted LDoS techniques. In: 2nd International Conference on Data, Engineering and Applications (IDEA), Bhopal, India, pp. 1–5 (2020). https://doi.org/10.1109/IDEA49133.2020.9170673
12. Kumar, R., Singh, D.K., Mishra, A.K.: An approach to extract fine detail and unclear information by enhancing computed tomography image. In: 2nd International Conference on Data, Engineering and Applications (IDEA), Bhopal, India, pp. 1–6 (2020). https://doi.org/10.1109/IDEA49133.2020.9170735

Exploring the Metaverse: A Comprehensive Review and Future Perspectives

Rahul Kumar[1(✉)], Ashok Kumar[1], Saurabh Singh[2], and Rudramani Bhutia[1]

[1] Department of Computer Science and Engineering, K L University, Guntur, Andhra Pradesh 522302, India
rahulkumar1680@gmail.com

[2] Department of Computer Science and Engineering, Maharana Pratap Engineering College, Kanpur, Uttar Pradesh 209217, India

Abstract. The concept of the metaverse has emerged as an exciting vision for the future of the internet. It envisions a fully immersive and interconnected virtual shared space that transcends physical boundaries, providing individuals with opportunities to work, play, and socialize in a hyper spatiotemporal environment. The metaverse promises a new era of digital experiences, where people can engage in real-time, multi-dimensional interactions, collaborating and communicating with others.

This research paper aims to delve into the fundamental aspects of the metaverse, exploring its definition, potential applications, and the transformative impact it may have across various domains. It seeks to provide a comprehensive review of the metaverse, analyzing its historical development, key technologies, and implications for creativity, commerce, and social interaction. The paper also acknowledges the challenges and opportunities associated with the metaverse and presents future perspectives for its advancement.

By examining the metaverse from multiple angles, this research paper intends to contribute to the understanding of this evolving paradigm and shed light on its potential implications for the future of human interaction in virtual environments.

Keywords: Metaverse · security · privacy · distributed virtual worlds · extended reality · artificial intelligence · blockchain

1 Introduction

The concept of the Metaverse has indeed been influenced by literary works, and one notable precursor is the science fiction novel "Snow Crash" published in 1992. This novel presented a realistic virtual world known as the Metaverse, which can be considered a prototype of the concept. The development of the Metaverse represents the advancement of virtual worlds, going beyond a simple mapping of the real world [1].

Unlike traditional concepts of space and time, the Metaverse offers a multidimensional and immersive virtual space. It provides users with an environment that transcends the limitations of physical reality, allowing for unique and interactive experiences. In this sense, the Metaverse represents a departure from traditional notions of spatial and temporal boundaries, offering a new paradigm of digital interaction and exploration [2].

S. Singh et al. (Eds.): AIDSP 2023, CCIS 2440, pp. 216–226, 2025.
https://doi.org/10.1007/978-3-031-88762-8_21

1.1 Background and Motivation

The metaverse has generated interest with advancements in virtual reality and immersive technologies. It aims to create a fully interconnected digital universe. This overview discusses the motivations behind studying the metaverse and the importance of integrating the real and virtual worlds. Technologies like the Internet of Things (IoT), digital twins, and blockchain play crucial roles in facilitating this integration. The IoT brings advancements to various aspects of life and production, digital twins enable real-time monitoring, and blockchain ensures secure management within the metaverse. [4]. The growing proliferation of IoT devices facilitates real-time information exchange and interaction between diverse devices. This phenomenon generates a vast amount of data that finds applications in various domains such as industrial production, city planning, healthcare, and more. [5]. The Internet of Things (IoT) plays [6, 7] a critical role as an enabling technology for digital twins. Digital twins involve creating virtual representations of physical objects or systems that can be monitored in real-time [8]. By combining IoT and digital twin technology, it becomes possible to map the Metaverse onto the real world, effectively bridging the gap between physical and virtual environments. This integration of IoT and digital twins contributes to the seamless connection and interaction between the physical and virtual realms within the Metaverse [7, 9, 10].

Blockchain technology, originating from Bitcoin, has had a profound impact on various domains, including business, society, and politics. It offers decentralized, immutable, transparent, and verifiable distributed databases. Blockchain technology provides secure transactions and data integrity by offering advantages in terms of speed, security, and availability. It has found applications beyond the financial industry and has the potential to disrupt existing business models and create new ones. With its ability to provide secure and anonymous services without relying on thirdparty control, blockchain can contribute to the integrity and security of transactions within the Metaverse [11–14].

1.2 Research Objectives

The research objectives outline the specific goals of the research paper. They serve as a roadmap for the subsequent sections and provide a clear direction for the study. The objectives may include:

1. Defining the metaverse: The paper aims to provide a clear and comprehensive definition of the metaverse, considering its conceptual origins, historical development, and key characteristics.
2. Examining enabling technologies: The research paper intends to explore the technologies that enable the creation and functioning of the metaverse, such as virtual reality, augmented reality, blockchain, artificial intelligence, and more.
3. Exploring applications and implications: The study seeks to investigate the potential applications and implications of the metaverse across various domains, including gaming, social interaction, commerce, education, healthcare, and architecture.
4. Identifying challenges and opportunities: The research paper aims to identify the challenges and opportunities associated with the metaverse, including technological hurdles, ethical considerations, privacy and security issues, legal and regulatory aspects, and the economic impact.

5. Presenting future perspectives: Based on the current understanding, the research paper aims to present future perspectives for the development and advancement of the metaverse, including potential directions for research, technological advancements, user experience improvements, and societal implications.

By addressing these research objectives, this paper seeks to contribute to the existing knowledge base, foster discussion, and provide insights for researchers, industry professionals, policymakers, and anyone interested in understanding the metaverse and its potential impact on our digital future.

2 Definition and Evolution of the Metaverse

2.1 Origins and Conceptualization

In this section, the research paper explores the origins and conceptualization of the metaverse. It delves into the early roots of the concept and examines influential works, ideas, and pioneers that contributed to its formation. The discussion may include references to science fiction literature, virtual reality experiments, early online communities, and visionary thinkers who envisioned a shared digital universe.

2.2 Evolution of the Metaverse Concept

This subsection focuses on the evolution of the metaverse concept over time. It traces the development of the term "metaverse" and its popularization within the fields of technology, gaming, and entertainment. The paper may discuss key milestones, influential projects, and significant advancements that have shaped the understanding and scope of the metaverse. Examples might include virtual worlds like Second Life, the emergence of multiplayer online games, and the integration of augmented reality in mobile applications.

2.3 Key Characteristics and Defining Features

This subsection explores the key characteristics and defining features of the metaverse. It aims to establish a set of common attributes that distinguish the metaverse from other virtual or augmented reality experiences. Some potential key characteristics and defining features of the metaverse might include:

1. Interconnectivity: The metaverse is characterized by a network of interconnected virtual worlds and environments, allowing users to seamlessly move between different spaces and experiences.
2. User-driven content creation: Users have the ability to create and contribute to the metaverse's content, shaping and influencing the virtual environment.
3. Persistence: The metaverse is designed to be persistent, meaning that it continues to exist and evolve even when users are not actively present. Changes made by one user can have an impact on the experiences of others.

4. Immersion and realism: The metaverse aims to provide immersive and realistic experiences through technologies like virtual reality and augmented reality, engaging multiple senses and creating a sense of presence.
5. Social interaction: The metaverse facilitates social interaction among users, allowing them to communicate, collaborate, and socialize in real time within the virtual environment.
6. Cross-platform compatibility: The metaverse is accessible across different platforms and devices, ensuring compatibility and enabling users to engage regardless of their preferred technology.
7. Scalability: The metaverse is designed to support a large number of users simultaneously, providing a scalable infrastructure to accommodate a diverse range of activities and interactions.

By exploring these key characteristics and defining features, the research paper provides a foundation for understanding the unique nature of the metaverse and its potential implications across various domains.

3 Key Technologies Enabling the Metaverse

1. Virtual Reality (VR): Virtual reality is a technology that creates a simulated digital environment that immerses users in a computer-generated 3D world. VR headsets and controllers enable users to interact with and navigate within the virtual environment. VR plays a crucial role in the metaverse by providing a sense of presence and immersion, allowing users to explore and interact with virtual worlds and entities [15].
2. Augmented Reality (AR): augmented reality (AR) combines virtual elements with the real world to enhance user perception and interaction. AR technologies like smartphones, tablets, and smart glasses overlay digital content onto the physical environment in real time. In the metaverse, AR enables the seamless integration of virtual and real-world experiences by overlaying digital content and interactions onto physical spaces. [16].
3. Mixed Reality (MR): mixed reality combines virtual reality and augmented reality, enabling users to interact with and manipulate virtual and real-world objects simultaneously. MR technologies enable the integration of virtual content into the user's physical environment, creating a more seamless and interactive experience. In the metaverse, MR can enhance the sense of presence and enable users to interact with virtual objects within their real-world surroundings [17].
4. Blockchain and Distributed Ledger Technology (DLT): Blockchain and DLT have the potential to play a significant role in the metaverse by enabling secure and transparent transactions, ownership of digital assets, and decentralized governance. These technologies provide a distributed and immutable ledger that can facilitate the exchange, verification, and protection of virtual assets, such as digital currencies, virtual goods, and unique virtual identities.
5. Artificial Intelligence (AI): Artificial intelligence technologies, such as machine learning, computer vision, and natural language processing, can augment the metaverse by enabling intelligent virtual entities, personalized experiences, and lifelike

simulations. AI algorithms can generate and control virtual characters, enable natural language interactions, and provide personalized recommendations and content based on user preferences and behavior.

6. Internet of Things (IoT): The Internet of Things (IoT) encompasses a network of interconnected physical devices that are equipped with sensors, software, and connectivity. In the context of the metaverse, IoT can enable the integration of physical objects and environments into the virtual space. Smart devices, wearables, and environmental sensors can capture and transmit data, enabling real-time interactions and feedback between the physical and virtual realms.
7. Other Enabling Technologies: In addition to the aforementioned technologies, several other enabling technologies contribute to the development of the metaverse. These may include cloud computing for storage and processing power, 5G and future wireless networks for high-speed connectivity, haptic feedback systems for tactile interactions, motion tracking for precise user movement, and advanced graphics rendering for realistic visual representations.

These key technologies provide the foundation for creating immersive and interconnected digital experiences within the metaverse. The integration and advancement of these technologies continue to drive the evolution and realization of the metaverse concept.

4 Applications and Implications of the Metaverse

1. Gaming and Entertainment: The metaverse offers significant opportunities for gaming and entertainment experiences. Users can immerse themselves in virtual worlds, interact with other players, and participate in multiplayer games. The metaverse allows for the creation of dynamic and evolving game environments, where players can engage in collaborative or competitive activities, explore vast landscapes, and undertake quests and challenges.
2. Social Interaction and Communication: The metaverse enhances social interaction and communication by providing virtual spaces where individuals can meet, connect, and socialize with others in real time. Users can create and customize avatars, engage in virtual conversations, attend virtual parties or gatherings, and participate in shared activities and events. The metaverse can bridge geographical boundaries, enabling people from different locations to interact and form connections.
3. Virtual Commerce and Economy: The metaverse has the potential to revolutionize commerce and the economy. Users can engage in virtual commerce, buying and selling digital goods, virtual real estate, and unique virtual assets. Blockchain and decentralized technologies can ensure secure transactions, establish ownership rights, and enable the development of virtual economies. Businesses can create virtual storefronts, host virtual events, and offer immersive shopping experiences within the metaverse.
4. Education and Training: The metaverse holds promise for education and training by providing immersive and interactive learning environments. Virtual classrooms and training simulations can enhance traditional teaching methods, allowing students to engage in realistic and experiential learning. Virtual field trips, historical recreations,

and scientific experiments can be conducted within the metaverse, providing engaging educational experiences. Collaboration and knowledge-sharing among students and educators can also be facilitated.
5. Healthcare and Therapy: The metaverse can be applied in healthcare and therapy settings to provide innovative solutions. Virtual reality can be used for pain management, phobia treatment, and rehabilitation. Telemedicine and remote consultations can be enhanced by integrating augmented reality for real-time visualization and guidance. Virtual support groups and therapy sessions can be conducted within the metaverse, offering accessible and inclusive mental health services.
6. Architecture and Design: Architects and designers can utilize the metaverse to create and visualize virtual architectural spaces and designs. Virtual environments allow for immersive walkthroughs, enabling clients to experience and provide feedback on proposed designs. Collaboration among architects, engineers, and clients can be enhanced through shared virtual workspaces, facilitating real-time design iterations and project coordination.
7. Virtual Events and Conferences: The metaverse offers opportunities for hosting virtual events and conferences. Participants can attend virtual conferences, trade shows, and exhibitions from anywhere in the world, eliminating the need for physical travel. Virtual event spaces can be designed to simulate real-world venues, enabling networking opportunities, presentations, and interactive sessions. Attendees can engage with virtual exhibitors, access digital content, and participate in immersive experiences.

These applications and implications of the metaverse highlight its potential to transform various domains, providing new and innovative ways for people to interact, learn, and conduct business, and experience entertainment and events. The metaverse opens up exciting possibilities for creating immersive, interconnected, and inclusive digital experiences.

5 Challenges and Opportunities

1. Technological Challenges: The metaverse faces several technological challenges, including the need for advanced hardware and infrastructure to support high-quality graphics, real-time interactions, and seamless experiences. Improvements in processing power, network bandwidth, and device capabilities are required to create a smooth and immersive metaverse. Additionally, ensuring interoperability among different platforms, devices, and technologies poses a significant technological challenge.
2. Privacy and Security: Privacy and security are major concerns in the metaverse. As users interact and share personal information within virtual environments, there is a need to protect their privacy and data. Ensuring secure transactions, safeguarding personal identities, and addressing potential vulnerabilities, such as hacking or identity theft, are critical challenges. Establishing robust security measures and user control over data privacy are essential for building trust in the metaverse.

3. Ethical Considerations: The metaverse raises ethical considerations related to user behavior, content moderation, and virtual experiences. Issues such as cyberbullying, harassment, and inappropriate content can arise within the metaverse. Ethical guidelines and responsible governance are necessary to address these concerns. Ensuring that the metaverse promotes inclusivity, diversity, and respectful interactions is crucial.
4. Accessibility and Inclusivity: Creating an accessible and inclusive metaverse is a significant challenge. Ensuring that individuals with disabilities can fully engage and participate in virtual experiences requires considering factors such as assistive technologies, user interfaces, and accommodating different sensory needs. Language barriers, affordability, and digital literacy also contribute to accessibility challenges that need to be addressed to avoid excluding certain populations.
5. Legal and Regulatory Aspects: The metaverse operates in a complex legal and regulatory landscape. Issues such as intellectual property rights, virtual asset ownership, taxation, and jurisdictional disputes need to be carefully addressed. Establishing clear legal frameworks and regulations that govern virtual transactions, user rights, and data protection are essential for the development and adoption of the metaverse.
6. Business and Economic Challenges: The metaverse presents both opportunities and challenges for businesses and the economy. Establishing sustainable business models, monetization strategies, and fair economic systems within the metaverse can be challenging. Ensuring that creators and content contributors are appropriately rewarded and that virtual economies are stable and transparent requires careful consideration. The balance between commercial interests and user experiences is crucial.
7. Environmental Impact: The metaverse's growth and increasing energy demands can have significant environmental implications. Data centers, servers, and highperformance computing required to support the metaverse consume substantial amounts of energy. Balancing the expansion of the metaverse with environmentally sustainable practices, such as energy-efficient infrastructure and responsible data management, is essential to mitigate its environmental impact.

These challenges present opportunities for innovation, collaboration, and research. Addressing these challenges can lead to the development of more robust and inclusive metaverse experiences while ensuring privacy, security, and ethical considerations are upheld. Overcoming these challenges will contribute to the responsible and sustainable growth of the metaverse.

6 Future Perspectives and Directions

1. Advancements in Hardware and Infrastructure: Advancements in hardware and infrastructure will play a crucial role in the future development of the metaverse. Improvements in processing power, graphics capabilities, network bandwidth, and storage capacity will enable more immersive and realistic virtual experiences. Advancements in wearable devices, haptic feedback systems, and sensory interfaces will further enhance user immersion and interaction within the metaverse.

2. Interoperability and Standards: Interoperability and the development of common standards will be key for the future of the metaverse. Seamless integration and communication between different platforms, devices, and technologies will enable users to traverse and interact within the metaverse more easily. The establishment of open standards for data formats, communication protocols, and virtual asset interoperability will foster a more connected and unified metaverse ecosystem.
3. User Experience and Interface Design: User experience and interface design will continue to evolve to create more intuitive and user-friendly interactions within the metaverse. Advancements in natural language processing, gesture recognition, and emotion detection will enable more natural and immersive user experiences. User interface design will focus on reducing complexity, improving accessibility, and enhancing the ease of navigation within the metaverse.
4. Integration with Physical Environments: The future of the metaverse will involve increased integration with physical environments. Augmented reality technologies will enable the overlay of virtual content onto the real world, blurring the boundaries between physical and virtual spaces. The metaverse will seamlessly connect virtual and physical objects, enabling users to interact with their surroundings in new and innovative ways.
5. Collaborative Efforts and Industry Partnerships: The development of the metaverse will require collaborative efforts and industry partnerships. Companies from various sectors, including technology, gaming, entertainment, education, healthcare, and architecture, will collaborate to create a unified and interconnected metaverse. Partnerships between hardware manufacturers, content creators, platform developers, and service providers will drive innovation, enhance interoperability, and create diverse metaverse experiences.
6. Social and Cultural Implications: The metaverse's future will have profound social and cultural implications. It will shape new forms of social interaction, community building, and identity expression. The metaverse can foster cross-cultural connections, enable global collaboration, and provide a platform for cultural exchange and creative expression. Societal norms, ethics, and values within the metaverse will need to be addressed to ensure inclusivity, diversity, and responsible digital citizenship.
7. Impact on Work and Society: The metaverse's future will impact various aspects of work and society. It will transform how we collaborate, learn, conduct business, and entertain ourselves. Remote work and telecommuting may become more prevalent, leveraging the immersive and collaborative capabilities of the metaverse. New job roles and industries may emerge, focusing on metaverse development, content creation, virtual events, and virtual economy management. The metaverse's impact on social dynamics, privacy, mental health, and the redistribution of economic opportunities will need to be studied and addressed.

Considering these future perspectives and directions, the development and advancement of the metaverse will continue to shape our digital future. It offers exciting possibilities for innovation, creativity, collaboration, and interconnectedness, while also posing challenges that require careful consideration and proactive solutions.

7 Conclusion

7.1 Summary of Key Findings

In this research paper, we have explored the concept of the metaverse, its evolution, key technologies enabling it, applications and implications, challenges and opportunities, and future perspectives. The key findings can be summarized as follows:

- The metaverse is a virtual universe where people can interact with each other and digital content in real-time, blurring the boundaries between the physical and digital worlds.
- The concept of the metaverse has evolved over time, driven by advancements in technology and the growing demand for immersive digital experiences.
- Virtual reality (VR), augmented reality (AR), mixed reality (MR), blockchain, artificial intelligence (AI), internet of things (IoT), and other enabling technologies play a critical role in realizing the metaverse.
- The metaverse has applications in gaming, entertainment, social interaction, commerce, education, healthcare, architecture, and virtual events, among others.
- Challenges such as technological limitations, privacy and security concerns, ethical considerations, accessibility, legal and regulatory aspects, business and economic factors, and environmental impact need to be addressed for the responsible and sustainable development of the metaverse.
- The future of the metaverse will see advancements in hardware and infrastructure, interoperability and standards, user experience and interface design, integration with physical environments, collaborative efforts and industry partnerships, social and cultural implications, and its impact on work and society.

7.2 Recommendations for Future Research

Based on the findings, the following recommendations for future research can be made:

- Further exploration of the technical aspects of the metaverse, including advancements in hardware, network infrastructure, and immersive technologies.
- Investigation into the development of interoperability standards and protocols to enable seamless connectivity and interaction within the metaverse.
- Research on user experience design principles, interface design, and interaction paradigms that enhance immersion, accessibility, and ease of use within the metaverse.
- Examination of the integration of the metaverse with physical environments, exploring the potential for augmented reality and the Internet of Things to create seamless and interactive experiences.
- Exploration of the social, cultural, and psychological impacts of the metaverse on individuals, communities, and society at large, considering issues of identity, social dynamics, privacy, and digital citizenship.
- Investigation into the economic implications of the metaverse, including sustainable business models, virtual economies, and the redistribution of economic opportunities.
- Research on the environmental impact of the metaverse and the development of strategies to minimize energy consumption and promote eco-friendly practices.

By focusing on these research areas, we can advance our understanding of the metaverse and its potential implications, ensuring its responsible and inclusive development in the future. The metaverse holds immense promise as a transformative digital space, and continued research and exploration will contribute to unlocking its full potential.

References

1. Dionisio, J.D.N., III, W.G.B., Gilbert, R.: 3D virtual worlds and the Metaverse: current status and future possibilities. ACM Comput. Surv. (CSUR) **45**(3), 1–38 (2013)
2. Mystakidis, S.: Metaverse. Encyclopedia **2**(1), 486–497 (2022)
3. Xianyi Yang, X.N.: Brief History of Communications. The Posts and Telecommunications Press (2020)
4. Liu, M., Fang, S., Dong, H., Xu, C.: Review of digital twin about concepts, technologies, and industrial applications. J. Manuf. Syst. **58**, 346–361 (2021)
5. Lee, I., Lee, K.: The Internet of Things (IoT): applications, investments, and challenges for enterprises. Bus. Horiz. **58**(4), 431–440 (2015)
6. Reyna, A., Martín, C., Chen, J., Soler, E., Díaz, M.: On blockchain and its integration with IoT. challenges and opportunities. Futur. Gener. Comput. Syst. **88**, 173–190 (2018)
7. Da Xu, L., Lu, Y., Li, L.: Embedding blockchain technology into IoT for security: a survey. IEEE Internet Things J. **8**(13), 10452–10473 (2021)
8. Shafique, K., Khawaja, B.A., Sabir, F., Qazi, S., Mustaqim, M.: Internet of things (IoT) for next-generation smart systems: a review of current challenges, future trends and prospects for emerging 5G-IoT scenarios. IEEE Access **8**, 23022–23040 (2020)
9. Wang, J., Lim, M.K., Wang, C., Tseng, M.L.: The evolution of the Internet of Things (IoT) over the past 20 years. Comput. Ind. Eng. **155**, 107174 (2021)
10. Fuller, A., Fan, Z., Day, C., Barlow, C.: Digital twin: enabling technologies, challenges and open research. IEEE Access **8**, 108952–108971 (2020)
11. Yli-Huumo, J., Ko, D., Choi, S., Park, S., Smolander, K.: Where is current research on blockchain technology?—A systematic review. PLoS One **11**(10), e0163477 (2016)
12. Nofer, M., Gomber, P., Hinz, O., Schiereck, D.: Blockchain. Bus. Inf. Syst. Eng. **59**, 183–187 (2017)
13. Monrat, A.A., Schelén, O., Andersson, K.: A survey of blockchain from the perspectives of applications, challenges, and opportunities. IEEE Access **7**, 117134–117151 (2019)
14. Atzori, M.: Blockchain technology and decentralized governance: Is the state still necessary? (2015). Available at SSRN 2709713
15. Leenes, R.: Privacy in the Metaverse: regulating a complex social construct in a virtual world. In: IFIP International Summer School on the Future of Identity in the Information Society, pp. 95-112. Springer US, Boston, MA (2007)
16. Falchuk, B., Loeb, S., Neff, R.: The social Metaverse: battle for privacy. IEEE Technol. Soc. Mag. **37**(2), 52–61 (2018)
17. Shang, J., Chen, S., Wu, J., Yin, S.: ARSpy: breaking location-based multiplayer augmented reality application for user location tracking. IEEE Trans. Mob. Comput. **21**(2), 433–447 (2020)
18. Yang, Q., Zhao, Y., Huang, H., Xiong, Z., Kang, J., Zheng, Z.: Fusing blockchain and AI with Metaverse: a survey. IEEE Open J. Comput. Soc. **3**, 122–136 (2022)
19. Jiang, W., Han, B., Habibi, M.A., Schotten, H.D.: The road towards 6G: a comprehensive survey. IEEE Open J. Commun. Soc. **2**, 334–366 (2021)
20. Pham, Q.V., et al.: A survey of multi-access edge computing in 5G and beyond: fundamentals, technology integration, and state-of-the-art. IEEE Access **8**, 116974–117017 (2020)

21. Dogra, A., Jha, R.K., Jain, S.: A survey on beyond 5G network with the advent of 6G: architecture and emerging technologies. IEEE Access **9**, 67512–67547 (2020)
22. Ahmad, W.S.H.M.W., et al.: 5G technology: towards dynamic spectrum sharing using cognitive radio networks. IEEE Access **8**, 14460–14488 (2020)
23. You, X., et al.: Towards 6G wireless communication networks: vision, enabling technologies, and new paradigm shifts. Sci. China Inf. Sci. **64**, 1–74 (2021)
24. Kumar, R., Mishra, A.K., Singh, D.K.: Packet loss avoidance in mobile Adhoc network by using trusted LDoS techniques. In: 2nd International Conference on Data, Engineering and Applications (IDEA), Bhopal, India, pp. 1–5 (2020). https://doi.org/10.1109/IDEA49133.2020.9170673
25. Kumar, R., Singh, D.K., Mishra, A.K.: An approach to extract fine detail and unclear information by enhancing computed tomography image. In: 2nd International Conference on Data, Engineering and Applications (IDEA), Bhopal, India, pp. 1–6 (2020).https://doi.org/10.1109/IDEA49133.2020.9170735

Author Index

S. Singh et al. (Eds.): AIDSP 2023, CCIS 2440, pp. 227–228, 2025.
https://doi.org/10.1007/978-3-031-88762-8

The manufacturer's authorised representative in the EU is Springer Nature Customer Service Centre GmbH, Europaplatz 3, 69115 Heidelberg, Germany. If you have any concerns regarding our products, please contact ProductSafety@springernature.com

Printed and bound by CPI Group (UK) Ltd, Croydon, CR0 4YY

31/07/2026

02182982-0003

Claudia Ratti • Rene Bellwied

The Deconfinement Transition of QCD

Theory Meets Experiment

Springer

Claudia Ratti
Department of Physics
University of Houston
Houston, TX, USA

Rene Bellwied
Department of Texas
University of Houston
Houston, TX, USA

ISSN 0075-8450 ISSN 1616-6361 (electronic)
Lecture Notes in Physics
ISBN 978-3-030-67234-8 ISBN 978-3-030-67235-5 (eBook)
https://doi.org/10.1007/978-3-030-67235-5

This Springer imprint is published by the registered company Springer Nature Switzerland AG
The registered company address is: Gewerbestrasse 11, 6330 Cham, Switzerland

To Carlo, Rita, and Mario

Preface

Over the past decade, both experiment and theory have made great strides in conjuring a deeper understanding of the physics of strong interactions. The parallel programs at the Relativistic Heavy Ion Collider and the Large Hadron Collider, in concert with the new supercomputing facilities in the USA and Europe, have enabled us to develop not only theoretical predictions based on accurate, continuum extrapolated lattice QCD calculations, but also to obtain experimental verifications to many of these predictions. The landscape of the phase diagram of nuclear matter under extreme conditions has thus come into full focus, and with this publication, we are trying to capture the latest developments on gaining the ultimate knowledge of the early evolution of matter shortly after the big bang.

This book is based on a graduate *Physics of strong interactions* course that we teach in the Physics Department at the University of Houston and on a series of lectures that we have delivered at summer schools for advanced Ph.D. students and postdocs and at doctoral training programs. Based on the aforementioned recent developments in theory and experiment, this is a very dynamic subject which continuously evolves, driven by the latest results. We therefore tried to capture the essence of new insights without losing track of the long established foundation of quantum chromodynamics.

The phase diagram of strongly interacting matter is still vastly unexplored, due to the fact that quantum chromodynamics cannot be solved at finite baryonic density. The focus is on the characterization of strongly interacting matter in the non-perturbative regime, based on the interplay between first principle fundamental theory and experimental observations. We present the state of the art in the field and discuss the main open questions and current methods to approach them. We stress that given the complexity of the problem, a synergy between fundamental theory, phenomenology, and experiment is crucial. This book is written in this spirit, as it reports on several complementary approaches to study different regions of the phase diagram. On the experimental side, the accelerator-based experiments are being complemented by astrophysical results of systems at higher density and lower temperature. In this regard, gravitational wave measurements and neutron star measurements have opened a new window of opportunity to link future finite density results to new theoretical insights, which are based on ever improving extrapolation methods of the underlying dynamics.

The book is meant for an audience at the advanced graduate/postdoctoral researcher level. It is divided into nine chapters, which cover a range of topics from equation of state at zero and finite density to heavy-flavors and quarkonia. We hope that it helps to train the next generation of scientists working on strongly interacting matter. The book will not only give them a broad overview of the field, but also expose them to the latest results and the plans for future investigations on the experimental and theoretical side.

We definitely benefitted from several enlightening discussions with students and colleagues in the field whose names are too numerous to list. Nevertheless, we would like to stress that input from theorists and experimentalists was crucial to arrive at a balanced manuscript, which shows that the future of quantum chromodynamics and the phase diagram of strong interactions is firmly rooted in the close collaboration between theory and experiment. Most of the recent predictions will be almost immediately verifiable through experiment, which prepares us for an exciting few decades of new physics to come.

Houston, TX, USA
September 2020

Claudia Ratti
Rene Bellwied

Acknowledgments

The authors acknowledge fruitful discussions with Gert Aarts, Jörg Aichelin, Paolo Alba, Marcus Bluhm, Szabolcs Borsanyi, Elena Bratkovskaya, Zoltan Fodor, Vincenzo Greco, Joaquin Grefa, Jana Günther, Volker Koch, Valentina Mantovani Sarti, Debora Mroczek, Marlene Nahrgang, Angel Nava Acuña, Jacquelyn Noronha-Hostler, Jorge Noronha, Paolo Parotto, Attila Pasztor, Israel Portillo, Krishna Rajagopal, Thomas Schaefer, Jamie Stafford, and Mikhail Stephanov. R.B. would also like to thank several students and postdoctoral fellows in his research group who directly contributed to the experimental results and discussions shown here, namely Fernando Flor, Anders Knospe, Nalinda Kulathunga, Corey Myers, Gabby Olinger, and Ejiro Umaka. This material is based upon work supported by the National Science Foundation under grant no. PHY-1654219 and by the U.S. Department of Energy, Office of Science, Office of Nuclear Physics, within the framework of the Beam Energy Scan Theory (BEST) Topical Collaboration as well as the US-DOE Grant No.DEFG02-07ER4152.

Contents

About the Authors

Prof. Claudia Ratti received her Ph.D. in theoretical physics from the University of Torino (Italy) in 2003. She has been a postdoctoral researcher at the Technical University of Munich (Germany), ECT* in Trento (Italy), State University of New York at Stony Brook (USA), and Wuppertal University (Germany). In 2010, she became junior professor and group leader at Torino University, thanks to an FIRB grant funded by the Italian Ministry of Education, University and Research. In 2014, she became an assistant professor in the Physics Department at the University of Houston. Since 2017, she is a tenured associate professor. Author of more than 70 publications in peer-reviewed international journals, she has presented the results of her research in more than a hundred seminars at international conferences and universities. In light of her achievements, she has been awarded the 2011 International Zonta Prize for Women in Science, the 2012 Prize "Giuseppe Borgia" for Best Italian Physicist below 35 granted by the Italian Academy of Science (Accademia Nazionale dei Lincei), and the NSF Career Award in 2017. Her research is mainly focused on the study of strongly interacting matter under extreme conditions of temperature and density, such as the one created at the Large Hadron Collider (LHC) at CERN and the Relativistic Heavy Ion Collider (RHIC) at BNL.

Prof. Rene Bellwied is a distinguished MD Anderson professor of physics at the University of Houston, TX, USA. He received his Ph.D. from the Johannes Gutenberg University in Mainz, Germany, in 1989, and was awarded a Feodor Lynen fellowship from the Humboldt Foundation to continue his research as a postdoctoral fellow at Stony Brook University on Long Island, NY, USA. In 1991, he became an assistant professor at Wayne State University in Detroit, where he remained as a professor until 2010. At the University Houston, he became the group leader of the Experimental Nuclear Physics effort, which is involved in the STAR experiment at the Relativistic Heavy Ion Collider (RHIC) at Brookhaven National Laboratory (BNL) on Long Island, New York, and the ALICE experiment at the Large Hadron Collider (LHC) at CERN in Geneva, Switzerland.

Dr. Bellwied was a founding member of the STAR collaboration. He served as one of eight project managers of STAR from 1992 to 2007, he was the deputy spokesperson of STAR from 2001 to 2002. He joined the ALICE experiment at CERN in 2010 and served as coordinator of the ALICE-USA collaboration from 2015 to 2018.

In addition, he was the chair of the U.S. National User Facility Organization (NUFO) for 4 years (2008–2012) and served as an appointed member on the Physics Policy Committee (PPC) of the American Physical Society (APS). He is presently chair of the Texas Section of the APS. He is also the lead organizer of an annual international conference in his field, he serves on two peer-reviewed journal editorial boards, and he is a reviewer for the four highest-impact nuclear physics journals, as well as the grant offices of the Department of Energy (DOE) and the National Science Foundation (NSF). He is a member of the Sigma-Xi Honors Society.

Prof. Bellwied has published over 1,000 peer-reviewed papers that gathered over 85,000 citations. His h-index is 151 according to Google Scholar.

Acronyms

AdS/CFT	Anti-de-Sitter/Conformal Field Theory
ALICE	A Large Ion Collision Experiment
AMPT	A Multi-Phase Transport Model
BES	Beam Energy Scan
BR	Bayesian Reconstruction
CBWC	Centrality Bin Width Correction
CEE	CSR External-Target Experiment
CEP	Critical End Point
CP	Critical Point
CSR	Cooler Storage Ring
DSE	Dyson-Schwinger Equations
EMCal	Electro-Magnetic Calorimeter
EoS	Equation of State
EPOS	Energy conserving quantum mechanical multiple scattering approach, based on Partons, parton ladders, Off-shell remnants, and Splitting of parton ladders
FAIR	Facility for Antiproton and Ion Research
FO	Freeze-Out
FRG	Functional Renormalization Group
FXT	Fixed-Target
GSI	Gesellschaft für SchwerIonenforschung mbh
HADES	High-Acceptance Di-Electron Spectrometer
HIJING	Heavy-Ion Jet INteraction Generator
HISQ	Highly Improved Staggered Quark
HMC	Hybrid Monte Carlo
HQET	Heavy Quark Effective Theory
hQM	Hypercentral Quark Model
HRG	Hadron Resonance Gas
HTL	Hard Thermal Loop
ITS	Inner Tracking Silicon
LHC	Large Hadron Collider

LO	Leading Order
LQCD	Lattice QCD
MC	Monte Carlo
MEM	Maximum Entropy Method
NBD	Negative Binomial Distribution
NICA	Nuclotron-based Ion Collider fAcility
NJL	Nambu Jona-Lasinio
NLO	Next-to-Leading Order
NRQCD	Non-Relativistic Quantum Chromodynamics
PDG	Particle Data Group
PNJL	Polyakov-loop-extended Nambu Jona-Lasinio
pNRQCD	potential Non-Relativistic Quantum Chromodynamics
pQCD	Perturbative Quantum Chromodynamics
PQM	Polyakov-loop-extended Quark Meson
QCD	Quantum Chromodynamics
QED	Quantum Electrodynamics
QGP	Quark-Gluon Plasma
QM	Quark Model
RHIC	Relativistic Heavy Ion Collider
RW	Roberge-Weiss
SB	Stefan-Boltzmann
SIS	Schwer-Ionen-Synchrotron
SPS	Super Proton Synchrotron
STAR	Solenoidal Tracker At RHIC
TOF	Time of Flight
TPC	Time Projection Chamber
UrQMD	Ultra-relativistic Quantum Molecular Dynamics
UV	Ultraviolet
VdW	van der Waals
WB	Wuppertal-Budapest

Part I

Bulk Properties of Strongly Interacting Matter

Introduction to Lattice QCD

1

Abstract

In this chapter we give a brief introduction to Quantum Chromodynamics, the fundamental theory of strong interactions, and its discretization on a four-dimensional hypercubic grid. We discuss the requirements imposed by gauge symmetry and translate them on the lattice. We identify the fermion fields as living on the lattice sites and the gauge fields on the lattice links. We briefly introduce the problem of fermion doubling on the lattice. Finally, we discuss Monte Carlo techniques and give an example of importance sampling algorithms. For books and reviews on the topic we recommend (Rothe (1992) World Sci Lect Notes Phys 43:1–381; Smit (2002) Camb Lect Notes Phys 15:1–271).

1.1 QED Action, QCD Action and Gauge Invariance

The derivation in this section follows the one described in Ref. [1]. Quantum Electrodynamics (QED) is the theory which describes the interaction between Dirac fermion fields (e.g. electrons) and the electromagnetic field. The QED action in Euclidean spacetime is given by

$$S_{QED} = \int d^4x \mathcal{L}_{QED}(x) \tag{1.1}$$

with

$$\mathcal{L}_{QED}(x) = \frac{1}{4}\mathcal{F}_{\mu\nu}(x)\mathcal{F}^{\mu\nu}(x) + \bar{\psi}(x)\gamma^{\mu}\left[\partial_{\mu} + ieA_{\mu}(x)\right]\psi(x) + m\bar{\psi}(x)\psi(x) \tag{1.2}$$

C. Ratti, R. Bellwied, *The Deconfinement Transition of QCD*, Lecture Notes in Physics 981, https://doi.org/10.1007/978-3-030-67235-5_1

where $\psi(x)$ is the Dirac field, $A_\mu(x)$ is the photon field, $\mathcal{F}_{\mu\nu}(x) = \partial_\mu A_\nu(x) + \partial_\nu A_\mu(x)$ is the electromagnetic field strength tensor, e is the elementary charge (with $e > 0$), m the fermion mass and γ^μ are Dirac matrices. Usually, one re-defines $A_\mu \to \frac{1}{e}A_\mu$, so that the Lagrangian reads

$$\mathcal{L}_{QED} = \frac{1}{4e^2}\mathcal{F}_{\mu\nu}(x)\mathcal{F}^{\mu\nu}(x) + \bar{\psi}(x)\gamma^\mu\left[\partial_\mu - iqA_\mu(x)\right]\psi(x) + m\bar{\psi}(x)\psi(x) \tag{1.3}$$

and $q = -1$ for the electron by convention. The QED action is invariant under gauge transformations of the form

$$\psi'(x) = e^{i\omega(x)q}\psi(x) \tag{1.4}$$

$$\bar{\psi}'(x) = e^{-i\omega(x)q}\bar{\psi}(x) \tag{1.5}$$

$$A'_\mu(x) = A_\mu(x) + \partial_\mu\omega(x), \tag{1.6}$$

namely we have $S(\bar{\psi}, \psi, A) = S(\bar{\psi}', \psi', A')$. The phase factors $\Omega(x) = e^{i\omega(x)}$ form an abelian $U(1)$ group for each x.

The covariant derivative $D_\mu\psi(x)$ transforms like $\psi(x)$ under gauge transformations

$$\begin{aligned} D'_\mu\psi'(x) &= \left[\partial_\mu - iqA'_\mu(x)\right]\psi'(x) = \qquad (1.7)\\ &= e^{iq\omega(x)}\left[\partial_\mu + iq\partial_\mu\omega(x) - iqA_\mu - iq\partial_\mu\omega(x)\right]\psi(x) = e^{iq\omega(x)}D_\mu\psi(x). \end{aligned}$$

Quantum Chromodynamics is the fundamental theory, which describes the strong interaction between quarks and gluons. Its action has the following form:

$$S_{QCD} = \int d^4x\,\mathcal{L}_{QCD}(x) \tag{1.8}$$

where the QCD Lagrangian reads

$$\mathcal{L}_{QCD}(x) = \frac{1}{4g^2}\mathcal{G}^k_{\mu\nu}(x)\mathcal{G}^{\mu\nu}_k(x) + \bar{\psi}(x)\gamma^\mu D_\mu\psi(x) + \bar{\psi}(x)m\psi(x). \tag{1.9}$$

The gauge group in this case is $SU(3)$, the group of unitary 3×3 matrices with determinant =1. An element of $SU(3)$ can be written as

$$\Omega = \exp[i\omega^k t_k] \tag{1.10}$$

where t_k is a complete set of Hermitian traceless 3×3 matrices. We have $\Omega^{-1} = \Omega^\dagger$ and

$$\mathrm{tr}(t_k) = 0 \quad \Rightarrow \quad \det(\Omega) = \exp[\mathrm{tr}(\ln \Omega)] = 1. \tag{1.11}$$

The matrices t_k are the generators of the group. The standard choice for them is $t_k = \frac{1}{2}\lambda_k$, where λ_k are the eight Gell-Mann matrices defined as follows:

$$\lambda_1 = \begin{pmatrix} 0 & 1 & 0 \\ 1 & 0 & 0 \\ 0 & 0 & 0 \end{pmatrix} \quad \lambda_2 = \begin{pmatrix} 0 & -i & 0 \\ i & 0 & 0 \\ 0 & 0 & 0 \end{pmatrix} \quad \lambda_3 = \begin{pmatrix} 1 & 0 & 0 \\ 0 & -1 & 0 \\ 0 & 0 & 0 \end{pmatrix} \quad \lambda_4 = \begin{pmatrix} 0 & 0 & 1 \\ 0 & 0 & 0 \\ 1 & 0 & 0 \end{pmatrix}$$

$$\lambda_5 = \begin{pmatrix} 0 & 0 & -i \\ 0 & 0 & 0 \\ i & 0 & 0 \end{pmatrix} \quad \lambda_6 = \begin{pmatrix} 0 & 0 & 0 \\ 0 & 0 & 1 \\ 0 & 1 & 0 \end{pmatrix} \quad \lambda_7 = \begin{pmatrix} 0 & 0 & 0 \\ 0 & 0 & -i \\ 0 & i & 0 \end{pmatrix} \quad \lambda_8 = \frac{1}{\sqrt{3}}\begin{pmatrix} 1 & 0 & 0 \\ 0 & 1 & 0 \\ 0 & 0 & -2 \end{pmatrix}.$$

They obey the following properties:

$$\begin{aligned} \mathrm{tr}[t_k t_l] &= \frac{1}{2}\delta_{kl} \\ [t_k, t_l] &= i f_{klm} t^m, \end{aligned} \tag{1.12}$$

where f_{klm} are the real structure constants of $SU(3)$, completely anti-symmetric in $k,\ l,\ m$.

The fermion fields ψ and $\bar{\psi}$ carry 3 discrete indices:

$$\psi(x) = \psi^{\alpha a f}(x) \tag{1.13}$$

where $\alpha = 1, \ldots, 4$ is the Dirac index, $a = 1,\ 2,\ 3\ (r,\ g,\ b)$ is the color index and $f = 1, \ldots,\ N_f$ is the flavor index. Quarks come in six different flavors (up u, down d, strange s, charm c, bottom b, top t). In the following, the heaviest quarks (b, t and sometimes c) will be mainly neglected. Gauge transformations and covariant derivatives act on the index a, Dirac γ matrices act on α, the diagonal mass matrix acts on f: $m = \mathrm{diag}(m_u,\ m_d,\ m_s,\ m_c,\ m_b,\ m_t)$.

The gauge transformation can be written as

$$\Omega = \exp[i\omega^k T_k], \tag{1.14}$$

where T_k are the $SU(3)$ generators in the chosen representation (e.g. the t_k correspond to the fundamental representation) and they satisfy the following

$$\begin{aligned} [T_k, T_l] &= i f_{klm} T^m \\ \mathrm{tr}(T_k T_l) &= \rho\delta_{kl}, \end{aligned} \tag{1.15}$$

where $\rho = \frac{1}{2}$ in the fundamental representation and $\rho = 3$ in the adjoint one.

In this case, the covariant derivative D_μ becomes a matrix, but we assume that it has a form similar to the QED one:

$$D_{\mu ab} = \delta_{ab}\partial_\mu - i(G_\mu)_{ab}, \tag{1.16}$$

where G_μ is the gluon field. We require the following properties under gauge transformations:

$$\begin{aligned}\psi'(x) &= \Omega(x)\psi(x)\\ \bar{\psi}'(x) &= \bar{\psi}(x)\Omega^\dagger(x)\\ D'_\mu(x)\psi'(x) &= \Omega(x)D_\mu\psi(x) = \Omega(x)D_\mu[\Omega^\dagger\psi'(x)],\end{aligned} \tag{1.17}$$

from which we get

$$D'_\mu(x) = \Omega(x)D_\mu(x)\Omega^\dagger(x) \tag{1.18}$$

or more explicitly

$$\partial_\mu - iG'_\mu = \Omega(\partial_\mu - iG_\mu)\Omega^\dagger = \Omega\partial_\mu\Omega^\dagger + \partial_\mu - i\Omega G_\mu\Omega^\dagger, \tag{1.19}$$

from which we get a transformation rule for G_μ:

$$G'_\mu = \Omega G_\mu\Omega^\dagger + i\Omega\partial_\mu\Omega^\dagger, \tag{1.20}$$

which reduces to (1.6) for an abelian group. In general, we can write $G_\mu(x) = G^m_\mu(x)T_m$, namely a function of x times the generator T_m.

The expression for the field tensor $\mathcal{G}_{\mu\nu}$ is chosen such that it transforms homogeneously under gauge transformations

$$\mathcal{G}'_{\mu\nu} = \Omega\mathcal{G}_{\mu\nu}\Omega^\dagger. \tag{1.21}$$

We can write it as

$$\begin{aligned}\mathcal{G}_{\mu\nu} = D_\mu G_\nu - D_\nu G_\mu &= \partial_\mu G_\nu - \partial_\nu G_\mu - i\left[G_\mu, G_\nu\right] =\\ &= \partial_\mu G_\nu - \partial_\nu G_\mu - iG^m_\mu G^n_\nu\left[T_m, T_n\right],\end{aligned} \tag{1.22}$$

or more specifically:

$$\mathcal{G}_{\mu\nu} = \mathcal{G}^k_{\mu\nu}T_k \tag{1.23}$$

with

$$\mathcal{G}^k_{\mu\nu} = \partial_\mu G^k_\nu - \partial_\nu G^k_\mu + f^k_{mn}G^m_\mu G^n_\nu. \tag{1.24}$$

The combination

$$\mathcal{G}^k_{\mu\nu}\mathcal{G}^{\mu\nu}_k = \frac{1}{\rho}\mathrm{tr}(\mathcal{G}_{\mu\nu}\mathcal{G}^{\mu\nu}) \tag{1.25}$$

is gauge invariant. We now can re-write the QCD Lagrangian in more detail:

$$\mathcal{L}_{QCD}(x) = \frac{1}{4g^2}\mathcal{G}^k_{\mu\nu}(x)\mathcal{G}^{\mu\nu}_k(x) + \bar{\psi}(x)\gamma^\mu(\partial_\mu - iT_m G^m_\mu)\psi(x) + \bar{\psi}(x)m\psi(x). \tag{1.26}$$

1.2 Action Discretization and Lattice Gauge Theory

The QCD partition function can be formally defined as follows

$$Z = \int [D\bar{\psi}][D\psi][DG]\exp[-S_{QCD}(\bar{\psi}, \psi, G)]. \tag{1.27}$$

Path integral expressions only have a meaning for systems with a finite number of degrees of freedom. For an action in the continuum space-time like the one above, the integrals are only formally defined, since we are dealing with an infinite number of degrees of freedom (labeled by the coordinates x and the discrete indices for the gauge and fermion fields). To give the above integral a precise meaning, one usually discretizes space-time by introducing a four-dimensional lattice, of which the one in Fig. 1.1 is a two-dimensional cartoon. The calculations are then performed on this discretized version of QCD, but eventually the lattice structure needs to be removed by taking the continuum limit, which is a non-trivial task.

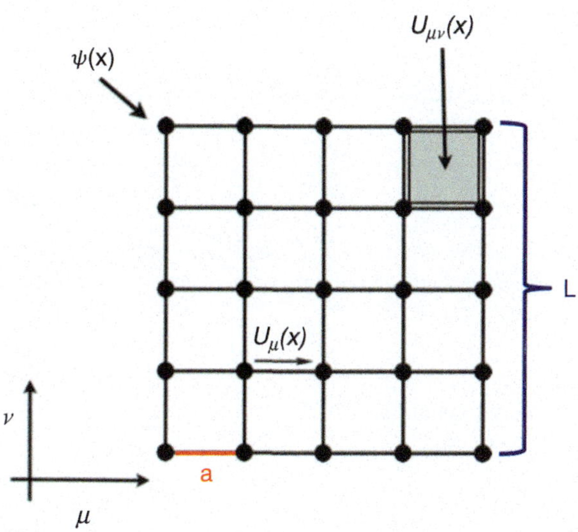

Fig. 1.1 Schematics of lattice QCD. The quark fields live on the sites, the gluon fields on the links. The lattice size is L and a indicates the lattice spacing

As Fig. 1.1 shows, we associate the fermion field $\psi(x)$ with the sites of the lattice. From now on, we will label the fields defined on the discrete lattice sites as $\psi_x = \psi(x)$. We consider a four-dimensional hypercubic lattice of size N^4, where N is the number of lattice spacings along a given direction. The lattice sites are labeled as

$$x_\mu = m_\mu a, \tag{1.28}$$

where a is the lattice spacing, usually measured in fm, and the discrete number m_μ labels the lattice site in the μ direction. In our case, $m_\mu = 0, \ldots, N$ and $\mu = 1, \ldots, 4$. The physical size of the hypercubic box is $L = Na$. The space-time volume is $V = L^4$. We will use the following notation

$$\sum_x = a^4 \sum_{m_1=0}^{N} \ldots \sum_{m_4=0}^{N} = a^4 \sum_m . \tag{1.29}$$

For a smooth function $f(x)$ we have, in the continuum limit

$$\sum_x f(x) \to \int_0^L f(x) d^4x \tag{1.30}$$

where the above limit is intended as $N \to \infty$, $a = \frac{L}{N} \to 0$ with L fixed. We choose $x = 0$ at the edge of the box, and N even.

The notation ψ_x means that the fermion field is assigned to the site x. Under local gauge transformations, it transforms as

$$\psi'_x = \Omega_x \psi_x . \tag{1.31}$$

The continuum derivative becomes a finite difference on the lattice

$$\partial_\mu \psi_x = \frac{1}{a}\left[\psi_{x+a\hat{\mu}} - \psi_x\right]. \tag{1.32}$$

We need to define a covariant derivative. We try the following form

$$D_\mu \psi_x = \frac{1}{a}\left[\psi_{x+a\hat{\mu}} - \psi_x\right] - i\left[\tilde{C}_{\mu,x}\psi_x + C_{\mu,x}\psi_{x+a\hat{\mu}}\right]. \tag{1.33}$$

In the above equation, $C_{\mu,x}$ and $\tilde{C}_{\mu,x}$ are supposed to compensate for the lack of gauge covariance, analogously to $G_\mu(x)$. We want the covariant derivative to satisfy

$$D'_\mu \psi'_x = \Omega_x D_\mu \psi_x \tag{1.34}$$

or, more explicitly,

$$
\begin{aligned}
&\frac{1}{a}\left[\psi'_{x+a\hat{\mu}}-\psi'_x\right]-i\left[\tilde{C}'_{\mu,x}\psi'_x+C'_{\mu,x}\psi'_{x+a\hat{\mu}}\right]=\\
&=\Omega_x\left[\frac{1}{a}\left[\psi_{x+a\hat{\mu}}-\psi_x\right]-i\left[\tilde{C}_{\mu,x}\psi_x+C_{\mu,x}\psi_{x+a\hat{\mu}}\right]\right]= \qquad (1.35)\\
&=\Omega_x\left[\frac{1}{a}\left[\Omega^\dagger_{x+a\hat{\mu}}\psi'_{x+a\hat{\mu}}-\Omega^\dagger_x\psi'_x\right]-i\left[\tilde{C}_{\mu,x}\Omega^\dagger_x\psi'_x+C_{\mu,x}\Omega^\dagger_{x+a\hat{\mu}}\psi_{x+a\hat{\mu}}\right]\right].
\end{aligned}
$$

Comparing coefficients between the left- and right-hand-side of the equation we get

$$
\tilde{C}'_{\mu,x}=\Omega_x\tilde{C}_{\mu,x}\Omega^\dagger_x \qquad (1.36)
$$

$$
\begin{aligned}
C'_{\mu,x}&=\Omega_x C_{\mu,x}\Omega^\dagger_{x+a\hat{\mu}}+\frac{i}{a}\left[\Omega_x\Omega^\dagger_{x+a\hat{\mu}}-1\right]=\\
&=\Omega_x C_{\mu,x}\Omega^\dagger_{x+a\hat{\mu}}+\frac{i}{a}\left[\Omega_x\Omega^\dagger_{x+a\hat{\mu}}-\Omega_x\Omega^\dagger_x\right]=\\
&=\Omega_x C_{\mu,x}\Omega^\dagger_{x+a\hat{\mu}}+i\Omega_x\partial_\mu\Omega^\dagger_x. \qquad (1.37)
\end{aligned}
$$

From Eq. (1.36) it is clear that we can set $\tilde{C}_{\mu,x}=0$. The transformation rule for $C_{\mu,x}$ is very similar to the one for the continuum gauge potentials. We try to define the field strength as follows

$$
\begin{aligned}
C_{\mu\nu,x}=D_\mu C_{\nu,x}-D_\nu C_{\mu,x}&=\frac{1}{a}[C_{\nu,x+a\hat{\mu}}-C_{\nu,x}]+\\
-iC_{\mu,x}C_{\nu,x+a\hat{\mu}}-\frac{1}{a}[C_{\mu,x+a\hat{\nu}}-C_{\mu,x}]&+iC_{\nu,x}C_{\mu,x+a\hat{\nu}}. \qquad (1.38)
\end{aligned}
$$

One can show that $C_{\mu\nu,x}$ transforms homogeneously under gauge transformations:

$$
C'_{\mu\nu,x}=\Omega_x C_{\mu\nu,x}\Omega^\dagger_{x+a\hat{\mu}+a\hat{\nu}} \qquad (1.39)
$$

and therefore $\mathrm{tr}(C_{\mu\nu,x}C^\dagger_{\mu\nu,x})$ is gauge invariant. How do we parametrize $C_{\mu,x}$? We can try with the following form

$$
C_{\mu,x}=\frac{i}{a}\left(U_{\mu x}-1\right) \qquad (1.40)
$$

with $U_{\mu x}$ unitary matrix. This parametrization of $C_{\mu,x}$ is consistent with the transformation rule for $C_{\mu,x}$

$$
U'_{\mu x}=\Omega_x U_{\mu x}\Omega^\dagger_{x+a\hat{\mu}}. \qquad (1.41)
$$

In order to establish a relationship with the gluon fields $G^k_\mu(x)$ in the continuum, we write

$$U_{\mu x} = e^{-iaG_{\mu x}}, \qquad \text{with} \qquad G_{\mu x} = G^k_{\mu x} T_k \tag{1.42}$$

and identify $G^k_{\mu x} = G^k_\mu(x)$. Therefore G^k_μ is identified with the smooth gauge potential in the continuum, evaluated at the lattice points $x_\mu = m_\mu a$. Then $aG_\mu(x) \to 0$ as $a \to 0$, and by construction:

$$C_{\mu,x} \to G_\mu(x), \qquad C_{\mu\nu,x} \to \mathcal{G}_{\mu\nu}(x); \tag{1.43}$$

in fact we have

$$\lim_{a\to 0}(C_{\mu,x}) = \frac{i}{a}\left[1 - iaG_{\mu x} - 1\right] = G_\mu(x). \tag{1.44}$$

A straight forward lattice-regulated gauge theory action is now given by

$$S = \sum_x \left[\frac{1}{2}(\bar{\psi}_x\gamma^\mu D_\mu\psi_x - \bar{\psi}_x\gamma^\mu D^\dagger_\mu\psi_x) + \bar{\psi}_x m\psi_x + \frac{1}{4\rho g^2}\mathrm{tr}(C_{\mu\nu,x}C^{\dagger\mu\nu,x})\right]. \tag{1.45}$$

In the limit $a \to 0$, this action reduces to the continuum one if we identify $\psi_x = \psi(x)$, $\bar{\psi}_x = \bar{\psi}(x)$, $G^k_{\mu x} = G^k_\mu(x)$. However, this action contains too many fermions. This is the well-known fermion doubling problem. We will come back to this later.

With a change of notation, the transformation property (1.41) becomes

$$U_{x,x+a\hat{\mu}} = U_{\mu x}, \qquad U_{x+a\hat{\mu},x} = U^\dagger_{\mu x} \tag{1.46}$$

which provides an easy interpretation for the gauge fields, as one can write

$$U'_{x,y} = \Omega_x U_{x,y}\Omega^\dagger_y \qquad \text{with} \qquad y = x + a\hat{\mu}. \tag{1.47}$$

It is natural to think of $U_{x,x+a\hat{\mu}}$ as belonging to the link between x and $x + a\hat{\mu}$. This is illustrated in the figure below.

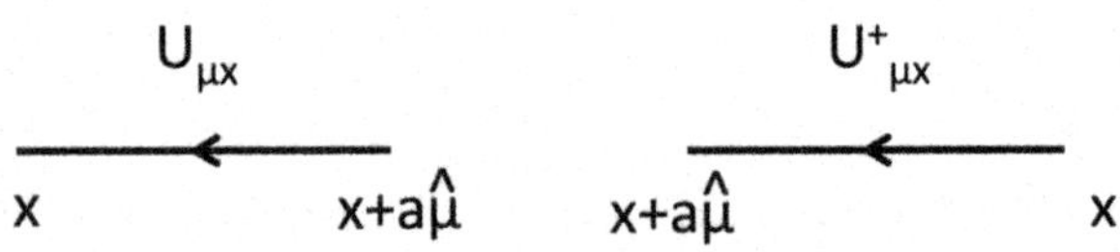

In general, we can identify $U_{x,y}$ with the parallel transporter from y to x along the link (x, y). The parallel transporter $U(p_{xy})$ along a path p_{xy} from y to x is defined by the path-ordered product (in the continuum):

$$U(p_{xy}) = \mathcal{P}\exp\left[-i\int_{p_{xy}} dz^\mu G_\mu(z)\right], \tag{1.48}$$

where $\mathcal{P}$ indicates a path-ordered product, which can be defined by dividing the path into N segments $(z_n, z_n + dz_n)$ with $n = 0, \ldots, N$, and taking the ordered product:

$$U(p_{xy}) = \lim_{N\to\infty} \exp\left[-i\int_{z_{N-1}}^{x} dz^\mu G_\mu(z)\right]\ldots\exp\left[-i\int_{z_n}^{z_{n+1}} dz^\mu G_\mu(z)\right]\ldots$$
$$\ldots\exp\left[-i\int_{y}^{z_1} dz^\mu G_\mu(z)\right] = \lim_{N\to\infty}\left[1-idz_0^\mu G_\mu(z_0)\right]\ldots\left[1-idz_N^\mu G_\mu(z_N)\right]. \tag{1.49}$$

Under a gauge transformation $G'_\mu(z) = \Omega(z)G_\mu(z)\Omega^\dagger(z)+\Omega(z)i\partial_\mu\Omega^\dagger(z)$ we have

$$\begin{aligned} 1 - idz_n^\mu G'_\mu(z_n) &= \Omega(z_n)\left[\Omega^\dagger(z_n) - idz_n^\mu G_\mu(z_n)\Omega^\dagger(z_n) + dz_n^\mu\partial_\mu\Omega^\dagger(z_n)\right] = \\ &= \Omega(z_n)\left[1 - idz_n^\mu G_\mu(z_n)\right]\Omega^\dagger(z_n) + O(dz^2) \end{aligned} \tag{1.50}$$

so that

$$U'(p_{xy}) = \Omega(x)U(p_{xy})\Omega^\dagger(y). \tag{1.51}$$

We can now re-write everything in terms of $U_{x,x+a\hat{\mu}}$:

$$D_\mu\psi_x = \frac{1}{a}\psi_{x+a\hat{\mu}} - iC_{\mu,x}\psi_{x+a\hat{\mu}} - \frac{1}{a}\psi_x. \tag{1.52}$$

We saw that we can write $C_{\mu,x} = \frac{i}{a}(U_{\mu x} - 1)$, so that

$$D_\mu\psi_x = \frac{1}{a}\psi_{x+a\hat{\mu}} + \frac{1}{a}U_{\mu x}\psi_{x+a\hat{\mu}} - \frac{1}{a}\psi_{x+a\hat{\mu}} - \frac{1}{a}\psi_x = \frac{1}{a}\left[U_{x,x+a\hat{\mu}}\psi_{x+a\hat{\mu}} - \psi_x\right]. \tag{1.53}$$

We then write

$$\begin{aligned} C_{\mu\nu,x} &= \frac{1}{a}\left[C_{\nu,x+a\hat{\mu}} - C_{\nu,x}\right] - iC_{\mu,x}C_{\nu,x+a\hat{\mu}} - \frac{1}{a}\left[C_{\mu,x+a\hat{\nu}} - C_{\mu,x}\right] + \\ &\quad + iC_{\nu,x}C_{\mu,x+a\hat{\nu}} = \\ &= \frac{i}{a^2}\left[U_{x+a\hat{\mu},x+a\hat{\mu}+a\hat{\nu}} - 1 - U_{x,x+a\hat{\nu}} + 1\right] + \\ &\quad + \frac{i}{a^2}\left[U_{x,x+a\hat{\mu}} - 1\right]\left[U_{x+a\hat{\mu},x+a\hat{\mu}+a\hat{\nu}} - 1\right] + \end{aligned}$$

$$
\begin{aligned}
& - \frac{i}{a^2}\left[U_{x+a\hat{\nu},x+a\hat{\nu}+a\hat{\mu}} - 1 - U_{x,x+a\hat{\mu}} + 1\right] - \frac{i}{a^2}\left[U_{x,x+a\hat{\nu}} - 1\right] \\
& \quad \left[U_{x+a\hat{\nu},x+a\hat{\nu}+a\hat{\mu}} - 1\right] = \\
& = \frac{i}{a^2}\left[U_{x,x+a\hat{\mu}}U_{x+a\hat{\mu},x+a\hat{\mu}+a\hat{\nu}} - U_{x,x+a\hat{\nu}}U_{x+a\hat{\nu},x+a\hat{\mu}+a\hat{\nu}}\right], \qquad (1.54)
\end{aligned}
$$

and from it we can calculate

$$
\begin{aligned}
\mathrm{tr}(C_{\mu\nu,x}C^{\dagger}_{\mu\nu,x}) &= \frac{1}{a^4}\mathrm{tr}\left\{\left[U_{x,x+a\hat{\mu}}U_{x+a\hat{\mu},x+a\hat{\mu}+a\hat{\nu}} - U_{x,x+a\hat{\nu}}U_{x+a\hat{\nu},x+a\hat{\mu}+a\hat{\nu}}\right]\right. \\
& \left.\left[U^{\dagger}_{x+a\hat{\mu},x+a\hat{\mu}+a\hat{\nu}}U^{\dagger}_{x,x+a\hat{\mu}} - U^{\dagger}_{x+a\hat{\nu},x+a\hat{\mu}+a\hat{\nu}}U^{\dagger}_{x,x+a\hat{\nu}}\right]\right\} = \\
&= \frac{1}{a^4}\mathrm{tr}\left\{1 - U_{x,x+a\hat{\nu}}U_{x+a\hat{\nu},x+a\hat{\mu}+a\hat{\nu}}U_{x+a\hat{\mu}+a\hat{\nu},x+a\hat{\mu}}U_{x+a\hat{\mu},x} + \right. \\
& \left. -U_{x,x+a\hat{\mu}}U_{x+a\hat{\mu},x+a\hat{\mu}+a\hat{\nu}}U_{x+a\hat{\mu}+a\hat{\nu},x+a\hat{\nu}}U_{x+a\hat{\nu},x} + 1\right\} \\
&= \frac{1}{a^4}\mathrm{tr}\left\{2 - U_{\mu\nu,x} - U^{\dagger}_{\mu\nu,x}\right\}, \qquad (1.55)
\end{aligned}
$$

where we have defined

$$
U_{\mu\nu,x} = U^{\dagger}_{\nu\mu,x} = U_{x,x+a\hat{\mu}}U_{x+a\hat{\mu},x+a\hat{\mu}+a\hat{\nu}}U_{x+a\hat{\mu}+a\hat{\nu},x+a\hat{\nu}}U_{x+a\hat{\nu},x}. \qquad (1.56)
$$

We can now re-write the lattice QCD action as

$$
\begin{aligned}
S_{QCD} &= \sum_{x\mu\nu}\frac{1}{2g^2\rho a^4}\mathrm{tr}\left\{1 - \Re U_{\mu\nu,x}\right\} + \\
& + \sum_{x\mu}\frac{1}{2a}\left[\bar{\psi}_x\gamma^{\mu}U_{\mu x}\psi_{x+a\hat{\mu}} - \bar{\psi}_{x+a\hat{\mu}}\gamma^{\mu}U^{\dagger}_{\mu x}\psi_x\right] + \\
& + \sum_{x}\bar{\psi}_x m\psi_x = \sum_{x\mu\nu}\frac{2N_c}{2g^2\rho a^4}\left\{1 - \frac{1}{2N_c}\mathrm{tr}\Re U_{\mu\nu,x}\right\} + \\
& + \sum_{x\mu}\frac{1}{2a}\left[\bar{\psi}_x\gamma^{\mu}U_{\mu x}\psi_{x+a\hat{\mu}} - \bar{\psi}_{x+a\hat{\mu}}\gamma^{\mu}U^{\dagger}_{\mu x}\psi_x\right] + \sum_{x}\bar{\psi}_x m\psi_x. \qquad (1.57)
\end{aligned}
$$

1.2.1 Pure Gauge Theory

The elementary square of a hypercubic lattice is called a *plaquette*. It is denoted as P. The product of the Us around P is usually denoted as U_P, so that

$$S_G = \frac{1}{g^2 \rho} \sum_P \Re(\mathrm{tr} U_P) + const. \tag{1.58}$$

is the way in which one can re-write the gauge-field part of the action in lattice units ($a = 1$). It depends on the representation chosen for the U fields.

We can define a lattice path integral:

$$Z = \int [DU] \exp[-S[U]], \qquad \text{with} \qquad DU = \prod_{x,\mu} dU_{\mu x}; \tag{1.59}$$

Z is well-defined for a finite lattice. We want Z to be gauge invariant. For this to be true, the integration measure DU has to satisfy

$$DU = DU^{\Omega} \qquad U^{\Omega}_{\mu x} = \Omega_x U_{\mu x} \Omega^{\dagger}_{x+a\hat{\mu}}. \tag{1.60}$$

A measure which is invariant under such transformations is the Haar measure

$$dU = \nu \sqrt{detg} \prod_k d\alpha_k \tag{1.61}$$

where α_k are the coordinates in group space, $U = U(\alpha)$ and g_{kl} is a metric on this space, of the form

$$g_{kl} = \frac{1}{\rho} \mathrm{tr} \left[\frac{\partial U}{\partial \alpha^k} \frac{\partial U^{\dagger}}{\partial \alpha^l} \right] \qquad \text{and} \qquad \rho = \frac{1}{2}; \tag{1.62}$$

ν is a normalization constant, chosen such that

$$\int dU = 1. \tag{1.63}$$

For the group $U(1)$ (e.g. in the case of QED) we have

$$U = \exp[i\alpha], \qquad g_{kl} = 1, \qquad \int dU = \int_{-\pi}^{\pi} \frac{d\alpha}{2\pi} = 1. \tag{1.64}$$

For the $(N^2 - 1)$–dimensional group $SU(N)$ we have

$$U = \exp\left[i\alpha^k t_k\right], \qquad g_{kl} = S_{km} S_{lm}, \qquad S_{kl} = \left(\frac{1 - e^{-i\alpha}}{i\alpha}\right)_{kl},$$
$$\alpha = \alpha^p F_p, \qquad (F_p)_{mn} = -i f_{pmn}. \tag{1.65}$$

This completes the definition of the partition function Z. Expectation values of observables are defined in the expected way as

$$\langle \mathrm{tr}U_p \mathrm{tr}U_{p'} \rangle = Z^{-1} \int dU \exp\left[-S[U]\right] \mathrm{tr}U_p \mathrm{tr}U_{p'}. \tag{1.66}$$

1.2.2 Fermions on the Lattice

In continuous Minkowski space-time, the free Dirac action can be written as

$$S = \int d^4x \left[\frac{1}{2} \left(\bar{\psi}(x)\gamma^\mu \partial_\mu \psi(x) - \partial_\mu \bar{\psi}(x)\gamma^\mu \psi(x) \right) - m\bar{\psi}(x)\psi(x) \right]. \tag{1.67}$$

We obtain a lattice version of the above action by replacing the derivatives by discrete differences:

$$\partial_\mu \psi(x) \to \frac{1}{a} \left[\psi(x + a\hat{\mu}) - \psi(x) \right] \tag{1.68}$$

so that we get

$$S = \sum_{x,\mu} \frac{1}{2a_\mu} \left[\bar{\psi}_x \gamma^\mu \psi_{x+a\hat{\mu}} - \bar{\psi}_{x+a\hat{\mu}} \gamma^\mu \psi_x \right] - m \sum_x \bar{\psi}_x \psi_x. \tag{1.69}$$

The path integral for fermions is tentatively defined as

$$Z = \int [D\bar{\psi}][D\psi] \exp[iS], \quad \text{with } [D\bar{\psi}][D\psi] = \prod_{x,\alpha} d\bar{\psi}_{x\alpha} d\psi_{x\alpha} = \prod_{x,\alpha} d\psi^\dagger_{x\alpha} d\psi_{x\alpha}. \tag{1.70}$$

In the last equality above, we used the rule $d(T\psi) = (\det T)^{-1} d\psi$ and $\det(\gamma^0) = 1$. Notice that we re-wrote everything in terms of the dimensionless ψ_x and $\bar{\psi}_x$:

$$\psi_x = a^{3/2}\psi(x); \qquad \bar{\psi}_x = a^{3/2}\bar{\psi}(x) \tag{1.71}$$

and also $d\psi^\dagger_{x\alpha}$ and $d\psi_{x\alpha}$ are dimensionless. We now want to define the rotation to imaginary (Euclidean) time. We have seen that the numbers m_μ are a group of integers specifying the lattice site x:

$$x_0 = m_0 a_0; \qquad \mathbf{x} = \mathbf{m}a, \qquad \sum_x = a_0 a^3 \sum_m \tag{1.72}$$

where in principle we allow for a different lattice spacing in the time direction and we label $\psi_m = \psi_x$; $\bar{\psi}_m = \bar{\psi}_x$. More explicitly, the action reads

$$S = \sum_m \left[\frac{1}{2} \left(\bar{\psi}_m \gamma^0 \psi_{m+\hat{0}} - \bar{\psi}_{m+\hat{0}} \gamma^0 \psi_m \right) + \sum_{k=1}^{3} \frac{a_0}{2a} \left(\bar{\psi}_m \gamma^k \psi_{m+\hat{k}} - \bar{\psi}_{m+\hat{k}} \gamma^k \psi_m \right) + \right.$$
$$\left. - \, a_0 m \left(\bar{\psi}_m \psi_m \right) \right]. \tag{1.73}$$

An analytical continuation to imaginary time can be easily written as

$$a_0 = |a_0| e^{-i\varphi} \qquad \varphi : 0 \to \frac{\pi}{2} \qquad a_0 \to -i a_4, \tag{1.74}$$

with $a_4 = |a_0|$. This transforms the path-integral into its Euclidean version:

$$Z = \int [D\bar{\psi}][D\psi] \exp[-S],$$
$$S = \sum_m \left[\sum_\mu \frac{a_4}{2a_\mu} \left(\bar{\psi}_m \gamma_\mu \psi_{m+\hat{\mu}} - \bar{\psi}_{m+\hat{\mu}} \gamma_\mu \psi_m \right) + a_4 m \bar{\psi}_m \psi_m \right] \tag{1.75}$$

where now $\mu = 1, \ldots, 4$ and $\gamma_4 = i\gamma^0$.

1.2.3 Fermion Doubling and Staggered Fermions

The model described by the discretized action above yields $2^4 = 16$ Dirac particles instead of 1. We can understand this by looking at the fermion propagator. We start from the partition function

$$Z = \int [D\bar{\psi}][D\psi] \exp[-\bar{\psi} M \psi] \tag{1.76}$$

where (setting $a = a_4 = 1$ for simplicity)

$$M_{xy} = \sum_{z\mu} \gamma^\mu \frac{1}{2} \left(\delta_{x,z} \delta_{y,z+\hat{\mu}} - \delta_{x,z+\hat{\mu}} \delta_{y,z} \right) + m \sum_z \delta_{x,z} \delta_{y,z}. \tag{1.77}$$

$M^{-1}_{xy} = S_{xy}$ is the fermion propagator. The path is easily integrated to give $Z = \det M$. Assuming an infinite space-time, S can be calculated in momentum space:

$$M(k, -l) = \sum_{xy} e^{-ikx+ily} M_{xy} = S(k)^{-1} \delta(k - l). \tag{1.78}$$

One thus finds

$$S(k)^{-1} = \sum_{\mu} i\gamma^{\mu} \sin k_{\mu} + m, \tag{1.79}$$

from which follows

$$S(k) = \frac{m - i\gamma^{\mu} S_{\mu}}{m^2 + S^2}, \qquad \text{with} \qquad S_{\mu} = \sin k_{\mu}. \tag{1.80}$$

Reverting to non-lattice units, we get

$$S(k) = \frac{m - i\sum_{\mu} \gamma^{\mu} \frac{\sin(ak_{\mu})}{a}}{m^2 + \sum_{\mu} \frac{\sin^2(ak_{\mu})}{a^2}} \tag{1.81}$$

which, for $a \to 0$ (continuum limit) yields

$$S(k) = \frac{m - i\gamma^{\mu} k_{\mu}}{m^2 + k^2} + O(a^2). \tag{1.82}$$

The above expression has a pole at $k_4 = i\sqrt{\mathbf{k}^2 + m^2}$, corresponding to a Dirac particle. The pole is near the zeroes of the sin functions at the origin $ak_{\mu} = 0$. We write $k = k_A + p$, where $k_A = \frac{\pi_A}{a}$. We can then write

$$\frac{\sin(ak_A + ap)}{a} = \sin(ak_A)\frac{\cos(ap)}{a} + \cos(ak_A)\frac{\sin(ap)}{a}. \tag{1.83}$$

Therefore, there are fifteen more regions in the four-dimensional torus with $-\pi \leq ak_{\mu} \leq \pi$ where the sin functions vanish, sixteen in total. The index A defines the following vectors:

$$\begin{aligned}
&\pi_0 = (0,0,0,0), \quad &&\pi_1 = (\pi,0,0,0), \quad &&\pi_2 = (0,\pi,0,0), \quad &&\pi_3 = (0,0,\pi,0) \\
&\pi_4 = (0,0,0,\pi), \quad &&\pi_{12} = (\pi,\pi,0,0), \quad &&\pi_{13} = (\pi,0,\pi,0), \quad &&\pi_{14} = (\pi,0,0,\pi) \\
&\pi_{23} = (0,\pi,\pi,0), \quad &&\pi_{24} = (0,\pi,0,\pi), \quad &&\pi_{34} = (0,0,\pi,\pi), \quad &&\pi_{123} = (\pi,\pi,\pi,0) \\
&\pi_{124} = (\pi,\pi,0,\pi), \quad &&\pi_{134} = (\pi,0,\pi,\pi), \quad &&\pi_{234} = (0,\pi,\pi,\pi), \quad &&\pi_{1234} = (\pi,\pi,\pi,\pi).
\end{aligned} \tag{1.84}$$

The zeroes of the sine function at the edges of the Brillouin zone destroy the continuum limit. Thus, there exist sixteen regions of integration over the momentum k, where $\tilde{p}_{\mu} = \sin(k_{\mu}a)/a$ takes a finite value in the limit $a \to 0$. We have

$$\frac{\sin(ak_A + ap)}{a} = \cos(ak_A)p \tag{1.85}$$

and $\cos(ak_A) = \pm 1$. Thus we can re-write the propagator as

$$S(k) = \frac{m - i\gamma^\mu \cos \pi_A p_\mu}{m^2 + p^2} = \frac{m \pm i\gamma^\mu p_\mu}{m^2 + p^2}. \tag{1.86}$$

Therefore, for each one of the values of π_A (and correspondingly of k_A) listed above, we get a propagator like the one in Eq. (1.86), corresponding to a particle. This gives rise to fifteen additional fermions, with dispersion relation defined by the relationship $k = k_A + p$ and Eq. (1.86) above. In the continuum limit, the partition function receives contributions from sixteen fermion-like excitations in momentum space, of which fifteen are pure lattice artifacts. More generally, in d space-time dimensions, the number of spurious fermions is 2^d. This is a well-known problem in lattice QCD. One way of mitigating the problem is to resort to the so-called staggered fermions.

The idea is to mitigate the fermion doubling problem, due to the vanishing of $\tilde{p}_\mu$ at the corners of the Brillouin zone, by reducing this zone, e.g. by doubling the effective lattice spacing. This can be accomplished in two steps: we need to distribute the fermionic degrees of freedom over the lattice in such a way that the effective lattice spacing for each fermion type is twice the lattice spacing. We then need to check that, in the continuum limit, the action reduces to the desired continuum form.

We start by considering a d-dimensional space-time lattice, and divide it into d-dimensional hypercubes of unit length. At each site in the elementary hypercube, we place a different kind of fermion and repeat this structure periodically throughout the lattice. By doing this, we effectively double the lattice spacing for each degree of freedom. Within a hypercube there are 2^d sites, but only $2^{d/2}$ components of the Dirac field. For this reason, we need $2^{d/2}$ different Dirac fields to reduce the Brillouin zone by a factor two. Therefore, in our standard four-dimensional space-time we need 2^2 different flavors of quarks.

We make the unitary transformation of variables on the fermionic fields defined at lattice site x:

$$\psi_x = \gamma^x \chi_x, \qquad \bar{\psi}_x = \bar{\chi}_x (\gamma^x)^\dagger \tag{1.87}$$

with

$$\gamma^x = (\gamma_1)^{m_1} (\gamma_2)^{m_2} (\gamma_3)^{m_3} (\gamma_4)^{m_4}. \tag{1.88}$$

We recall that the lattice site is identified by the four integer numbers m_i above (see Eq. (1.72)). The matrices γ^x defined above are chosen so that

$$\left[(\gamma^x)^\dagger \gamma_\mu \gamma^{x+\hat{\mu}} \right]_{\alpha\beta} = \left[\gamma^x \gamma_\mu \left(\gamma^{x+\hat{\mu}} \right)^\dagger \right]_{\alpha\beta} = \eta_{\mu x} \delta_{\alpha\beta} \tag{1.89}$$

with

$$\eta_{1x} = 1, \qquad \eta_{2x} = (-1)^{x_1}, \qquad \eta_{3x} = (-1)^{x_1 + x_2}, \qquad \eta_{4x} = (-1)^{x_1 + x_2 + x_3}. \tag{1.90}$$

This unitary transformation removes the γ matrices from the naïve fermionic action, which acquires the form

$$S = \sum_{\alpha=1}^{4} \left[\sum_{x,\mu} \eta_{\mu x} \frac{1}{2} \left(\bar{\chi}_x^\alpha \chi_{x+\hat{\mu}}^\alpha - \bar{\chi}_{x+\hat{\mu}}^\alpha \chi_x^\alpha \right) + m \sum_x \bar{\chi}_x^\alpha \chi_x^\alpha \right]. \tag{1.91}$$

Notice that the sum over α is the addition of four identical terms, one for each value of μ. This reduces the fermion doubling from sixteen to four. Adding the gluonic fields we thus have

$$S_F = \sum_{x,\mu} \eta_{\mu x} \frac{1}{2} \left[\bar{\chi}_x^a \left(U_{\mu x}\right)_{ab} \chi_{x+\hat{\mu}}^b - \bar{\chi}_{x+\hat{\mu}}^a \left(U_{\mu x}^\dagger\right)_{ab} \chi_x^b \right] + m \sum_x \bar{\chi}_x^a \chi_x^a \tag{1.92}$$

with a, b color indices. This action leads to four degenerate flavors in the continuum limit. At finite lattice spacing, these flavors are not degenerate, they are an artifact of the discretization and they are called *tastes*.

1.3 Monte Carlo Methods

Here we give a brief overview of Monte Carlo methods, including a couple of examples. For more details we refer the readers to e.g. the book in Ref. [2]. When one computes an observable in lattice QCD, one has a problem: having to perform a very large number of integrations. Suppose we want to compute the ensemble average

$$\langle O \rangle = \frac{\int DU\, O[U] e^{-S[U]}}{\int DU e^{-S[U]}}. \tag{1.93}$$

Let us choose a relatively small lattice, with $N = 10$ in each direction. The number of link variables $U_{\mu x}$ is $\sim 4 \times 10^4$, because the index μ takes four values and the index x corresponds to 10^4 values. In the case of $SU(3)$, each $U_{\mu x}$ is a function of 8 real parameters: there are 320,000 integrations to be done. Using a mesh of 10 points per integration, the multiple integral will be approximated by a sum of 10^{320000} terms. The important point to keep in mind, is that most configurations have a very large action and therefore only a small fraction will yield a significant contribution. Therefore, one usually resorts to importance sampling techniques.

They consist of generating a sequence of link variable configurations with a probability distribution given by the Boltzmann factor $\exp[-S(U)]$. Then the ensemble average $\langle O \rangle$ will be approximated as

$$\langle O \rangle \simeq \frac{1}{N} \sum_{i=1}^{N} O\left(\{U\}_i\right) \tag{1.94}$$

where $\{U\}_i$, $(i = 1, \ldots N)$ are the generated link configurations. The procedure for calculating the ensemble average goes as follows

- given a rule for generating a sequence of configurations, one first updates them until thermalization is achieved, namely until they are distributed with the desired probability;
- as a second step, one selects configurations which are uncorrelated (usually one every ten generated configurations is saved);
- finally, the ensemble average will be given by the above sum with an error $\sim 1/\sqrt{N}$.

1.3.1 The Metropolis Algorithm

We will now spend a few words to describe the Metropolis algorithm, which is frequently used to generate the configurations [3]. Let C be any configuration which is to be updated. We suggest a new configuration C', with a transition probability (probability for suggesting C' as the new configuration) $P_0(C \to C')$ which only satisfies

$$P_0(C \to C') = P_0(C' \to C). \tag{1.95}$$

We give an example for a pure $U(1)$ gauge theory. The configuration C is specified by the values of the link variables:

$$U_\mu(n) = \exp(i\theta_\mu(n)) \tag{1.96}$$

for $\mu = 1, \ldots 4$ and all lattice sites n. As a new configuration, we choose one of the link variables and multiply it by $\exp(i\chi)$, where χ is a random number between $-\pi$ and π. The transition probability has to satisfy the detailed balance

$$e^{-S(C)} P(C \to C') = e^{-S(C')} P(C' \to C), \tag{1.97}$$

where $S(C)$ is the action calculated on the configuration C, and analogously for C'. This choice satisfies Eq. (1.95), but is C' an acceptable configuration? This depends on $S(C)$ and $S(C')$. The decision is made as follows: if $e^{-S(C')} > e^{-S(C)}$, the configuration C' is accepted, because the action has been lowered and therefore it is closer to equilibrium. If not, one accepts C' with probability $\exp(-S(C'))/\exp(-S(C))$. One generates a random number $R \in [0, 1]$ and accepts C' if

$$R \leq \frac{\exp(-S(C'))}{\exp(-S(C))}. \tag{1.98}$$

Otherwise, C' is rejected and the old configuration is kept. This conditional acceptance allows the system to increase its action and the algorithm builds in the quantum fluctuations (the classical configurations correspond to the minima of the action). This algorithm is usually used to update a single variable at a time. It becomes very slow for non-local actions (e.g. with fermions).

1.3.2 The Hybrid Monte Carlo (HMC) Algorithm

The HMC algorithm involves parallel updates of fields at all lattice sites, followed by an accept/reject decision for the whole configuration [4]. There are no truncation errors. The requirement is a method for choosing candidate configurations which can be computed efficiently, and whose reverse probability $P_0(C' \to C)$ is easy to obtain. The acceptance rate should be large and should not depend too strongly on the size of the system. Finally, the correlation between successive configurations should be minimized.

One introduces a "computer" time parameter τ and a Hamiltonian that describes the evolution of the field $\phi(\tau)$ in this time. The conjugate momenta $\pi(\tau)$ are then introduced, and one gets

$$H'(\phi, \pi) = \frac{1}{2}\pi^2 + S'(\phi) \tag{1.99}$$

where S' is an arbitrary action. The equations of motion are then

$$\dot{\phi} = \pi, \qquad \dot{\pi} = -\delta S'/\delta\phi. \tag{1.100}$$

One generates some initial momenta π at random from a Gaussian distribution of zero mean and unit variance:

$$P_G(\pi) \propto \exp(-\pi^2/2) \tag{1.101}$$

and lets the system evolve deterministically for some time τ_0 according to Hamilton's equations. The evolution defines a mapping as $(\phi, \pi) = (\phi(0), \pi(0)) \to (\phi(\tau_0), \pi(\tau_0))$ if the space trajectory $(\phi(\tau), \pi(\tau))$ is a solution of Hamilton's equations.

The probability P_0 of selecting the candidate phase space configuration is therefore (notice that, in the previous section, we indicated the configuration with $C = (\phi, \pi)$)

$$P_0((\phi, \pi) \to (\phi', \pi')) = \delta[(\phi', \pi') - (\phi(\tau_0), \pi(\tau_0))]. \tag{1.102}$$

Finally, the probability of accepting this candidate is

$$P_A((\phi, \pi) \to (\phi', \pi')) = \min[1, \exp[\delta H]] \tag{1.103}$$

where $\delta H = H(\phi', \pi') - H(\phi, \pi)$. The transition probability for the ϕ field is then

$$P(\phi \to \phi') = \int [d\pi][d\pi'] P_G(\pi) P_0((\phi, \pi) \to (\phi', \pi')) P_A((\phi, \pi) \to (\phi', \pi')), \tag{1.104}$$

which satisfies the detailed balance condition (1.97) exactly. For this to be true, the condition (1.95) has to be satisfied. From the identity

$$\begin{aligned} &\exp[-H(\phi, \pi)] \min(1, \exp(-\delta H)) = \min(\exp(-H(\phi, \pi)), \exp(-H(\phi', \pi'))) = \\ &= \exp[-H(\phi', \pi')] \min(\exp(\delta H), 1) \end{aligned} \tag{1.105}$$

and noticing that $P_S(\phi) P_G(\pi) \propto \exp[H(\phi, \pi)]$ (where $P_S(\phi)$ is the unique fixed point distribution to which any Markov process will converge to) and H is invariant under $\pi \to -\pi$ we get

$$\begin{aligned} &P_S(\phi) P_G(\pi) P_A((\phi, \pi) \to (\phi', \pi')) = P_S(\phi') P_G(\pi') P_A((\phi', \pi') \to (\phi, \pi)) = \\ &= P_S(\phi') P_G(-\pi') P_A((\phi', -\pi') \to (\phi, -\pi)). \end{aligned} \tag{1.106}$$

Multiplying the above equation by P_0, integrating over π and π' and using condition (1.95) we get

$$\begin{aligned} &\int [d\pi][d\pi'] P_S(\phi) P_G(\pi) P_0((\phi, \pi) \to (\phi', \pi')) P_A((\phi, \pi) \to (\phi', \pi')) = \\ &= \int [d\pi][d\pi'] P_S(\phi') P_G(-\pi') P_0((\phi', -\pi') \to (\phi, -\pi)) P_A((\phi', -\pi') \to (\phi, -\pi)) \end{aligned} \tag{1.107}$$

which yields the detailed balance and the invariance of the measure. This algorithm will therefore generate $\phi - field$ configurations with the correct distribution for any δH. Therefore, we now summarize the steps of the HMC algorithm

- Choose the coordinates $\{\phi_i\}$ in some arbitrary way
- Choose the conjugate momenta $\{\pi_i\}$ from a Gaussian ensemble
- Calculate $\tilde{\pi}_i$ according to

$$\tilde{\pi}_i(n) = \pi_i(n) - \frac{\epsilon}{2} \frac{\partial S[\phi]}{\partial \phi_i(n)} \tag{1.108}$$

- Iterate the following equations to let the fields evolve over a certain number of time steps

$$\phi_i(n+1) = \phi_i(n) + \epsilon \tilde{\pi}_i(n)$$

$$\tilde{\pi}_i(n+1) = \tilde{\pi}_i(n) - \epsilon \frac{\partial S[\phi]}{\partial \phi_i(n+1)} \tag{1.109}$$

where $\epsilon = \tau_n/n$ is the time step. We call $\{\phi_i', \pi_i'\}$ the last configuration generated in the molecular dynamics chain.

- Accept $\{\phi_i', \pi_i'\}$ as the new configuration with probability

$$p = \min\{1, \exp(-H(\phi', \pi'))/\exp(-H(\phi, \pi))\}. \tag{1.110}$$

- If the new configuration is rejected, keep the old one and repeat the previous steps. If it is accepted, use it to generate a new configuration following the previous steps.

References

1. Smit, J.: Camb. Lect. Notes Phys. **15**, 1–271 (2002)
2. Rothe, H.J.: World Sci. Lect. Notes Phys. **43**, 1–381 (1992)
3. Bhanot, G.: Rep. Prog. Phys. **51**, 429 (1988)
4. Duane, S., Kennedy, A.D., Pendleton, B.J., Roweth, D.: Phys. Lett. B **195**, 216–222 (1987)

Phase Transitions in QCD

2

Abstract

In this chapter we introduce the concept of phase transition and we discuss the chiral and deconfinement phase transitions of QCD. We discuss the infinite quark limit, for which we have a deconfinement phase transition due to the breaking of the $Z(3)$ symmetry, and for which the Polyakov loop is the order parameter. We also consider the opposite, massless quark limit, for which the chiral condensate is the order parameter. We finish the chapter by discussing the Columbia plot, which is the quark-mass dependence of the order of the phase transition.

2.1 QCD Partition Function on the Lattice

The study of finite temperature QCD (or any field theory) starts with the Grand Canonical partition function

$$Z = \mathrm{tr}\exp\left[-\beta\left(\hat{H} - \mu\hat{N}\right)\right] \tag{2.1}$$

where $\hat{H}$ is the Hamiltonian of the system, $\hat{N}$ is the number operator for some conserved charge and μ is the chemical potential corresponding to that conserved charge; $\beta = 1/T$ and T is the temperature of the system. We can re-write Z in terms of a path-integral as follows

$$Z = \int_{b.c.} D\psi D\bar{\psi} DG_\mu \exp\left[-\int_0^\beta d\tau \int d^3\mathcal{L}\right] \tag{2.2}$$

where $b.c.$ are the boundary conditions: antiperiodic for ψ and $\bar{\psi}$, periodic for G_μ. The thermal expectation value of a physical observable O can be written as

C. Ratti, R. Bellwied, *The Deconfinement Transition of QCD*, Lecture Notes in Physics 981, https://doi.org/10.1007/978-3-030-67235-5_2

$$\langle O\rangle = \frac{\int_{b.c.} D\psi D\bar{\psi} DG_\mu O \exp\left[-\int_0^\beta d\tau \int d^3\mathcal{L}\right]}{\int_{b.c.} D\psi D\bar{\psi} DG_\mu \exp\left[-\int_0^\beta d\tau \int d^3\mathcal{L}\right]}. \tag{2.3}$$

We now proceed to define the discrete, lattice QCD version of the above. In the following, we will indicate the number of lattice points in the spatial and temporal directions as N_s and N_t, respectively. The physical volume of the lattice box is $V = (N_s a)^3$, while the temperature T is defined as $T = \frac{1}{N_t a}$. The staggered fermion action (1.92) can be re-written in the following way at finite temperature and chemical potential:

$$S_F = \sum_{x,\nu=1}^{3} \eta_{\nu x}\frac{1}{2}\left[\bar{\chi}_x^a (U_{\nu x})_{ab}\, \chi_{x+\hat{\nu}}^b - \bar{\chi}_{x+\hat{\nu}}^a \left(U_{\nu x}^\dagger\right)_{ab} \chi_x^b\right] +$$
$$+ \sum_x \frac{1}{2}\left[e^{\mu a} \bar{\chi}_x^a (U_{4x})_{ab}\, \chi_{x+\hat{4}}^b - e^{-\mu a} \bar{\chi}_x^a \left(U_{4x}^\dagger\right)_{ab} \chi_{x-\hat{4}}^b\right] + m \sum_x \bar{\chi}_x^a \chi_x^a; \tag{2.4}$$

μ is the chemical potential, which will be discussed later. The partition function can be integrated over the fermionic fields and becomes

$$Z = \int_{b.c.} DU\, [\det M]\, e^{-S_G[U]} \tag{2.5}$$

where M is the fermionic matrix. In the staggered fermion case we have

$$M = \sum_{\nu=1}^{3} \eta_{\nu x}\frac{1}{2}\left[(U_{\nu x})_{ab}\, \delta_{x,z}\delta_{x',z+\hat{\nu}} - \left(U_{\nu x}^\dagger\right)_{ab} \delta_{x,z+\hat{\nu}}\delta_{x',z}\right] +$$
$$+ \frac{1}{2}\left[e^{\mu a} (U_{4x})_{ab}\, \delta_{x',x+\hat{4}} - e^{-\mu a} \left(U_{4x}^\dagger\right)_{ab} \delta_{x',x-\hat{4}}\right] + m \sum_z \delta_{x,z}\delta_{x',z}. \tag{2.6}$$

2.2 Phase Transitions

We want to study the phase transition of QCD, from a confined hadronic system to deconfined quarks and gluons. For this reason, we will now define what a phase transition is, and classify them according to the behavior of some thermodynamic quantities.

- A phase transition is a transformation of a thermodynamic system from one phase or state of matter to another.

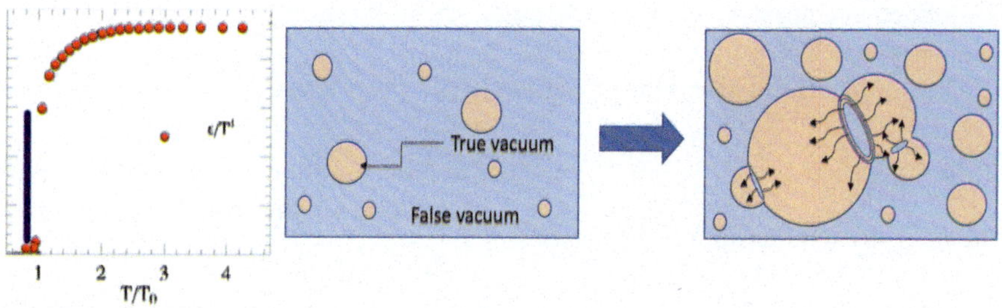

Fig. 2.1 Left: example of energy density as a function of the temperature for a first-order phase transition: the blue line indicates the latent heat. Right: bubble formation and phase coexistence for a first-order phase transition

- During a phase transition, the properties of the medium change, often discontinuously, as a result of some external conditions.
- The measurement of the external conditions at which the transformation occurs (e.g. T, μ_B) is called the phase transition point.
- An order parameter is some observable, which is able to distinguish between the two phases.

Paul Ehrenfest classified phase transitions based on the behavior of the thermodynamic free energy.

- A **first order** phase transition is characterized by a discontinuity in the first derivative of the free energy with respect to some thermodynamic variable; it involves a latent heat, and at the phase boundary the system is in a mixed phase regime, in which some parts of the system have completed the transition and others have not. This case is shown in Fig. 2.1.
- In the case of a **second order** phase transition, the free energy and its first derivative are continuous at T_c. A new state grows continuously out of the previous one. For $T \to T_c$, the two states become quantitatively the same. The second derivative of the free-energy can be discontinuous or diverge at T_c. One expects a power-law behavior in $(1 - \frac{T}{T_c})$ at T_c.
- In the case of an **analytic crossover**, the free energy and all its derivatives are continuous at T_c. In this case, the system changes smoothly from one phase to the other. The phase transition point in this case is identified by the peak in the derivative of some observable (susceptibilities), as we will see later.

2.3 Polyakov Loop

We will now introduce the concept of Polyakov loop. As we will see, in the case of a purely gluonic system, the Polyakov loop is the order parameter for the deconfinement phase transition, associated to the breaking of the $Z(3)$ discrete symmetry (see Ref. [1] for more details).

The pure-gauge field action is invariant under gauge transformations of the form

$$G'_\mu = \Omega(G_\mu + i\partial_\mu)\Omega^\dagger. \tag{2.7}$$

The field strength tensor transforms as

$$\mathcal{G}'_{\mu\nu} = \Omega\mathcal{G}_{\mu\nu}\Omega^\dagger \tag{2.8}$$

under gauge transformations. Besides, the field $G_\mu(x)$ obeys periodic boundary conditions in the Euclidean time direction:

$$G_\mu(\mathbf{x}, x_4 + \beta) = G_\mu(\mathbf{x}, x_4). \tag{2.9}$$

In order to maintain these conditions, we initially consider strictly periodic gauge transformations in Euclidean time:

$$\Omega(\mathbf{x}, x_4 + \beta) = \Omega(\mathbf{x}, x_4); \tag{2.10}$$

however, there are topologically non-trivial transformations (twisted gauge transformations) that are periodic up to a constant matrix $h \in SU(N)$:

$$\Omega(\mathbf{x}, x_4 + \beta) = h\Omega(\mathbf{x}, x_4). \tag{2.11}$$

When a transformation of this kind is applied to a strictly periodic vector potential G_μ, the latter behaves as follows

$$\begin{aligned} G'_\mu(\mathbf{x}, x_4 + \beta) &= \Omega(\mathbf{x}, x_4 + \beta)\left[G_\mu(\mathbf{x}, x_4 + \beta) + i\partial_\mu\right]\Omega(\mathbf{x}, x_4 + \beta)^\dagger = \\ &= h\Omega(\mathbf{x}, x_4)\left[G_\mu(\mathbf{x}, x_4) + i\partial_\mu\right]\Omega(\mathbf{x}, x_4)^\dagger h^\dagger = hG'_\mu(\mathbf{x}, x_4)h^\dagger. \end{aligned} \tag{2.12}$$

We must thus have that

$$hG'_\mu(\mathbf{x}, x_4)h^\dagger = G'_\mu(\mathbf{x}, x_4), \tag{2.13}$$

which is only true if h commutes with G'_μ. This limits us to consider twist matrices h in the center $Z(N)$ of the gauge group $SU(N)$. The elements of the center commute with all the group elements, and are therefore multiples of the identity matrix:

$$h = z\mathbb{1}, \qquad z = \exp(\frac{2\pi i n}{N}), \qquad n = 1, \ldots, N \tag{2.14}$$

and $N = 3$ in our case. The $Z(3)$ symmetry gets explicitly broken by dynamical fields that transform in the fundamental representation of $SU(3)$. In fact, quark fields transform as

$$\psi' = \Omega\psi \tag{2.15}$$

and are anti-periodic in Euclidean time:

$$\psi(\mathbf{x}, x_4 + \beta) = -\psi(\mathbf{x}, x_4). \tag{2.16}$$

Under a twisted transformation, they therefore transform as

$$\psi'(\mathbf{x}, x_4 + \beta) = \Omega(\mathbf{x}, x_4 + \beta)\psi(\mathbf{x}, x_4 + \beta) = -z\Omega(\mathbf{x}, x_4)\psi(\mathbf{x}, x_4). \tag{2.17}$$

To maintain the boundary condition (2.16) for fermions, we must restrict ourselves to $z = 1$, so that the center symmetry disappears. Notice that $Z(3)$ is a symmetry of the Euclidean action, not the Hamiltonian: it is characterized by non-trivial boundary conditions in Euclidean time.

We now define the Polyakov loop

$$\Phi(\mathbf{x}) = \frac{1}{3}\mathrm{tr}\mathcal{P}\exp\left[-\int_0^\beta dx_4 G_4(\mathbf{x}, x_4)\right], \tag{2.18}$$

where $\mathcal{P}$ is the path-ordering operator. As clear from the definition above, the Polyakov loop is a complex scalar field that depends on the spatial coordinate $\mathbf{x}$. It transforms non-trivially under twisted $Z(3)$ gauge transformations:

$$\begin{aligned}\Phi'(\mathbf{x}) &= \frac{1}{3}\mathrm{tr}\mathcal{P}\exp\left[-\int_0^\beta dx_4 G'_4(\mathbf{x}, x_4)\right]\\ &= \frac{1}{3}\mathrm{tr}\left[\Omega(\mathbf{x}, \beta)\mathcal{P}\exp\left[-\int_0^\beta dx_4 G_4(\mathbf{x}, x_4)\right]\Omega^\dagger(\mathbf{x}, 0)\right]\\ &= \frac{1}{3}\mathrm{tr}\left[z\Omega(\mathbf{x}, 0)\mathcal{P}\exp\left[-\int_0^\beta dx_4 G_4(\mathbf{x}, x_4)\right]\Omega^\dagger(\mathbf{x}, 0)\right] = z\Phi(\mathbf{x}).\end{aligned} \tag{2.19}$$

The Polyakov loop is invariant under these transformations when $z = 1$ (strictly periodic gauge transformations: it is a physical quantity). Therefore, a non-zero expectation value $\langle\Phi\rangle$ means that the $Z(3)$ symmetry is spontaneously broken. This happens at high T. Instead, in the case of $\langle\Phi\rangle = 0$, we have a confined phase in which the $Z(3)$ symmetry is exact. In the pure-gauge sector of QCD (in the limit of infinitely-heavy quarks), the Polyakov loop is therefore an order parameter for the confinement-deconfinement phase transition. As $\beta \to 0$, or $T \to \infty$, we have $\langle\Phi\rangle \to 1$.

We now demonstrate the transformation property (2.19) of the Polyakov loop under gauge transformations. The path ordering in the Polyakov loop definition is to be intended as follows: we divide the path into n infinitesimal segments

$x_1, x_2, \ldots x_{n+1}$ going from 0 to β. On each one, we can approximate the exponential by its Taylor expansion, keeping only the first few terms;

$$\Phi(\mathbf{x}) = \frac{1}{3}\mathrm{tr} \lim_{dx_l \to 0} [1 - G_4(x_1)d(x_1)_4] \ldots [1 - G_4(x_n)d(x_n)_4]\,. \tag{2.20}$$

We recall that, under a gauge transformation we have (setting $ix_0 = x_4$):

$$G_4(x) \to \Omega(x)G_4(x)\Omega^\dagger(x) - \Omega(x)\partial_4\Omega^\dagger(x) \quad \text{with} \quad \Omega(x) = \exp[i\theta(x)] \tag{2.21}$$

and θ is a hermitian matrix belonging to the Lie algebra of $SU(N)$. An infinitesimal gauge transformation can be written as

$$\Omega \simeq 1\!\!1 + i\theta(x) + \ldots \tag{2.22}$$

so that we have

$$\begin{aligned}
\Omega(x)G_4(x)\Omega^\dagger(x) &= [1\!\!1 + i\theta(x)]\, G_4(x)\, [1\!\!1 - i\theta(x)] = G_4(x) + i\, [\theta(x), G_4(x)] \\
\Omega\partial_4\Omega^\dagger &= [1\!\!1 + i\theta(x)][-i\partial_4\theta(x)]
\end{aligned} \tag{2.23}$$

and finally

$$G_4(x) \to -i\partial_4\theta(x) + G_4(x) + i\, [\theta, G_4]\,. \tag{2.24}$$

Up to linear terms in dx_l and $\theta(x)$ we therefore get

$$\begin{aligned}
&1\!\!1 - G_4(x_l)d(x_l)_4 \to 1\!\!1 - G_4(x_l)d(x_l)_4 + \\
&-i\, [\theta(x_l), G_4(x_l)]\, d(x_l)_4 + i\, [\partial_4\theta(x_l)]\, d(x_l)_4
\end{aligned} \tag{2.25}$$

and

$$\partial_4\theta(x_l)d(x_l)_4 = \theta(\mathbf{x}, (x_{l+1})_4) - \theta(\mathbf{x}, (x_l)_4). \tag{2.26}$$

Therefore, up to leading order in θ, we have

$$1\!\!1 - G_4(x_l)d(x_l)_4 \to e^{i\theta(x_{l+1})}\, [1\!\!1 - G_4(x_l)d(x_l)_4]\, e^{-i\theta(x_l)}. \tag{2.27}$$

We thus conclude that

$$\Phi'(\mathbf{x}) = \Omega(\mathbf{x}, T)\Phi(\mathbf{x})\Omega^\dagger(\mathbf{x}, 0). \tag{2.28}$$

2.3.1 Physical Interpretation of the Polyakov Loop

Consider an infinitely heavy quark coupled to a fluctuating gauge potential. We start from the static Dirac equation in imaginary time [2]:

$$[\partial_\tau - igG_4]\,\psi(\mathbf{x},\tau) = 0. \tag{2.29}$$

The above equation can be solved as follows: we re-write it as

$$\frac{\partial_\tau \psi(\mathbf{x},\tau)}{\psi(\mathbf{x},\tau)} = +igG_4(\mathbf{x},\tau), \tag{2.30}$$

which can be integrated to give

$$\ln[\psi(\mathbf{x},\tau)] = ig\int_0^\tau d\tau' G_4(\mathbf{x},\tau') + \ln[\psi(\mathbf{x},0)], \tag{2.31}$$

or equivalently

$$\psi(\mathbf{x},\tau) = \mathcal{T}\exp\left[ig\int_0^\tau d\tau' G_4(\mathbf{x},\tau')\right]\psi(\mathbf{x},0). \tag{2.32}$$

The exponential of the free energy is defined as

$$e^{-\beta F} = \frac{1}{N_c}\sum_{i,n}\langle n|\psi_i(\mathbf{x})e^{-\beta H}\psi_i^\dagger(\mathbf{x})|n\rangle \tag{2.33}$$

where $|n\rangle$ are purely gluonic states, i is an index which corresponds to the quark color, $\psi_i^\dagger(\mathbf{x})$ creates a quark of color i at point $\mathbf{x}$ and $\psi_i(\mathbf{x})$ destroys a quark of color i at point $\mathbf{x}$.

We now give a temperature-dependence to the fields by using the evolution operator in imaginary time:

$$e^{\beta H}\psi_i(\mathbf{x})e^{-\beta H} = \psi_i(\mathbf{x},\beta), \qquad \text{with} \qquad \psi_i(\mathbf{x}) = \psi_i(\mathbf{x},0). \tag{2.34}$$

We can thus re-write

$$\begin{aligned} e^{-\beta F} &= \frac{1}{N_c}\sum_{i,n}\langle n|e^{-\beta H}\psi_i(\mathbf{x},\beta)\psi_i^\dagger(\mathbf{x},0)|n\rangle \\ &= \frac{1}{N_c}\sum_{i,n}e^{-\beta E_n}\langle n|\psi_i(\mathbf{x},\beta)\psi_i^\dagger(\mathbf{x},0)|n\rangle. \end{aligned} \tag{2.35}$$

Now we explicitly write the solution of the static Dirac equation found above:

$$\psi_i(\mathbf{x},\beta) = \sum_j \mathcal{T}\exp\left[ig\int_0^\beta d\tau G_4(\mathbf{x},\tau)\right]_{ij}\psi_j(\mathbf{x},0) \tag{2.36}$$

so that

$$\begin{aligned}
e^{-\beta F} &= \sum_n e^{-\beta E_n}\langle n|\frac{1}{N_c}\sum_{i,j}\mathcal{T}\exp\left[ig\int_0^\beta d\tau G_4(\mathbf{x},\tau)\right]_{ij}\psi_j(\mathbf{x},0)\psi_i^\dagger(\mathbf{x},0)|n\rangle = \\
&= \sum_n e^{-\beta E_n}\langle n|\frac{1}{N_c}\sum_{i,j}\mathcal{T}\exp\left[ig\int_0^\beta d\tau G_4(\mathbf{x},\tau)\right]_{ij}\delta_{i,j}|n\rangle = \\
&= \sum_n e^{-\beta E_n}\langle n|\frac{1}{N_c}\mathrm{tr}_{spin}\mathcal{T}\exp\left[ig\int_0^\beta d\tau G_4(\mathbf{x},\tau)\right]|n\rangle = \\
&= \sum_n e^{-\beta E_n}\langle n|\Phi(\mathbf{x})|n\rangle,
\end{aligned} \tag{2.37}$$

from which we read

$$e^{-\beta(F-F_0)} = \langle\Phi(\mathbf{x})\rangle, \tag{2.38}$$

where F_0 is the free-energy of gluons. From the above definition it is clear that, if $\langle\Phi(\mathbf{x})\rangle = 0$, then we have $F \to \infty$: it costs an infinite amount of energy to put or pull out a quark in the system (confined phase, corresponding to a restored $Z(3)$ symmetry). If $\langle\Phi(\mathbf{x})\rangle \neq 0$, then F is finite: states with a single quark are possible (deconfined phase, corresponding to a spontaneously broken $Z(3)$ symmetry). The left panel of Fig. 2.2 shows the behavior of the Polyakov loop as a function of T/T_c, calculated on the lattice for a pure gauge system in Ref. [4]. The transition temperature T_c in this case is $\simeq$270 MeV and the Polyakov loop is a good order parameter for the deconfinement phase transition of QCD, which is of the first order in this case. The right panel of Fig. 2.2 shows the Polyakov loop as a function of the temperature, calculated on the lattice for a system with 2+1 dynamical quark flavors. The phase transition in this case is a smooth crossover (we will come back to this point later), and the Polyakov loop is no longer an order parameter for the deconfinement phase transition: quarks break the $Z(3)$ symmetry explicitly.

2.3.2 Polyakov Loop on the Lattice

We now want to define the discretized version of the Polyakov loop. In the classical Maxwell theory, an external current J^μ appears in the weight factor in the real-time path integral as

$$e^{iS} \to e^{iS+i\int d^4x J^\mu G_\mu}. \tag{2.39}$$

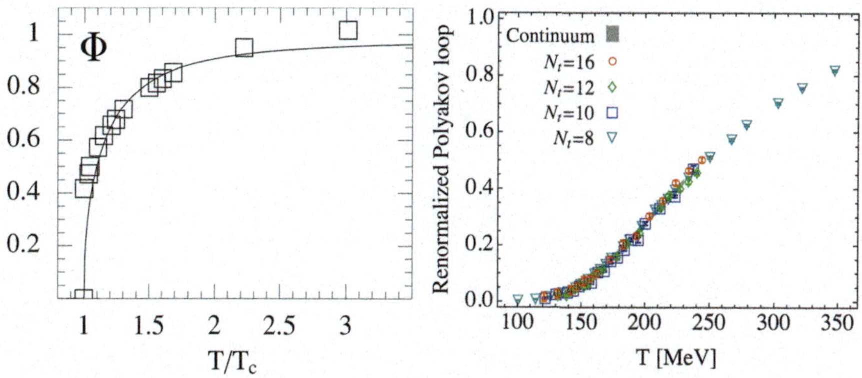

Fig. 2.2 Left: from Ref. [3]. Lattice QCD results for the Polyakov loop as a function of T/T_c, for a purely gluonic system. $T_c \simeq 270$ MeV in this case (lattice results from Ref. [4]). Right: from Ref. [5]. Renormalized Polyakov loop as a function of the temperature, for a system of $2+1$ quark flavors

For a line current along a path $z^\mu(\tau)$ we have

$$J^\mu(x) = \int d\tau \frac{dz^\mu(\tau)}{d\tau} \delta^4(x - z(\tau)). \tag{2.40}$$

The phase $\exp[i \int J^\mu G_\mu]$ takes the form

$$\exp\left[i \int dz^\mu G_\mu\right], \tag{2.41}$$

where the integral is taken along the path specified by $z(\tau)$. The current is conserved ($\partial_\mu J^\mu$=0) for a closed path. In Minkowski space we have

$$J^0 = -iJ_4, \qquad dz^0 = -idz_4 \qquad G_0 = iG_4 \tag{2.42}$$

thus, the phase remains a phase:

$$\exp\left[i \int_C \sum_{\mu=1}^{4} dz_\mu G_\mu(z)\right] \tag{2.43}$$

where C is a path specified by $z(\tau)$. This is what we call Wilson line. We now need to define the lattice version of the Wilson/Polyakov line. We recall the definition of the lattice link variables

$$U_\mu(x) = U_{x,x+\hat{\mu}} = e^{iaG_\mu(x)}. \tag{2.44}$$

We can see that the exponential in Eq. (2.43), corresponds to the product of the link variables along the contour C on the lattice:

$$W_C[U] = \mathrm{tr} \prod_{l \in C} U_l, \tag{2.45}$$

where U_l is the link variable on the contour C. The thermal expectation value is given by

$$\langle W_C \rangle = \frac{\int DU\, W_C[U] e^{-S_G[U]}}{\int DU e^{-S_G[U]}}. \tag{2.46}$$

For our purposes, we need the Polyakov loop, and we want it to depend only on T: the lattice definition of the Polyakov loop reads therefore

$$\langle \Phi(T) \rangle = \frac{1}{V N_c} \sum_{\mathbf{x}} \langle \mathrm{tr} \prod_{x_4=1}^{N_\tau} U_{(\mathbf{x},x_4),(\mathbf{x},x_4+a)} \rangle \tag{2.47}$$

where we took an average over space.

2.4 Chiral Symmetry

Consider the QCD Lagrangian with 2 flavors of quarks (u and d) and $m_u = m_d$ (for an introduction to chiral symmetry see e.g. Ref. [6])

$$\mathcal{L}_{QCD} = -\frac{1}{4} \mathcal{G}_{\mu\nu} \mathcal{G}^{\mu\nu} + i\bar{\psi} \not{D} \psi - m\bar{\psi}\psi \tag{2.48}$$

with $\not{D} = \gamma^\mu D_\mu$. We can decompose the spinors into left and right components by using the projectors as follows

$$\begin{aligned} \psi_L &= P_L \psi = \frac{1}{2} [\mathbb{1} - \gamma_5] \psi \\ \psi_R &= P_R \psi = \frac{1}{2} [\mathbb{1} + \gamma_5] \psi, \end{aligned} \tag{2.49}$$

with $P_L + P_R = \mathbb{1}$, $P_L P_R = P_R P_L = 0$ and $\gamma_5 = i\gamma_0\gamma_1\gamma_2\gamma_3$. The γ_5 matrix satisfies the following properties

$$\gamma_5^\dagger = \gamma_5; \qquad (\gamma_5)^2 = \mathbb{1}; \qquad \{\gamma_\mu, \gamma_5\} = 0. \tag{2.50}$$

Written in terms of ψ_L and ψ_R, the QCD Lagrangian becomes

$$\mathcal{L}_{QCD} = -\frac{1}{4}\mathcal{G}_{\mu\nu}\mathcal{G}^{\mu\nu} + i\left[\bar{\psi}_L + \bar{\psi}_R\right]\not{D}\left[\psi_L + \psi_R\right] - m\left[\bar{\psi}_L + \bar{\psi}_R\right]\left[\psi_L + \psi_R\right]$$
$$= -\frac{1}{4}\mathcal{G}_{\mu\nu}\mathcal{G}^{\mu\nu} + i\bar{\psi}_L\not{D}\psi_L + i\bar{\psi}_R\not{D}\psi_R - m\bar{\psi}_L\psi_R - m\bar{\psi}_R\psi_L, \qquad (2.51)$$

because $\psi = \psi_L + \psi_R$, $\bar{\psi}_L\not{D}\psi_R = \bar{\psi}_R\not{D}\psi_L = 0$ and $\bar{\psi}_L\psi_L = \bar{\psi}_R\psi_R = 0$. Let us consider a few transformations on the QCD Lagrangian and see if all terms are invariant under them.

- The first transformation we consider is $U(1)_V$, which acts on the spinors as follows:

$$\psi \to e^{-i\alpha}\psi \qquad (2.52)$$

  with α real and constant. Both $\bar{\psi}\not{D}\,\psi$ and $\bar{\psi}m\psi$ are invariant under this transformation: $U(1)_V$ is a symmetry of the QCD Lagrangian.
- The second transformation we consider is $U(1)_A$, which acts on the spinors as follows:

$$\psi \to e^{-i\alpha\gamma_5}\psi \qquad (2.53)$$

  with α real and constant. In this case, $\bar{\psi}\not{D}\psi$ is invariant under this transformation, but $\bar{\psi}m\psi$ is not invariant.
- The third transformation we consider is $SU(2)_V$, which acts on the spinors as follows:

$$\psi \to e^{-i\frac{\tau^a}{2}\theta_a}\psi \qquad (2.54)$$

  with θ_a real and constant. Both $\bar{\psi}\not{D}\,\psi$ and $\bar{\psi}m\psi$ are invariant under this transformation: $SU(2)_V$ is a symmetry of the QCD Lagrangian (under the assumption that $m_u = m_d$).
- The fourth and last transformation we consider is $SU(2)_A$, which acts on the spinors as follows:

$$\psi \to e^{-i\frac{\tau^a}{2}\theta_a\gamma_5}\psi \qquad (2.55)$$

  with θ_a real and constant. In this case, $\bar{\psi}\not{D}\,\psi$ is invariant under this transformation, but $\bar{\psi}m\psi$ is not invariant.

Therefore, the symmetry group of $\mathcal{L}_{QCD}$ with $m_u = m_d \neq 0$ is $SU(2)_V \times U(1)_V$.

Let us now consider the QCD Lagrangian in the chiral limit: $m_u = m_d = 0$. In this case, $\mathcal{L}_{QCD}$ is invariant under all four *global* transformations listed above (*global* means that α and θ_a are constant and do not depend on $\mathbf{x}$), and the corresponding conserved currents are:

- The $U(1)_V$ symmetry corresponds to baryon number conservation, with current $j^\mu = \bar{\psi}\gamma^\mu\psi$.
- The $U(1)_A$ symmetry would correspond to the conservation of the current $j^{5\mu} = \bar{\psi}\gamma^\mu\gamma_5\psi$, but it is broken at the quantum level (quantum anomaly).
- The $SU(2)_V$ symmetry corresponds to isospin conservation, with current $j_k^\mu = \bar{\psi}\gamma^\mu\tau_k\psi$.
- The $SU(2)_A$ symmetry corresponds to chirality conservation, with current $j_k^{5\mu} = \bar{\psi}\gamma^\mu\gamma_5\tau_k\psi$.

Therefore, the symmetry group of QCD with $m_u = m_d = 0$ is $SU(2)_V \times SU(2)_A \times U(1)_V$.

Equivalent transformations involve the left- and right-handed spinors seen above:

$$\psi_L \to \psi_L' = e^{-i\frac{\tau^a}{2}\theta_{La}}\psi_L, \qquad SU(2)_L \qquad j_{L\mu}^k = \bar{\psi}_L\gamma_\mu\tau^k\psi_L \tag{2.56}$$

$$\psi_R \to \psi_R' = e^{-i\frac{\tau^a}{2}\theta_{Ra}}\psi_R, \qquad SU(2)_R \qquad j_{R\mu}^k = \bar{\psi}_R\gamma_\mu\tau^k\psi_R. \tag{2.57}$$

These transformations are completely equivalent to the ones seen above, in particular: $SU(2)_V$ is a simultaneous transformation of ψ_L and ψ_R with $\boldsymbol{\theta}_L = \boldsymbol{\theta}_R$, and $SU(2)_A$ is a simultaneous transformation of ψ_L and ψ_R with $\boldsymbol{\theta}_L = -\boldsymbol{\theta}_R$. So, the symmetry group of QCD with $m_u = m_d = 0$ can also be written as $SU(2)_R \times SU(2)_L \times U(1)_V$.

2.4.1 Experimental Observation

Experimentally, baryon-number-violating processes have never been observed: $U(1)_V$, associated to baryon number conservation, is indeed an exact symmetry of QCD. The conserved charge, baryon number, is defined as

$$B = \frac{1}{3}\int d^3x\bar{\psi}\gamma^\mu\psi. \tag{2.58}$$

The $SU(2)_V$ symmetry is a "good" symmetry: the conserved charge, isospin, is defined as

$$Q_V^k = \int d^3x\psi^\dagger\frac{\tau^k}{2}\psi. \tag{2.59}$$

The vacuum state of QCD, $|0\rangle$, is an isospin singlet (i.e. invariant under isospin rotations $Q_V^k|0\rangle = |0\rangle$). The energy levels (hadrons) form degenerate isospin multiplets. We have to point out that this is only an approximate symmetry, because $m_u \simeq m_d$ and $m_p \simeq m_n$.

It turns out that the symmetry $SU(2)_A$ is spontaneously broken: the conserved charge would be

$$Q_A^k = \int d^3x \psi^\dagger \frac{\tau^k}{2} \gamma_5 \psi. \tag{2.60}$$

However, the QCD vacuum is not invariant under chiral rotations: $Q_A^k|0\rangle \neq |0\rangle$. No degenerate multiplets with opposite parity have been observed. $\langle\bar{\psi}\psi\rangle$ in the vacuum is called chiral condensate. It is the order parameter for chiral symmetry. Chiral symmetry implies that every isospin multiplet has a partner with the same mass, the same quantum numbers but opposite parity. However, the experimental observation is very different. For example, the ρ meson is a $J^\pi = 1^-$ particle, namely a vector meson, and its mass is $m_\rho = 0.77$ GeV. Its chiral partner would be the a_1 particle, with $J^\pi = 1^+$ (axial vector meson); however, we have $m_{a_1} = 1.23$ GeV. This is the most striking experimental evidence that chiral symmetry is NOT realized.

Chiral symmetry is spontaneously broken: it means that the Lagrangian of QCD has the symmetry, but the vacuum does NOT have it. In the limit of $m_u = m_d = 0$ (chiral limit), $\mathcal{L}_{QCD}$ has the $SU(2)_A$ symmetry, but the vacuum is not symmetric. A typical visualization of spontaneous symmetry breaking is the Mexican hat potential.

Figure 2.3a shows a parabolic potential, which is symmetric for rotations around the vertical axis. The ground state of the system, represented by the location of the sphere, is symmetric for rotations around the vertical axis as well. Figure 2.3b shows the Mexican hat potential. It is symmetric under rotations around the vertical axis, but the ground state is not: the ball in the figure sits at $y = 0$; a rotation would move it to a different position in the (x, y) plane.

The chiral condensate $\langle\bar{\psi}\psi\rangle = \langle\bar{u}u\rangle + \langle\bar{d}d\rangle = \langle\bar{\psi}_L\psi_R\rangle + \langle\bar{\psi}_R\psi_L\rangle \neq 0$ in the ground state means that the vacuum has lost the symmetry of $\mathcal{L}_{QCD}$. $\bar{\psi}\psi$ is not invariant under $SU(2)_A$ transformations. A non-zero expectation value for this quantity means that the symmetry is spontaneously broken. As a consequence, there are three massless pseudoscalar bosons (Goldstone bosons), identified with the pions. In reality, the pions are not exactly massless, even though they are very light: this is due to the fact that the u and d quark masses are small, but not exactly zero.

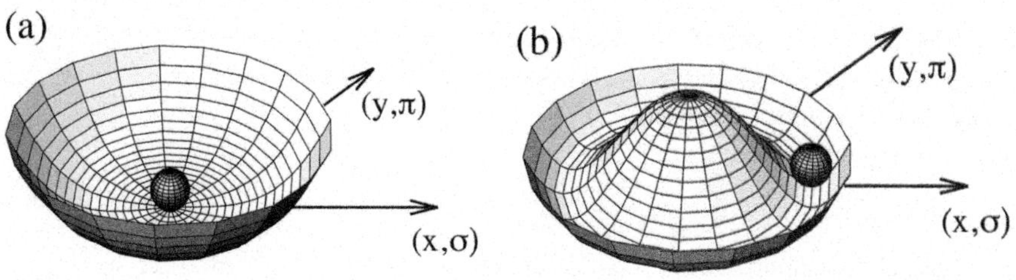

Fig. 2.3 **(a)** Mexican hat potential: it is symmetric under rotations around the vertical axis, but the ground state is not: the sphere in the figure sits at $y = 0$; a rotation would move it to a different position in the (x, y) plane. **(b)** Parabolic potential: it is symmetric under rotations around the vertical axis, and so is the ground state.

2.5 Chiral Phase Transition

2.5.1 Chiral Limit and Transition Temperature

The definition of the chiral condensate of a flavor f is given by

$$\langle\bar{\psi}\psi\rangle_f = \frac{T}{V}\frac{\partial \ln Z}{\partial m_f}. \tag{2.61}$$

On the lattice, it can be calculated as follows

$$\frac{\partial \ln Z}{\partial m_f} = \frac{\partial \ln Z}{\partial M_f}\frac{\partial M_f}{\partial m_f} = \frac{1}{Z}\frac{\partial Z}{\partial M_f}\frac{\partial M_f}{\partial m_f} = \frac{1}{Z}\int DU\,D\bar{\psi}\,D\psi\,(\bar{\psi}_f\psi_f)\,e^{-S_G}e^{-\bar{\psi}M\psi} \tag{2.62}$$

where $M_f^{-1} = \psi_f\bar{\psi}_f$ is the quark propagator with flavor f. From the above it follows that

$$\langle\bar{\psi}\psi\rangle_f = \frac{T}{V}\frac{1}{Z}\int DU\,D\bar{\psi}\,D\psi\left(\mathrm{tr}M_f^{-1}\right)e^{-S_G-S_F}. \tag{2.63}$$

In the vicinity of the chiral phase transition, we can express the free energy density as the sum of a singular and a regular contribution:

$$F = -\frac{T}{V}\ln Z = F_{sing}(t,h) + F_{reg}(T, m_l, m_s, \mu) \tag{2.64}$$

where $h = \frac{1}{h_0}\frac{m_l}{m_s}$ is the explicit chiral symmetry breaking term, while $t = \frac{1}{t_0}\left[\frac{T-T_c^0}{T_c^0} + \kappa_q\left(\frac{\mu_q}{T}\right)^2\right]$ contains all the thermal variables that do not explicitly break chiral symmetry and $\mu_q = \mu_u = \mu_d$. T_c^0 is the phase transition temperature in the chiral limit and for $\mu_q = 0$, t_0 and h_0 are unknown normalization parameters, and κ_q is a parameter that determines the chiral transition temperature change with μ_q. Therefore, T_c^0, κ_q, t_0, h_0 are four unknown quantities unique to QCD. They can be determined by analyzing the scaling behavior of the chiral condensate and susceptibilities from lattice QCD.

To summarize:

- in the limit $m_l \to 0$ QCD is expected to undergo a phase transition at some critical temperature T_c^0, at which chiral symmetry gets restored;
- $\langle\bar{\psi}\psi\rangle_l$ will vanish at T_c^0;
- the dependence of $\langle\bar{\psi}\psi\rangle_l$ in the vicinity of the critical point $(T, m_l) = (T_c^0, 0)$ is controlled by a scaling function that arises from the singular part of the partition function;

- this transition is supposed to be of second order in the continuum limit and to belong to the same universality class as the $3d - O(4)$ model. The latter is a spin model of general interest in condensed matter physics;
- however, in lattice QCD the situation is more complicated: for staggered fermions away from the continuum limit there is only one Goldstone boson. For this reason, the universality class is expected to be the one of $3d - O(2)$ spin model.

In the vicinity of a critical point, the regular contributions are negligible and the universal behavior of the order parameter, which we generically call M, is defined through a scaling function f_G that arises from the singular contribution to the logarithm of the partition function:

$$F_{sing}(T, m_q) = h_0 h^{1+1/\delta} f_s(z) \tag{2.65}$$

so that

$$M(t, h) = -\frac{\partial F_{sing}}{\partial H} = h^{1/\delta} f_G(z) \tag{2.66}$$

with $z = t/h^{1/(\beta\delta)}$, $H = h_0 h$, $f_G(z) = -\left(1 + \frac{1}{\delta}\right) f_s(z) + \frac{z}{\beta\delta} \frac{\partial f_s(z)}{\partial z}$. β and δ are critical exponents describing how the order parameter M approaches the critical point, when one of the variables is set to zero

$$\begin{aligned} M &= (-t)^\beta \qquad h = 0, \qquad t < 0 \\ M &= h^{1/\delta} \qquad t = 0. \end{aligned} \tag{2.67}$$

The parameters t_0 and h_0 are thus fixed from the above equations, and the critical exponents uniquely determine the universality class of the phase transition.

The chiral condensate is an order parameter for the chiral phase transition only in the chiral limit. At finite light quark mass values, additive and multiplicative renormalization is needed to define the order parameter in the continuum limit. For example, one can introduce as an order parameter for chiral symmetry restoration in (2 + 1)-flavor QCD

$$M = \hat{m}_s \left(\langle \bar{\psi}\psi \rangle_l - \frac{m_l}{m_s} \langle \bar{\psi}\psi \rangle_s \right) N_t^4, \tag{2.68}$$

where $\hat{m}_s$ is the strange quark mass in lattice units. The non-subtracted order parameter M_b is defined as

$$M_b = N_t^4 \hat{m}_s \langle \bar{\psi}\psi \rangle_l \tag{2.69}$$

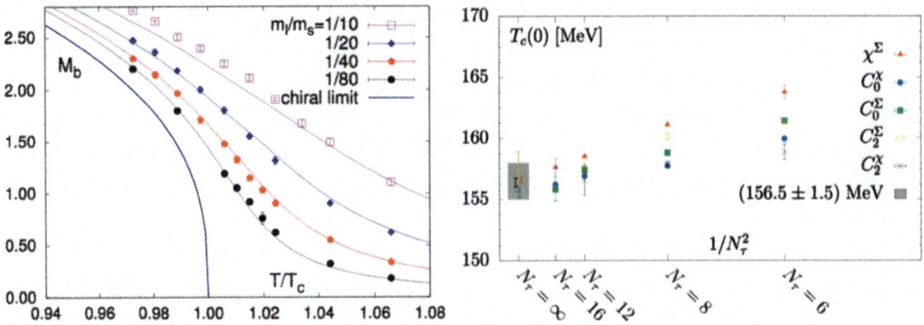

Fig. 2.4 Left: from Ref. [7]. M_b defined in Eq. (2.69) as a function of the temperature, for different values of the light quark mass. Right: from Ref. [8]. Continuum extrapolations of pseudo-critical temperatures T_c using five different chiral observables defined in Ref. [8]

so that M and M_b agree in the chiral limit. M_b has been studied on the lattice as a function of the temperature for different values of the light quark mass. This is shown in the left panel of Fig. 2.4. The curves correspond to the 3d-$O(N)$ scaling behavior, which agree with the data as expected.

The chiral order parameter M can be used to define two susceptibilities, obtained by either deriving M with respect to t, which yields the mixed susceptibility

$$\chi_t \sim \frac{\partial M}{\partial t} \sim \frac{\partial^2 f_s}{\partial t \partial h} \tag{2.70}$$

or with respect to h, which yields the chiral susceptibility:

$$\chi_h \sim \frac{\partial M}{\partial h} \sim \frac{\partial^2 f_s}{\partial h^2}. \tag{2.71}$$

These two quantities are the only two second-order derivatives of the free energy of QCD, that diverge at the chiral critical point, namely at $T = T_c^0$ for $m_l = 0$. Their singular behavior is controlled by the scaling functions $f'_G(z) = \frac{df_G(z)}{dz}$ and $f_\chi(z) = \frac{1}{\delta}\left(f_G(z) - \frac{z}{\beta} f'_G(z)\right)$. At finite quark masses, these two functions have maxima at universal values $z = z_t$ and $z = z_p$, through which one can define pseudo-critical temperatures. The scaling behavior of the chiral susceptibility for two light, degenerate flavors is given by

$$\chi_l(T) = \frac{2\partial\langle\bar{\psi}\psi\rangle_l}{\partial m_l} \tag{2.72}$$

and it is related to that of χ_h. The product of χ_l and m_s^2 is related to the scaling function by

$$\frac{m_s^2 \chi_l}{T^4} = \frac{1}{h_0} h^{1/\delta - 1} f_\chi(z) + \text{regular terms.} \tag{2.73}$$

Thus, the chiral susceptibility diverges at the chiral critical temperature T_c^0 in the chiral limit $h \to 0$ at $\mu_B = 0$. The peak location of the chiral susceptibility is associated with the peak of $f_\chi(z)$, located at $z = z_p$, and identifies the chiral pseudo-critical temperature T_c for $h > 0$.

$$T_c = T_c^0 \left[1 + \frac{z_p}{z_0} \left(\frac{m_l}{m_s} \right)^{1/(\beta\delta)} \right] + \text{regular terms,} \tag{2.74}$$

where $z_0 = t_0 / h_0^{1/(\beta\delta)}$. Recently, the authors of Ref. [10] performed a thorough analysis of the chiral phase transition in QCD with $N_f = 2 + 1$, in a range of light quark masses corresponding to $58 \leq m_\pi \leq 163$ MeV. After the thermodynamic, continuum and chiral limit they reported a chiral transition temperature of $T_c^0 = 132^{+3}_{-6}$ MeV at vanishing light quark masses.

In a system of dynamical quarks with physical masses, the chiral transition temperature increases with respect to the one in the chiral limit. The value of the chiral transition temperature of QCD for a system of 2 + 1 quark flavors quoted by the HotQCD collaboration is $T_c = 156.5 \pm 1.5$ MeV [8], which is in agreement with previous estimates [5]. The most recent estimate for the chiral phase transition temperature was presented in Ref. [9], and it yields a value of $T_c = 158.0 \pm 0.6$ MeV. In Ref. [9], a thorough discussion on the width of the crossover transition is also presented. A natural definition of the width of the susceptibility peak is its second derivative at T_c: $(\Delta T)^2 = -\chi_l(T_c) \left[\frac{d^2}{dT^2} \chi_l \right]^{-1}_{T=T_c}$. The reported value for ΔT is 15 ± 1 MeV.

The authors of Ref. [11] studied the finite size scaling of the lattice chiral susceptibility χ_l. For a real (first-order) phase transition, the finite size scaling is determined by the geometric dimension: the height of the peak is proportional to the volume V and the width to $1/V$. For a second-order transition, the singular behavior is given by some power of V with appropriate critical exponents. For an analytic crossover, there is no singular behavior and the susceptibility peak does not get sharper when the volume is increased. Its height and width will be volume-independent for large volumes.

Figure 2.5 shows the behavior of the chiral susceptibility as a function of $6/g^2$ (related to the temperature) for $N_t = 4$ (left panel) and $N_t = 6$ (right panel) and with $N_s/N_t = 3, 4, 5, 6$. The volume increases by up to a factor eight between the smallest and largest cases. The figure clearly shows that no critical behavior develops as the volume increases: the QCD phase transition is an analytic crossover at $\mu_B = 0$ for $N_f = 2 + 1$.

The left panel of Fig. 2.6 shows the continuum extrapolated results for another definition of the chiral condensate,

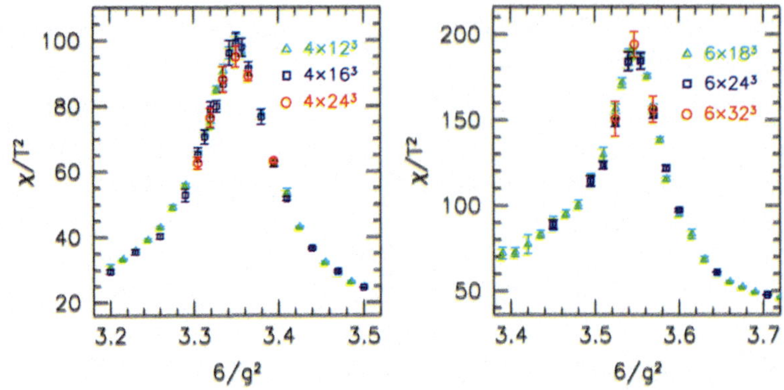

Fig. 2.5 From Ref. [11]. Chiral susceptibility as a function of $6/g^2$ (related to the temperature) for $N_t = 4$ (left panel) and $N_t = 6$ (right panel) and with $N_s/N_t = 3, 4, 5, 6$

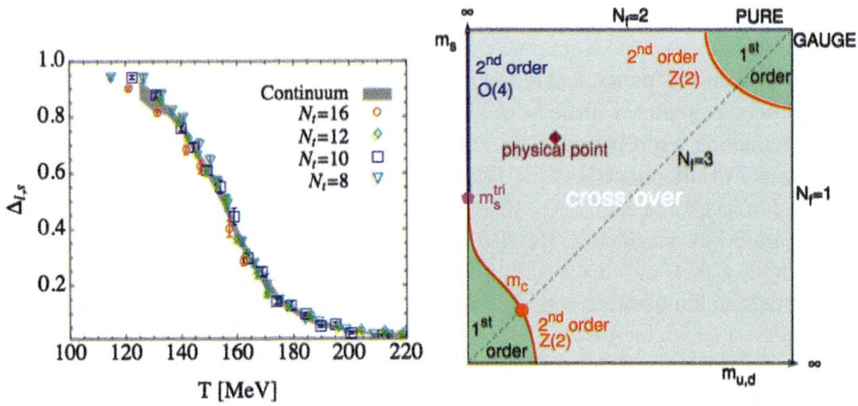

Fig. 2.6 Left: from Ref. [5]. Continuum extrapolated results for $\Delta_{l,s}$ as a function of the temperature. Right: cartoon of the nature of the QCD transition as a function of the light $(m_{u,d})$ and strange (m_s) quark masses, at zero baryonic chemical potential

$$\Delta_{l,s} = \frac{\langle\bar{\psi}\psi\rangle_{l,T} - \frac{m_l}{m_s}\langle\bar{\psi}\psi\rangle_{s,T}}{\langle\bar{\psi}\psi\rangle_{l,0} - \frac{m_l}{m_s}\langle\bar{\psi}\psi\rangle_{s,0}}, \tag{2.75}$$

as a function of the temperature. The crossover nature of the QCD chiral transition for physical values of the quark masses is evident from this plot.

It is clear from the discussion above, that the nature of the QCD phase transition depends crucially on the values of the quark masses. This is summarized in the right panel of Fig. 2.6, which shows the so-called Columbia plot. On the axis there are the two degenerate light quark masses $(m_{u,d})$ and the strange quark mass (m_s). The plane is divided into first-order (green) and crossover (grey) regions, separated by second-order phase transition lines (red). The diamond indicates the physical values of the quark masses, situated in the crossover region as discussed above. The $N_f = 3$ chiral limit was postulated to have a first order phase transition in Ref. [12], on the basis of the fact that any effective theory for the chiral condensate would contain its cubic power. This limit was studied e.g. in Ref. [13], in which an upper bound for the critical pion mass (corresponding to the point labeled as m_c in the figure), was found to be $m_\pi^c \sim 50\,\text{MeV}$. The separation line between crossover and first order phase transition at $N_f = 3$ is in the $Z(2)$ universality class, since the only massless field at the chiral critical point is a sigma meson, with the universality class of the Ising model [14].

2.5.2 Lattice QCD Predictions on Parity Doubling and Chiral Symmetry Restoration

In the past decades, chiral symmetry restoration at high temperature has been studied extensively in the mesonic sector [15]. More recently, the baryonic sector has also received attention, due to the fact that the combination of chiral symmetry restoration and parity leads to testable predictions. In particular, when both symmetries are realized, the phenomenon of parity doubling takes place, namely a degeneracy between baryonic channels of opposite parity. This topic was at first investigated in the quenched approximation in Ref. [16], and more recently in the case of $2+1$ dynamical flavors [17–19].

Results of these lattice QCD simulations show an interesting feature of chiral symmetry restoration, namely that the positive parity state is largely unaffected by the increase in temperature in all channels, whereas the mass of the negative parity state decreases with increasing temperature, leading to parity doubling in the quark-gluon plasma. Figure 2.7 shows the temperature dependence of the ground state masses for the octet (top) and decuplet (bottom) baryons [19].

These results, albeit obtained so far for heavier than physical values of the quark masses, are extremely promising as they lead to a clear experimental signature for chiral symmetry restoration.

2.6 QCD in an External Magnetic Field

The study of QCD in an external magnetic field is an active field of research (for a recent review see e.g. [20]). The study of strong magnetic fields plays an important role in at least three areas of high-energy physics: noncentral heavy-ion collisions, compact stars and the early universe. In the case of noncentral collisions, very strong and time-dependent fields are created due to the fact that the two colliding nuclei are effectively electric currents which produce a magnetic field $\mathbf{B}$ according to Maxwell's equations. The intensity of the generated fields can reach up to $|\mathbf{B}| \simeq 10^{19}$ Gauss, but their life-time is estimated to be relatively short [21–23].

Magnetars are a specific category of neutron stars, characterized by very high magnetic fields and relatively low rotation frequency. The magnetic field depends

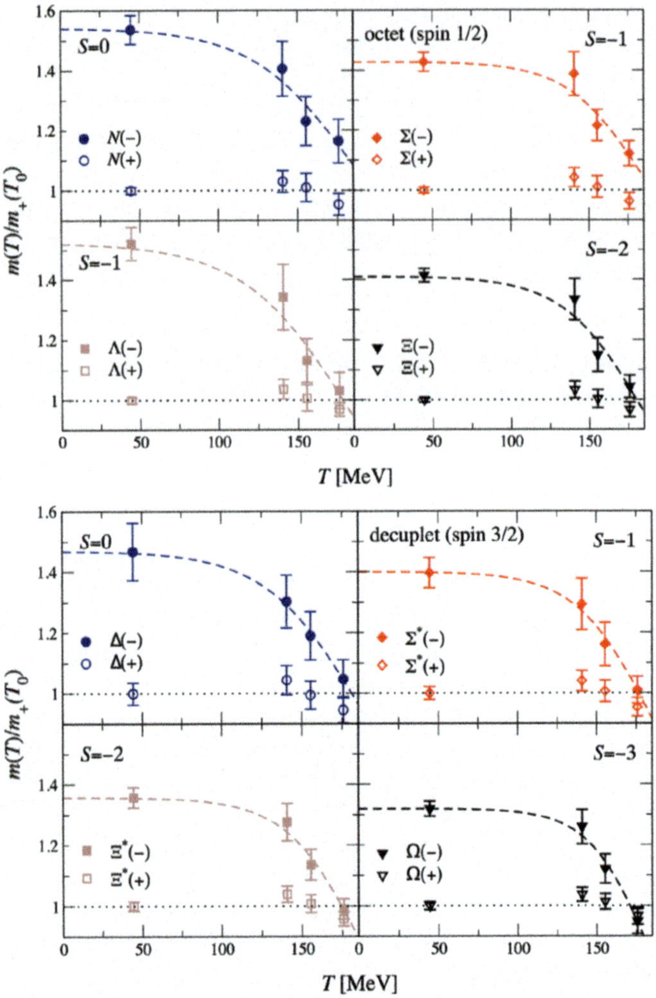

Fig. 2.7 From Ref. [19]. Temperature dependence of the ground state masses, normalized with the mass of the positive parity state at the lowest temperature, in the hadronic phase. Positive (negative) parity states are indicated with open (closed) symbols. The top panels show the octet baryon masses, while the bottom panels show the decuplet baryon states

on the density achieved in the star, and can be as large as $|\mathbf{B}| \simeq 10^{14} - 10^{15}$ Gauss on the surface and $|\mathbf{B}| \simeq 10^{16} - 10^{19}$ Gauss in the core. In order to extract the magnetar properties such as mass and radius, the equation of state of dense matter in the presence of a magnetic field is therefore needed.

Finally, if primordial magnetic fields were generated from bubble collisions during the electroweak transition, they are of the order $|\mathbf{B}|/T^2 \lesssim 0.5$ and they

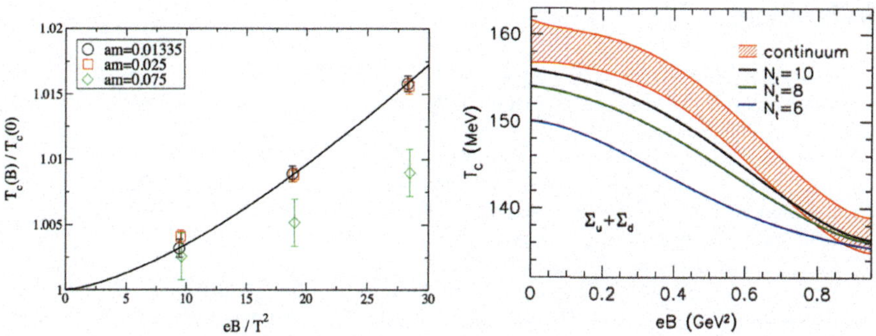

Fig. 2.8 Left: from Ref. [35]. Transition temperature, normalized to the value at zero magnetic field, as a function of $|e\mathbf{B}|/T^2$. Right: from Ref. [36]. Transition temperature as a function of $|e\mathbf{B}|$ at different lattice spacings (solid lines) and in the continuum limit (red band)

might be responsible for a first order electroweak phase transition, thus potentially being relevant for baryogenesis [24].

2.6.1 Magnetic Catalysis

We call magnetic catalysis the effect according to which either the magnitude of a condensate is increased by an external magnetic field (if the condensate is already non-zero at $|\mathbf{B}| = 0$), or an external magnetic field induces the breaking of a symmetry, and consequently the appearance of a non-zero condensate, if the symmetry was not broken at $|\mathbf{B}| = 0$.

In the case of QCD, the chiral condensate is non-zero already for $|\mathbf{B}| = 0$ at low temperatures (chiral symmetry is broken) and its fate in an external magnetic field has been investigated in several effective models [25–34].

QCD at zero chemical potential, in the presence of an abelian background field A_μ from which the magnetic field stems, is not affected by the sign problem and can therefore be simulated on the lattice. The first simulations in $SU(3)_c$ lattice gauge theory with two flavors were performed in Ref. [35]. In this pioneering work, the authors used quark masses corresponding to a pion mass in the range (200 MeV)$<$ $m_\pi <$(480 MeV) and found a slow increase of the transition temperature with increasing magnetic field. This behavior is shown in the left panel of Fig. 2.8 and was later confirmed in Refs. [36, 37].

In Refs. [36, 37], results were obtained also using physical quark masses. In this case, the transition temperature decreases with increasing magnetic field, as shown in the right panel of Fig. 2.8. This suggests that the transition temperature is a complicated function of the magnetic field and the quark masses. For recent results on the value of the quark mass above which inverse catalysis does not take place, see [38]. This value corresponds to a pion mass of $m_\pi = 497(4)$ MeV. For other recent results for QCD in a magnetic field, see Refs. [39–42].

References

1. Holland, K., Wiese, U.J.: [arXiv:hep-ph/0011193 [hep-ph]]
2. McLerran, L.D., Svetitsky, B.: Phys. Lett. B **98**, 195–198 (1981). https://doi.org/10.1016/0370-2693(81)90986-2
3. Ratti, C., Thaler, M.A., Weise, W.: Phys. Rev. D **73**, 014019 (2006)
4. Kaczmarek, O., Karsch, F., Petreczky, P., Zantow, F.: Phys. Lett. B **543**, 41 (2002)
5. Borsanyi, S., et al.: [Wuppertal-Budapest Collaboration] JHEP **1009**, 073 (2010)
6. Koch, V.: [arXiv:nucl-th/9512029 [nucl-th]]
7. Ejiri, S., et al.: Phys. Rev. D **80**, 094505 (2009)
8. Bazavov, A., et al.: [HotQCD Collaboration] Phys. Lett. B **795**, 15 (2019)
9. Borsanyi, S., Fodor, Z., Guenther, J.N., Kara, R., Katz, S.D., Parotto, P., Pasztor, A., Ratti, C., Szabo, K.K.: Phys. Rev. Lett. **125**(5), 052001 (2020)
10. Ding, H.T., Hegde, P., Kaczmarek, O., Karsch, F., Lahiri, A., Li, S.T., Mukherjee, S., Ohno, H., Petreczky, P., Schmidt, C., et al.: Phys. Rev. Lett. **123**(6), 062002 (2019)
11. Aoki, Y., Endrodi, G., Fodor, Z., Katz, S.D., Szabo, K.K.: Nature **443**, 675 (2006)
12. Pisarski, R.D., Wilczek, F.: Phys. Rev. D **29**, 338 (1984)
13. Bazavov, A., Ding, H.-T., Hegde, P., Karsch, F., Laermann, E., Mukherjee, S., Petreczky, P., Schmidt, C.: Phys. Rev. D **95**(7), 074505 (2017)
14. Gavin, S., Gocksch, A., Pisarski, R.D.: Phys. Rev. D **49**, R3079 (1994)
15. Rapp, R., Wambach, J.: Adv. Nucl. Phys. **25**, 1 (2000)
16. Datta, S., Gupta, S., Padmanath, M., Maiti, J., Mathur, N.: JHEP **02**, 145 (2013)
17. Aarts, G., Allton, C., Hands, S., Jäger, B., Praki, C., Skullerud, J.I.: Phys. Rev. D **92**(1), 014503 (2015)
18. Aarts, G., Allton, C., De Boni D., Hands, S., Jäger B., Praki, C., Skullerud, J.I.: JHEP **06**, 034 (2017)
19. Aarts, G., Allton, C., De Boni, D., Jäger, B.: Phys. Rev. D **99**(7), 074503 (2019)
20. Andersen, J.O., Naylor, W.R., Tranberg, A.: Rev. Mod. Phys. **88**, 025001 (2016)
21. Kharzeev, D.E., McLerran, L.D., Warringa, H.J.: Nucl. Phys. A **803**, 227–253 (2008)
22. Skokov, V., Illarionov, A., Toneev, V.: Int. J. Mod. Phys. A **24**, 5925–5932 (2009)
23. Bzdak, A., Skokov, V.: Phys. Lett. B **710**, 171–174 (2012)
24. De Simone, A., Nardini, G., Quiros, M., Riotto, A.: JCAP **10**, 030 (2011)
25. Klimenko, K.: Z. Phys. C **54**, 323–330 (1992)
26. Klimenko, K.: Theor. Math. Phys. **89**, 1161–1168 (1992)
27. Klimenko, K.: Theor. Math. Phys. **90**, 1–6 (1992)
28. Ebert, D., Klimenko, K., Vdovichenko, M., Vshivtsev, A.: Phys. Rev. D **61**, 025005 (2000)
29. Klevansky, S., Lemmer, R.H.: Phys. Rev. D **39**, 3478–3489 (1989)
30. Klimenko, K., Zhukovsky, V.: Phys. Lett. B **665**, 352–355 (2008)
31. Gusynin, V., Miransky, V., Shovkovy, I.: Phys. Rev. Lett. **73**, 3499–3502 (1994)
32. Miransky, V., Shovkovy, I.: Phys. Rev. D **66**, 045006 (2002)
33. Ozaki, S.: Phys. Rev. D **89**(5), 054022 (2014)
34. Haber, A., Preis, F., Schmitt, A.: Phys. Rev. D **90**(12), 125036 (2014)
35. D'Elia, M., Mukherjee, S., Sanfilippo, F.: Phys. Rev. D **82**, 051501 (2010)
36. Bali, G., Bruckmann, F., Endrodi, G., Fodor, Z., Katz, S., Krieg, S., Schafer, A., Szabo, K.: JHEP **02**, 044 (2012)
37. Bali, G., Bruckmann, F., Endrodi, G., Fodor, Z., Katz, S., Schafer, A.: Phys. Rev. D **86**, 071502 (2012)
38. Endrodi, G., Giordano, M., Katz, S.D., Kovács, T.G., Pittler, F.: JHEP **07**, 007 (2019)
39. Bali, G.S., Brandt, B.B., Endrődi, G., Gläßle, B.: Phys. Rev. Lett. **121**(7), 072001 (2018)
40. Endrodi, G., Kaminski, M., Schafer, A., Wu, J., Yaffe, L.: JHEP **09**, 070 (2018)
41. Endrődi, G., Markó, G.: JHEP **08**, 036 (2019)
42. Bali, G.S., Endrődi, G., Piemonte, S.: JHEP **07**, 183 (2020)

Equation of State of QCD at Finite Temperature and $\mu_B = 0$

3

Abstract

In this chapter we discuss the main techniques to calculate the equation of state of QCD at finite temperature and $\mu_B = 0$ in lattice QCD. In particular, we describe the differential and integral methods. We also discuss the Stefan-Boltzmann ideal gas limit, as well as the comparison with perturbative QCD results. The most recent lattice QCD results are introduced and discussed.

3.1 Equation of State of QCD at $\mu_B = 0$

Once the lattice action is defined, the finite temperature lattice QCD formalism can be extended to the numerical evaluation of the Equation of State (EoS) of QCD. The energy density and the pressure are defined as

$$\epsilon = \frac{T^2}{V} \left. \frac{\partial \ln Z(V,T)}{\partial T} \right|_V \qquad p = \left. \frac{T \partial \ln Z(V,T)}{\partial V} \right|_T . \tag{3.1}$$

All other quantities of interest can be evaluated making use of standard thermodynamic relations:

$$I = \epsilon - 3p; \qquad s = \frac{\epsilon + p}{T}; \qquad c_s^2 = \frac{dp}{d\epsilon}, \tag{3.2}$$

where I is the interaction measure or trace anomaly and it vanishes in the ideal gas limit where $\epsilon = 3p$, s is the entropy density and c_s^2 is the speed of sound squared. We first consider the pure gauge case, to illustrate the method. The partition function in this case is given by

C. Ratti, R. Bellwied, *The Deconfinement Transition of QCD*, Lecture Notes in Physics 981, https://doi.org/10.1007/978-3-030-67235-5_3

$$Z(V,T) = \int DUe^{-S_G[U]}. \tag{3.3}$$

The two most popular methods to determine the EoS of QCD are the differential and the integral methods.

3.1.1 Differential Method

The differential method was introduced in Ref. [1], but soon it was noticed that it can yield a negative pressure near the critical temperature [2]. A proposal to improve the method has been presented in Ref. [3] (for more details see [4]). In order to distinguish between the derivatives with respect to T and V, in the differential method the theory is defined on a $(3+1)$-dimensional anisotropic lattice with different lattice spacings in the spatial (a_s) and temporal (a_t) directions. The variables $\{T, V\}$ are then replaced by the anisotropy parameter ξ and a scale a, in the following way

$$\xi = \frac{a_s}{a_t}, \qquad a = a_s. \tag{3.4}$$

The partial derivatives with respect to T and V are written in terms of the variables ξ and a, at fixed N_s and N_t:

$$T\left.\frac{\partial}{\partial T}\right|_V = \xi\left.\frac{\partial}{\partial \xi}\right|_a \quad \text{and} \quad 3V\left.\frac{\partial}{\partial V}\right|_T = a\left.\frac{\partial}{\partial a}\right|_\xi + \xi\left.\frac{\partial}{\partial \xi}\right|_a. \tag{3.5}$$

We have seen in Chap. 1 that the pure gauge action can be written as

$$S_G = \sum_{x\mu\nu}\frac{1}{g^2a^4}\mathrm{tr}\left\{1 - \Re U_{\mu\nu,x}\right\} \tag{3.6}$$

where we had the same lattice spacing a in the spatial and temporal directions. Now we need two different ones. In the equation above, we have

$$U_{\mu\nu,x} = U^\dagger_{\nu\mu,x} = U_{x,x+a\hat{\mu}}U_{x+a\hat{\mu},x+a\hat{\mu}+a\hat{\nu}}U_{x+a\hat{\mu}+a\hat{\nu},x+a\hat{\nu}}U_{x+a\hat{\nu},x}. \tag{3.7}$$

We define $P_{\mu\nu}(x) = 1 - \frac{\Re\mathrm{tr}U_{\mu\nu,x}}{N_c}$. We introduce the following notation for the average spatial and temporal plaquettes

$$P_s = \sum_{x,j>i,i,j=1}^{3}\frac{P_{ij}(x)}{3N_s^3N_t}, \qquad P_t = \sum_{x,i=1}^{3}\frac{P_{0i}(x)}{3N_s^3N_t}. \tag{3.8}$$

The Wilson action that we defined in Chap. 1 can be written as

$$S_G[U] = 6N_cN_s^3N_t\,[\kappa_s P_s + \kappa_t P_t] \tag{3.9}$$

with

$$\kappa_s = \frac{1}{\xi g_s^2} \quad \text{and} \quad \kappa_t = \frac{\xi}{g_t^2} \tag{3.10}$$

where g_s is the gauge coupling along the spatial direction and g_t is the gauge coupling along the temporal direction. Using these definitions we obtain

$$\epsilon = \frac{T^2}{V}\frac{\partial \ln Z}{\partial T} = \frac{T}{V}\xi\frac{\partial \ln Z}{\partial \xi} = -\frac{T}{V}\frac{\xi\int DU\frac{\partial S_G[U]}{\partial \xi}e^{-S_G[U]}}{Z}. \tag{3.11}$$

Recalling that

$$T = \frac{\xi}{N_t a}, \qquad V = N_s^3 a^3 \tag{3.12}$$

we can re-write

$$\begin{aligned}\epsilon &= \frac{\xi^2}{N_tN_s^3a^4}\frac{\int DU\left(-\frac{\partial S_G[U]}{\partial \xi}\right)e^{-S_G[U]}}{Z} = -\frac{\xi^2 6N_cN_s^3N_t}{N_s^3N_ta^4}\left[\frac{\partial \kappa_s}{\partial \xi}\langle P_s\rangle + \frac{\partial \kappa_t}{\partial \xi}\langle P_t\rangle\right]\\ &= -\frac{\xi^2 6N_c}{a^4}\left[\frac{\partial \kappa_s}{\partial \xi}\langle P_s\rangle + \frac{\partial \kappa_t}{\partial \xi}\langle P_t\rangle\right],\end{aligned} \tag{3.13}$$

where we used the following identity

$$\frac{\partial S_G[U]}{\partial \xi} = \left[\frac{\partial \kappa_s}{\partial \xi}P_s + \frac{\partial \kappa_t}{\partial \xi}P_t\right]6N_cN_s^3N_t. \tag{3.14}$$

This energy density contains a contribution from the vacuum, similar to the zero-point energy of the continuum theory, which can be eliminated by subtracting $\epsilon(T = 0)$. The latter can be approximated by ϵ evaluated on the symmetric N_s^4 lattice for sufficiently large N_s; thus we get

$$a^4\epsilon = -6N_c\xi^2\left[\frac{\partial \kappa_s}{\xi}D_s + \frac{\partial \kappa_t}{\partial \xi}D_t\right] \tag{3.15}$$

with $D_i = \langle P_i\rangle - \langle P_0\rangle$ and $\langle P_0\rangle$ is the average plaquette at $T = 0$.

In the same way we have

$$p = T\frac{\partial \ln Z}{\partial V} = \frac{T}{3V}\left[a\frac{\partial \ln Z}{\partial a} + \xi\frac{\partial \ln Z}{\partial \xi}\right]$$

$$= -\frac{\frac{T}{3V}a\left[\int DU \frac{\partial S_G[U]}{\partial a} e^{-S_G[U]}\right]}{Z} - \frac{\frac{T}{3V}\xi\left[\int DU \frac{\partial S_G[U]}{\partial \xi} e^{-S_G[U]}\right]}{Z}. \quad (3.16)$$

Using the identities

$$\frac{\partial S_G[U]}{\partial a} = 6N_c N_s^3 N_t \left[\frac{\partial \kappa_s}{\partial a} P_s + \frac{\partial \kappa_t}{\partial a} P_t\right]$$
$$\frac{\partial S_G[U]}{\partial \xi} = 6N_c N_s^3 N_t \left[\frac{\partial \kappa_s}{\partial \xi} P_s + \frac{\partial \kappa_t}{\partial \xi} P_t\right]. \quad (3.17)$$

and subtracting the zero-point pressure we get

$$a^4 p = -2N_c \xi^2 \left[\frac{\partial \kappa_s}{\partial \xi} D_s + \frac{\partial \kappa_t}{\partial \xi} D_t\right] - 2N_c \xi a \left[\frac{\partial \kappa_s}{\partial a} D_s + \frac{\partial \kappa_t}{\partial a} D_t\right]. \quad (3.18)$$

We also define the quantity $\Delta I = \epsilon - 3p$, the so-called interaction measure. Another name for it is the "trace anomaly", as it is generated by the breaking of the conformal invariance of QCD at the quantum level. We get

$$a^4 \Delta I = 6N_c \xi a \left[\frac{\partial \kappa_s}{\partial a} D_s + \frac{\partial \kappa_t}{\partial a} D_t\right]. \quad (3.19)$$

All these expressions involve derivatives of the couplings g_s and g_t with respect to a and ξ. In the weak coupling limit, the two g_i^{-2} can be expanded around their symmetric lattice value $g^{-2}(a)$ [5] ($\xi = 1$ in the following)

$$g_i^{-2}(a, \xi) = g^{-2}(a) + c_i(\xi) + O(g^2(a)) \quad (3.20)$$

with the condition $c_i(\xi = 1) = 0$. The numbers $c_i(\xi)$ are known as the Karsch coefficients. We need

$$b(a) = a \frac{\partial g^{-2}(a)}{\partial a}. \quad (3.21)$$

With the definition of the β-function calculated in perturbative QCD at one loop,

$$B(\alpha_s) = \frac{\mu}{2}\frac{\partial \alpha_s}{\partial \mu} = -\frac{33 - 2N_f}{12\pi}\alpha_s^2 + \ldots \qquad (\alpha_s = \frac{g^2}{4\pi}) \quad (3.22)$$

we get $b(a) = \frac{B(\alpha_s)}{2\pi\alpha_s^2}$ because

$$2B(\alpha_s) = a\frac{\partial \alpha_s}{\partial a} = a\frac{\partial \alpha_s}{\partial g^{-2}}\frac{\partial g^{-2}}{\partial a} \quad (3.23)$$

from which follows

$$
\begin{aligned}
a\frac{\partial \kappa_s}{\partial a} &= \frac{a}{\xi}\frac{\partial g_s^{-2}}{\partial a} = \frac{B}{2\pi\alpha_s^2\xi} \\
\frac{\partial \kappa_s}{\partial \xi} &= -\frac{g_s^{-2}}{\xi^2} + \frac{1}{\xi}\frac{\partial c_s}{\partial \xi} \\
a\frac{\partial \kappa_t}{\partial a} &= a\xi\frac{\partial g_t^{-2}}{\partial a} = \frac{\xi B}{2\pi\alpha_s^2} \\
\frac{\partial \kappa_t}{\partial \xi} &= -g_t^{-2} + \xi\frac{\partial c_t}{\partial \xi}.
\end{aligned}
\tag{3.24}
$$

Putting everything together, in the $\xi \to 1$ limit one finds

$$
\begin{aligned}
\frac{\epsilon}{T^4} &= 2\frac{N_c N_t^3}{N_s^3}\left[\frac{D_s - D_t}{g^2} - \left(\frac{\partial c_s}{\partial \xi}D_s + \frac{\partial c_t}{\partial \xi}D_t\right)\right] \\
\frac{p}{T^4} &= \frac{2}{3}\frac{N_c N_t^3}{N_s^3}\left[\frac{D_s - D_t}{g^2} - \left(\frac{\partial c_s}{\partial \xi}D_s + \frac{\partial c_t}{\partial \xi}D_t\right)\right] - \frac{2}{3}\frac{N_c N_t^3}{N_s^3}\frac{B}{2\pi\alpha_s^2}\left[D_s + D_t\right] \\
\frac{\Delta I}{T^4} &= 2\frac{N_c N_t^3}{N_s^3}\frac{B}{2\pi\alpha_s^2}\left[D_s + D_t\right].
\end{aligned}
\tag{3.25}
$$

It was argued that the problem encountered in the differential method, namely that the pressure turns negative near the transition temperature, is due to the use of perturbative formulas for the derivatives of the coupling. To cure this problem, the integral method was introduced [6–8].

3.1.2 Integral Method

The integral method is defined on an isotropic lattice ($a_s = a_t = a$) and uses the fact that the following expression for the pressure holds in the case of a homogeneous system,

$$
p = \frac{T}{V}\ln Z(T, V). \tag{3.26}
$$

Therefore, evaluating the pressure is equivalent to evaluating the partition function Z. For $\xi = 1$, the Wilson action takes the form

$$
-\frac{\partial \ln Z}{\partial \beta} = S_G[U] = 3N_s^3 N_t\left[P_s + P_t\right], \text{ with } \beta = 2\frac{N_c}{g^2} \text{ and } g_s = g_t = g. \tag{3.27}
$$

Therefore we can get

$$\ln Z|_\beta - \ln Z|_{\beta_0} = -3N_s^3 N_t \int_{\beta_0}^{\beta} d\beta' \left[\langle P_s(\beta')\rangle + \langle P_t(\beta')\rangle\right]. \tag{3.28}$$

This procedure gives us the pressure up to an additive constant

$$\left.\frac{p}{T^4}\right|_\beta - \left.\frac{p}{T^4}\right|_{\beta_0} = -3N_t^4 \int_{\beta_0}^{\beta} d\beta' \left[\langle P_s(\beta')\rangle + \langle P_t(\beta')\rangle\right]. \tag{3.29}$$

Below $T \simeq 200\,\text{MeV}$, the pressure $p \sim \exp[-m_g/T]$ is dominated by the lightest glueball states, which have a mass $m_g \sim 1$ GeV. For this reason, the pressure is small for $T \leq T_0$. The additive constant is therefore usually chosen to be zero at some coupling β_0 corresponding to a small temperature $T \leq T_0$. Making this choice and performing the usual vacuum subtraction, we get

$$\frac{p(\beta)}{T^4} = -3N_t^4 \int_{\beta_0}^{\beta} d\beta' \left[D_s(\beta') + D_t(\beta')\right]. \tag{3.30}$$

From the pressure, one can obtain the other observables through thermodynamic identities

$$\begin{aligned} \left.\frac{\partial p}{\partial T}\right|_V &= \left.\frac{\partial S}{\partial V}\right|_T = s = \frac{\epsilon + p}{T} \\ \frac{\Delta I}{T^4} &= \frac{\epsilon - 3p}{T^4} = T\left.\frac{\partial(p/T^4)}{\partial T}\right|_V . \end{aligned} \tag{3.31}$$

For an isotropic lattice we get

$$T = \frac{1}{N_t a}; \qquad \frac{\partial}{\partial T} = \frac{\partial a}{\partial T}\frac{\partial}{\partial a} = -\frac{1}{N_t T^2}\frac{\partial}{\partial a} = -\frac{N_t^2 a^2}{N_t}\frac{\partial}{\partial a} \tag{3.32}$$

so that

$$T\left.\frac{\partial}{\partial T}\right|_V = -a\frac{\partial}{\partial a}. \tag{3.33}$$

It is then easy to see that

$$\frac{\Delta I}{T^4} = N_t^4 \left(a\frac{\partial \beta}{\partial a}\right)[D_s + D_t] = 2N_c N_t^4 \frac{B}{2\pi\alpha_s^2}[D_s + D_t]. \tag{3.34}$$

We can use a non-perturbative β-function for $B(\alpha_s)$. High precision results for the trace anomaly for the pure gauge sector of QCD are shown in the left panel of Fig. 3.1. The right panel shows a comparison with the trace anomaly obtained with the glueball resonance model, estimated from the twelve lightest glueballs [9].

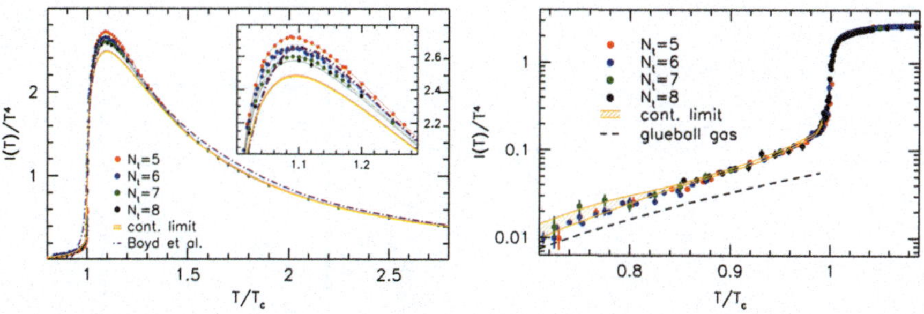

Fig. 3.1 From Ref. [9]. Left: the trace anomaly on $N_s/N_t = 8$ lattices for various lattice spacings in the transition region. The result of a combined spline fit for each lattice spacing, together with the continuum extrapolation is shown by the colored lines and the yellow band, respectively. For comparison, results with the standard Wilson action [8] are also shown by the dashed-dotted line. Right: the trace anomaly in the confined phase measured with various lattice spacings and the continuum extrapolation (yellow band). The dashed line corresponds to the glueball resonance model, estimated from the twelve lightest glueballs

In the case of a system containing dynamical quarks, one can think of other derivatives of the pressure with respect to the bare parameters of the action

$$p^{lat}(\beta, m_q) - p^{lat}(\beta_0, m_{q0}) = \frac{1}{N_t N_s^3} \int_{\beta_0, m_{q0}}^{\beta, m_q} \left[d\beta \frac{\partial \ln Z}{\partial \beta} + \sum_q dm_q \frac{\partial \ln Z}{\partial m_q} \right]. \tag{3.35}$$

We have seen that

$$\begin{aligned} \langle -S_G \rangle &= \frac{1}{N_t N_s^3} \frac{\partial \ln Z}{\partial \beta} \\ \langle \bar{\psi}_q \psi_q \rangle &= \frac{1}{N_t N_s^3} \frac{\partial \ln Z}{\partial m_q}. \end{aligned} \tag{3.36}$$

Subtracting the vacuum contribution one gets

$$\begin{aligned} \langle S_G \rangle^{sub} &= \langle S_G \rangle_{N_t, N_s} - \langle S_G \rangle_{N_t^{sub}, N_s} \\ \langle \bar{\psi}_q \psi_q \rangle^{sub} &= \langle \bar{\psi}_q \psi_q \rangle_{N_t, N_s} - \langle \bar{\psi}_q \psi_q \rangle_{N_t^{sub}, N_s} \end{aligned} \tag{3.37}$$

and therefore

$$\frac{p(T)}{T^4} - \frac{p(T_0)}{T_0^4} = N_t^4 \int_{(\beta_0, m_{q0})}^{(\beta, m_q)} \left[d\beta \langle -S_G \rangle^{sub} + \sum_q dm_q \langle \bar{\psi}_q \psi_q \rangle^{sub} \right]. \tag{3.38}$$

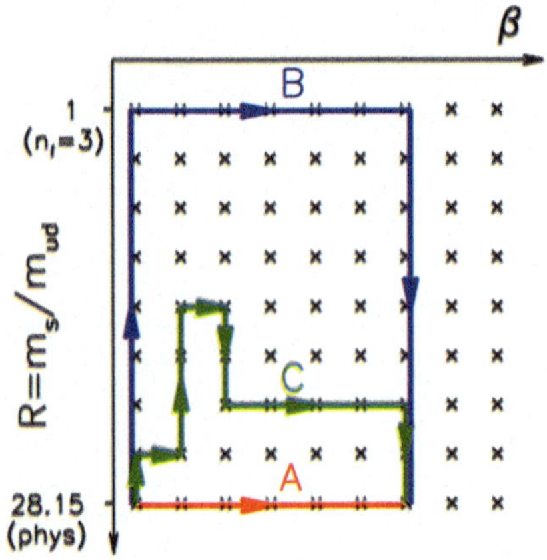

Fig. 3.2 From Ref. [10]. Illustration of possible integration paths in the (β, R) plane, which can be used to obtain the pressure at a certain point

In Ref. [10] it was noticed that $\langle -S_G \rangle^{sub}$ decreases as $(N_t)^{-4}$, while $\langle \bar{\psi}_q \psi_q \rangle^{sub}$ only decreases as $(N_t)^{-2}$: the term containing the gauge action has $(N_t)^2$ larger errors. The standard integral method consists of calculating the trace anomaly for several temperatures, then integrate to get the pressure up to an integration constant. This modified method sets m_s (the strange quark mass) to its physical value and changes R from 1 to 28.15 (the physical value), where the light quark mass is $m_{u,d} = \frac{m_s}{R}$. One then calculates the derivatives of the pressure at several values of β and R:

$$D_\beta = \left.\frac{\partial}{\partial \beta}\right|_R \frac{p}{T^4} = N_t^4 \left[\langle -S_G \rangle^{sub} + \frac{\partial m_s^{phys}}{\partial \beta} \left(\langle \bar{\psi}_s \psi_s \rangle^{sub} + \frac{1}{R} \langle \bar{\psi}_{ud} \psi_{ud} \rangle^{sub} \right) \right]$$

$$D_R = \left.\frac{\partial}{\partial R}\right|_\beta \frac{p}{T^4} = -N_t^4 \frac{m_s^{phys}}{R^2} \langle \bar{\psi}_{ud} \psi_{ud} \rangle^{sub}. \tag{3.39}$$

Different paths of integration can be considered in the (β, R) plane. The integral of the total derivative only depends on the initial and final points. An example of integration paths is shown in Fig. 3.2.

3.1.3 High Temperature, Ideal Gas Limit

Since the effective QCD coupling tends to zero logarithmically at short distance, it is reasonable to attempt a perturbative expansion of the thermodynamic potential at high energy density or temperature. To zeroth-order in the coupling, the QCD plasma is an ideal gas of gluons and quarks. The pressure can be written down immediately in the most general way

$$p_0 = \frac{T}{V} \ln Z \tag{3.40}$$

where

$$\ln Z = Vd \int \frac{d^3p}{(2\pi)^3} \ln\left(1 \pm e^{-\beta(|\mathbf{p}|-\mu)}\right)^{\pm 1} \tag{3.41}$$

where the sign "+" is for fermions and "−" is for bosons, and d indicates the number of degrees of freedom. In the specific case of QCD we have

$$\begin{aligned} p^{QCD} &= \frac{T}{V} V d_g \int \frac{d^3p}{(2\pi)^3} \ln\left(\left(1 - e^{-\beta(|\mathbf{p}|)}\right)^{-1}\right) + \\ &+ \frac{T}{V} V d_q \int \frac{d^3p}{(2\pi)^3} \ln\left(1 + e^{-\beta(|\mathbf{p}|-\mu)}\right) + \\ &+ \frac{T}{V} V d_{\bar{q}} \int \frac{d^3p}{(2\pi)^3} \ln\left(1 + e^{-\beta(|\mathbf{p}|+\mu)}\right), \end{aligned} \tag{3.42}$$

where $d_g = 2 \cdot (N_c^2 - 1) = 16$ (two possible polarizations times eight gluon colors) and $d_q = d_{\bar{q}} = 2 \cdot N_c = 6$. The gluonic contribution is

$$\begin{aligned} p_g &= \frac{T}{V} V d_g \frac{4\pi}{(2\pi)^3} \int_0^\infty |\mathbf{p}|^2 d|\mathbf{p}| \ln\left[\frac{1}{1 - e^{-\beta|\mathbf{p}|}}\right] = \\ &= T d_g \frac{1}{2\pi^2} \frac{|\mathbf{p}|^3}{3} \ln\left[\frac{1}{1 - e^{-\beta|\mathbf{p}|}}\right] + T^4 d_g \frac{1}{6\pi^2} \int_0^\infty \frac{|\mathbf{p}|^3}{T^3} \frac{d(|\mathbf{p}|/T)}{e^{|\mathbf{p}|/T} - 1} = \\ &= \frac{T^4 d_g}{6\pi^2} \frac{\pi^4}{15} = \frac{\pi^2 T^4}{90} d_g. \end{aligned} \tag{3.43}$$

The contribution of a quark of flavor f is

$$\begin{aligned} p_q &= T^4 d_q \frac{4\pi}{(2\pi)^3} \int_0^\infty \frac{|\mathbf{p}|^3}{3T^3} \left[\frac{d(|\mathbf{p}/T)}{e^{\beta(|\mathbf{p}|-\mu)}} + \frac{d(|\mathbf{p}/T)}{e^{\beta(|\mathbf{p}|+\mu)}}\right] = \\ &= \frac{d_q}{6\pi^2} \left[\frac{7\pi^4 T^4}{60} + \frac{\mu_f^2 \pi^2 T^2}{2} + \frac{\mu_f^4}{4}\right]. \end{aligned} \tag{3.44}$$

The total Stefan-Boltzmann limit for the QCD pressure is

$$p^{QCD} = \frac{\pi^2}{45} T^4 (N_c^2 - 1) + \sum_f \frac{N_c}{3\pi^2} \left[\frac{7\pi^4 T^4}{60} + \frac{\mu_f^2 \pi^2 T^2}{2} + \frac{\mu_f^4}{4}\right]. \tag{3.45}$$

In the limit $T \to \infty$, $\frac{p^{QCD}}{T^4}$ tends to a constant

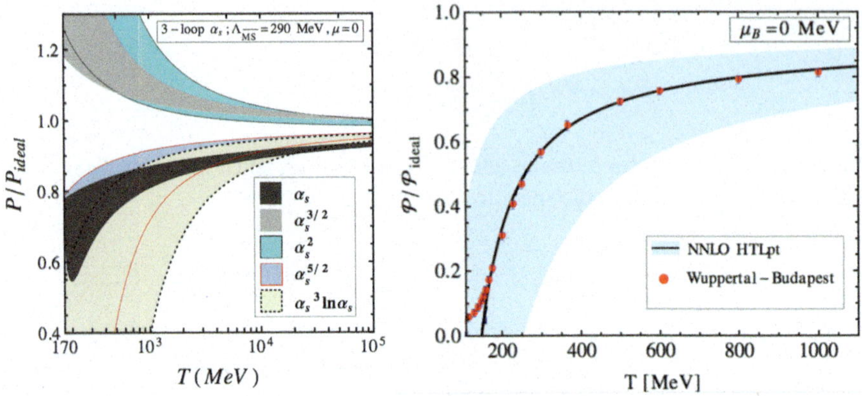

Fig. 3.3 Left: from Ref. [34]. Weak-coupling expansion for the scaled QCD pressure with $N_f = 3$. Shaded bands show the result of varying the renormalization scale μ by a factor of 2 around $\mu = 2\pi T$. Right: modified from Ref. [36]. Three-loop HTL perturbative resummation of the QCD pressure, rescaled by the ideal gas limit, as a function of the temperature (shaded band). The perturbative results are compared to lattice QCD data from the WB (red) [10] collaboration

$$\frac{p^{QCD}}{T^4} = \frac{8\pi^2}{45} + 7\frac{N_f \pi^2}{60}. \tag{3.46}$$

For $N_f = 3$ we have $\lim_{T\to\infty} \frac{p^{QCD}}{T^4} = 5.20896$.

Going beyond the zeroth-order in the coupling, a perturbative series can be systematically constructed and truncated at a given order of the QCD coupling constant. This has been a long-standing activity [11–17], and the pressure is now known through order $g^6 \log g$ [18]. The convergence of the standard perturbative QCD series turns out to be slow: the total pressure calculated at different orders in the coupling, divided by the ideal gas limit, is shown in the left panel of Fig. 3.3. From the figure it is clear that the results of the different orders are very different from each other: the various approximations oscillate widely and show no signs of convergence in the temperature range shown. For this reason, a reorganization of the perturbation series for thermal QCD was introduced, called Hard Thermal Loop (HTL) [19–35], which systematically shifts the perturbative expansion from being around an ideal gas of massless particles, which is the physical picture of the naive weak-coupling expansion, to being around a gas of massive quasiparticles. This shift incorporates the classical physics of the high temperature quark-gluon plasma, so that loop corrections correspond to true quantum and thermal corrections to the classical high temperature limit. The pressure, calculated at three-loop HTL order, is shown in the right panel of Fig. 3.3. From the figure it is evident that this technique brings the agreement between lattice and perturbative QCD down to lower temperatures.

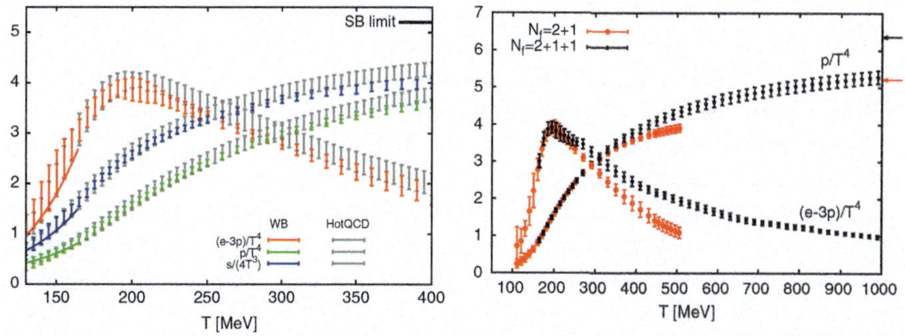

Fig. 3.4 Left: continuum extrapolated results for trace anomaly, entropy density and pressure. The gray points are from the HotQCD collaboration [39], while the colored ones are from the WB collaboration [37]. The figure also shows the Stefan-Boltzmann limit for the pressure and the scaled entropy; the curves at low temperature correspond to the HRG model predictions. Right: the trace anomaly and pressure in the $2 + 1$ and $2 + 1 + 1$ flavor theories. (From Ref. [42])

3.1.4 Results

The Equation of State of QCD for a system of $2 + 1$ quark flavors with physical masses is known since a few years from first principles. The Wuppertal Budapest (WB) collaboration [10, 37] published continuum extrapolated results for pressure, energy density, entropy density, speed of sound and interaction measure as functions of the temperature in a tree-level Symanzik improved gauge, and stout-improved staggered fermion action with two levels of smearing (2stout) [38]. These results were later confirmed by the HotQCD collaboration [39] with a different kind of staggered fermion discretization: the highly improved staggered quark (HISQ) action introduced in [40]. The left panel of Fig. 3.4 shows scaled interaction measure, pressure and entropy as functions of the temperature. The colored bands correspond to results from the WB collaboration, while the gray bands correspond to the HotQCD collaboration results. The Stefan-Boltzmann limit for pressure and scaled entropy is indicated by the horizontal line at high temperatures.

Some observables show a good agreement with resummation [26] or dimensional reduction [41] techniques already at $T \sim 400$ MeV, while for others this is not the case. For example, in the case of the trace anomaly, the results from Hard Thermal Loop perturbation theory to three-loop order [26] show a large uncertainty, corresponding to varying the renormalization scale. The results from Electrostatic QCD are in good agreement with the HTL ones, and the lattice results approach the perturbative ones at $T \sim 2 - 3T_c$.

Simulations with Wilson, overlap or domain wall fermions are notoriously more time consuming. For this reason, results for the QCD Equation of State in these more rigorous formulations are not yet available with physical quark masses. They will serve as important cross-checks in the coming years. It is worth mentioning that, recently, first results for the equation of state obtained from other approaches to lattice QCD are becoming available: these include the gradient flow method

[43, 44], which extracts the thermodynamic quantities from the energy-momentum tensor, and twisted mass fermions [45]; the former are limited so far to the quenched approximation [43] or heavier than physical quark masses [44], the latter to two flavors with heavier-than-physical quark masses.

The effect of the charm quark on QCD thermodynamics has been recently investigated [42]. Results are shown in the right panel of Fig. 3.4. This analysis indicates that the charm quark is a relevant degree of freedom already at $T \sim 250$ MeV, and its effects should not be neglected. For a recent study of the effect of including the charm quark in the EoS for hydrodynamic simulations of heavy ion collisions at the LHC energies see e.g. Ref. [46]. Reference [42] actually contains the QCD EoS for cosmology, which extends the flavor content to the bottom and top quarks, relevant at cosmologically high temperatures.

Recently, an important validation of the lattice QCD Equation of State has been obtained from a Bayesian analysis [47]. This framework, based on a comparison of data from RHIC and the LHC to theoretical models, has applied state-of-the-art statistical techniques to the combined analysis of a large number of observables while varying the model parameters. The posterior distribution over possible equations of states turned out to be consistent with results from lattice QCD simulations, as shown in Fig. 3.5. This analysis has also been successfully applied to infer the behavior of other quantities, such as the shear viscosity of the QGP at zero [48] and finite density [49].

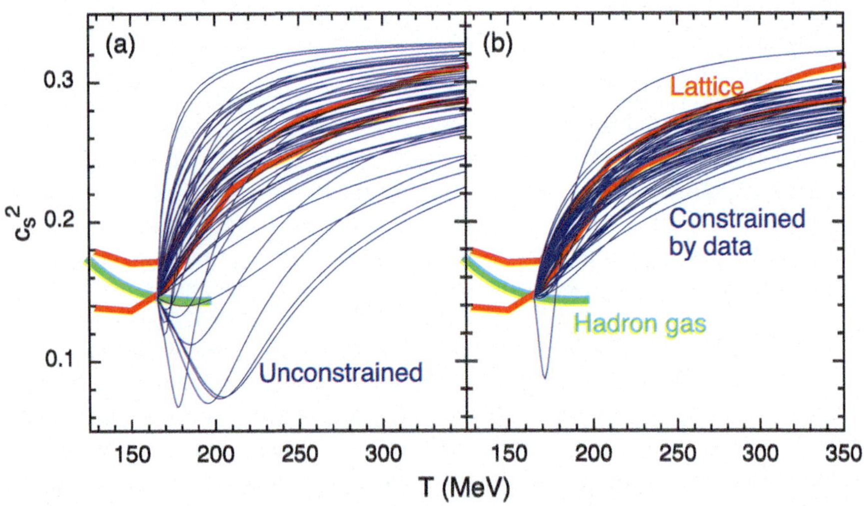

Fig. 3.5 From Ref. [47]. Constraints on the QCD equation of state from the Bayesian analysis. (**a**) Fifty equations of state were generated by randomly choosing the parameters from the prior distribution and weighted by the posterior likelihood (**b**). The two red lines in each figure represent the range of lattice equations of state shown in [39], and the green line shows the equation of state of a non-interacting hadron gas. This suggests that the matter created in heavy-ion collisions at RHIC and at the LHC has a pressure that is similar to that expected from equilibrated matter

References

1. Engels, J., Karsch, F., Satz, H., Montvay, I.: Nucl. Phys. B **205**, 545–577 (1982)
2. Deng, Y.: Nucl. Phys. B (Proc. Suppl.) **9**, 334 (1989)
3. Gavai, R.V., Gupta, S., Mukherjee, S.: Pramana **71**, 487–508 (2008)
4. Mukherjee, S.: QCD Thermodynamics from Lattice Gauge Theory. Ph. D. Thesis, Tata Institute (2006).
5. Hasenfratz, A., Hasenfratz, P.: Nucl. Phys. B **193**, 210 (1981)
6. Engels, J., Fingberg, J., Karsch, F., Miller, D., Weber, M.: Phys. Lett. B **252**, 625–630 (1990)
7. Boyd, G., Engels, J., Karsch, F., Laermann, E., Legeland, C., Lutgemeier, M., Petersson, B.: Phys. Rev. Lett. **75**, 4169–4172 (1995)
8. Boyd, G., Engels, J., Karsch, F., Laermann, E., Legeland, C., Lutgemeier, M., Petersson, B.: Nucl. Phys. B **469**, 419–444 (1996)
9. Borsanyi, S., Endrodi, G., Fodor, Z., Katz, S.D., Szabo, K.K.: JHEP **1207**, 056 (2012)
10. Borsanyi, S., Endrodi, G., Fodor, Z., Jakovac, A., Katz, S.D., Krieg, S., Ratti, C., Szabo, K.K.: JHEP **1011**, 077 (2010)
11. Shuryak, E.V.: Sov. Phys. JETP **47**, 212 (1978) [Zh. Eksp. Teor. Fiz. **74**, 408 (1978)]
12. Kapusta, J.I., Nucl. Phys. B **148**, 461 (1979).
13. Toimela, T.: Int. J. Theor. Phys. **24**, 901 (1985) Erratum: [Int. J. Theor. Phys. **26**, 1021 (1987)]
14. Arnold, P.B., Zhai, C.X.: Phys. Rev. D **51**, 1906 (1995)
15. Braaten, E., Nieto, A.: Phys. Rev. Lett. **76**, 1417 (1996)
16. Braaten, E., Nieto, A.: Phys. Rev. D **53**, 3421 (1996)
17. Zhai, C.X., Kastening, B.M.: Phys. Rev. D **52**, 7232 (1995)
18. Kajantie, K., Laine, M., Rummukainen, K., Schroder, Y.: Phys. Rev. D **67**, 105008 (2003)
19. Braaten, E., Pisarski, R.D.: Phys. Rev. D **45**(6), R1827 (1992)
20. Andersen, J.O., Braaten, E., Petitgirard, E., Strickland, M.: Phys. Rev. D **66**, 085016 (2002)
21. Andersen, J.O., Petitgirard, E., Strickland, M.: Phys. Rev. D **70**, 045001 (2004)
22. Andersen, J.O., Strickland, M., Su, N.: Phys. Rev. Lett. **104**, 122003 (2010)
23. Andersen, J.O., Strickland, M., Su, N.: JHEP **1008**, 113 (2010)
24. Andersen, J.O., Leganger, L.E., Strickland, M., Su, N.: Phys. Lett. B **696**, 468 (2011)
25. Andersen, J.O., Leganger, L.E., Strickland, M., Su, N.: Phys. Rev. D **84**, 087703 (2011)
26. Andersen, J.O., Leganger, L.E., Strickland, M., Su, N.: JHEP **1108**, 053 (2011)
27. Chakraborty, P., Mustafa, M.G., Thoma, M.H.: Eur. Phys. J. C **23**, 591 (2002)
28. Chakraborty, P., Mustafa, M.G., Thoma, M.H.: Phys. Rev. D **68**, 085012 (2003)
29. Haque, N., Mustafa, M.G., Thoma, M.H.: Phys. Rev. D **84**, 054009 (2011)
30. Haque, N., Mustafa, M.G.: Nucl. Phys. A **862–863**, 271 (2011)
31. Haque, N., Mustafa, M.G.: arXiv:1007.2076 [hep-ph]
32. Andersen, J.O., Mogliacci, S., Su, N., Vuorinen, A.: Phys. Rev. D **87**(7), 074003 (2013)
33. Mogliacci, S., Andersen, J.O., Strickland, M., Su, N., Vuorinen, A.: JHEP **1312**, 055 (2013)
34. Haque, N., Mustafa, M.G., Strickland, M.: Phys. Rev. D **87**(10), 105007 (2013)
35. Haque, N., Mustafa, M.G., Strickland, M.: JHEP **1307**, 184 (2013)
36. Haque, N., Andersen, J.O., Mustafa, M.G., Strickland, M., Su, N.: Phys. Rev. D **89**(6), 061701 (2014)
37. Borsanyi, S., Fodor, Z., Hoelbling, C., Katz, S.D., Krieg, S., Szabo, K.K.: Phys. Lett. B **730**, 99 (2014)
38. Aoki, Y., Fodor, Z., Katz, S.D., Szabo, K.K.: JHEP **0601**, 089 (2006)
39. Bazavov, A., et al.: [HotQCD Collaboration] Phys. Rev. D **90**, 094503 (2014)
40. Follana, E., et al.: [HPQCD and UKQCD Collaborations] Phys. Rev. D **75**, 054502 (2007)
41. Laine, M., Schroder, Y.: Phys. Rev. D **73**, 085009 (2006)
42. Borsanyi, S., et al.: Nature **539**(7627), 69 (2016)
43. Asakawa, M., et al.: [FlowQCD Collaboration] Phys. Rev. D **90**(1), 011501 (2014). Erratum: [Phys. Rev. D **92**(5), 059902 (2015)]

44. Kanaya, K., Ejiri, S., Iwami, R., Kitazawa, M., Suzuki, H., Taniguchi, Y., Umeda, T., Wakabayashi, N.: PoS LATTICE **2016**, 063 (2016)
45. Burger, F., et al.: [tmfT Collaboration] Phys. Rev. D **91**(7), 074504 (2015)
46. Alba, P., Mantovani Sarti, V., Noronha, J., Noronha-Hostler, J., Parotto, P., Portillo Vazquez, I., Ratti, C.: Phys. Rev. C **98**(3), 034909 (2018)
47. Pratt, S., Sangaline, E., Sorensen, P., Wang, H.: Phys. Rev. Lett. **114**, 202301 (2015)
48. Bernhard, J.E., Moreland, J.S., Bass, S.A., Liu, J., Heinz, U.: Phys. Rev. C **94**(2), 024907 (2016)
49. Auvinen, J., Bernhard, J.E., Bass, S.A., Karpenko, I.: Phys. Rev. C **97**(4), 044905 (2018)

QCD at Finite Chemical Potential

4

Abstract

In this chapter we discuss current methods to extend the QCD equation of state and transition temperature to finite chemical potential. Due to the sign problem, direct simulations are not currently possible and are replaced by approximated methods, which allow to expand the observables to small values of the chemical potential. We introduce the Taylor expansion and imaginary chemical potential simulation methods, and discuss the most recent results on the QCD equation of state and transition line extended to finite μ_B. We also discuss current limits on the position of the QCD critical point from first principles.

4.1 Sign Problem

The chemical potential μ appears in the fermionic part of the QCD action as follows

$$S = \int_0^\beta d\tau \int d^3x\bar{\psi}\left[\gamma_\mu D^\mu + \mu\gamma_4 + m\right]\psi = \int d^4x\bar{\psi}M\psi. \tag{4.1}$$

From the above equation, it is evident that μ appears in the QCD action in the same way as the imaginary part of the fourth-component of an abelian vector field (same as $iG_4(x)$). With a non-zero chemical potential, the QCD action becomes complex. This can be seen by the absence of γ_5 hermiticity. At $\mu = 0$ we have

$$(\gamma_5 M)^\dagger = \gamma_5 M, \tag{4.2}$$

which implies $M^\dagger = \gamma_5 M\gamma_5$. This leads to

$$\det(M^\dagger) = \det(\gamma_5 M\gamma_5) = \det M = (\det M)^* . \tag{4.3}$$

C. Ratti, R. Bellwied, *The Deconfinement Transition of QCD*, Lecture Notes in Physics 981, https://doi.org/10.1007/978-3-030-67235-5_4

However, when $\mu \neq 0$ we find

$$M^{\dagger}(\mu) = \gamma_5 M(-\mu^*)\gamma_5, \tag{4.4}$$

resulting in

$$[\det M(\mu)]^* = \det\left[M(-\mu^*)\right] \tag{4.5}$$

which is in general a complex number. For real μ we get

$$[\det M(\mu)]^* = \det\left[M(-\mu)\right] \tag{4.6}$$

while for imaginary μ we get

$$[\det M(\mu)]^* = \det\left[M(\mu)\right] \tag{4.7}$$

which is a real number. Therefore, simulations at imaginary chemical potential are allowed and do not generate a sign problem. However, when the chemical potential is real, the QCD action becomes complex, and cannot be used as a weight in the Monte Carlo sampling techniques detailed in Sect. 1.3.

4.2 Equation of State of QCD at Finite Chemical Potential

The equation of state of QCD at finite density is an extremely important quantity, which can be used to describe the matter created in low-energy ultrarelativistic heavy-ion collisions, or the core of compact stellar objects, or neutron star mergers. Recently, results from perturbative QCD at extreme density have been obtained and used to constrain the matter in the core of neutron stars [1]. However, the densities needed for perturbation theory to be valid are too large to be useful in the vicinity of the QCD transition line, hence the need for non-perturbative techniques. Unfortunately, due to the sign problem, the equation of state is currently only available in a limited range of chemical potentials $\mu_B/T \leq 2$.

The thermodynamic quantities at finite chemical potential μ_B are related through the following identities

$$\begin{aligned} s(T,\mu_B) &= \left.\frac{\partial p}{\partial T}\right|_{\mu_B}, & n_B(T,\mu_B) &= \chi_1^B = \left.\frac{\partial p}{\partial \mu_B}\right|_T, \\ \epsilon(T,\mu_B) &= Ts - p + \mu_B n_B, & c_s^2(T,\mu_B) &= \left.\frac{\partial p}{\partial \epsilon}\right|_{s/n_B}, \end{aligned} \tag{4.8}$$

where s is the entropy density, p the pressure, ϵ the energy density, n_B the baryonic density and c_s^2 the speed of sound squared.

While several methods have been proposed to actually solve QCD at finite μ_B, either by finding appropriate dual variables [2] or by the use of Lefschetz thimbles [3–5] or complex Langevin equation [6–14], none of them has reached the level in which quantitatively reliable results can be expected soon. The currently available results are obtained through two main methods, the Taylor expansion of thermodynamic quantities in powers of μ_B/T [15–19] or their simulation at imaginary chemical potentials followed by an analytical continuation to real μ_B [20–27]. While other methods have also been employed, e.g. reweighting of the generated configurations [28–31], use of the canonical ensemble [32–34] and density of state methods [35, 36], here we focus on the first two.

4.2.1 Taylor Expansion

The pressure of QCD can be written as a Taylor series in powers of μ_B/T around $\mu_B = 0$:

$$\frac{p(T,\mu_B)}{T^4} = \frac{p(T,0)}{T^4} + \sum_{n=1}^{\infty} \frac{1}{n!} \frac{\partial^n (p/T^4)}{\partial(\frac{\mu_B}{T})^n}\bigg|_{\mu_B=0} \left(\frac{\mu_B}{T}\right)^n = \sum_{n=0}^{\infty} c_n(T) \left(\frac{\mu_B}{T}\right)^n, \tag{4.9}$$

where we defined

$$c_n = \frac{1}{n!} \frac{\partial^n (p/T^4)}{\partial(\mu_B/T)^n}. \tag{4.10}$$

On the lattice, the c_n coefficients can be defined in terms of the quark chemical potentials. It is then straightforward to perform a change of basis, through the following relationships:

$$\begin{aligned} \mu_u &= \frac{1}{3}\mu_B + \frac{2}{3}\mu_Q \\ \mu_d &= \frac{1}{3}\mu_B - \frac{1}{3}\mu_Q \\ \mu_s &= \frac{1}{3}\mu_B - \frac{1}{3}\mu_Q - \mu_S, \end{aligned} \tag{4.11}$$

so that

$$\begin{aligned} \frac{\partial}{\partial \mu_B} &= \frac{1}{3}\partial_u + \frac{1}{3}\partial_d + \frac{1}{3}\partial_s \\ \frac{\partial}{\partial \mu_Q} &= \frac{2}{3}\partial_u - \frac{1}{3}\partial_d - \frac{1}{3}\partial_s \end{aligned}$$

$$\frac{\partial}{\partial \mu_S} = -\partial_s. \tag{4.12}$$

In the above equation we have used the abbreviation $\partial_i = \frac{\partial}{\partial \mu_i}$ for the derivative with respect to the chemical potential of quark flavor i.

The coefficients c_n can be calculated on the lattice, since they are defined at $\mu_B = 0$. To see how to do it in practice, we recall that $p = T \left.\frac{\partial \ln Z}{\partial V}\right|_T$ and that we can write the partition function as

$$Z = \int DU \, (\det M)^{N_f} \, e^{-S_G[U]} = \int DU e^{-S_G[U]} e^{N_f \ln \det M(\mu)} \tag{4.13}$$

so that we can write

$$\frac{\partial \ln Z}{\partial \mu} = \frac{\int DU e^{-S_G[U]} e^{N_f \ln \det M(\mu)}}{Z} N_f \frac{\partial [\ln \det M(\mu)]}{\partial \mu} = \left\langle N_f \frac{\partial \ln \det M(\mu)}{\partial \mu} \right\rangle. \tag{4.14}$$

Analogously we have

$$\begin{aligned}
\frac{\partial^2 \ln Z}{\partial \mu^2} &= \frac{\partial}{\partial \mu} \left[\frac{\int DU e^{-S_G[U]} e^{N_f \ln \det M(\mu)}}{Z} N_f \frac{\partial [\ln \det M(\mu)]}{\partial \mu} \right] = \\
&= \frac{\int DU e^{-S_G[U]} e^{N_f \ln \det M(\mu)}}{Z} \left(N_f \frac{\partial [\ln \det M(\mu)]}{\partial \mu} \right)^2 + \\
&+ \frac{\int DU e^{-S_G[U]} e^{N_f \ln \det M(\mu)}}{Z} \left(N_f \frac{\partial^2 [\ln \det M(\mu)]}{\partial \mu^2} \right) + \\
&- \frac{1}{Z^2} \int DU e^{-S_G[U]} e^{N_f \ln \det M(\mu)} \frac{N_f \partial [\ln \det M(\mu)]}{\partial \mu} \times \\
&\times \int DU e^{-S_G[U]} e^{N_f \ln \det M(\mu)} \frac{N_f \partial [\ln \det M(\mu)]}{\partial \mu} = \\
&= \left\langle N_f \frac{\partial^2 \ln \det M(\mu)}{\partial \mu^2} \right\rangle + \left\langle \left(N_f \frac{\partial \ln \det M(\mu)}{\partial \mu} \right)^2 \right\rangle - \left\langle N_f \frac{\partial \ln \det M(\mu)}{\partial \mu} \right\rangle^2
\end{aligned} \tag{4.15}$$

and so on. Writing $\ln \det M = \operatorname{tr} \ln M$, we can express these as

$$\begin{aligned}
\frac{\partial}{\partial \mu} \ln \det M &= \operatorname{tr} M^{-1} \frac{\partial M}{\partial \mu}; \\
\frac{\partial^2}{\partial \mu^2} \ln \det M &= \operatorname{tr} M^{-1} \frac{\partial^2 M}{\partial \mu^2} - \operatorname{tr} M^{-1} \frac{\partial M}{\partial \mu} M^{-1} \frac{\partial M}{\partial \mu}.
\end{aligned} \tag{4.16}$$

Notice that these coefficients can be written as:

$$c_n^i = \frac{1}{n!}\chi_n^i, \quad \text{where} \quad \chi_n^i = \frac{\partial^n (p/T^4)}{\partial(\mu_i/T)^n}. \tag{4.17}$$

The χ_n^i are called susceptibilities of conserved charges (or quark flavors) and they have important physical meanings. For example, the first-order up quark susceptibility

$$\chi_1^u = \frac{\partial(p/T^4)}{\partial(\mu_u/T)} = \frac{1}{T^3}n_u \tag{4.18}$$

represents the density of u quarks, while the second-order one

$$\chi_2^u = \frac{\partial}{\partial(\mu_u/T)}\left(\frac{n_u}{T^3}\right) \tag{4.19}$$

measures the response of the up quark density to an infinitesimal change in the up quark chemical potential. Finally, the non-diagonal second-order susceptibility

$$\chi_{11}^{us} = \frac{\partial}{\partial(\mu_s/T)}\left(\frac{n_u}{T^3}\right) \tag{4.20}$$

measures the response of the up quark density to an infinitesimal change in the strange quark chemical potential, namely it measures the correlation between up and strange quarks. These observables, and their conserved-charge equivalents, are relevant to experiment, as we will see in Chap. 5.

We note here that susceptibilities share the same Taylor expansion coefficients as the pressure:

$$\begin{aligned}
\frac{p(T,\ \mu_B)}{T^4} &= \frac{p(T,\ \mu_B{=}0)}{T^4} + c_2(T)\left(\frac{\mu_B}{T}\right)^2 + c_4(T)\left(\frac{\mu_B}{T}\right)^4 + c_6(T)\left(\frac{\mu_B}{T}\right)^6 + \ldots \\
\frac{T\chi_1(T,\ \mu_B)}{\mu_B} &= 2c_2(T) + 4c_4(T)\left(\frac{\mu_B}{T}\right)^2 + 6c_6(T)\left(\frac{\mu_B}{T}\right)^4 + \ldots \\
\chi_2(T,\ \mu_B) &= 2c_2(T) + 12c_4(T)\left(\frac{\mu_B}{T}\right)^2 + 30c_6(T)\left(\frac{\mu_B}{T}\right)^4 + \ldots \\
\frac{T\chi_3(T,\ \mu_B)}{\mu_B} &= 24c_4(T) + 120c_6(T)\left(\frac{\mu_B}{T}\right)^2 + \ldots
\end{aligned} \tag{4.21}$$

and so on. Therefore, if we obtain the first few coefficients, we can extrapolate both the pressure and the susceptibilities to finite μ_B. Notice that, since QCD is symmetric with respect to a change of sign in chemical potential (excess of matter vs. anti-matter), these observables are all even functions of μ_B.

4.2.2 Simulations at Imaginary Chemical Potential

4.2.2.1 Simulation Setup

The fact that the higher order coefficients appear in the Taylor series of the lower order ones in power of μ_B/T allows one to determine them more precisely through simulations at imaginary chemical potential, than what can be achieved with the direct simulation method. In fact, the direct simulation method at $\mu_B = 0$ yields very noisy results for the higher order $c_n(T)$. In the case of imaginary μ_B, μ_B^2 and μ_B^6 will be negative in Eqs. (4.21). A combined fit of these lower-order coefficients at finite, imaginary μ_B can then be performed. The coefficients of this fit, namely $c_2(T)$, $c_4(T)$ and $c_6(T)$, can thus be extracted. When pursuing this strategy, one has to choose values for μ_S and μ_Q as well (or equivalently, for μ_d and μ_s). This is not an issue when simulating at $\mu_B = 0$, because in that case one usually sets $\mu_S = 0$ and $\mu_Q = 0$ as well.

The two main choices when simulating at finite imaginary μ_B are to either set $\mu_S = 0$ and $\mu_Q = 0$ (corresponding to $\mu_u = \mu_d = \frac{1}{3}\mu_B$) or fix μ_S and μ_Q to match the experimental situation in a heavy-ion collision, namely

$$\begin{aligned} \langle n_S \rangle &= 0 \\ \langle n_Q \rangle &= 0.4 \langle n_B \rangle . \end{aligned} \tag{4.22}$$

These two relationships stem from the initial conditions of the ions being collided in the accelerator, which do not contain any valence strange quark and for which the number of protons is approximately 0.4 times the atomic number. From the first condition we get

$$\langle n_S \rangle = 0 \quad \leftrightarrow \quad \left\langle \frac{\partial \ln Z}{\partial \mu_S} \right\rangle = 0 \tag{4.23}$$

from which, deriving with respect to μ_B, we get a relationship between μ_S, μ_B and T:

$$\frac{d}{d\mu_B}\left[\frac{\partial \ln Z}{\partial \mu_S}\right] = \frac{\partial^2 \ln Z}{\partial \mu_S \partial \mu_B} + \frac{\partial^2 \ln Z}{\partial \mu_S^2}\frac{\partial \mu_S}{\partial \mu_B} = 0 \tag{4.24}$$

which means

$$\frac{\partial \mu_S}{\partial \mu_B} = -\frac{\chi_{11}^{BS}(T)}{\chi_2^S(T)} \tag{4.25}$$

to lowest order. The higher order conditions are achieved by further deriving $\frac{\partial \ln Z}{\partial \mu_S}$ with respect to μ_B. By solving these differential equations, we get the $\mu_S(\mu_B, T)$ that satisfies the desired conditions. An example is shown in the top panel of Fig. 4.1.

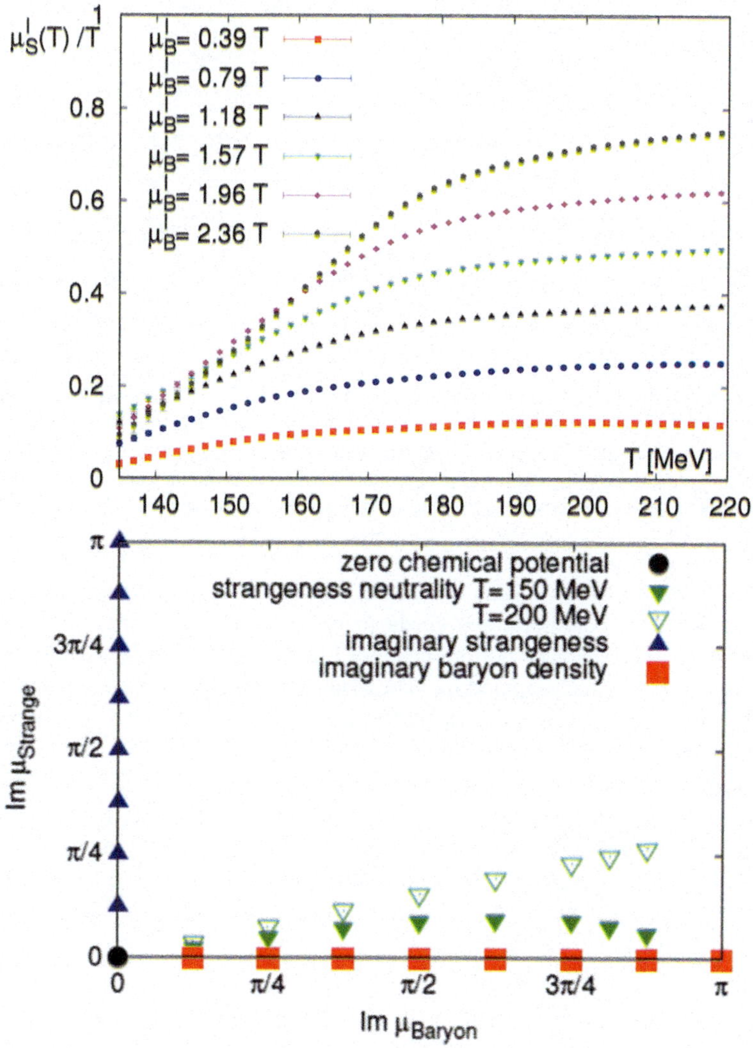

Fig. 4.1 Top: from Ref. [37]. Continuum extrapolation of the imaginary strangeness chemical potentials that realize strangeness neutrality. Results are shown as functions of the temperature, for different values of imaginary μ_B. Bottom: landscape of simulation points in the $(\Im(\mu_B), \Im(\mu_S))$ plane: the black circle corresponds to $\Im(\mu_B) = \Im(\mu_S) = 0$, the blue triangles to simulations at finite imaginary μ_S and zero baryon chemical potential, the red squares to simulations at finite imaginary μ_B and zero strangeness chemical potential, and the two sets of green triangles correspond to simulations at finite $\Im(\mu_B)$ and $\Im(\mu_S)$ that satisfy the experimental constraints (4.22) at two different temperature values

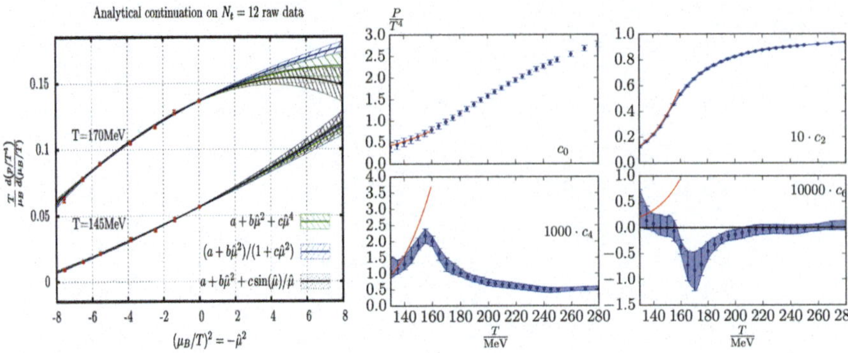

Fig. 4.2 From Ref. [41]. Left: analytical continuation of $T\chi_1^B/\mu_B$ from negative to positive $(\mu_B/T)^2$ for $T = 145$ MeV (lower curves) and $T = 170$ MeV (upper curves). The different colors correspond to different fitting functions. Right: coefficients $c_0 \ldots c_6$ for the Taylor expansion of the pressure around $\mu_B = 0$. The data are continuum extrapolated and they are shown as functions of the temperature, in comparison to the Hadron Resonance Gas model prediction (red lines)

Therefore, the drawback of this method is that, for each value of the temperature, a different set of configurations needs to be generated, with specific values of imaginary μ_B and μ_S. The bottom panel of Fig. 4.1 shows possible landscapes for simulations at imaginary μ_B and μ_S. The black dot corresponds to direct simulation of all coefficients at $\mu_B = 0$, as performed e.g. by the HotQCD collaboration. The red squares correspond to finite μ_B and $\mu_S = 0$, while the green triangles are trajectories which ensure the strangeness-neutrality condition at $T = 150$ MeV (full) and $T = 200$ MeV (empty). The idea is to simulate lower order fluctuations at imaginary μ_B and fit them to extract the higher order fluctuations at $\mu_B = 0$.

In Ref. [41], the rescaled baryonic density $T\chi_1^B/\mu_B$ was fitted with three different functions:

$$\begin{aligned} B_1(\hat{\mu}) &= a + b\hat{\mu}^2 + c\hat{\mu}^4 \\ B_2(\hat{\mu}) &= (a + b\hat{\mu}^2)/(1 + c\hat{\mu}^2) \\ B_3(\hat{\mu}) &= a + b\hat{\mu}^2 + c\sin(\hat{\mu})/\hat{\mu}. \end{aligned} \tag{4.26}$$

These fits are shown in the left panel of Fig. 4.2.

More recently, the authors of Ref. [38] have performed a new, combined fit of lower-order fluctuations up to order four. This exploits the fact that all lower-order fluctuations share the same set of Taylor coefficients, which are typically a numerical factor times the T-dependent higher-order fluctuations at $\mu_B = 0$. The formulas used for the combined fit of $\chi_1, \ldots \chi_4$ are

$$\chi_1^B(\hat{\mu}_B) = 2c_2\hat{\mu}_B + 4c_4\hat{\mu}_B^3 + 6c_6\hat{\mu}_B^5 + \frac{4!}{7!}c_4\epsilon_1\hat{\mu}_B^7 + \frac{4!}{9!}c_4\epsilon_2\hat{\mu}_B^9$$

$$\chi_2^B(\hat{\mu}_B) = 2c_2 + 12c_4\hat{\mu}_B^2 + 30c_6\hat{\mu}_B^4 + \frac{4!}{6!}c_4\epsilon_1\hat{\mu}_B^6 + \frac{4!}{8!}c_4\epsilon_2\hat{\mu}_B^8$$

$$\chi_3^B(\hat{\mu}_B) = 24c_4\hat{\mu}_B + 120c_6\hat{\mu}_B^3 + \frac{4!}{5!}c_4\epsilon_1\hat{\mu}_B^5 + \frac{4!}{7!}c_4\epsilon_2\hat{\mu}_B^7$$

$$\chi_4^B(\hat{\mu}_B) = 24c_4 + 360c_6\hat{\mu}_B^2 + c_4\epsilon_1\hat{\mu}_B^4 + \frac{4!}{6!}c_4\epsilon_2\hat{\mu}_B^6 \quad (4.27)$$

where ϵ_1 and ϵ_2 are drawn randomly from a normal distribution with mean -1.25 and variance 2.75. This is done to eliminate the ambiguity coming from the fitting function and just assumes that $8!c_8 \leq 4!c_4$ and $10!c_{10} \leq 4!c_4$, or equivalently that $\chi_8^B \leq \chi_4^B$ and $\chi_{10}^B \leq \chi_4^B$.

4.2.3 Results

After the early results for c_2, c_4 and c_6 [16], the first continuum extrapolated results for c_2 were published in Ref. [39]; in Ref. [40] c_4 was shown, but only at finite lattice spacing. The continuum limit for c_6 was published for the first time in [41] in the case of strangeness neutrality.

The first four Taylor coefficients from Ref. [41] are shown in the right panel of Fig. 4.2. From these coefficients, the pressure can be reconstructed at finite temperature and chemical potential, and from the pressure all other quantities can be obtained through the thermodynamic relationships listed in Eqs. (4.8). The top panel of Fig. 4.3 shows the isentropic trajectories in the (T, μ_B)-plane obtained in Ref. [41]. They correspond to the values of T and μ_B at which the entropy per baryon takes a constant value, determined at the freeze-out points (in black) for each collision energy at RHIC. These trajectories are phenomenologically relevant, as the system created in a heavy-ion collision would follow them in the absence of dissipation (no viscosity). The pressure and interaction measure obtained along the highest and lowest trajectories are shown in the bottom panel of Fig. 4.3. The details of the freeze-out points will be discussed in Chap. 5. Results for c_6 were later published in Ref. [43], both in the case of strangeness neutrality and at $\mu_S = \mu_Q = 0$. A continuum estimate was performed, based on $N_t = 6$ and 8 lattices. The three panels of Fig. 4.4 show c_2, c_4 and c_6 in the case of strangeness neutrality (called P_2, P_4 and P_6 in Ref. [43]), while the three panels of Fig. 4.5 show χ_2^B, χ_4^B/χ_2^B and χ_6^B/χ_2^B calculated at $\mu_S = \mu_Q = 0$.

In [44], a first determination of c_8, at two values of the temperature and $N_t = 8$ was presented. More recently, diagonal and non-diagonal coefficients up to c_8 have been calculated at $N_t = 12$ in Ref. [38] at $\mu_S = \mu_Q = 0$. The diagonal ones are shown in Fig. 4.6. It is worth mentioning that, while the results from Ref. [43] have been obtained through direct simulation of the higher order coefficients at $\mu_B = 0$, the results from Refs. [41] and [38] have been obtained by simulating the lower-

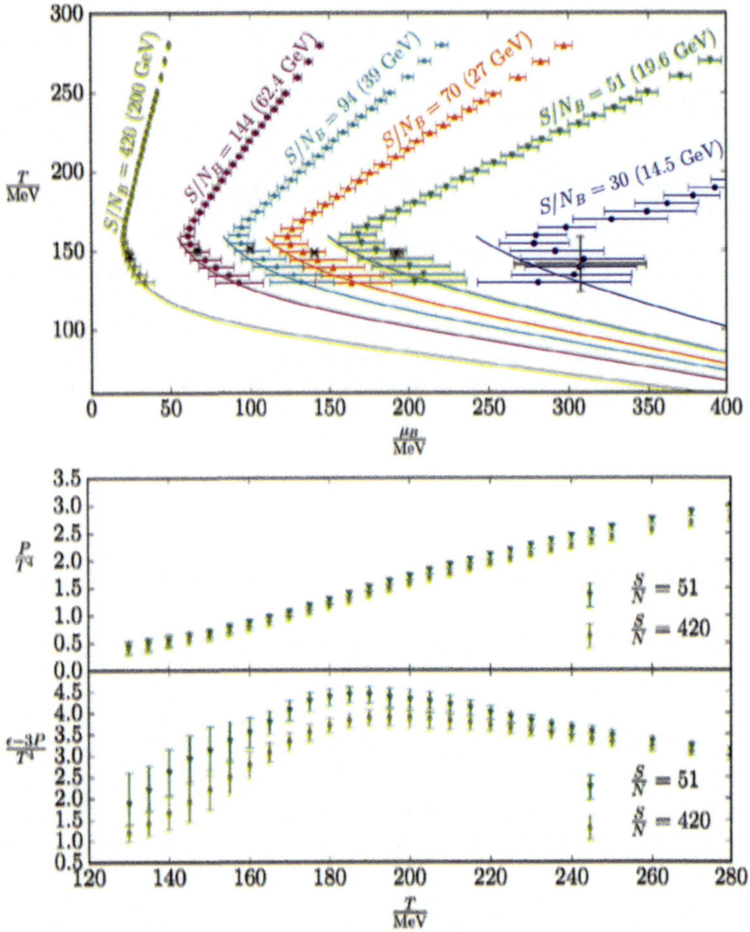

Fig. 4.3 From Ref. [41]. Top: the QCD phase diagram in the $(T,\ \mu_B)$ plane with the isentropic trajectories: the contours calculated at fixed S/N_B values. The black points are the chemical freeze-out parameters from Ref. [42]. Bottom: pressure (top) and interaction measure (bottom) as functions of the temperature, calculated along the highest and lowest isentropic trajectories from the top panel

order coefficients at finite imaginary chemical potential and fitting them to obtain the higher-order ones at $\mu_B = 0$.

The diagonal and off-diagonal coefficients from Ref. [38] have been used to construct a lattice-based equation of state at finite μ_B, μ_S and μ_Q in Ref. [45] through the following Taylor expansion:

$$\frac{p(T, \mu_B, \mu_S, \mu_Q)}{T^4} = \sum_{i,j,k} \frac{1}{i!j!k!} \chi_{ijk}^{BQS} \left(\frac{\mu_B}{T}\right)^i \left(\frac{\mu_S}{T}\right)^j \left(\frac{\mu_Q}{T}\right)^k, \tag{4.28}$$

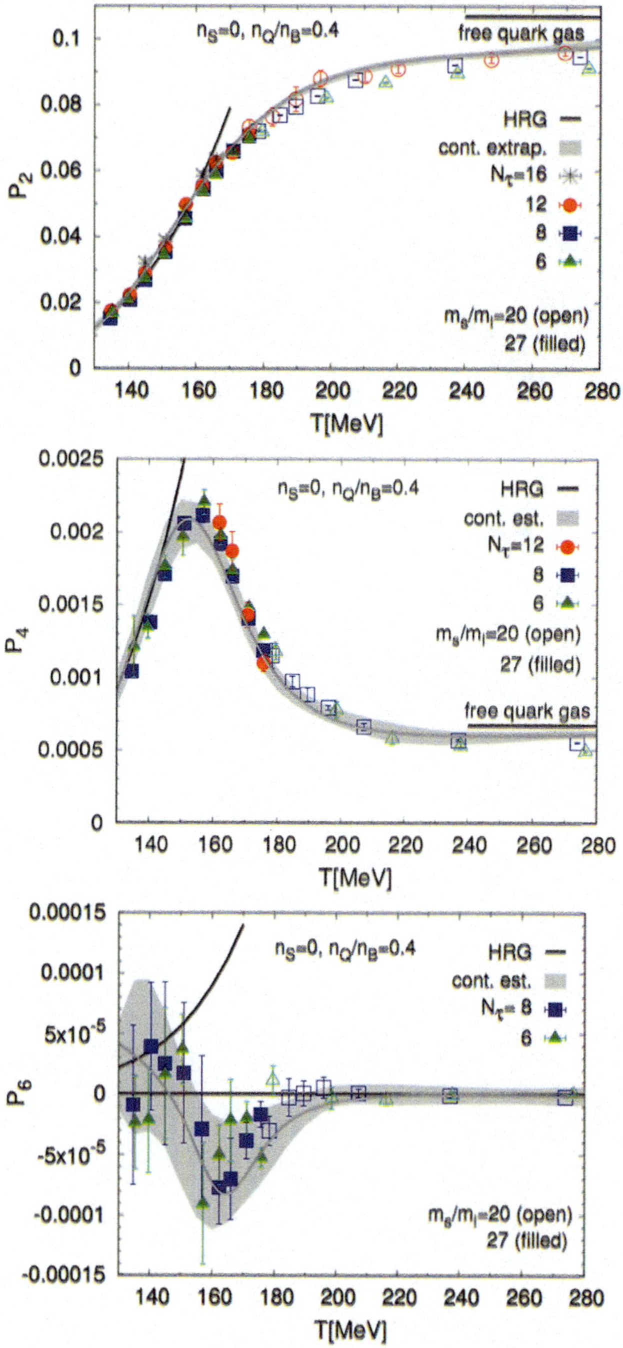

Fig. 4.4 From Ref. [43]. Expansion coefficients of the pressure as functions of the temperature in the case of strangeness neutrality. The broad gray bands show the continuum extrapolation. At low temperature, lines for HRG model calculations based on hadron resonances listed by the Particle Data Group are shown

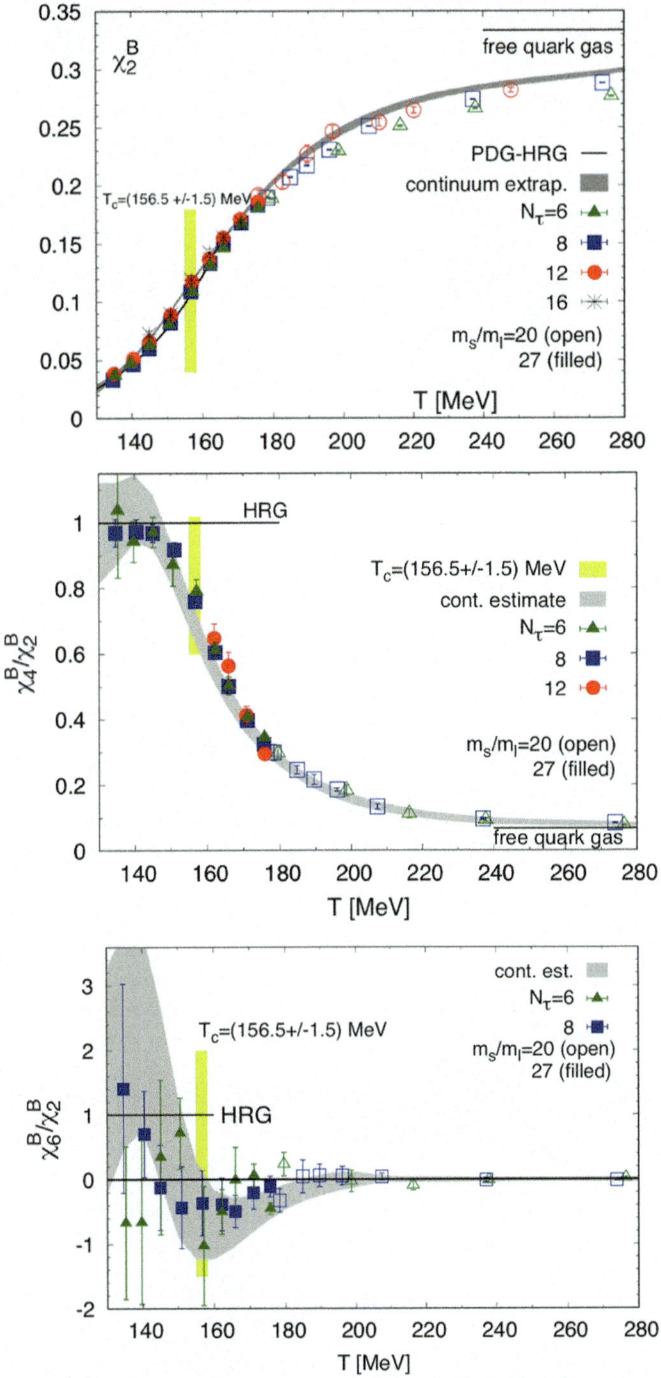

Fig. 4.5 From Ref. [43]. Expansion coefficients of the pressure as functions of the temperature in the case of $\mu_S = \mu_Q = 0$. The broad gray bands show the continuum extrapolation. At low temperature, lines for HRG model calculations based on hadron resonances listed by the Particle Data Group are shown

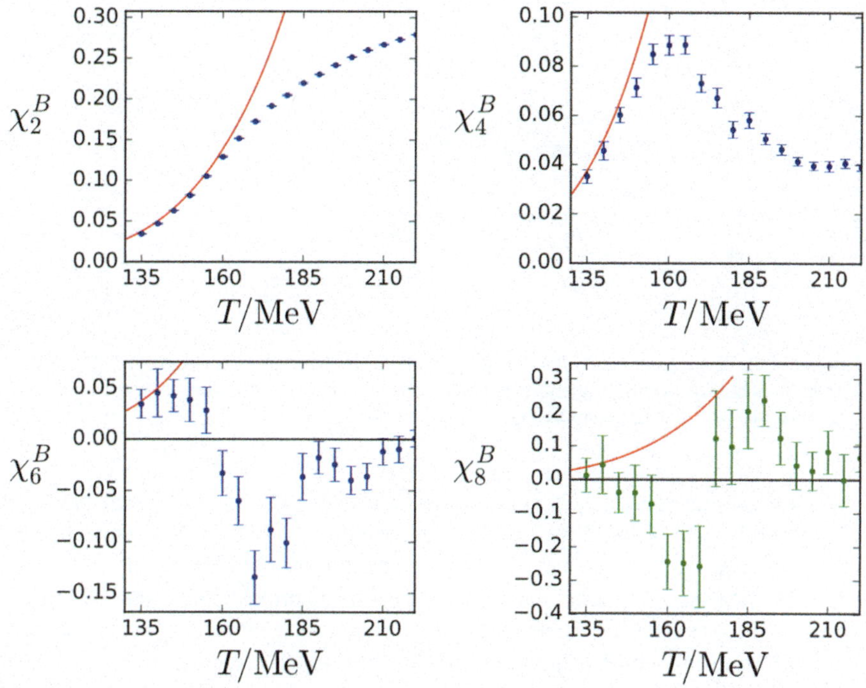

Fig. 4.6 From Ref. [38]. Expansion coefficients of the pressure as functions of the temperature at $\mu_S = \mu_Q = 0$, obtained at finite lattice spacing for $N_t = 12$

where

$$\chi_{ijk}^{BQS} = \frac{\partial^{i+j+k}(p/T^4)}{\partial(\frac{\mu_B}{T})^i \partial(\frac{\mu_Q}{T})^j \partial(\frac{\mu_S}{T})^k}. \tag{4.29}$$

This is necessary as, even if strangeness is globally conserved in heavy-ion collisions, the fluid cells that enter hydrodynamic simulations can have large local fluctuations that violate strangeness neutrality and electric charge conservation. Similar results were obtained in Ref. [46].

4.3 QCD Phase Diagram at Imaginary Chemical Potential

QCD at imaginary chemical potential is a rich topic, because there is an interplay between the chemical potential and the $Z(3)$ symmetry. For a detailed review on this specific topic, see Ref. [47]. We have seen that, in the pure gauge sector of QCD, if $\langle\Phi\rangle \neq 0$ (deconfined phase), a center transformation will change the expectation value of the Polyakov loop:

$$\langle\Phi\rangle \to z\langle\Phi\rangle \to z^2\langle\Phi\rangle \tag{4.30}$$

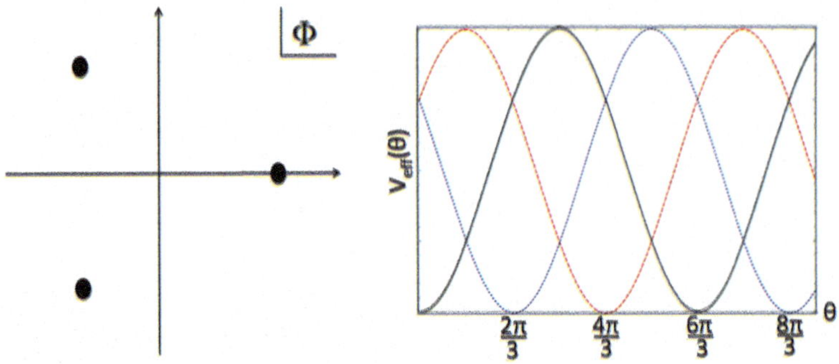

Fig. 4.7 Left: Equivalent vacua for the SU(3) pure gauge theory, in the case in which the $Z(3)$ symmetry is broken. Right: The effective potential as a function of the phase θ at the three $Z(3)$ images of $\langle\Phi\rangle = 1$, for the group SU(3)

and, as a consequence, the $Z(3)$ symmetry is broken. The vacuum corresponds to $\langle\Phi\rangle$=1 perturbatively, but there are three equivalent vacua in the SU(3) gauge theory, corresponding to $z = 1,\ e^{i2\pi/3},\ e^{-i2\pi/3}$ (see the left panel of Fig. 4.7).

In the presence of quarks, the center symmetry is broken explicitly. The trivial vacuum $\langle\Phi\rangle = 1$ is preferred in this case: quarks act as an external symmetry breaking field, choosing a preferred direction. However, in the presence of an imaginary chemical potential μ_I, something non-trivial happens: the center transformation can be undone by a shift in μ_I. Recall that the lattice gauge field, U_μ, is related to G_μ by: $U_\mu = e^{-iG_\mu}$. Since the chemical potential appears in the QCD action in the same way as the imaginary part of the fourth-component of an abelian vector field, we have that U_4 gets multiplied by a phase factor in the presence of a finite μ_I: $U_4 \to e^{i\mu_I} U_4$. Therefore, the following combination appears under a twisted gauge transformation

$$z^k e^{i\frac{\mu_I}{T}} = e^{i(\frac{\mu_I}{T}+\frac{2\pi k}{N})}. \tag{4.31}$$

This leads to a new symmetry, called Roberge-Weiss symmetry: the partition function becomes periodic in μ as follows:

$$Z\left(\frac{\mu}{T}\right) = Z\left(\frac{\mu}{T} + \frac{2\pi i k}{3}\right). \tag{4.32}$$

The partition function and the phase structure are therefore periodic in the μ_I direction with period $\frac{2\pi T}{3}$ and the range of $\frac{\mu_I}{T}$ is therefore limited by $\frac{\pi}{3}$ starting at $\mu_I = 0$. Another way to interpret the Roberge-Weiss symmetry is to note that an increase in μ_I is equivalent to a center transformation: since the Polyakov loop is not invariant under it, the choice of preferred vacuum will change as μ_I increases. This is illustrated by the right panel of Fig. 4.7, which shows the effective Polyakov loop potential as a function of the phase $\theta = \mu_I/T$. The three curves correspond to the three values of $\langle\Phi\rangle$ listed below. When $\langle\Phi\rangle \neq 0$, the Polyakov loop expectation value cycles through the three different possibilities: the preferred vacuum is given by

$$\begin{aligned} \langle\Phi\rangle &\simeq 1 \quad \text{at} \quad \mu_I/T \simeq 0 \quad \text{black curve} \\ \langle\Phi\rangle &\simeq z \quad \text{at} \quad \mu_I/T \simeq \frac{2\pi}{3} \quad \text{blue curve} \\ \langle\Phi\rangle &\simeq z^2 \quad \text{at} \quad \mu_I/T \simeq \frac{4\pi}{3} \quad \text{red curve.} \end{aligned} \tag{4.33}$$

As a consequence, exactly at the boundaries (given by $\mu_I/T = (2r+1)\pi/3$, with $r = 1, 2, 3$) we find a proper first order phase transition, with the Polyakov loop as the order parameter. The first order boundary lines separate areas of the phase diagram in which the Polyakov loop expectation value takes one of the three values discussed above. This is shown in the left panel of Fig. 4.8. The structure shown in the figure repeats itself periodically for $\mu_I/T > 4\pi/3$. The arrows in the figure point in the direction of the three minima shown in the left panel of Fig. 4.7. The dotted lines indicate the thermal transition line. For real μ_B, we can write a Taylor expansion of the transition line $T_c(\mu_B)$ in powers of $\mu_B/T_c(\mu_B)$ as follows:

$$\frac{T_c(\mu_B)}{T_c(0)} = 1 - \kappa_2 \left(\frac{\mu_B}{T_c(\mu_B)}\right)^2 + \ldots . \tag{4.34}$$

Therefore, for small μ_I the transition increases quadratically with μ_I and can be connected to the Roberge-Weiss vertical line at $\mu_I/T = \pi/3$ (see the right panel of Fig. 4.8). The point where the lines meet is known as the Roberge-Weiss endpoint. For larger μ_I, the phase structure is determined by the periodicity discussed above.

It is interesting to discuss how the Columbia Plot discussed in Chap. 3 extends to non-zero μ_I. Several cases need to be discussed:

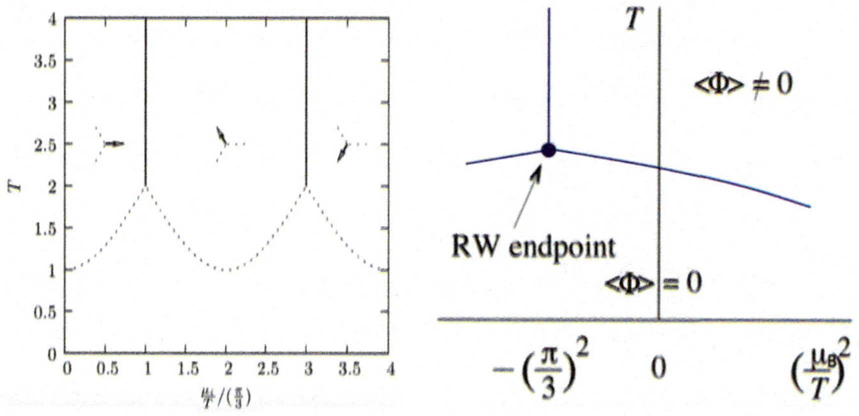

Fig. 4.8 Left: Phase structure in the (T, μ_I) plane. Right: Phase structure in the $(T, (\mu_B/T)^2)$ plane

1. *heavy or light quarks*: the first order transition remains first order for all values of μ_I. At the Roberge-Weiss endpoint, three first order lines come together:

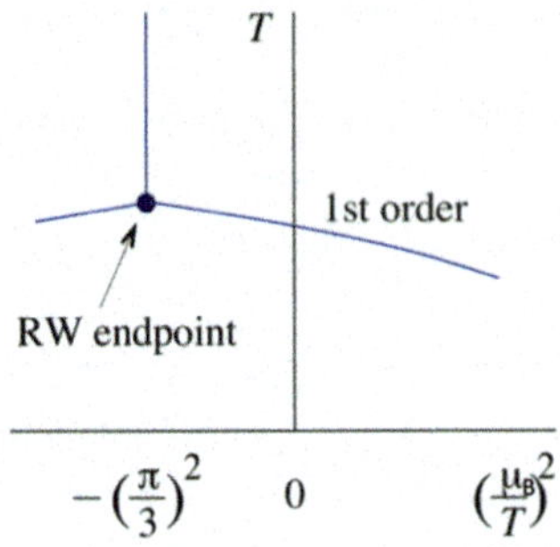

2. *quarks with intermediate mass*: the crossover at $\mu_B^2 = 0$ turns into a first-order transition at some value of μ_I and maybe also at some value of real μ_B (the latter would correspond to the QCD critical point)

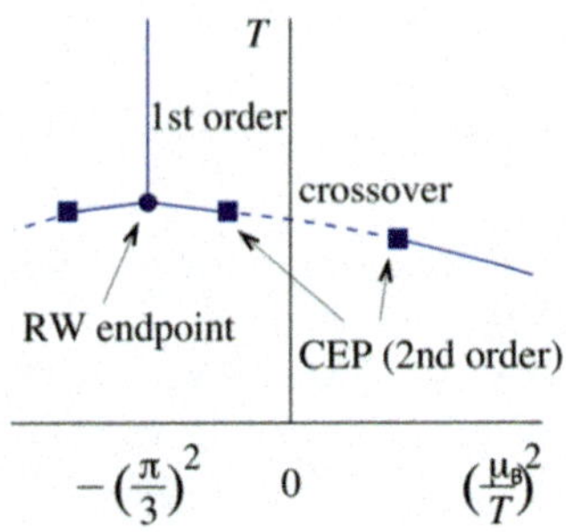

3. *quarks with specific mass value*: there will exist one value of quark masses for which the critical point at finite μ_I coincides with the Roberge-Weiss point. The transition in this case is a crossover for all values of μ_I. There might still be a critical point for real μ_B.

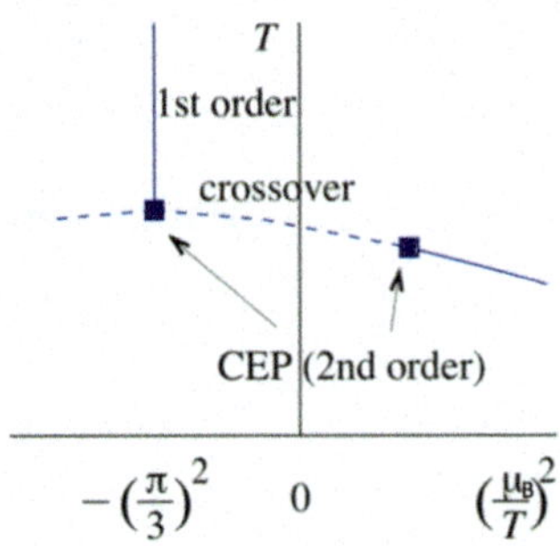

4.4 QCD Phase Diagram at Real Chemical Potential

The fact that direct lattice simulations at real μ_B cannot currently be performed makes the knowledge of the QCD phase diagram relatively limited. The definition of transition temperature that we have seen in Chap. 2 can be extended to finite chemical potential through a Taylor expansion as follows

$$\frac{T_c(\mu_B)}{T_c(0)} = 1 - \kappa_2 \left(\frac{\mu_B}{T_c(\mu_B)}\right)^2 + \kappa_4 \left(\frac{\mu_B}{T_c(\mu_B)}\right)^4 + \dots . \tag{4.35}$$

For the first few coefficients in the above equation, it is sufficient to study QCD at small μ_B, for which there are several methods. For example, one can calculate the derivative of the chiral condensate with respect to μ_B, on $\mu_B = 0$ ensembles [19, 48]. In this way, we can study the chiral condensate and chiral susceptibility as Taylor expansions in μ_B/T and check how the transition temperature defined before moves when increasing μ_B. The chiral condensate can be calculated on the lattice as (l indicates light quark flavors):

$$\langle\bar{\psi}\psi\rangle_l = \frac{T}{V}\frac{\partial \ln Z}{\partial m_l} = N_l \frac{T}{V}\langle X_l\rangle, \qquad \text{where} \qquad \langle X_l\rangle = \langle\frac{1}{2}\mathrm{tr}M_l^{-1}\rangle \tag{4.36}$$

and M_l is the light flavor fermionic matrix defined in Eq. (1.77). The chiral susceptibility is given by

$$\langle\chi\rangle_l = \frac{T}{V}\frac{\partial^2 \ln Z}{\partial m_l^2} = \frac{T}{V}N_l^2\left[\langle X_l X_l\rangle - \langle X_l\rangle\langle X_l\rangle\right] + \frac{T}{V}N_l\langle Y_l\rangle,$$

$$\text{where} \quad \langle Y_l\rangle = -\langle\frac{1}{2}\mathrm{tr}(M_l^{-1}M_l^{-1})\rangle \tag{4.37}$$

and Y_l is the connected part of the chiral susceptibility, while the term in square brackets is its disconnected part. We calculate

$$P_l = \frac{\partial X_l}{\partial \mu_l} = -\frac{1}{2}\mathrm{tr}(M_l^{-1}M_l' M_l^{-1}) \tag{4.38}$$

and from this we get the derivative of the chiral condensate with respect to μ_l:

$$\begin{aligned}\frac{\partial\langle\bar{\psi}\psi\rangle}{\partial\mu_l} = \frac{T}{V}\Big[&\langle-\frac{1}{2}\mathrm{tr}M_l^{-1}\mathrm{tr}(M_l^{-1}M_l')\rangle - \langle\frac{1}{2}\mathrm{tr}M_l^{-1}\rangle\langle\mathrm{tr}(M_l^{-1}M_l')\rangle \\ &+\langle-\frac{1}{2}\mathrm{tr}(M_l^{-1}M_l' M_l^{-1})\rangle\Big]\end{aligned} \tag{4.39}$$

and more explicitly (recalling that $\mathrm{tr}\ln M = \ln\det M$):

$$
\begin{aligned}
\frac{\partial\langle\bar{\psi}\psi\rangle_l}{\partial\mu_l} &= N_l\frac{T}{V}\frac{\partial}{\partial\mu_l}\frac{1}{Z}\int DUe^{-S_G}\frac{1}{2}\mathrm{tr}M_l^{-1}e^{N_f\ln\det M_f}\\
&= N_l\frac{T}{V}\frac{1}{Z}\int DUe^{-S_G}\frac{1}{2}\mathrm{tr}M_l^{-1}\left(\mathrm{tr}M_l^{-1}M_l'e^{N_f\ln\det M_f}\right)+\\
&- N_l\frac{T}{V}\frac{1}{Z^2}\int DUe^{-S_G}\frac{1}{2}\mathrm{tr}M_l^{-1}e^{N_f\ln\det M_f}\int DUe^{-S_G}\mathrm{tr}M_l^{-1}M_l'e^{N_f\ln\det M_f}+\\
&+ N_l\frac{T}{V}\frac{1}{Z}\int DU\frac{\partial X_l}{\partial\mu_l}e^{-S_G}e^{N_f\ln\det M_f}
\end{aligned}
\tag{4.40}
$$

Analogously, we have

$$
\frac{\partial\langle\bar{\psi}\psi\rangle_l}{\partial\mu_s} = \langle-\frac{1}{2}\mathrm{tr}M_l^{-1}\mathrm{tr}M_s^{-1}M_s'\rangle - \langle\frac{1}{2}\mathrm{tr}M_l^{-1}\rangle\langle\mathrm{tr}M_s^{-1}M_s'\rangle. \tag{4.41}
$$

We define $A_i = \mathrm{tr}M_i^{-1}M_i'$. We now can easily write the derivative of the disconnected part of the chiral susceptibility with respect to the chemical potentials

$$
\frac{\partial}{\partial\mu_l}\langle\chi\rangle_{l,disc} \propto \frac{\partial}{\partial\mu_l}\left[\langle X_lX_l\rangle - \langle X_l\rangle^2\right] = \tag{4.42}
$$

$$
= \langle X_lX_lA_l\rangle - \langle X_lX_l\rangle\langle A_l\rangle - 2\langle X_l\rangle\langle X_lA_l\rangle + 2\langle X_l\rangle^2\langle A_l\rangle + 2\langle P_lX_l\rangle - 2\langle P_l\rangle\langle X_l\rangle
$$

and

$$
\frac{\partial}{\partial\mu_s}\langle\chi\rangle_{l,disc} \propto \langle X_lX_lA_s\rangle - \langle X_lX_l\rangle\langle A_s\rangle - 2\langle X_l\rangle\langle X_lA_s\rangle + 2\langle X_l\rangle^2\langle A_s\rangle \tag{4.43}
$$

where we wrote $\propto$ instead of $=$ because we neglected a factor $\frac{T}{V}N_l^2$. This approach is however unpractical beyond μ_B^2 order, since the signal-to-noise ratio of higher μ_B derivatives is suppressed with powers of the volume. Another way is to set $\mu_B = i\mu_B^I$, and perform simulations at imaginary chemical potential. Several simulations have been performed to extract κ_2, both with the imaginary-μ_B formalism and Taylor expansion. Details of the simulations also vary with respect to the choice of μ_S and μ_Q. The easiest possibility is to set $\mu_S = \mu_Q = 0$ (corresponding to $\mu_u = \mu_d$ and $\mu_s = 0$) or $\mu_u = \mu_d = \mu_s$. A more complicated choice, which is however closer to the situation in heavy-ion collision experiments, is to solve the two constraints $\langle n_S\rangle$=0 and $\langle n_Q\rangle = 0.4\langle n_B\rangle$, obtaining $\mu_S(T,\mu_B)$ and $\mu_Q(T,\mu_B)$. In the case in which the quark chemical potentials are all equal ($\mu_s = \mu_{u,d}$), and for $m_s/m_{u,d} = 20$, the curvature obtained from the disconnected part of the renormalized susceptibility of the light quark chiral condensate is $\kappa = 0.020(4)$ [49]. Using the chiral susceptibility and two different ways to extract the inflection point of the chiral condensate, the authors of Ref. [50] found

$\kappa = 0.0135(20)$ at $\mu_S = 0$. In Ref. [37], a value of $\kappa = 0.0149(21)$ was found from the chiral condensate, chiral susceptibility and strange quark susceptibility, in the case of strangeness neutrality ($\langle n_S \rangle = 0$). This result was further refined recently [51], leading to a more precise value of $\kappa_2 = 0.0153(18)$. Other recent results were obtained in Ref. [52], in which a comparison between the results from Taylor expansion and analytic continuation from imaginary chemical potential was performed, and in Ref. [53], in which the Taylor expansion of chiral observables was used to extract the curvature and the first estimate of κ_4. A compilation of values of κ_2 in the literature is shown in the top panel of Fig. 4.9. A more recent estimate for κ_4 was presented in Ref. [51], in which the error-bar was considerably reduced. The two results for κ_4 available in the literature so far are shown in the bottom panel of Fig. 4.9.

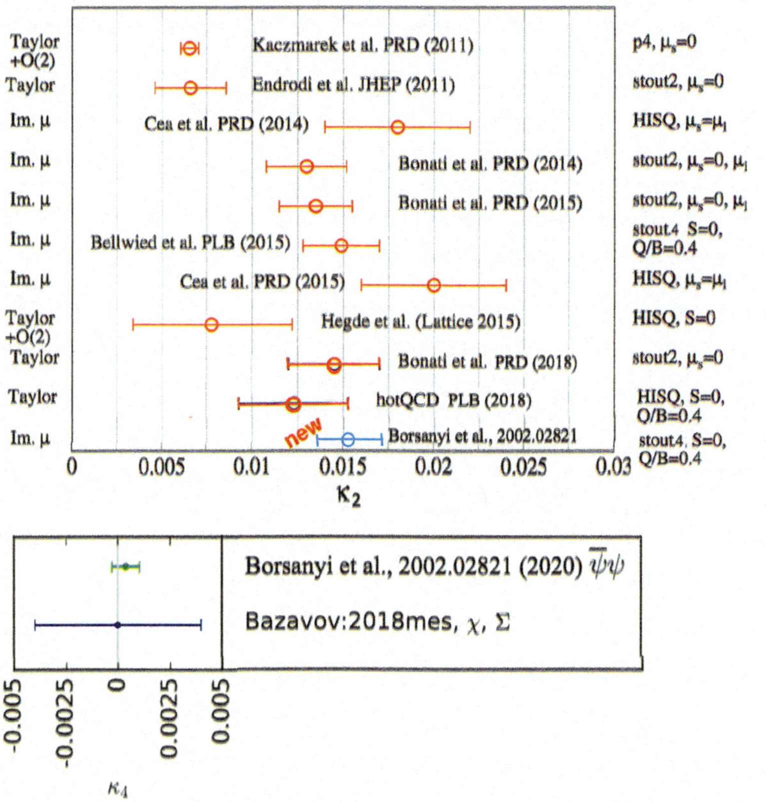

Fig. 4.9 Top: compilation of values of κ_2 in the literature. The values of the curvature have been obtained with different methods, different actions and different conditions on μ_S and μ_Q [19, 37, 48–52, 52–56]. Bottom: Compilation of values of κ_4 in the literature [51, 53]

4.4.1 Limits on the Critical Point Location

One of the most exciting open questions in the field of strongly interacting matter under extreme conditions is whether the transition to deconfined quarks and gluons, which we have seen to be a crossover at $\mu_B = 0$, becomes a first order phase transition at large values of the chemical potential. This would lead to the existence of a critical point (CP) on the QCD phase diagram. While the theory cannot be solved in this regime, several models predict the existence and position of the CP [57, 58]. Experimental verification should become possible through several low-energy heavy-ion programs. In order to create a high-density system in experiments, due to baryon number conservation, some of the baryons from the colliding nuclei must be transported to the mid-rapidity region. This can be achieved by systematically decreasing the energy of the colliding nuclei, which leads to a systematic scan of the phase diagram. The first part of this so-called Beam Energy Scan was started at RHIC in 2010. The Second Beam Energy Scan (BESII) is running between 2019 and 2021, with improved beam quality and detector capability. Other facilities are exploring the high density region of the QCD phase diagram, such as NA61/SHINE and HADES, and the future NICA in Dubna, FAIR at the GSI (Darmstadt) and the CSR in Lanzhou (CEE). A discussion of the main experimental results will be presented in Chap. 7. For a recent review on mapping the QCD phase diagram through the Beam Energy Scan, see Ref. [59].

While QCD cannot be solved at high baryon density due to the sign problem, several attempts have been made to at least constrain the location of the critical point to a reduced portion of the QCD phase diagram. The Taylor expansion coefficients introduced in Sect. 4.2.1 carry information on the CP position [60]. In fact, the Taylor series breaks down at the nearest non-analytic point to the series origin, which corresponds to the critical point on the QCD phase diagram if it is on the real μ_B axis. The distance between the series origin and this singularity is called radius of convergence of the Taylor series, and it can be obtained from ratios of subsequent expansion coefficients of thermodynamic observables it the limit of infinite order in the Taylor expansion. In particular, in the case of the free energy, the radius of convergence reads [61]

$$r^f_{n,m}(T) = \left| \frac{\frac{m!}{n!}\chi_n(T)}{\chi_m(T)} \right|^{1/(m-n)} \tag{4.44}$$

while from the second-order baryon susceptibility we get

$$r^\chi_{n,m}(T) = \left| \frac{\frac{(m-2)!}{(n-2)!}\chi_n(T)}{\chi_m(T)} \right|^{1/(m-n)} . \tag{4.45}$$

These two definitions coincide in the limit $n \to \infty$. While the number of coefficients currently available is not enough to reliably calculate the radius of convergence, an estimate has been presented in Refs. [43, 44, 62]. An example from Ref. [43], based on $r^\chi_{6,4}$, is shown in the top panel of Fig. 4.10. The bottom panel shows a recent analysis from Ref. [62].

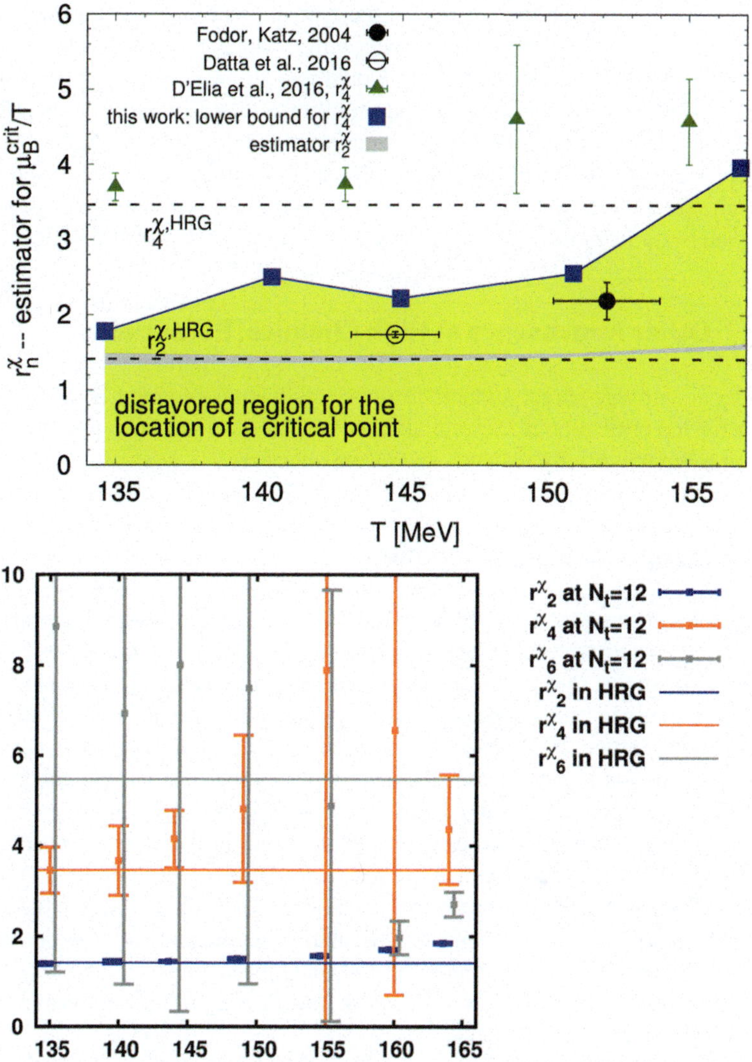

Fig. 4.10 Top: from Ref. [43]. Estimators for the radius of convergence of the Taylor series for net baryon-number fluctuations, $\chi_2^B(T, \mu_B)$, in the case of vanishing electric charge and strangeness chemical potentials obtained on lattices with $N_t = 8$. Shown are lower bounds for $r_{4,6}^\chi$ from Ref. [43] (squares) and from calculations with an imaginary chemical potential (triangles) [44]. The gray band shows an estimator from $r_{2,4}^\chi$. The figure also shows an estimate for the location of the critical point obtained from calculations using Taylor expansion (empty circle) [64] and unimproved staggered fermions with a reweighting technique (full circle) [31]. The two dashed lines are estimates from the HRG model. Bottom: from Ref. [62]. Estimators for the radius of convergence of the Taylor series for net-baryon number fluctuations, $\chi_2^B(T, \mu_B)$ on lattices with $N_t = 12$. The plot shows $r_{2,4}^\chi, r_{4,6}^\chi, r_{6,8}^\chi$ on the lattice (points with error-bar) and in the HRG model (horizontal lines)

More recent limits on the position of the critical point have been determined by looking at properties of the chiral susceptibility, which is supposed to diverge at the critical point. For this reason, one would expect an increase of the height of the peak of the T-dependent chiral susceptibility, and a decrease in its width with increasing chemical potential, if the critical point is approached. The top panel of Fig. 4.11 shows the width of the peak for this quantity as a function of the chemical potential. The bottom panel shows the height of the peak as a function of $(\mu_B/T)^2$. Both panels are from Ref. [51]. Similar results were presented in Ref. [63].

4.5 Other Approaches at High Chemical Potential

In order to provide direct support to the experimental program at RHIC in the search for the critical point, several alternative methods have been proposed in the literature. We will briefly review a few recent ones here.

4.5.1 Dyson-Schwinger Equation

Dyson-Schwinger equations (DSE) and the functional renormalization group (FRG) represent an alternative approach to study QCD at finite density, which relies on approximations that are less controlled than the lattice QCD ones. These approximations can however be controlled by constraints such as symmetries and by comparison to lattice QCD results when available. This approach has an advantage with respect to other effective models such as the Nambu Jona-Lasinio model or the Polyakov-loop-extended Nambu Jona-Lasinio or Quark Meson model: the gluon degrees of freedom, which are usually integrated out in the latter approaches and therefore have no effect on the medium, are instead directly accessible in DSE.

The authors of Ref. [65] have considered a system of 2 and 2 + 1 quark flavors and solved the quark and gluon propagator DSE at finite temperature and quark chemical potential. They found a critical point in the QCD phase diagram, and investigated how its position moves from 2 to 2 + 1 quark flavors. This result is shown in the left panel of Fig. 4.12.

In Ref. [66] it was pointed out that baryonic degrees of freedom, which had been omitted in the quark-gluon interaction used in Ref. [65], might have an impact on the location of the critical point. The effect of these contributions was considered only in the case of $N_f = 2$ quark flavors, and it was found that the critical point moves to higher temperatures and lower chemical potentials, compared to the case in which these effects are neglected. This is shown in the right panel of Fig. 4.12.

The results obtained in these publications exclude the presence of a critical point for $\mu_B/T < 1$.

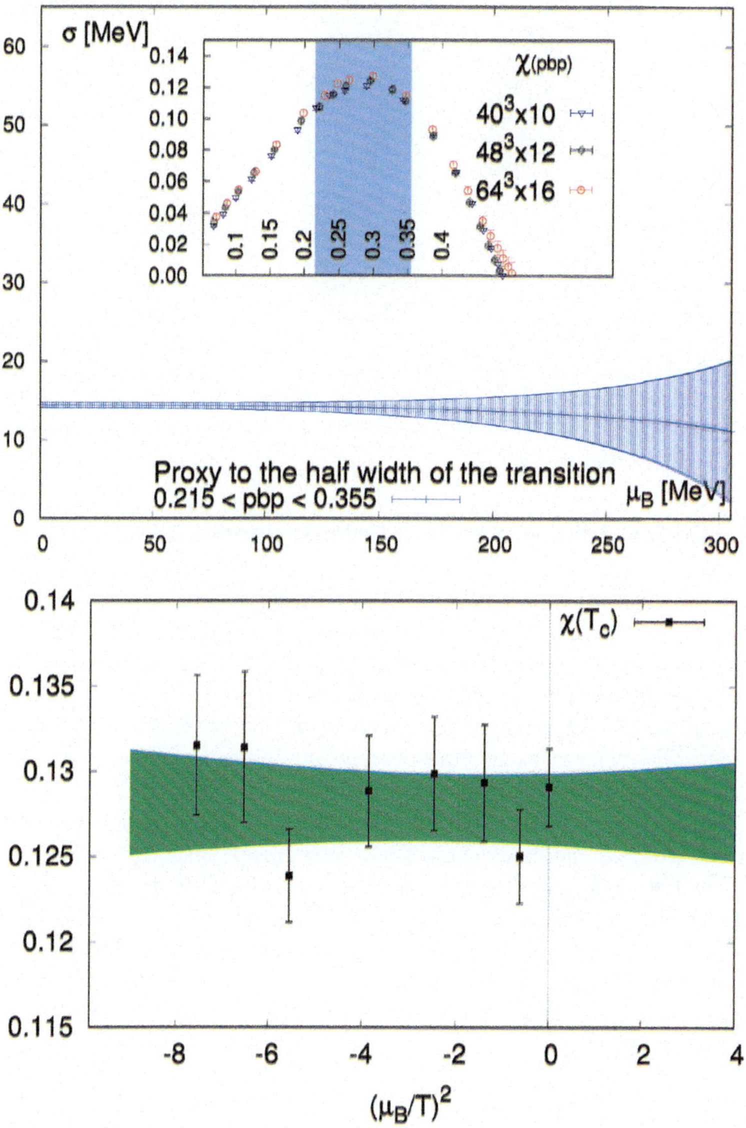

Fig. 4.11 From Ref. [51]. Top: half width σ of the transition, corresponding to the width of the chiral susceptibility (χ) peak, simulated at imaginary chemical potential and extrapolated to real μ_B. Bottom: result of a μ_B-by-μ_B analysis for the value of the chiral susceptibility at the crossover temperature after continuum extrapolation. The green band shows a linear extrapolation in $(\mu_B/T)^2$

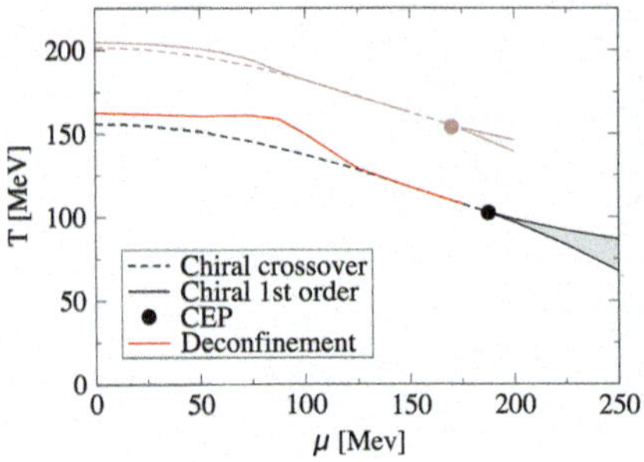

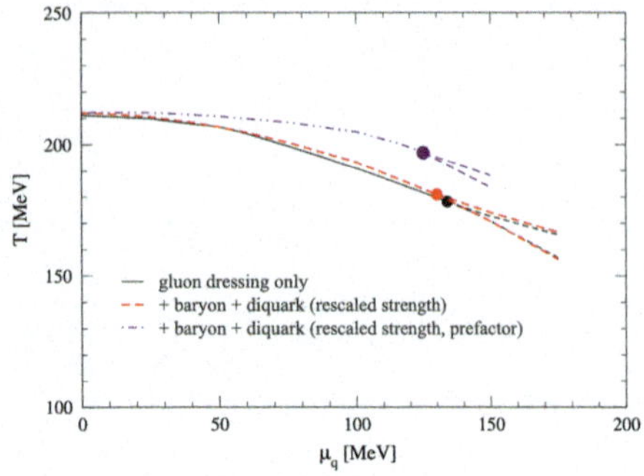

Fig. 4.12 Left: from Ref. [65]. The phase diagram for $N_f = 2$ (light colors, top lines) and $N_f = 2 + 1$ (dark colors, bottom lines) from Dyson-Schwinger equations. Right: from Ref. [66]. Comparison of the phase diagram for $N_f = 2$ including different types of self-energy contributions

4.5.2 Critical Point in the Black Hole Engineering Approach

Among the approaches to extend QCD to higher baryon densities, black hole engineering [67, 68] has been one of the most successful ones. It is based on the holographic correspondence [69], a well-known tool developed in string theory, and it fulfills the following necessary requirements: it reproduces the thermodynamics of QCD in the crossover region at zero and small baryon density, and the system that it describes behaves as an almost perfect liquid for the temperatures probed in heavy-ion collisions.

In Ref. [70], the most realistic black hole engineering model was shown to reproduce the QCD thermodynamic Taylor coefficients displayed in Sect. 4.2.3 up to χ_8^B. Such a model predicts the existence of a critical point on the QCD phase diagram at $T = 89$ MeV and $\mu_B = 724$ MeV, corresponding to a collision energy $\sqrt{s} = 2.5 - 4.1$ GeV. This prediction was made on the basis of the divergence of χ_2^B in the phase diagram, shown in the top panel of Fig. 4.13. The bottom panel of Fig. 4.13 shows the predicted critical point, and the limits on the forbidden regions from different approaches.

4.5.3 Lattice-Based Approach with a 3D-Ising Model Critical Point

The QCD critical point, if it exists, is in the 3D-Ising model universality class [57, 72–75]. This information has been exploited in Ref. [76] to construct a family of Equations of State for QCD, which match lattice QCD results up to order μ_B^4 and exhibit the correct scaling behavior in the vicinity of the critical point.

This is achieved through

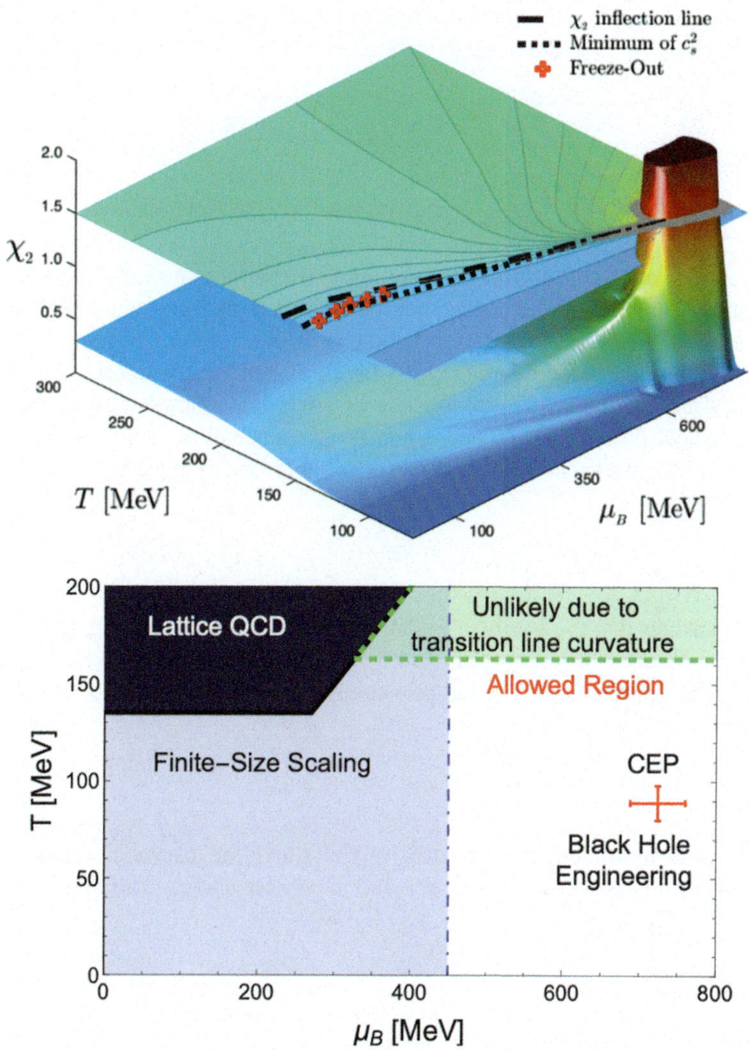

Fig. 4.13 From Ref. [70]. Top: baryon susceptibility χ_2^B in the (T, μ_B) plane determined from black hole engineering. The upper plane in the top plot shows the phase diagram, with the chemical freeze-out points in red. The dashed line corresponds to the location of the inflection point of χ_2^B. The dotted line gives the location of the minimum of the speed of sound squared in the phase diagram. Bottom: regions in the QCD phase diagram where the presence of a critical point has been excluded by current lattice QCD constraints [43] and a finite-size scaling analysis [71]. Temperatures above 155 MeV are also unlikely due to constraints from the curvature of the transition line. The location of the critical point in the phase diagram found in the black hole engineering approach, taking into account systematic uncertainties, is also shown

- a non-universal mapping between the Ising model phase diagram in the variables $r = (T - T_c)/T_c$ (reduced temperature) and h (magnetic field) and the QCD one

$$\frac{T - T_C}{T_C} = w(r\rho \sin\alpha_1 + h \sin\alpha_2)$$
$$\frac{\mu_B - \mu_{BC}}{T_C} = w(-r\rho \cos\alpha_1 - h \cos\alpha_2), \tag{4.46}$$

  where T_C and μ_{BC} are the temperature and chemical potential of the critical point, α_1 and α_2 are the angles that the r and h axes form with the T =const. lines, and (w, ρ) are scale factors for the variables r and h (see Fig. 4.14);
- the assumption that the lattice QCD Taylor expansion coefficients discussed in Sect. 4.2.3 can be written as the sum of an "Ising" contribution coming from the critical point of QCD, and a non-Ising contribution, which would contain the regular part as well as any other possible criticality present in the region of interest:

$$T^4 c_n^{LAT}(T) = T^4 c_n^{Non-Ising}(T) + T_C^4 c_n^{Ising}(T). \tag{4.47}$$

  This ensures that the family of Equations of State constructed with this prescription reproduces the lattice QCD results for the pressure and its Taylor coefficients up to the desired order. The full pressure is then reconstructed as

$$P(T, \mu_B) = T^4 \sum_n c_{2n}^{Non-Ising}(T) \left(\frac{\mu_B}{T}\right)^{2n} + P_{crit}^{QCD}(T, \mu_B). \tag{4.48}$$

Figure 4.15 shows the entropy density (left) and the baryonic density (right) obtained in this approach for one specific parameter choice. The hope is that the

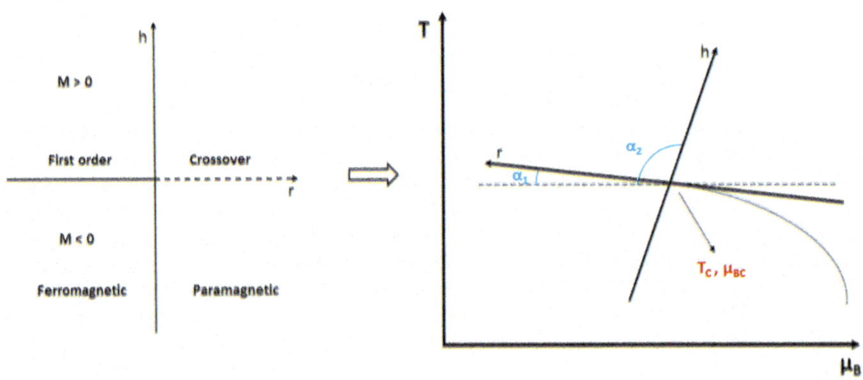

Fig. 4.14 From Ref. [76]. Non-universal map from Ising variables (r, h) to QCD coordinates (T, μ_B)

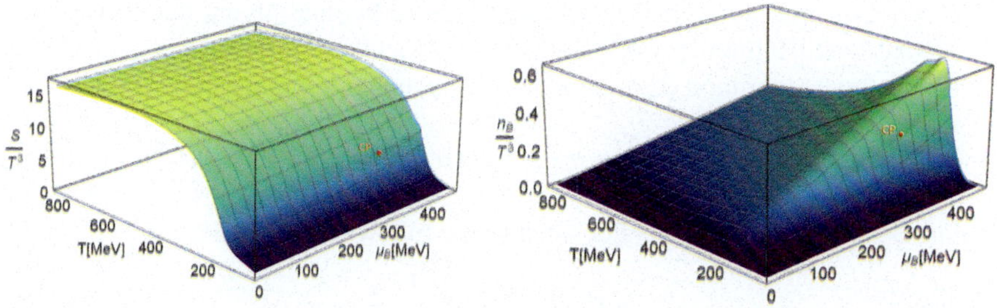

Fig. 4.15 From Ref. [76]. Entropy density (left) and baryonic density (right) as functions of the temperature and chemical potential. The critical point is marked with a red dot. Both quantities exhibit a discontinuity for $\mu_B > \mu_{BC}$, as expected

experimental results and thermodynamic stability conditions will help to constrain the parameter space, including the position of the critical point.

4.6 QCD at Finite Isospin Chemical Potential

Numerical simulations of QCD at finite isospin chemical potential are possible on the lattice. The fermionic part of the Euclidean QCD Lagrangian gets modified as follows by a finite μ_I:

$$\begin{aligned} S_{ud} &= \bar{\psi} M_{ud} \psi \\ M_{ud} &= \gamma^\mu(\partial_\mu + iG_\mu)\mathbb{1} + m_{ud}\mathbb{1} + \mu_I \gamma_4 \tau_3 + i\lambda\gamma_5\tau_2, \end{aligned} \tag{4.49}$$

where G_μ is the gluon field and τ_a the Pauli matrices. The term proportional to λ is an explicit symmetry breaking term that couples to the charged pion field; λ is referred to as pionic source and a non-zero value is needed in a finite volume to trigger the spontaneous breaking of the $U_{\tau_3}(1)$ symmetry due to the chemical potential. Results are subsequently extrapolated to $\lambda \to 0$.

The partition function on the lattice can be written as

$$Z = \int [DU] e^{-\beta S_G} (\det M_{ud})^{1/4} (\det M_s)^{1/4}, \tag{4.50}$$

where

$$M_{ud} = \begin{pmatrix} \not{D}(\mu_I) + m_{ud} & \lambda\eta_5 \\ -\lambda\eta_5 & \not{D}(-\mu_I) + m_{ud} \end{pmatrix} \tag{4.51}$$

with $\eta_5 = (-1)^{m_x+m_y+m_z+m_t}$ and the m_i are the lattice site coordinates (see Eqs. (1.72) and (1.90)). From the above formula for M_{ud} it is evident that the chemical potential μ_I is introduced on the lattice by multiplying the forward and backward timelike links by $e^{a\mu_I}$ and $e^{-a\mu_I}$, respectively.

The staggered fermion equivalent of chiral symmetry implies that

$$\not{D}(\mu_I)\eta_5 + \eta_5 \not{D}(\mu_I) = 0 \tag{4.52}$$

holds. Besides, the Dirac operator satisfies the following condition

$$\eta_5 \not{D}(\mu_I)\eta_5 = \not{D}(\mu_I)^\dagger . \tag{4.53}$$

Therefore, the light fermion determinant obeys

$$\tau_1 \eta_5 M_{ud} \eta_5 \tau_1 = M_{ud}^\dagger . \tag{4.54}$$

Taking the determinant of both sides shows that $\det(M_{ud})$ is real. Besides, it is also positive because if we consider

$$M'_{ud} = B M_{ud} B = \begin{pmatrix} \not{D}(\mu_I) + m_{ud} & \lambda \\ -\lambda & [\not{D}(\mu_I) + m_{ud}]^\dagger \end{pmatrix} \tag{4.55}$$

where $B = \mathrm{diag}(1, \eta_5)$, since $\det B = 1$ we get

$$\det M_{ud} = \det[(\not{D}(\mu_I) + m_{ud})\,(\not{D}(\mu_I) + m_{ud})^\dagger + \lambda^2] > 0. \tag{4.56}$$

This shows that lattice QCD simulations based on importance sampling techniques can be performed. The authors of Ref. [77] considered a system of $N_f = 2 + 1$ on $24^3 \times 6$ lattices and simulated the chiral condensate and the pion condensate as functions of temperature and μ_I. The results are shown in the two panels of Fig. 4.16.

The corresponding phase diagram is shown in the left panel of Fig. 4.17. The chiral phase transition temperature at $\mu_I = 0$ is consistent with the one reported

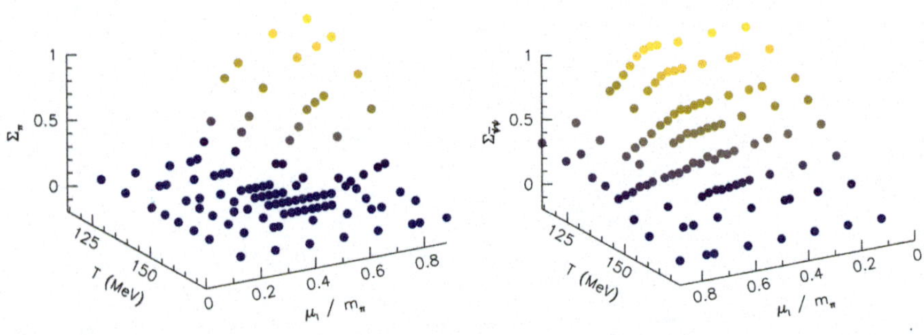

Fig. 4.16 From Ref. [77]. Pion condensate (left) and quark condensate (right) as functions of temperature and isospin chemical potential simulated on $24^3 \times 6$ lattices

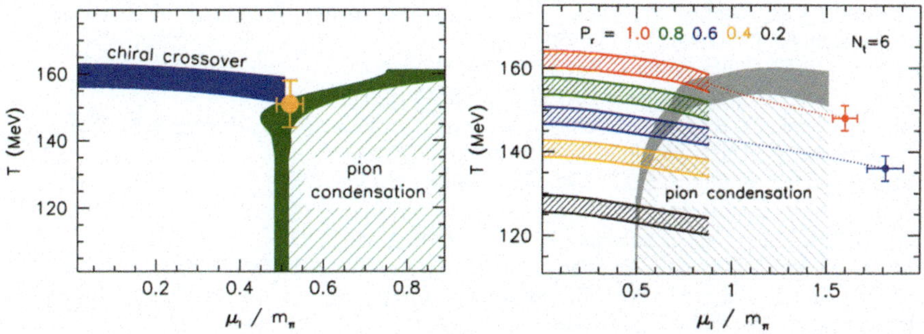

Fig. 4.17 From Ref. [77]. Left: phase diagram of QCD in the (T, μ_I) plane. The blue band is the chiral crossover transition temperature, while the green line is the boundary of the pion condensation phase. The yellow point marks the triple point. Right: the colored bands indicate contour lines of constant renormalized Polyakov loop

in Sect. 2.5.1; the pion condensate starts at $\mu_i = m_\pi/2$. A triple point is found at $(T_C, \mu_{IC}) = (151(7), 70(5))$ MeV. This is consistent with expectations from the NJL model [78–82].

The right panel of Fig. 4.17 shows the contours of constant renormalized Polyakov loop. Since this observable does not show a pronounced inflection point as a function of the temperature, these curves capture the dependence of deconfinement on the isospin chemical potential: e.g. the contour for which the Polyakov loop is 1 can be chosen to identify a transition temperature for deconfinement. It is interesting to notice that, contrary to the chiral phase transition line shown in the left panel of Fig. 4.17, these contour lines are insensitive to the pion condensation phase.

References

1. Fraga, E.S., Kurkela, A., Vuorinen, A.: Astrophys. J. **781**, L25 (2014)
2. Gattringer, C.: PoS LATTICE **2013**, 002 (2014)
3. Alexandru, A., Basar, G., Bedaque, P.F., Ridgway, G.W., Warrington, N.C.: JHEP **1605**, 053 (2016)
4. Alexandru, A., Basar, G., Bedaque, P.: Phys. Rev. D **93**(1), 014504 (2016)
5. Scorzato, L.: PoS LATTICE **2015**, 016 (2016)
6. Parisi, G., Wu, Y.S.: Sci. Sin. **24**, 483 (1981)
7. Damgaard, P.H., Huffel, H.: Phys. Rep. **152**, 227 (1987)
8. Aarts, G., Seiler, E., Stamatescu, I.O.: Phys. Rev. D **81**, 054508 (2010)
9. Aarts, G., James, F.A., Seiler, E., Stamatescu, I.O.: Eur. Phys. J. C **71**, 1756 (2011)
10. Mollgaard, A., Splittorff, K.: Phys. Rev. D **88**(11), 116007 (2013)
11. Aarts, G., Seiler, E., Sexty, D., Stamatescu, I.O.: Phys. Rev. D **90**(11), 114505 (2014)
12. Mollgaard, A., Splittorff, K.: Phys. Rev. D **91**(3), 036007 (2015)
13. Splittorff, K.: Phys. Rev. D **91**(3), 034507 (2015)
14. Nishimura, J., Shimasaki, S.: Phys. Rev. D **92**(1), 011501 (2015)
15. Allton, C.R., Ejiri, S., Hands, S.J., Kaczmarek, O., Karsch, F., Laermann, E., Schmidt, C., Scorzato, L.: Phys. Rev. **D66**, 074507 (2002)

16. Allton, C.R., Doring, M., Ejiri, S., Hands, S.J., Kaczmarek, O., Karsch, F., Laermann, E., Redlich, K.: Phys. Rev. **D71**, 054508 (2005)
17. Gavai, R.V., Gupta, S.: Phys. Rev. **D78**, 114503 (2008)
18. Basak, S., et al.: (MILC) PoS **LATTICE2008**, 171 (2008)
19. Kaczmarek, O., Karsch, F., Laermann, E., Miao, C., Mukherjee, S., Petreczky, P., Schmidt, C., Soeldner, W., Unger, W.: Phys. Rev. **D83**, 014504 (2011)
20. de Forcrand, P., Philipsen, O.: Nucl. Phys. **B642**, 290–306 (2002)
21. D'Elia, M., Lombardo, M.P.: Phys. Rev. **D67**, 014505 (2003)
22. Wu, L.K., Luo, X.Q., Chen, H.S.: Phys. Rev. **D76**, 034505 (2007)
23. D'Elia, M., Di Renzo, F., Lombardo, M.P.: Phys. Rev. **D76**, 114509 (2007)
24. Conradi, S., D'Elia, M.: Phys. Rev. **D76**, 074501 (2007)
25. de Forcrand, P., Philipsen, O.: JHEP **11**, 012 (2008)
26. D'Elia, M., Sanfilippo, F.: Phys. Rev. **D80**, 014502 (2009)
27. Moscicki, J.T., Wos, M., Lamanna, M., de Forcrand, P., Philipsen, O.: Comput. Phys. Commun. **181**, 1715–1726 (2010)
28. Fodor, Z., Katz, S.D.: Phys. Lett. **B534**, 87–92 (2002)
29. Fodor, Z., Katz, S.D.: JHEP **03**, 014 (2002)
30. Csikor, F., Egri, G.I., Fodor, Z., Katz, S.D., Szabo, K.K., Toth, A.I.: JHEP **05**, 046 (2004)
31. Fodor, Z., Katz, S.D.: JHEP **04**, 050 (2004)
32. Alexandru, A., Faber, M., Horvath, I., Liu, K.F.: Phys. Rev. **D72**, 114513 (2005)
33. Kratochvila, S., de Forcrand, P.: PoS **LAT2005**, 167 (2006)
34. Ejiri, S.: Phys. Rev. **D78**, 074507 (2008)
35. Fodor, Z., Katz, S.D., Schmidt, C.: JHEP **03**, 121 (2007)
36. Alexandru, A., Gattringer, C., Schadler, H.P., Splittorff, K., Verbaarschot, J.J.M.: Phys. Rev. **D91**, 074501 (2015)
37. Bellwied, R., Borsanyi, S., Fodor, Z., Günther J., Katz, S., Ratti, C., Szabo, K.: Phys. Lett. B **751**, 559–564 (2015)
38. Borsanyi, S., Fodor, Z., Guenther, J.N., Katz, S.K., Szabo, K.K., Pasztor, A., Portillo, I., Ratti, C.: JHEP **10**, 205 (2018)
39. Borsanyi, S., Endrodi, G., Fodor, Z., Katz, S.D., Krieg, S., Ratti, C., Szabo, K.K.: JHEP **08**, 053 (2012)
40. Hegde, P.: [BNL?Bielefeld?CCNU] Nucl. Phys. A **931**, 851–855 (2014)
41. Guenther, J.N., Bellwied, R., Borsanyi, S., Fodor, Z., Katz, S.D., Pasztor, A., Ratti, C., Szabó, K.K.: Nucl. Phys. A **967**, 720–723 (2017)
42. Alba, P., Alberico, W., Bellwied, R., Bluhm, M., Mantovani Sarti, V., Nahrgang, M., Ratti, C.: Phys. Lett. B **738**, 305–310 (2014)
43. Bazavov, A., Ding, H.T., Hegde, P., Kaczmarek, O., Karsch, F., Laermann, E., Maezawa, Y., Mukherjee, S., Ohno, H., Petreczky, P., Sandmeyer, H., Steinbrecher, P., Schmidt, C., Sharma, S., Soeldner, W., Wagner, M.: Phys. Rev. D **95**(5), 054504 (2017)
44. D'Elia, M., Gagliardi, G., Sanfilippo, F.: Phys. Rev. D **95**(9), 094503 (2017)
45. Noronha-Hostler, J., Parotto, P., Ratti, C., Stafford, J.M.: Phys. Rev. C **100**(6), 064910 (2019)
46. Monnai, A., Schenke, B., Shen, C.: Phys. Rev. C **100**(2), 024907 (2019)
47. Aarts, G.: J. Phys. Conf. Ser. **706**(2), 022004 (2016)
48. Endrodi, G., Fodor, Z., Katz, S.D., Szabo, K.K.: JHEP **04**, 001 (2011)
49. Cea, P., Cosmai, L., Papa, A.: Phys. Rev. D **93**(1), 014507 (2016)
50. Bonati, C., D'Elia, M., Mariti, M., Mesiti, M., Negro, F., Sanfilippo, F.: Phys. Rev. D **92**(5), 054503 (2015)
51. Borsanyi, S., Fodor, Z., Guenther, J.N., Kara, R., Katz, S.D., Parotto, P., Pasztor, A., Ratti, C., Szabo, K.K.: Phys. Rev. Lett. **125**(5), 052001 (2020)
52. Bonati, C., D'Elia, M., Negro, F., Sanfilippo, F., Zambello, K.: Phys. Rev. D **98**(5), 054510 (2018)
53. Bazavov, A., et al.: [HotQCD] Phys. Lett. B **795**, 15–21 (2019)
54. Cea, P., Cosmai, L., Papa, A.: Phys. Rev. D **89**(7), 074512 (2014)

55. Bonati, C., D'Elia, M., Mariti, M., Mesiti, M., Negro, F., Sanfilippo, F.: Phys. Rev. D **90**(11), 114025 (2014)
56. Hegde, P., et al.: [Bielefeld-BNL-CCNU] PoS **LATTICE2015**, 141 (2016)
57. Halasz, A.M., Jackson, A., Shrock, R., Stephanov, M.A., Verbaarschot, J.: Phys. Rev. D **58**, 096007 (1998)
58. Stephanov, M.A.: Prog. Theor. Phys. Suppl. **153**, 139–156 (2004)
59. Bzdak, A., Esumi, S., Koch, V., Liao, J., Stephanov, M., Xu, N.: Phys. Rep. **853**, 1–87 (2020)
60. Gavai, R., Gupta, S.: Phys. Rev. D **71**, 114014 (2005)
61. Karsch, F., Schaefer, B., Wagner, M., Wambach, J.: PoS **LATTICE2011**, 219 (2011)
62. Fodor, Z., Giordano, M., Günther, J.N., Kapás, K., Katz, S.D., Pásztor, A., Portillo, I., Ratti, C., Sexty, D., Szabó, K.K.: Nucl. Phys. A **982**, 843–846 (2019)
63. Steinbrecher, P.: [HotQCD] PoS **LATTICE2018**, 160 (2018)
64. Datta, S., Gavai, R.V., Gupta, S.: Phys. Rev. D **95**(5), 054512 (2017)
65. Fischer, C.S., Luecker, J., Phys. Lett. B **718** (2013), 1036–1043
66. Eichmann, G., Fischer, C.S., Welzbacher, C.A.: Phys. Rev. D **93**(3), 034013 (2016)
67. DeWolfe, O., Gubser, S.S., Rosen, C.: Phys. Rev. D **83**, 086005 (2011)
68. Rougemont, R., Critelli, R., Noronha-Hostler, J., Noronha, J., Ratti, C.: Phys. Rev. D **96**(1), 014032 (2017)
69. Maldacena, J.M.: Int. J. Theor. Phys. **38**, 1113–1133 (1999)
70. Critelli, R., Noronha, J., Noronha-Hostler, J., Portillo, I., Ratti, C., Rougemont, R.: Phys. Rev. D **96**(9), 096026 (2017)
71. Fraga, E.S., Palhares, L.F., Sorensen, P.: Phys. Rev. C **84**, 011903 (2011)
72. Rajagopal, K., Wilczek, F.: Nucl. Phys. B **399**, 395–425 (1993)
73. Berges, J., Rajagopal, K.: Nucl. Phys. B **538**, 215–232 (1999)
74. Karsch, F., Laermann, E., Schmidt, C.: Phys. Lett. B **520**, 41–49 (2001)
75. de Forcrand, P., Philipsen, O.: Nucl. Phys. B **673**, 170–186 (2003)
76. Parotto, P., Bluhm, M., Mroczek, D., Nahrgang, M., Noronha-Hostler, J., Rajagopal, K., Ratti, C., Schäfer, T., Stephanov, M.: Phys. Rev. C **101**(3), 034901 (2020)
77. Brandt, B.B., Endrodi, G., Schmalzbauer, S.: Phys. Rev. D **97**(5), 054514 (2018)
78. Ratti, C., Weise, W.: Phys. Rev. D **70**, 054013 (2004)
79. Andersen, J.O., Brauner, T.: Phys. Rev. D **81**, 096004 (2010)
80. Zhang, T., Brauner, T., Rischke, D.H.: JHEP **06**, 064 (2010)
81. Strodthoff, N., Schaefer, B.J., von Smekal, L.: Phys. Rev. D **85**, 074007 (2012)
82. Xia, T., He, L., Zhuang, P.: Phys. Rev. D **88**(5), 056013 (2013)

Fluctuations of Conserved Charges 5

Abstract

In this chapter we introduce one of the most promising set of observables for the comparison of lattice QCD results to experimental measurements: fluctuations and correlations of conserved charges. After their definition and relation to probability distributions, we discuss their relevance in determining chemical freeze-out parameters, critical point location and the chiral phase transition. The chapter ends with a compilation of lattice QCD results on these quantities. Their comparison to experimental data is discussed in Chap. 7.

5.1 Introduction and Definition

A system in thermal equilibrium, for a Grand-Canonical Ensemble, is characterized by its partition function

$$Z = \mathrm{tr}\left[\exp\left(-\frac{H - \sum_i \mu_i Q_i}{T}\right)\right], \tag{5.1}$$

where H is the Hamiltonian of the system, Q_i are the conserved charges and μ_i the chemical potential of charge i. In the case of QCD with $2+1+1$ quark flavors, they are charm C, strangeness S, baryon number B and electric charge Q, which can be expressed in terms of the four quark flavors $u,\ d,\ s$ and c. We can define the following quantities:

$$\text{Mean} \qquad \langle Q_i \rangle = T\frac{\partial}{\partial \mu_i}\ln Z$$

$$\text{Variance} \qquad \langle (\delta Q_i)^2 \rangle = T^2\frac{\partial^2}{\partial (\mu_i)^2}\ln Z \propto \chi_2^i, \qquad \text{with } \delta Q_i = Q_i - \langle Q_i \rangle$$

C. Ratti, R. Bellwied, *The Deconfinement Transition of QCD*, Lecture Notes in Physics 981, https://doi.org/10.1007/978-3-030-67235-5_5

$$\text{Co} - \text{variance} \qquad \langle \delta Q_i \delta Q_j \rangle = T^2 \frac{\partial^2}{\partial \mu_i \partial \mu_j} \ln Z \propto \chi_{11}^{ij}, \tag{5.2}$$

which are the first few central moments of the distribution of the conserved charge Q_i, as we will see below. We have introduced the susceptibilities χ_{ijk}^{BQS} in Chap. 4, but we repeat their general definition here for convenience:

$$\chi_{ijk}^{lmn} = \frac{\partial^{i+j+k}(p/T^4)}{\partial \left(\frac{\mu_l}{T}\right)^i \partial \left(\frac{\mu_m}{T}\right)^j \partial \left(\frac{\mu_n}{T}\right)^k}. \tag{5.3}$$

The definition of the susceptibilities in the above equation is chosen so that they are dimensionless. The diagonal ones are a measure for the fluctuations of the system, while the off-diagonal ones characterize the correlations between the conserved charges Q_i and Q_j.

5.2 Probability Functions

A real-valued function $F(x)$ is called cumulative distribution function if:

1. $F(x)$ is monotonically increasing or constant: $F(x_1) \leq F(x_2)$ if $x_1 \leq x_2$.
2. $F(x)$ is everywhere continuous from the right:

$$F(x) = \lim_{\epsilon \to 0^+} F(x + \epsilon) \tag{5.4}$$

3. $F(-\infty) = 0$, $F(+\infty) = 1$.

The function $F(x)$ represents the probability of the event $X \leq x$, where X is a random variable:

$$Pr\{X \leq x\} = F(x). \tag{5.5}$$

If $F(x)$ is continuous, it can be written as

$$F(x) = \int_{-\infty}^{x} f(t)dt, \tag{5.6}$$

where $f(t)$ is called the probability density function, or distribution. Given the distribution $f(x)$, the n-th moment about the origin is defined as

$$\mu_n' = \int_{-\infty}^{\infty} x^n f(x)dx. \tag{5.7}$$

In particular, the first moment is the mean

$$M = \mu_1' = \int_{-\infty}^{\infty} x f(x) dx. \tag{5.8}$$

For the second and higher moments, the central moments are usually used, rather than the moments about zero, because they provide clearer information about the distribution shape. They are defined as

$$\mu_n = \int_{-\infty}^{\infty} (x - M)^n f(x) dx. \tag{5.9}$$

In particular the variance is defined as:

$$\sigma^2 = \int_{-\infty}^{\infty} (x - M)^2 f(x) dx. \tag{5.10}$$

Another, very useful definition, is the normalized moments, defined as the n-th central moment divided by σ^n. In particular, the normalized third central moment is called skewness and it is equal to zero for any symmetric distribution

$$S = \frac{\mu_3}{\sigma^3}. \tag{5.11}$$

A negative skewness corresponds to a distribution skewed to the left, while a positive skewness corresponds to a distribution skewed to the right (see the top left and top right panels of Fig. 5.1). The fourth central moment is a measurement of the heaviness of the tail of a distribution, compared to the normal distribution having the same variance. It is always strictly positive, and for a normal distribution it is equal to $3\sigma^4$. The kurtosis κ is the normalized fourth central moment minus three:

$$\kappa = \frac{\mu_4}{\sigma^4} - 3. \tag{5.12}$$

If a distribution has heavy tails, the kurtosis will be large. Light-tailed distributions have a small kurtosis. Example of distributions having all the same mean and variance and different kurtosis values are shown in the bottom panel of Fig. 5.1. It follows from the definition, that both skewness and kurtosis are equal to zero for a gaussian distribution. For this reason, they are referred to as "non-gaussian moments".

We define a characteristic function, called moment generating function

$$\Phi(t) = \langle e^{tx} \rangle = \int_{-\infty}^{\infty} e^{tx} f(x) dx. \tag{5.13}$$

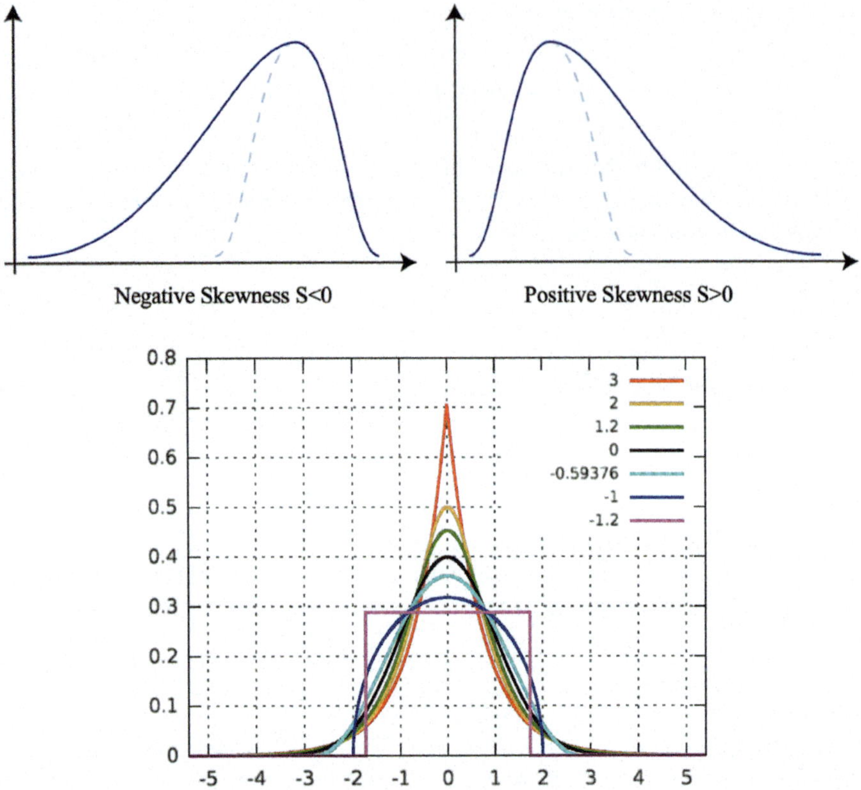

Fig. 5.1 Example of distributions with negative skewness (top left) and positive skewness (top right). The bottom panel shows distributions having all zero mean and unit variance and the different kurtosis values listed in the legend

It is easy to show that:

$$\mu_n' = \left.\frac{\mathrm{d}^n}{\mathrm{d}t^n}\Phi(t)\right|_{t=0}. \tag{5.14}$$

A set of quantities that provide an alternative to the moments of a probability distribution are the cumulants κ_n of the distribution. The cumulants of a random variable are defined using the cumulant generating function, which is the natural logarithm of the moment generating function defined above. The cumulants are defined through the power series of $\ln(\Phi(t))$:

$$\ln(\Phi(t)) = \sum_{n=0}^{\infty} \kappa_n \frac{t^n}{n!}. \tag{5.15}$$

It is easy to show that:

$$\kappa_1 = M; \qquad \kappa_2 = \sigma^2; \qquad \kappa_3 = \mu_3; \qquad \kappa_4 = \mu_4 - 3\mu_2^2; \tag{5.16}$$

here we show it for the first two cumulants and let the reader derive the higher order ones. As a first step we want to show that, using the definition of κ_1 as Taylor coefficient of the cumulant generating function, we obtain the mean M as defined in Eq. (5.9). We get:

$$\begin{aligned}\kappa_1 &= \left.\frac{\partial \ln \Phi(t)}{\partial t}\right|_{t=0} = \left.\frac{\partial \ln\left[\int_{-\infty}^{\infty} e^{tx} f(x)dx\right]}{\partial t}\right|_{t=0} \\ &= \left.\frac{\int_{-\infty}^{\infty} xe^{tx} f(x)dx}{\int_{-\infty}^{\infty} e^{tx} f(x)dx}\right|_{t=0} = \int_{-\infty}^{\infty} xf(x)dx.\end{aligned} \tag{5.17}$$

Now we proceed with the proof for the second cumulant:

$$\begin{aligned}\kappa_2 &= \left.\frac{\partial^2 \ln \Phi(t)}{\partial t^2}\right|_{t=0} = \left.\frac{\partial}{\partial t}\left[\frac{\int_{-\infty}^{\infty} xe^{tx} f(x)dx}{\int_{-\infty}^{\infty} e^{tx} f(x)dx}\right]\right|_{t=0} = \\ &= \left.\frac{\left(\int_{-\infty}^{\infty} x^2e^{tx} f(x)dx\right)\left(\int_{-\infty}^{\infty} e^{tx} f(x)dx\right) - \left(\int_{-\infty}^{\infty} xe^{tx} f(x)dx\right)^2}{\left(\int_{-\infty}^{\infty} e^{tx} f(x)dx\right)^2}\right|_{t=0} = \\ &= \int_{-\infty}^{\infty} x^2 f(x)dx - M^2,\end{aligned} \tag{5.18}$$

which is indeed the expression for the variance:

$$\begin{aligned}\sigma^2 &= \int_{-\infty}^{\infty} (x-M)^2 f(x)dx = \int_{-\infty}^{\infty} x^2 f(x)dx - 2M\int_{-\infty}^{\infty} xf(x)dx + M^2 \\ &= \int_{-\infty}^{\infty} x^2 f(x)dx - M^2.\end{aligned} \tag{5.19}$$

Coming back to statistical physics, in a large system an extensive quantity like the energy or the number of particles can be thought of as the sum of the energies associated with a number of nearly independent regions: these extensive quantities should be related to cumulants. In particular, we can identify the pressure with the cumulant generating function:

$$p = -T \ln\langle \exp[-\beta(E + \mu_i Q_i)]\rangle \tag{5.20}$$

so that we get:

$$
\begin{aligned}
\chi_1^i &= \frac{\partial(p/T^4)}{\partial(\mu_i/T)} = \frac{M_i}{VT^3} \\
\chi_2^i &= \frac{\partial^2(p/T^4)}{\partial(\mu_i/T)^2} = \frac{\sigma_i^2}{VT^3} \\
\chi_3^i &= \frac{\partial^3(p/T^4)}{\partial(\mu_i/T)^3} = \frac{\langle(\delta N_i)^3\rangle}{VT^3} = \frac{S_i\sigma_i^3}{VT^3} \\
\chi_4^i &= \frac{\partial^4(p/T^4)}{\partial(\mu_i/T)^4} = \frac{\langle(\delta N_i)^4\rangle - 3\langle(\delta N_i)^2\rangle^2}{VT^3} = \frac{\kappa_i\sigma_i^4}{VT^3}.
\end{aligned}
\tag{5.21}
$$

To eliminate the volume and temperature factors, one usually defines ratios as follows:

$$
\frac{\chi_1}{\chi_2} = \frac{M}{\sigma^2}; \quad \frac{\chi_3}{\chi_2} = S\sigma; \quad \frac{\chi_3}{\chi_1} = \frac{S\sigma^3}{M}; \quad \frac{\chi_4}{\chi_2} = \kappa\sigma^2. \tag{5.22}
$$

5.3 Chemical Freeze-Out Parameters

5.3.1 Canonical vs Grand Canonical Ensemble

The time-evolution of a heavy-ion collision is shown in Fig. 5.2. After the initial pre-equilibration phase, the system is in the deconfined, quark-gluon plasma phase until hadronization, where the phase transition to confined, hadronic matter occurs.

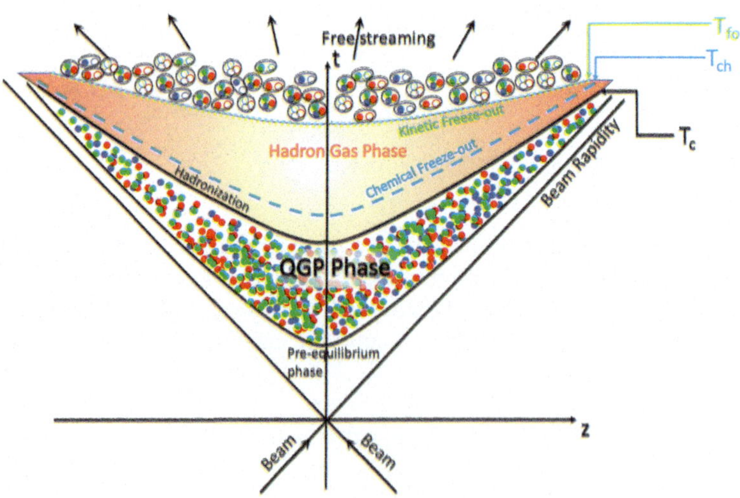

Fig. 5.2 Cartoon of the time evolution of a heavy-ion collision

When hadrons are formed, they continue to interact with each other, initially through inelastic scattering. The chemical freeze-out is the moment in the evolution of a heavy-ion collision that takes place after hadronization, when all inelastic collisions between particle species cease. All particle multiplicities and the higher order fluctuations of the multiplicity distributions are related to the thermal conditions at the chemical freeze-out. As the system keeps expanding and cooling down, it reaches the kinetic freeze-out, at which also all elastic collisions between hadrons cease. Afterwards, the hadrons that are stable for strong decays free stream to the detector.

We have seen that, in lattice QCD, one can calculate the cumulants of the QCD conserved charges, namely electric charge Q, baryon number B and strangeness S. The purpose is to compare the lattice QCD results, which depend on T and μ_B, to the cumulants of the distributions of conserved charges measured in experiments, which carry information on the chemical freeze-out. As we will see, this allows one to obtain the chemical freeze-out temperature and chemical potential from a comparison between first principle calculations and experimental measurements.

In order to perform a meaningful comparison, many factors need to be taken into account, as we will see. However, first we need to address the meaning of a distribution of conserved charges measured in experiment. As the name suggests, these charges are conserved by strong interactions, meaning that the same amount of net-baryons, net-electrically charged particles and net-strange particles that is present in the colliding nuclei should be measured by the detector at the end of the system evolution.

This is strictly true only if we look at the entire system. However, by studying a sufficiently small sub-system, fluctuations become meaningful. The small sub-system may exchange quanta with the rest of the system. This is similar to the assumptions which govern a thermal system in the Grand Canonical ensemble.

Consider Fig. 5.3 in the context of a heavy-ion collision. The total system corresponds to all particles distributed in rapidity Y over a range Y_{total}. The shaded area is the sub-system of particles within the accepted rapidity window Y_{accept}.

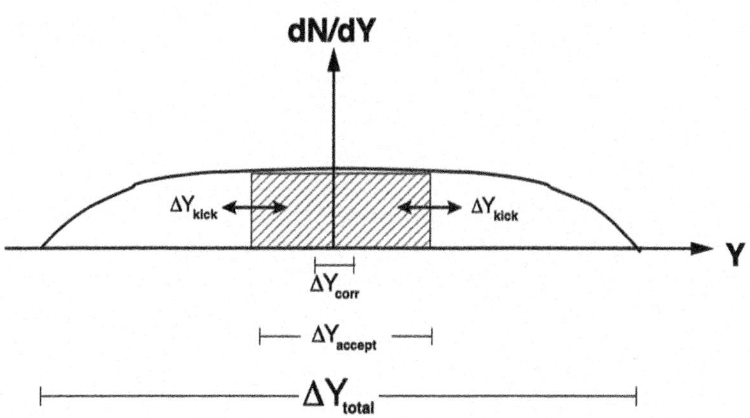

Fig. 5.3 From Ref. [1]. Cartoon of the various rapidity scales relevant for charge fluctuations

Fluctuations of conserved charges can be considered as meaningful observables if the following separation of scales occurs:

- $\Delta Y_{accept} \gg \Delta Y_{corr}$
- $\Delta Y_{total} \gg \Delta Y_{accept} \gg \Delta Y_{kick}$

where ΔY_{total} is the range for the total charge multiplicity distribution, ΔY_{accept} is the rapidity interval for the accepted charged particles, ΔY_{corr} is the charge correlation length characteristic to the physics of interest and ΔY_{kick} is the typical rapidity shift that charges receive during and after hadronization. This criterion ensures that total charge conservation does not suppress the signal and that the signal survives hadronization and the hadronic phase. For more details see Ref. [1].

5.3.2 Lattice QCD Observables

As we have seen, on the lattice we can calculate the fluctuations of conserved charges χ_{ijk}^{lmn}. Their ratios are related to the experimental cumulants of net-charge distributions according to Eqs. (5.22). The lattice QCD results are surfaces of these ratios as functions of T and μ_B: by comparing the lattice curves to two experimental points we can obtain, for each collision energy, the temperature T_{ch} and the chemical potential μ_{Bch} at the chemical freeze-out. For example, we can consider the following ratios of electric charge fluctuations, and the first terms of their Taylor expansion around $\mu_B = 0$:

$$\begin{aligned}
R_{31}^{Q}(T,\mu_B) &= \frac{\chi_3^Q(T,\mu_B)}{\chi_1^Q(T,\mu_B)} \\
&= \frac{\chi_{31}^{QB}(T,0)+\chi_4^Q(T,0)q_1(T)+\chi_{31}^{QS}(T,0)s_1(T)}{\chi_{11}^{QB}(T,0)+\chi_2^Q(T,0)q_1(T)+\chi_{11}^{QS}(T,0)s_1(T)}+O(\mu_B^2), \\
R_{12}^{Q}(T,\mu_B) &= \frac{\chi_1^Q(T,\mu_B)}{\chi_2^Q(T,\mu_B)} \\
&= \frac{\chi_{11}^{QB}(T,0)+\chi_2^Q(T,0)q_1(T)+\chi_{11}^{QS}(T,0)s_1(T)}{\chi_2^Q(T,0)}\frac{\mu_B}{T}+O(\mu_B^3),
\end{aligned} \tag{5.23}$$

where $q_1(T)=(\partial(\mu_Q/T)/\partial(\mu_B/T))|_{\mu_B=0}$ and $s_1(T)=(\partial(\mu_S/T)/\partial(\mu_B/T))|_{\mu_B=0}$. The leading order in R_{31}^Q is independent of μ_B; for this reason, this observable can be used to extract the freeze-out temperature. Once T_{ch} has been determined, one can use R_{12}^Q to determine μ_B. If we look at the fluctuations of baryon number we have

$$
\begin{aligned}
R_{31}^{B}(T,\mu_B) &= \frac{\chi_3^B(T,\mu_B)}{\chi_1^B(T,\mu_B)} \\
&= \frac{\chi_4^B(T,0)+\chi_{31}^{BQ}(T,0)q_1(T)+\chi_{31}^{BS}(T,0)s_1(T)}{\chi_2^B(T,0)+\chi_{11}^{BQ}(T,0)q_1(T)+\chi_{11}^{BS}(T,0)s_1(t)} + O(\mu_B^2), \\
R_{12}^{B}(T,\mu_B) &= \frac{\chi_1^B(T,\mu_B)}{\chi_2^B(T,\mu_B)} \\
&= \frac{\chi_2^B(T,0)+\chi_{11}^{BQ}(T,0)q_1(T)+\chi_{11}^{BS}(T,0)s_1(T)}{\chi_2^B(T,0)}\frac{\mu_B}{T} + O(\mu_B^3).
\end{aligned}
\tag{5.24}
$$

Similarly to the electric charge fluctuations, also R_{31}^B allows to extract T_{ch} and from R_{12}^B we can obtain μ_{Bch}. In this chapter we show lattice QCD results for various diagonal and off-diagonal fluctuations in Sect. 5.5. Several comparison between lattice QCD predictions and experimental data have been performed in the literature. We will discuss this in detail in Chap. 7.

5.4 Critical Fluctuations

5.4.1 Model Predictions for a Critical Point in the QCD Phase Diagram

Potential critical fluctuations, i.e. fluctuations of net-particle distributions beyond the expected Poissonian distribution, were first explained by strong correlations of pions and protons to the σ field near a critical point [2,3]. In QCD with two massless quarks, chiral symmetry is restored at finite temperature. This phase transition is likely second order and belongs to the O(4) universality class in three dimensional models (see Sect. 2.5). QCD with physical quark masses shows instead a smooth crossover transition at $\mu_B = 0$ [4].

Therefore, any tri-critical point at finite μ_B in the massless calculation turns into a critical end-point in QCD with physical quark masses. It can also be argued that the end-point is in the universality class of the Ising model in three dimensions, as discussed in Sect. 4.5.3. This is due to the fact that, for finite quark masses, no symmetry remains that would require the order parameter to have more than just one component. The field theory which describes the static critical behavior, the one-component ϕ^4 theory in three dimensions, has the critical exponents of the 3D-Ising model [5]. Using universality, one can predict the exponents for divergent susceptibilities at the end-point for any particle that couples strongly to the σ field.

The σ-meson is the most abundant particle at freeze-out, because it is (nearly) massless and so the equilibrium occupation number of the long-wavelength modes is large. Because pions are still massive at the critical endpoint, the σ mesons cannot immediately decay into $\pi\pi$. Instead, they persist as the density of the system further decreases.

For example, for particles like pions and protons that interact with the critical mode, the fluctuations in the number of particles in a given acceptance window will increase near the critical point as the critical mode becomes massless and develops large long-wavelength correlations. Protons are of special importance because their fluctuations are proxy to the fluctuations of the conserved baryon number and because their coupling to the critical mode is relatively large. The effects of long wavelength fluctuations can be quantified by measuring cumulants of event-by-event particle multiplicities. A proposed signature is their non-monotonic behavior as a function of the collision energy [6]. The contribution of critical fluctuations to these cumulants is proportional to some positive power of the correlation length ξ which, in the idealized thermodynamic limit, diverges at the critical point [7]:

$$\kappa_2 = \frac{T}{V}\xi^2; \qquad \kappa_3 = \frac{2\lambda_3 T}{V}\xi^6; \qquad \kappa_4 = \frac{6T}{V}\left[2\left(\lambda_3\xi\right)^2 - \lambda_4\right]\xi^8, \quad (5.25)$$

where λ_3 and λ_4 are coupling constants of the σ^3 and σ^4 terms in the effective action for the σ field, respectively. In reality, the correlation reaches only a maximum value due to finite time and size considerations in the vicinity of the critical point [8]. Higher moments depend on higher powers of the correlation length, which makes them more sensitive to the critical point.

A specific prediction regarding the kurtosis measurement of net-baryon fluctuations, based on the linear σ model, was presented in [6]. The author points out that the kurtosis of the order parameter is universally negative when the critical point is approached on the crossover side of the phase separation line. As a consequence, the kurtosis of the net-proton multiplicity may become smaller than the value given by independent Poisson statistics, before exhibiting a strong rise at the critical point itself. This leads to a unique shape of the kurtosis measurement as a function of beam energy, which is a reliable measure of the baryo-chemical potential. This is shown in the top and middle panels of Fig. 5.4. In particular, the central panel shows the expected qualitative behavior of the kurtosis as a function of the Ising model variable t, which is proportional to $\sqrt{s}$, along the hypothetical freeze-out curve sketched in green in the top and bottom panels. It was shown in Ref. [9] that this behavior is diminished when calculations take into account sub-leading contributions to the kurtosis beyond leading order.

Ideally, this type of critical fluctuations of particle multiplicities that can serve as a proxy for a conserved quantum number should also be reflected in the susceptibilities of the same quantum number on the lattice as a function of the baryo-chemical potential. Early studies [10, 11] hinted at potential evidence for a critical point, which was in line with numerous predictions from effective chiral models [12–23].

These early calculations had the purpose of illustrating the methods used to find a critical point in lattice simulations (see Sect. 4.4.1) but were performed on coarse lattices with unphysical parameter values. For this reason, the location of

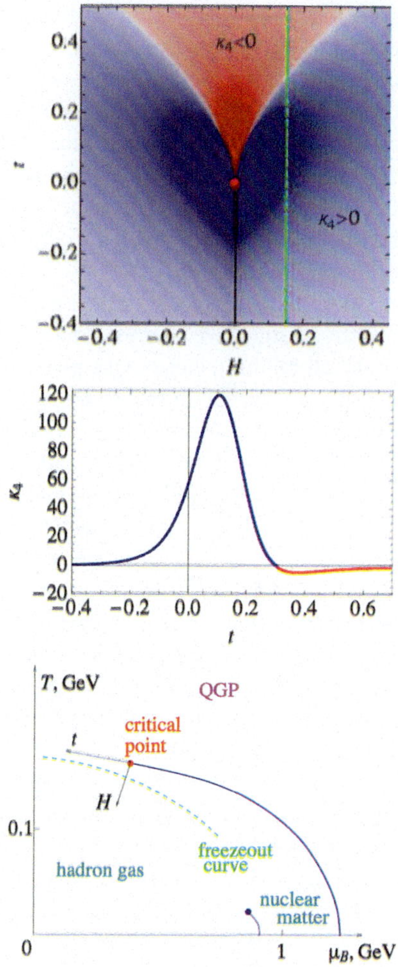

Fig. 5.4 From Ref. [6]. Top: the density plot of the function κ_4 in the Ising model variables. Middle: the expected behavior of the kurtosis as a function of t. Bottom: a sketch of the phase diagram of QCD with the freeze-out curve and a possible mapping of the Ising coordinates t and H

the corresponding critical points cannot be considered realistic. As of now, there has been no viable calculation on the lattice that predicts a critical point in the density regime that can be explored with the latest calculations (up to $\mu_B/T = 2$). Nevertheless the experimental program at RHIC and FAIR, which presently reaches out to μ_B/T = 4, covers a range that includes numerous predictions of effective model calculations [24–28] and even AdS/CFT applications, which perfectly link to non-critical lattice predictions in the low density regime [29].

5.4.2 Model Predictions on Chiral Criticality

Since any critical point on the QCD phase diagram will have to be at finite baryochemical potential and, based on the latest lattice results, above μ_B/T = 2, the question arises whether any fluctuation measurements at the LHC near $\mu_B = 0$ can contribute to the exploration of transition features in QCD. O(4) scaling functions enable us to study the generic structure of higher cumulants at vanishing baryon chemical potential. Under these conditions, the chiral transition is a true second order phase transition in the limit of vanishing light quark masses. For μ_B near zero, however, signatures of chiral criticality may exist also for physical values of the quark masses. Critical behavior is signaled by long-range correlations and increased fluctuations, owing to the appearance of massless modes near the second order phase transition.

As in the case of the critical point searches, fluctuations of baryon number and electric charge have been shown to be sensitive indicators for such critical behavior. In the chiral limit, close to the phase transition line $T_c(\mu_B, m_q = 0)$, fluctuations of e.g. the net-baryon number density are expected to reflect the universal properties of the three dimensional, O(4) symmetric spin model. The O(4) scaling relations for cumulants of net baryon number fluctuations will differ significantly from predictions based on statistical models. In comparison to the critical point searches at higher μ_B, chiral models, such as the Nambu-Jona-Lasinio (NJL) model, showed early on that the critical mode in this high-T, low-μ_B part of the phase diagram requires even higher moments of the net-baryon number distribution in order to measure significant non-statistical fluctuations. In particular, the ratios of the sixth/second and eighth/second cumulants of the net baryon number fluctuations feature a rapid sign change and rise near the transition temperature. These deviations from typical statistical fluctuations become even more pronounced at finite density towards a critical point.

Recent high precision lattice QCD calculations reaching out to the sixth and eighth order cumulants of baryon number fluctuations, performed in the transition region at vanishing baryon chemical potential and physical quark masses, confirm the chiral model predictions and do indeed show that these cumulants reflect the basic features expected close to the chiral phase transition. These results will be discussed in the next section. This suggests that, at physical values of the light and strange quark masses, these cumulants are sensitive to the critical dynamics in the chiral limit and may be employed to characterize also the crossover transition in strongly interacting matter.

Thus, also at vanishing baryon chemical potential, i.e. under the conditions approximately realized in the highest energy collisions at RHIC or LHC, the question arises to what extent a refined analysis of event-by-event particle multiplicity fluctuations can establish the existence of the chiral phase transition.

5.5 Results from Lattice QCD

First principle results for fluctuations of conserved charges have steadily become available over the last few years. Second order diagonal susceptibilities of conserved charges can be evaluated on the lattice as linear combinations of quark number susceptibilities (I is the isospin):

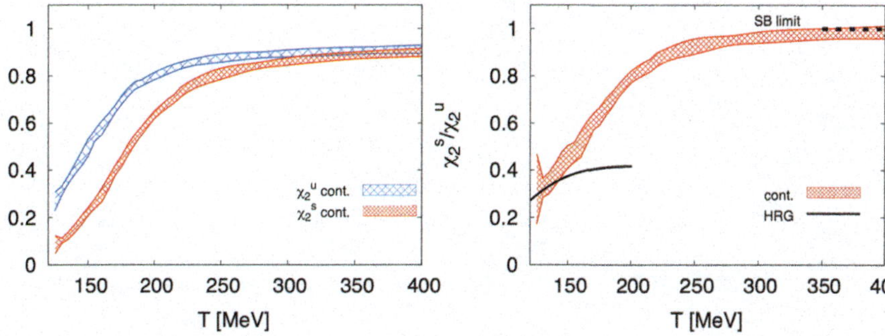

Fig. 5.5 From Ref. [30]. Left: comparison between the continuum limit of light and strange quark susceptibilities. Right: ratio χ_2^s/χ_2^u as a function of the temperature. The black, solid curve is the HRG model prediction. The dashed line indicates the ideal gas limit

$$
\begin{aligned}
\chi_2^B &= \frac{1}{9}\left[\chi_2^u + \chi_2^d + \chi_2^s + 2\chi_{11}^{us} + 2\chi_{11}^{ds} + 2\chi_{11}^{ud}\right] \\
\chi_2^Q &= \frac{1}{9}\left[4\chi_2^u + \chi_2^d + \chi_2^s - 4\chi_{11}^{us} + 2\chi_{11}^{ds} - 4\chi_{11}^{ud}\right] \\
\chi_2^I &= \frac{1}{4}\left[\chi_2^u + \chi_2^d - 2\chi_{11}^{ud}\right] \\
\chi_2^S &= \chi_2^s.
\end{aligned}
\tag{5.26}
$$

The first continuum extrapolated results for these quantities have been published in Ref. [30]. Figure 5.5 shows a comparison between the second order fluctuation for u and s quarks (left panel) and their ratio (right panel) as functions of the temperature. It is evident from the figure that, while the functional form of second-order fluctuations as functions of the temperature is similar, the two curves are shifted in temperature by about 15 MeV. This observation has triggered some discussion in the community about the possibility of a flavor hierarchy in the QCD transition, namely different flavors hadronizing at different temperatures (see the discussion in Sect. 6.3.2). The left panel of Fig. 5.6 shows the χ_{11}^{us} correlator between the u and s quark flavors, while the right panel shows a comparison between all different second order diagonal fluctuations, rescaled by their respective ideal gas limit. In Ref. [31], second-order diagonal and non-diagonal BQS correlators were calculated in the continuum limit. The off-diagonal correlators are shown in the three top panels of Fig. 5.7. More recently, these off-diagonal correlators were calculated in Ref. [32], which presents a detailed comparison to the HRG model to find the measured contribution of different hadrons and a connection to experiment (see the discussion in Sect. 6.3.4). These are shown in the three bottom panels of Fig. 5.7.

Continuum extrapolated results for the ratio of fourth-to-second order strangeness and light quark fluctuations at $\mu_B = 0$ were presented for the first time

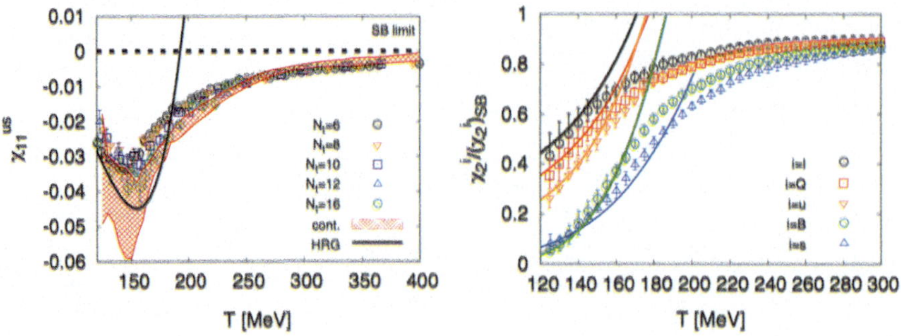

Fig. 5.6 From Ref. [30]. Left: non-diagonal $u-s$ correlator as a function of the temperature. The red band is the continuum extrapolation, the black curve is the HRG model prediction. The dashed line indicates the ideal gas limit. Right: comparison between all diagonal susceptibilities, rescaled by the corresponding ideal gas limit, as functions of the temperature

in Ref. [33], where it was pointed out that the shift in temperature between light and strange quarks persists in higher order fluctuations. These ratios are monotonically increasing functions of the temperature in a hadronic system, whereas the peak observed in the lattice results (shown in the left panel of Fig. 5.8) indicates a change in the degrees of freedom and the onset of quark liberation.

In Ref. [34], continuum extrapolated results for χ_3^Q/χ_1^Q (right panel of Fig. 5.8), χ_3^B/χ_1^B (left panel of Fig. 5.9) and χ_4^B/χ_2^B (right panel of Fig. 5.9) as functions of the temperature were published at $\mu_B = 0$. The latter is a very interesting quantity, as it is supposed to be identically equal to one in the non-interacting Hadron Resonance Gas model (see the discussion in Sect. 6.2). Any deviation from one, therefore, can be interpreted as a signal of change of degrees of freedom, or interactions that go beyond those captured by the non-interacting HRG model. As it is evident from the right panel of Fig. 5.9, such a deviation happens for $T \geq 150$ MeV.

Results for higher order fluctuations are also becoming available. We showed the results for diagonal higher order baryon number fluctuations from Ref. [35] in Fig. 4.6; they were obtained at finite lattice space, for $N_t = 12$. Similar results were recently published at $N_t = 8$ in Ref. [36]. They are shown in Fig. 5.10. Results for higher order fluctuations at finite lattice spacing were also presented in Ref. [37]. All non-diagonal BQS correlators up to fourth order were published in Ref. [35] for $N_t = 12$ and $\mu_B = 0$.

In order to compare to the experimental results at RHIC, several diagonal and off-diagonal fluctuations were extrapolated to finite chemical potential. Results for χ_1^B/χ_2^B as functions of the chemical potential for different values of the temperature were obtained in Ref. [40]. They are shown in Fig. 5.11; for a comparison of these results with the experimental data from the STAR collaboration, see Chap. 7.

Due to their divergence at the critical point, higher order baryon number fluctuations are particularly interesting and a lot of activity has been devoted to

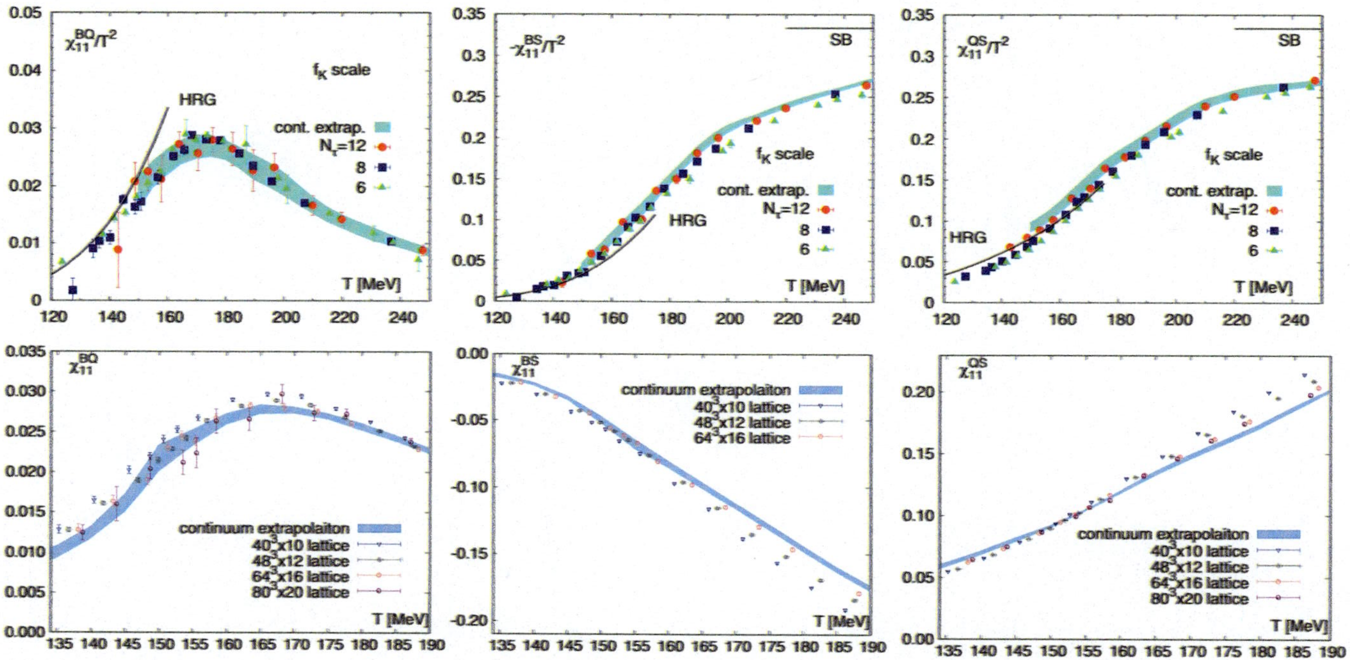

Fig. 5.7 Top: from Ref. [31]. Off-diagonal second-order $B-Q$ (left), $-(B-S)$ (center) and $Q-S$ (right) correlators as functions of the temperature. The cyan bands show the continuum extrapolated results, the black curve at low-T is the HRG model prediction and the one at high-T is the Stefan-Boltzmann limit. Bottom: from Ref. [32]. Off-diagonal $B-Q$ (left), $B-S$ (center) and $Q-S$ (right) correlators as functions of the temperature. The different points show the different N_t results, while the light blue band is the continuum extrapolation

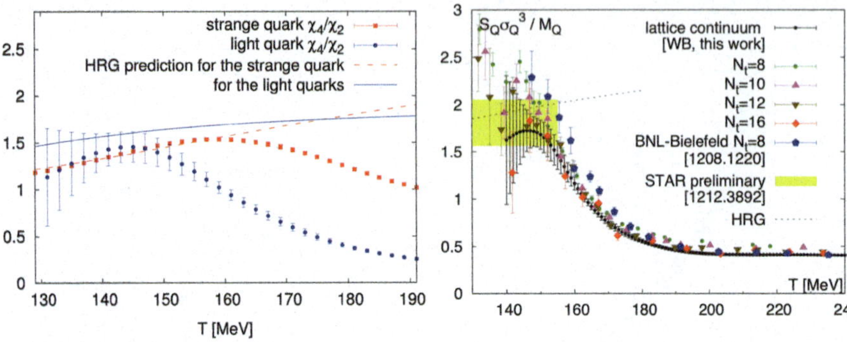

Fig. 5.8 Left: from Ref. [33]. Ratio of fourth-to-second order fluctuations of strange (red) vs light (blue) flavors as functions of the temperature. The dots show the lattice QCD results, while the lines are the corresponding HRG model curves. Right: from Ref. [34]. Ratio of third-to-first order fluctuations of electric charge. The black dots are the continuum extrapolated lattice QCD results from Ref. [34], the blue points are the $N_t = 8$ lattice QCD results from Ref. [38], the dashed line is the HRG model result, the yellow band is the experimental value for this observable from the STAR collaboration [39]

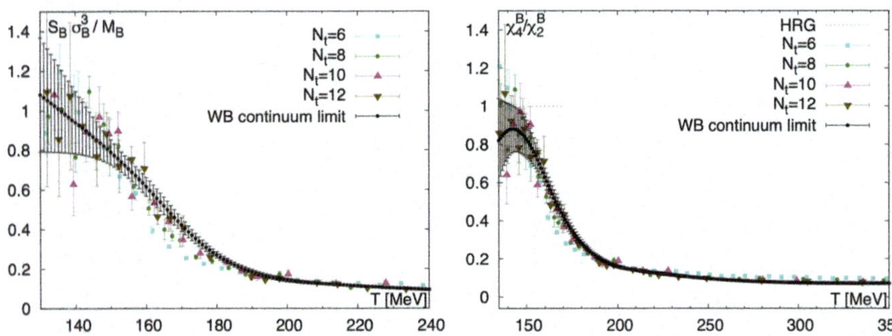

Fig. 5.9 From Ref. [34]. Left: third-to-first order cumulant of baryon number as a function of the temperature at $\mu_B = 0$. Right: fourth-to-second order cumulant of baryon number as a function of the temperature at $\mu_B = 0$. In both panels, the black points show the continuum extrapolated result. The dashed line on the right plot is the HRG model prediction

their extrapolation to finite density and their comparison to experimental results. χ_3^B/χ_1^B and χ_4^B/χ_2^B were extrapolated to finite μ_B in Ref. [35]. These results will be discussed in Chap. 7 in comparison to the experimental data from the STAR collaboration.

The most recent results for higher order fluctuations at finite T and μ_B were published in Ref. [36]. Figure 5.12 shows χ_3^B/χ_1^B (left) and χ_4^B/χ_2^B as functions of μ_B/T for $T = 155$ MeV and $T = 158$ MeV. These values were chosen because they are the lower and upper bound on the transition temperature value quoted in Ref. [41]. The bands do not correspond to an actual continuum extrapolation, but to a continuum estimate based on $N_t = 8$ and $N_t = 12$ lattices.

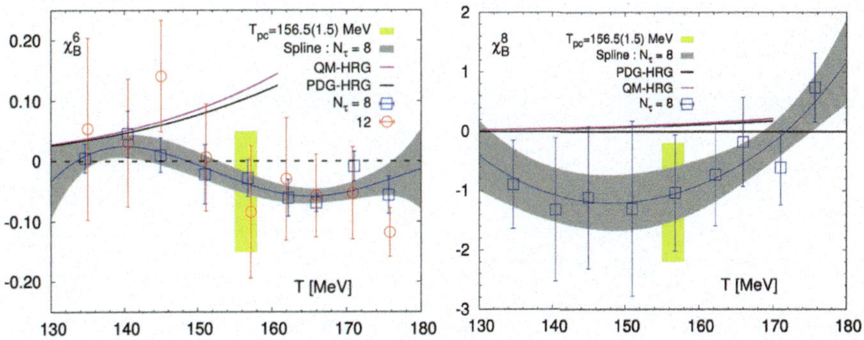

Fig. 5.10 From Ref. [36]. Sixth-order (left) and eighth-order (right) baryon number fluctuations as functions of the temperature. The gray band indicates a spline interpolation of the data obtained on the $32^3 \times 8$ lattices

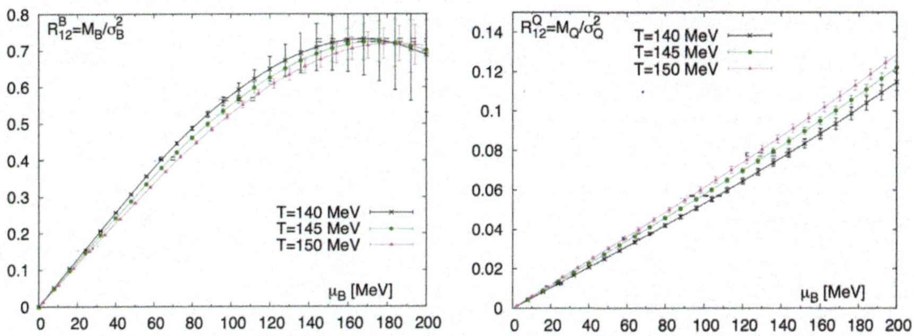

Fig. 5.11 From Ref. [40]. χ_1^B/χ_2^B (left) and χ_1^Q/χ_2^Q (right) as functions of the chemical potential for different values of the temperature

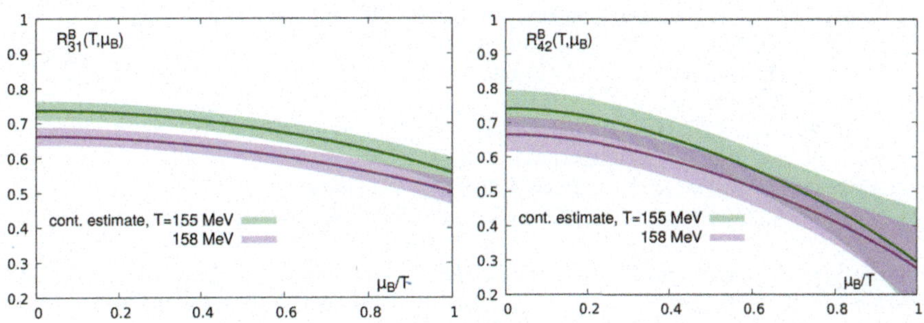

Fig. 5.12 From Ref. [36]. χ_3^B/χ_1^B (left) and χ_4^B/χ_2^B (right) extrapolated to finite chemical potential for two values of the temperature. The bands show a continuum estimate for these quantities, based on $N_t = 8$ and $N_t = 12$ lattices

The ratio of fifth-to-first order and sixth-to-second order fluctuations of baryon number were also published in Ref. [36] at $N_t = 8$ and extrapolated to finite μ_B. They are shown in the left and right panels of Fig. 5.13, respectively.

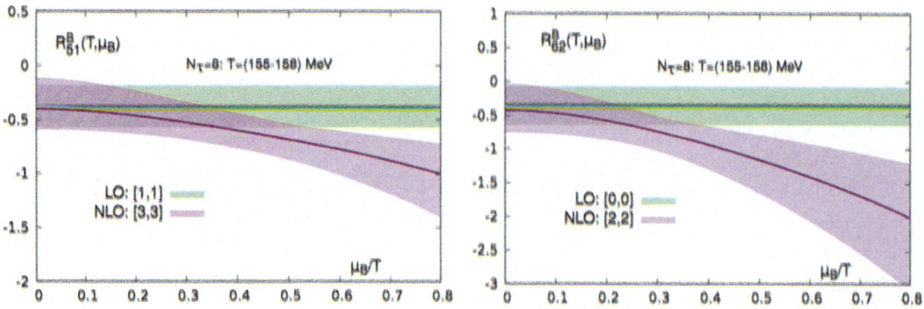

Fig. 5.13 From Ref. [36]. χ_5^B/χ_1^B (left) and χ_6^B/χ_2^B (right) extrapolated to finite chemical potential for two different extrapolation orders. The bands show the result obtained on $N_t = 8$ lattices

It is worth mentioning that the high-temperature behavior of diagonal and off-diagonal fluctuations was studied in Refs. [42, 43]. It was pointed out that these observables approach the ideal gas limit much faster than the global thermodynamic observables discussed in Chap. 3. Besides, the analyzed observables agree with perturbation theory predictions for $T \geq 250\,\text{MeV}$ [44, 45].

References

1. Koch, V.: [arXiv:0810.2520 [nucl-th]]
2. Stephanov, M.A., Rajagopal, K., Shuryak, E.V.: Phys. Rev. Lett. **81**, 4816–4819 (1998)
3. Athanasiou, C., Rajagopal, K., Stephanov, M.: Phys. Rev. D **82**, 074008 (2010)
4. Aoki, Y., Endrodi, G., Fodor, Z., Katz, S.D., Szabo, K.K.: Nature **443**, 675–678 (2006)
5. Stephanov, M.A.: Prog. Theor. Phys. Suppl. **153**, 139–156 (2004)
6. Stephanov, M.A.: Phys. Rev. Lett. **107**, 052301 (2011)
7. Stephanov, M.A.: Phys. Rev. Lett. **102**, 032301 (2009)
8. Berdnikov, B., Rajagopal, K.: Phys. Rev. D **61** (2000), 105017
9. Mroczek, D., Noronha-Hostler J., Acuna A.R.N., Ratti, C., Parotto, P., Stephanov, M.A.: [arXiv:2008.04022 [nucl-th]]
10. Fodor, Z., Katz, S.D.: JHEP **04**, 050 (2004)
11. Gavai, R.V., Gupta, S.: Phys. Rev. D **71**, 114014 (2005)
12. Asakawa, M., Yazaki, K.: Nucl. Phys. A **504**, 668–684 (1989)
13. Barducci, A., Casalbuoni, R., De Curtis S., Gatto, R., Pettini, G.: Phys. Lett. B **231**, 463–470 (1989)
14. Barducci, A., Casalbuoni, R., De Curtis S., Gatto, R., Pettini, G.: Phys. Rev. D **41**, 1610 (1990)
15. Barducci, A., Casalbuoni, R., Pettini, G., Gatto, R.: Phys. Rev. D **49**, 426–436 (1994)
16. Berges, J., Rajagopal, K.: Nucl. Phys. B **538**, 215–232 (1999)
17. Halasz, A.M., Jackson, A.D., Shrock, R.E., Stephanov, M.A., Verbaarschot, J.J.M.: Phys. Rev. D **58**, 096007 (1998)
18. Scavenius, O., Mocsy, A., Mishustin, I.N., Rischke, D.H.: Phys. Rev. C **64**, 045202 (2001)
19. Antoniou, N.G., Kapoyannis, A.S.: Phys. Lett. B **563**, 165–172 (2003)
20. Hatta, Y., Ikeda, T.: Phys. Rev. D **67**, 014028 (2003)
21. Roessner, S., Ratti, C., Weise, W.: Phys. Rev. D **75**, 034007 (2007)
22. Roessner, S., Hell, T., Ratti, C., Weise, W.: Nucl. Phys. A **814**, 118–143 (2008)
23. Sasaki, C., Friman, B., Redlich, K.: Phys. Rev. D **75**, 074013 (2007)

24. Fukushima, K.: Phys. Rev. D **78**, 114019 (2008)
25. Zhang, Z., Fukushima, K., Kunihiro, T.: Phys. Rev. D **79**, 014004 (2009)
26. Fan, W., Luo, X., Zong, H.: Chin. Phys. C **43**(3), 033103 (2019)
27. Herbst, T.K., Pawlowski, J.M., Schaefer, B.J.: Phys. Rev. D **88**(1), 014007 (2013)
28. Eichmann, G., Fischer, C.S., Welzbacher, C.A.: Phys. Rev. D **93**(3), 034013 (2016)
29. Critelli, R., Noronha, J., Noronha-Hostler, J., Portillo, I., Ratti, C., Rougemont, R.: Phys. Rev. D **96**(9), 096026 (2017)
30. Borsanyi, S., Fodor, Z., Katz, S.D., Krieg, S., Ratti, C., Szabo, K.: JHEP **01**, 138 (2012)
31. Bazavov, A., et al.: [HotQCD] Phys. Rev. D **86**, 034509 (2012)
32. Bellwied, R., Borsanyi, S., Fodor, Z., Guenther, J.N., Noronha-Hostler, J., Parotto, P., Pasztor, A., Ratti, C., Stafford, J.M.: Phys. Rev. D **101**(3), 034506 (2020)
33. Bellwied, R., Borsanyi, S., Fodor, Z., Katz, S.D., Ratti, C.: Phys. Rev. Lett. **111**, 202302 (2013)
34. Borsanyi, S., Fodor, Z., Katz, S.D., Krieg, S., Ratti, C., Szabo, K.K.: Phys. Rev. Lett. **111**, 062005 (2013)
35. Borsanyi, S., Fodor, Z., Guenther, J.N., Katz, S.K., Szabo, K.K., Pasztor, A., Portillo, I., Ratti, C.: JHEP **10**, 205 (2018)
36. Bazavov, A., Bollweg, D., Ding, H.T., Enns, P., Goswami, J., Hegde, P., Kaczmarek, O., Karsch, F., Larsen, R., Mukherjee, S., Ohno, H., Petreczky, P., Schmidt, C., Sharma, S., Steinbrecher, P.: Phys. Rev. D **101**(7), 074502 (2020)
37. D'Elia, M., Gagliardi, G., Sanfilippo, F.: Phys. Rev. D **95**(9), 094503 (2017)
38. Bazavov, A., Ding, H.T., Hegde, P., Kaczmarek, O., Karsch, F., Laermann, E., Mukherjee, S., Petreczky, P., Schmidt, C., Smith, D., Soeldner, W., Wagner, M.: Phys. Rev. Lett. **109**, 192302 (2012)
39. Sahoo, N.R.: [STAR] Acta Phys. Polon. Supp. **6**, 437–444 (2013)
40. Borsanyi, S., Fodor, Z., Katz, S.D., Krieg, S., Ratti, C., Szabo, K.K.: Phys. Rev. Lett. **113**, 052301 (2014)
41. Bazavov, A., et al.: [HotQCD] Phys. Lett. B **795**, 15–21 (2019)
42. Bellwied, R., Borsanyi, S., Fodor, Z., Katz, S.D., Pasztor, A., Ratti, C., Szabo, K.K.: Phys. Rev. D **92**(11), 114505 (2015)
43. Ding, H.T., Mukherjee, S., Ohno, H., Petreczky, P., Schadler, H.P.: Phys. Rev. D **92**(7), 074043 (2015)
44. Haque, N., Bandyopadhyay, A., Andersen, J.O., Mustafa, M.G., Strickland, M., Su, N.: JHEP **05**, 027 (2014)
45. Mogliacci, S., Andersen, J.O., Strickland, M., Su, N., Vuorinen, A.: JHEP **12**, 055 (2013)

The Hadron Resonance Gas Model

6

Abstract

This chapter is devoted to one of the most successful models for the description of strongly interacting matter in the hadronic phase: the Hadron Resonance Gas (HRG) model. We start by introducing the model in Hagedorn's original formulation, which includes an exponential hadron mass spectrum. We then present a more recent version of the model, in which the resonance spectrum is taken from the Particle Data Book. We discuss possible model extensions, which include excluded volume corrections and Van der Waals interaction. We then discuss applications of the model, ranging from comparison to lattice QCD predictions to thermal fits of experimental particle yields and fluctuations.

6.1 Introduction

In 1965, Rolf Hagedorn wrote a renowned paper [1] in which he had the idea that, if we consider a system of interacting hadrons in the ground state, the interactions among them can be well approximated by creating higher and higher resonances, that take part in the thermodynamics as if they were non-interacting particles [2]. By this time, after the construction of particle accelerators capable of higher and higher collision energy, the number of known particles was growing rapidly from a few dozens to hundreds.

The partition function that Hagedorn formulated was of the form

$$\ln Z(V,T) = \frac{V}{(2\pi)^3} 4\pi \int_0^\infty p^2 dp \left\{ \int_0^\infty dm \left[\rho_F(m) \ln \left(1 + e^{-\sqrt{\mathbf{p}^2+m^2}T} \right) + \right.\right.$$
$$\left.\left. - \; \rho_B(m) \ln \left(1 - e^{-\sqrt{\mathbf{p}^2+m^2}T} \right) \right] \right\}, \qquad (6.1)$$

C. Ratti, R. Bellwied, *The Deconfinement Transition of QCD*, Lecture Notes in Physics 981, https://doi.org/10.1007/978-3-030-67235-5_6

where $\rho_F(m)$ and $\rho_B(m)$ are the fermionic (baryonic) and bosonic (mesonic) mass spectra, respectively: $\rho_F(m)dm$ yields the number of excited baryons with mass between m and $m + dm$. Hagedorn obtained the following bootstrap equation for ρ:

$$\rho(m, V_0) = \delta(m - m_0) + \sum_N \frac{1}{N!} \left[\frac{V_0}{(2\pi)^3} \right]^{N-1}$$

$$\int \prod_{i=1}^{N} \left(dm_i \rho(m_i) d^3 p_i \right) \delta^4 \left(\sum_i p_i - p \right). \tag{6.2}$$

This formula shows that the heavier particles arise from interactions between the lighter ones. The first state is the lightest possible state (pion), and V_0 is the size of the overall system. Analytical solutions of this equation exist, provided that certain assumptions are made. They all have an exponential behavior

$$\rho(m) \propto \frac{1}{m^3} \exp(m/T_H), \tag{6.3}$$

where T_H is the so-called Hagedorn temperature. We can re-write the formula for the partition function in the following way, for fermions and bosons respectively:

$$\begin{aligned}
\ln Z_F(V, T) &= \frac{V}{2\pi^2} \int_0^\infty \frac{dm}{m^3} \exp\left(\frac{m}{T_H}\right) \\
&\sum_N \int_0^\infty p^2 dp \left[\frac{(-1)^{N+1}}{N} \exp\left(-\frac{\sqrt{\mathbf{p}^2 + m^2}}{T} \right)^N \right] = \\
&= \frac{VT}{2\pi^2} \int_0^\infty \frac{dm}{m^3} \exp\left(\frac{m}{T_H}\right) \sum_N \frac{(-1)^{N+1}}{N^2} m^2 K_2(N\frac{m}{T}) \\
\ln Z_B(V, T) &= -\frac{V}{2\pi^2} \int_0^\infty \frac{dm}{m^3} \exp\left(\frac{m}{T_H}\right) \\
&\sum_N \frac{(-1)}{N} \int_0^\infty p^2 dp \left[\exp\left(-\frac{\sqrt{\mathbf{p}^2 + m^2}}{T} \right)^N \right] = \\
&= \frac{VT}{2\pi^2} \int_0^\infty \frac{dm}{m^3} \exp\left(\frac{m}{T_H}\right) \sum_N \frac{m^2}{N^2} K_2(N\frac{m}{T}),
\end{aligned} \tag{6.4}$$

where $K_2(z)$ is the modified Bessel function of the second kind. If we keep only the first term in the series, this corresponds to the so-called Boltzmann approximation. In that case, the partition function takes the form

$$\ln Z(T, V) \simeq \frac{VT}{2\pi^2} \int dm \frac{m^2}{m^3} \exp\left[\frac{m}{T_H}\right] K_2(\frac{m}{T}) \sim V \left[\frac{T}{2\pi}\right]^{3/2}$$
$$\int \frac{dm}{m^{3/2}} \exp\left[\frac{m}{T_H} - \frac{m}{T}\right], \tag{6.5}$$

a result which is divergent at $T > T_H$. Therefore, in the Hadron Resonance Gas model an increase of energy leads to the following consequences:

- there is a fixed temperature limit, $T \to T_H$;
- the momenta of the constituents do not continue to increase;
- more and more species of heavier particles appear.

Therefore, this is a new non-kinetic way of using the energy: to increase the number of species and their masses, not the momentum per particle. The temperature is a measure of the momentum of the constituents: if it cannot continue to increase, there has to be a limiting temperature for hadronic systems. Hagedorn's interpretation of T_H was that of a limiting temperature for hadronic matter. Cabibbo and Parisi had the intuition of considering T_H as the transition temperature to a system with different degrees of freedom, namely deconfined strongly interacting matter [3]: T_H signals the transition from hadronic matter to a Quark-Gluon Plasma. Hagedorn's first estimate of the limiting temperature was $T_H \sim 158$ MeV, which is very close to the actual QCD transition temperature discussed in Chap. 2.

The recent increase in numerical accuracy of lattice QCD calculations and their extrapolation to the continuum limit makes it possible to confront the fundamental results of QCD with Hagedorn's concepts, which provide a theoretical scenario for the thermodynamics of strongly interacting hadronic matter. In particular, the equation of state calculated on the lattice at vanishing and finite chemical potential, and restricted to the hadronic phase, can be directly compared to the one obtained from the partition function of the HRG model.

In recent times, many more particles have been discovered: one usually considers a discrete mass spectrum, which accounts for all experimentally known hadrons and resonances. In this case, the partition function has $\rho(m)$ replaced by the spin degeneracy factor of the ith hadron, with a summation taken over all known resonance species listed by the Particle Data Group (PDG). The latter indicates different confidence levels on the existence of individual resonances. The most well-established states are denoted by **** (four stars), whereas * (one star) indicates states with the least experimental confirmation. Therefore, we will have a sum, over all hadrons and resonances, of one-particle, ideal gas partition functions:

$$\ln Z^{HRG}(T, V, \mu) = \sum_{i \in PDG} \ln Z_i(T, V, \mu) =$$
$$= \frac{V g_i}{2\pi^2} \int_0^\infty \pm p^2 dp \ln\left[1 \pm \lambda_i(T, \mu) \exp(-\beta \epsilon_i)\right] \tag{6.6}$$

where:

- $+$ is for fermions and $-$ is for bosons;
- $\epsilon_i = \sqrt{\mathbf{p}^2 + m_i^2}$;
- g_i is the spin degeneracy factor;
- $\lambda_i(T, \boldsymbol{\mu}) = \exp\left[\frac{B_i\mu_B + S_i\mu_S + Q_i\mu_Q}{T}\right]$ is the particle fugacity.

If we expand the logarithm like we did before we obtain

$$\ln Z_i(T, V, \boldsymbol{\mu}) = \frac{V g_i}{2\pi^2} \sum_{N=1}^{\infty} \frac{(\pm 1)^{N+1}\lambda_i^N}{N} \int p^2 dp \left[\exp\left(-\beta\epsilon_i\right)\right]^N . \tag{6.7}$$

Performing the integral, we get

$$\ln Z_i(T, V, \boldsymbol{\mu}) = \frac{V g_i T}{2\pi^2} \sum_{N=1}^{\infty} \frac{(\pm 1)^{N+1}\lambda_i^N}{N^2} m_i^2 K_2\left(N\frac{m_i}{T}\right). \tag{6.8}$$

Also in this case, keeping only the first term in the sum corresponds to the Boltzmann approximation. The particle density can be obtained from the partition function:

$$n_i(T, \boldsymbol{\mu}) = \frac{\langle N_i \rangle}{V} = \frac{T g_i}{2\pi^2} \sum_{N=1}^{\infty} \frac{(\pm 1)^{N+1}\lambda_i^N}{N} m_i^2 K_2\left(N\frac{m_i}{T}\right). \tag{6.9}$$

6.2 Boltzmann Approximation

Let us look again at the expression for the HRG model partition function:

$$\begin{aligned}\ln Z(T, \mu_B, \mu_S, \mu_Q, V) = &\sum_{i \in mesons} \ln Z_{m_i}^B(T, \mu_S, \mu_Q, V) + \\ &+ \sum_{i \in baryons} \ln Z_{m_i}^F(T, \mu_B, \mu_S, \mu_Q, V).\end{aligned} \tag{6.10}$$

It is worth mentioning that the partition function for baryons contains a separate contribution from each particle and its anti-particle. Let us first consider the case for which $\mu_S = \mu_Q = 0$. The baryonic pressure is given by

$$\frac{p_B}{T} = \sum_{i \in baryons} \frac{g_i}{2\pi^2} \int_0^{\infty} p^2 dp \left\{ \ln\left[1 + \exp\left(\frac{\mu_B}{T}\right) \exp\left(-\frac{\sqrt{\mathbf{p}^2 + m^2}}{T}\right)\right] + \right.$$

$$+ \ln\left[1+\exp\left(-\frac{\mu_B}{T}\right)\exp\left(-\frac{\sqrt{\mathbf{p}^2+m^2}}{T}\right)\right]\Bigg\}. \tag{6.11}$$

Making use of the fugacity expansion that we have seen before, we obtain

$$\frac{p_B}{T^4} = \sum_{i\in baryons} \frac{g_i}{\pi^2}\left(\frac{m_i}{T}\right)^2 \sum_{N=1}^{\infty}(-1)^{N+1}N^{-2}K_2(N\frac{m_i}{T})\cosh\left[N\frac{\mu_B}{T}\right]. \tag{6.12}$$

We are interested in temperatures $T \leq 200\,\text{MeV}$. The mass of the lightest baryon, $m_N \simeq 1\,\text{GeV}$, is about five times larger than the temperature. The Bessel function can be approximated by its asymptotic form, valid for large arguments: $K_2(x) \sim \sqrt{\frac{\pi}{2x}}\exp(-x)$. This shows that higher order terms are suppressed by factors $\exp[-N(m-\mu_B)/T]$. As long as $(m_N - \mu_B) \geq T$, the contribution of baryons to the partition function is well approximated by the leading term, which constitutes the Boltzmann approximation. In this case we have:

$$\frac{p_B}{T^4} = \sum_{i\in baryons} \frac{g_i}{\pi^2}\left(\frac{m_i}{T}\right)^2 K_2\left(\frac{m_i}{T}\right)\cosh\left(\frac{\mu_B}{T}\right) = F(T)\cosh\left(\frac{\mu_B}{T}\right), \tag{6.13}$$

where we defined $F(T) = \sum_{i\in baryons}\frac{g_i}{\pi^2}\left(\frac{m_i}{T}\right)^2 K_2\left(\frac{m_i}{T}\right)$. This simple relation is independent of the details of the mass spectrum, as long as all fermions are sufficiently heavy. It shows that the entire chemical potential dependence of the baryonic pressure is contained in the factor $\cosh\left(\frac{\mu_B}{T}\right)$.

From the above derivation, we can now read the behavior of fluctuations in the HRG model:

$$\frac{\partial^n\left(p/T^4\right)}{\partial\left(\mu_B/T\right)^n} = F(T)\cosh\left(\frac{\mu_B}{T}\right) \quad \text{for } n \text{ even,}$$

$$\frac{\partial^n\left(p/T^4\right)}{\partial\left(\mu_B/T\right)^n} = F(T)\sinh\left(\frac{\mu_B}{T}\right) \quad \text{for } n \text{ odd.} \tag{6.14}$$

Therefore, baryon number fluctuations in the HRG model have a trivial dependence on μ_B. When we consider ratios of fluctuations, like the ones introduced in Chap. 5, we get

$$\frac{\chi_n^B}{\chi_m^B} = 1 \qquad \text{for } n, m \text{ even or odd}$$

$$\frac{\chi_n^B}{\chi_m^B} = \frac{\sinh\left(\mu_B/T\right)}{\cosh(\mu_B/T)} \qquad \text{for } n \text{ odd, } m \text{ even}$$

$$\frac{\chi_n^B}{\chi_m^B} = \frac{\cosh(\mu_B/T)}{\sinh(\mu_B/T)} \quad \text{for } n \text{ even, } m \text{ odd.}$$

This behavior is confirmed by lattice QCD, as we have seen in Chap. 5; in particular, the ratio of even μ_B-fluctuations is equal to one below T_c. This observable has been proposed as an onset of deconfinement, since above T_c it rapidly decreases and tends to its asymptotic limit, which is not 1/9 contrary to expectations, due to quantum statistics. More recently, it was pointed out that a value smaller than one for, e.g., χ_4^B/χ_2^B can be obtained even in the hadronic phase in the interactive version of the HRG model.

Things become slightly more complicated if we consider $\mu_S \neq 0$ (and eventually also $\mu_Q \neq 0$). Let us consider only strange hadrons. In this case, the pressure can be written as

$$\begin{aligned} p_S(T, \mu_B, \mu_S) = {} & p_{|S|=1,M} \cosh(\mu_S/T) + p_{|S|=1,B} \cosh(\frac{\mu_B - \mu_S}{T}) + \\ & + p_{|S|=2,B} \cosh(\frac{\mu_B - 2\mu_S}{T}) + p_{|S|=3,B} \cosh(\frac{\mu_B - 3\mu_S}{T}), \end{aligned} \tag{6.15}$$

where

$$\frac{p_{|S|=1,M}}{T^4} = \sum_{i \in \, mesons \; with \; |S|=1} \frac{g_i}{\pi^2} \left(\frac{m_i}{T}\right)^2 K_2\left(\frac{m_i}{T}\right). \tag{6.16}$$

Summing over strange and non-strange contributions, we obtain the total pressure:

$$\begin{aligned} p(T, \mu_B, \mu_S) = {} & p_{00}^{BS} + p_{10}^{BS} \cosh\left(\frac{\mu_B}{T}\right) + p_{01}^{BS} \cosh\left(\frac{\mu_S}{T}\right) + \\ & + p_{11}^{BS} \cosh\left(\frac{\mu_B - \mu_S}{T}\right) + p_{12}^{BS} \cosh\left(\frac{\mu_B - 2\mu_S}{T}\right) \\ & + p_{13}^{BS} \cosh\left(\frac{\mu_B - 3\mu_S}{T}\right). \end{aligned} \tag{6.17}$$

This formula has been used on the lattice to identify the contribution of hadrons to the pressure, divided in families according to their strangeness and baryon number content [4,5]. The lattice analysis has been performed by simulating at $\mu_B = \mu_Q = 0$ and $\Im(\mu_S) > 0$. In this case, by differentiating the above expression, we get

$$\Im(\chi_1^B) = -p_{11}^{BS} \sin\left(\frac{\mu_I}{T}\right) - p_{12}^{BS} \sin\left(\frac{2\mu_I}{T}\right) - p_{13}^{BS} \sin\left(\frac{3\mu_I}{T}\right)$$

$$\chi_2^B = p_{10}^{BS} + p_{11}^{BS} \cos\left(\frac{\mu_I}{T}\right) + p_{12}^{BS} \cos\left(\frac{2\mu_I}{T}\right) + p_{13}^{BS} \cos\left(\frac{3\mu_I}{T}\right)$$

$$\Im(\chi_1^S) = (p_{01}^{BS}+p_{11}^{BS})\sin\left(\frac{\mu_I}{T}\right)+2p_{12}^{BS}\sin\left(\frac{2\mu_I}{T}\right)+3p_{13}^{BS}\sin\left(\frac{3\mu_I}{T}\right) \quad (6.18)$$

with $\mu_S = i\mu_I$. The left-hand side has been simulated on the lattice at finite imaginary μ_S. The above functions have then been fitted to the lattice results. Through this procedure it was possible to extract the contributions of the different hadronic sectors to the QCD pressure. This will be discussed further in Sect. 6.3.3.

6.3 Comparison to Lattice QCD Results

6.3.1 Early Days: Distorted Hadronic Spectrum

In many of the figures presented in the previous chapters, a HRG model prediction curve was shown together with the lattice QCD results for the thermodynamic observables. The comparison between lattice QCD results and HRG model has been very successful over the years, and has been used to study several quantities. In the early days, when lattice QCD simulations were performed for heavier-than-physical quark masses and on coarse lattices, it was shown that one can still reproduce the thermodynamic quantities by means of the HRG model if one uses a resonance spectrum which is distorted by heavy quark masses and discretization effects in the same way as lattice QCD simulations [6–8]. HRG results with physical quark masses were then used as a guidance to understand how far the distorted lattice QCD results were from the physical ones. In Fig. 6.1 we show an example of these early-day comparisons: the left panel shows the quark number susceptibility and the right panel shows $(\Delta p)/T^4 = (p(T, \mu_q/T) - p(T, \mu_q/T = 0))/T^4$ as functions of the temperature for $\mu_q/T = 1$. The dots correspond to lattice QCD results evaluated at heavy quark masses corresponding to $m_\pi = 770$ MeV. The dashed and dotted curves are the HRG model predictions, obtained with a resonance spectrum adequate to the heavy quark masses used in the lattice simulations. The full lines are HRG model results obtained with the physical spectrum, and for three values of the transition temperature. More recently, systematic studies of the effect of heavier-than-physical quark masses on the equation of state and fluctuations of conserved charges and on the chiral phase transition temperature have been presented in Refs. [9–11]. The left panel of Fig. 6.2 shows a comparison between the trace anomaly, evaluated on coarse lattices and for heavier-than-physical quark masses, and the HRG model predictions with the physical and the distorted hadronic spectrum. The right panel shows a similar comparison for the subtracted chiral condensate defined in Eq. (2.75).

The comparison between HRG model and lattice QCD simulations has also led to important physical insight and interpretation of the lattice QCD results. Recently, lattice QCD simulations with physical quark masses have become extremely precise. This has led to a few analyses in which it was pointed out that, to improve the agreement between the HRG model and lattice QCD, either an additional exponential mass spectrum at high temperatures or heavier, not-yet-detected resonances need to be included in the resonance spectrum (see the discussion in Sect. 6.3.3).

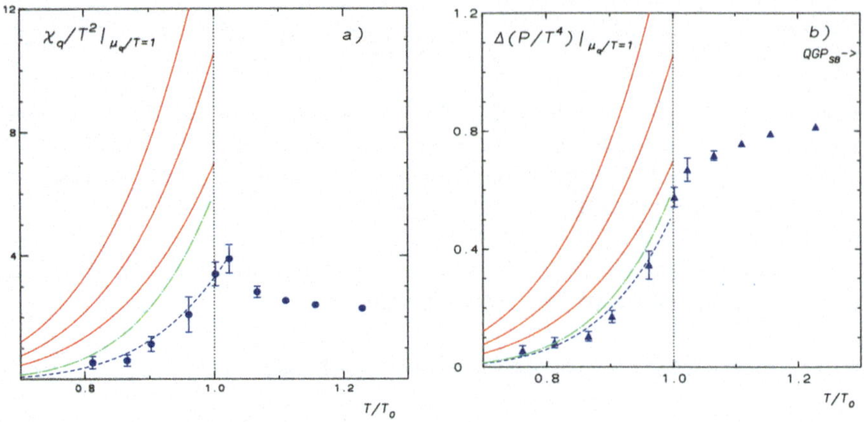

Fig. 6.1 From Ref. [7]. Quark number susceptibility (left) and change in pressure (right) for a fixed quark chemical potential $\mu_q/T = 1$ as functions of T/T_0. The lattice QCD results have been obtained with heavy quark masses, corresponding to $m_\pi = 770$ MeV. The dashed and dotted lines are HRG model predictions, obtained with a resonance mass spectrum consistent with the unphysical quark masses used in lattice QCD calculations. The full lines are resonance gas model results obtained with physical hadron masses and for three values of the transition temperature

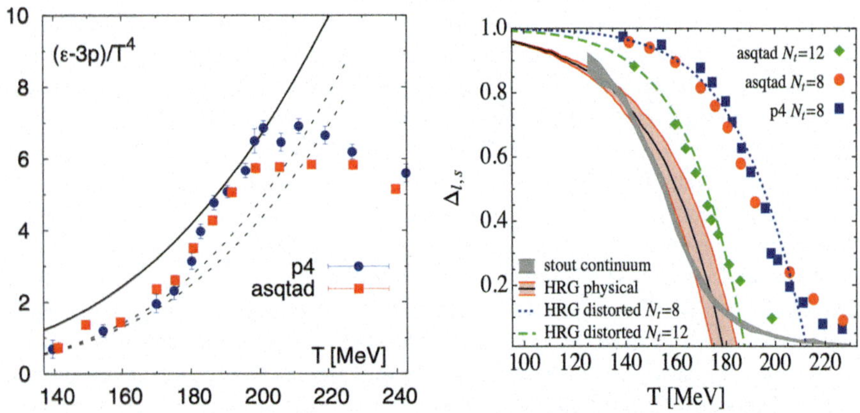

Fig. 6.2 Left: from Ref. [9]. Trace anomaly calculated in lattice QCD at quark masses $m_{u,d} = 0.2m_s$, compared to the HRG model predictions with physical hadron masses (solid line) and modified hadron masses (dashed lines). Right: from Ref. [11]. Subtracted chiral condensate, as defined in Eq. (2.75), as a function of the temperature. Full symbols are obtained with the asqtad and p4 actions at finite lattice spacing and for $m_{u,d} = 0.2m_s$ [12,13]. Grey bands are the continuum extrapolated results from Ref. [11]. The solid line is the HRG model result with physical masses. the dashed lines are the HRG+Chiral Perturbation Theory model results with distorted hadron masses, which take into account the discretization and unphysical quark mass effects

6.3.2 Investigation of a Possible Flavor Hierarchy

As discussed in Chap. 5, a clear separation in second- and fourth-order fluctuations between light and strange quarks was observed in lattice QCD. This seems to indicate a separation of the corresponding transition temperatures, suggesting that strange quarks might hadronize at higher temperature, compared to light quarks. Since no unambiguous temperature can be assigned to the QCD transition due to its crossover nature, the question arises whether hadrons of different flavors indeed form at different temperatures, and whether this is also translated to the freeze-out stage, as the thermal fits of particle yields at STAR and ALICE seem to suggest [14, 15].

In Ref. [4], a series of observables was introduced, based on combinations of up to fourth-order cumulants of net-strangeness fluctuations and their correlations with net-baryon number and electric charge, to probe the strangeness-carrying degrees of freedom at high temperatures. Using the two strangeness fluctuations χ_2^S and χ_4^S and the four baryon-strangeness correlations χ_{11}^{BS}, χ_{13}^{BS}, χ_{22}^{BS} and χ_{31}^{BS}, observables were introduced to project onto the four different partial pressures introduced in Eq. (6.15):

$$
\begin{aligned}
M(c_1, c_2) &= \chi_2^S - \chi_{22}^{BS} + c_1 v_1 + c_2 v_2 \\
B_1(c_1, c_2) &= \frac{1}{2}\left(\chi_4^S - \chi_2^S + 5\chi_{13}^{BS} + 7\chi_{22}^{BS}\right) + c_1 v_1 + c_2 v_2 \\
B_2(c_1, c_2) &= -\frac{1}{4}\left(\chi_4^S - \chi_2^S + 4\chi_{13}^{BS} + 4\chi_{22}^{BS}\right) + c_1 v_1 + c_2 v_2 \\
B_3(c_1, c_2) &= \frac{1}{18}\left(\chi_4^S - \chi_2^S + 3\chi_{13}^{BS} + 3\chi_{22}^{BS}\right) + c_1 v_1 + c_2 v_2
\end{aligned} \tag{6.19}
$$

where the combination $c_1 v_1 + c_2 v_2$ spans a two-dimensional plane in the 6-dimensional space of susceptibilities on which the partial pressure p_S^{HRG} vanishes identically when the strangeness degrees of freedom are described by a gas of uncorrelated hadrons, irrespective of their masses. Therefore, the two parameters c_1 and c_2 can be used to build observables that have an identical interpretation in the hadronic phase, while they differ in a medium in which the strangeness degrees of freedom are carried by quarks. The observables v_1 and v_2 are defined as

$$
\begin{aligned}
v_1 &= \chi_{31}^{BS} - \chi_{11}^{BS}, \\
v_2 &= \frac{1}{3}\left(\chi_2^S - \chi_4^S\right) - 2\chi_{13}^{BS} - 4\chi_{22}^{BS} - 2\chi_{31}^{BS}.
\end{aligned} \tag{6.20}
$$

From the above definitions and the discussion on the Boltzmann approximation in Sect. 6.2, it should be obvious that v_1 and v_2 both vanish in an uncorrelated hadron gas, yielding $M(c_1, c_2) \rightarrow p_{01}^{BS}$, $B_1(c_1, c_2) \rightarrow p_{11}^{BS}$, $B_2(c_1, c_2) \rightarrow p_{12}^{BS}$, $B_3(c_1, c_2) \rightarrow p_{13}^{BS}$. The four panels of Fig. 6.3 show the behavior of these four

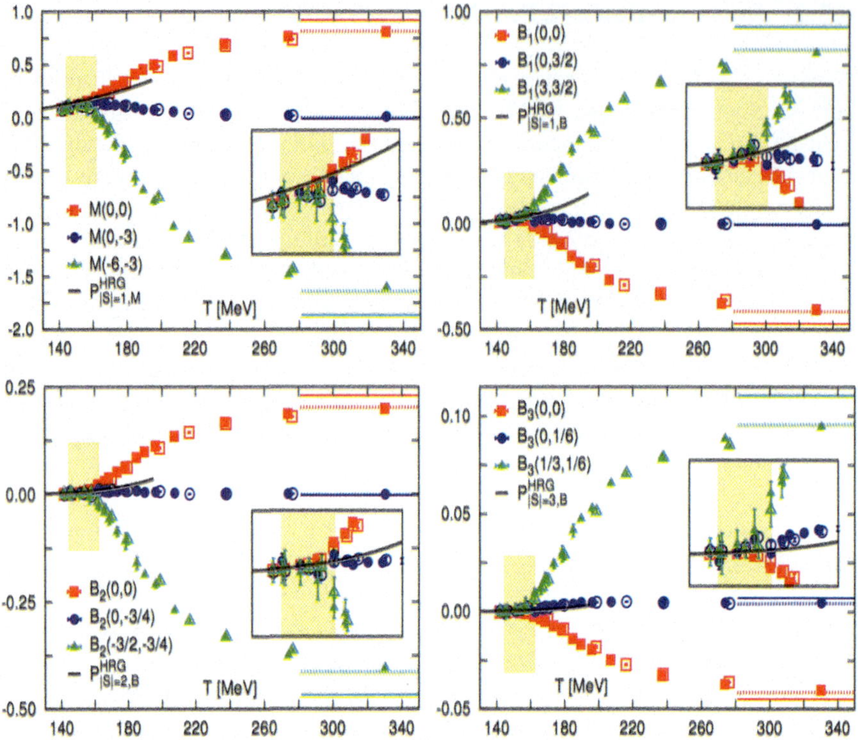

Fig. 6.3 From Ref. [4]. $M(c_1, c_2)$, $B_1(c_1, c_2)$ (top), $B_2(c_1, c_2)$, $B_3(c_1, c_2)$ (bottom) as functions of the temperature for different values of the parameters. The point at which different values for c_1 and c_2 cause a splitting of the results for a given observable can be identified as the onset of deconfinement for strange quarks, which in all panels happens around the upper limit for the chiral transition band

observables as functions of the temperature, for different values of c_1 and c_2. The temperature at which different values for these parameters cause a splitting of the lattice results can be identified as the onset of deconfinement for strange quarks. A consistent picture emerges, which indicates a temperature close to the upper bound of the chiral transition band (in orange).

This issue was further addressed in Ref. [16], which is based on the same concept but also considers a similar set of observables for light quarks:

$$\begin{aligned} v_1^f &= \chi_{11}^{Bf} - \chi_{31}^{Bf} \\ v_2^f &= \frac{1}{3}\left(\chi_2^f - \chi_4^f\right) + 2\chi_{13}^{Bf} - 4\chi_{22}^{Bf} + 2\chi_{31}^{Bf} \\ w^f &= \chi_{13}^{Bf} - \chi_{11}^{Bf}, \end{aligned} \tag{6.21}$$

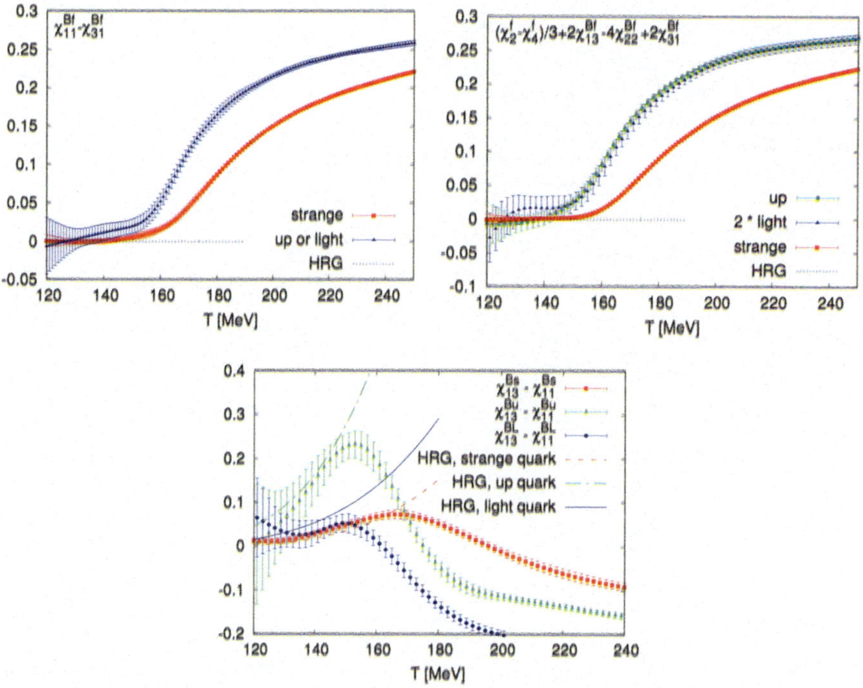

Fig. 6.4 From Ref. [16]. Continuum extrapolated lattice QCD results for v_1^f (top left), v_2^f (top right) and w^f (bottom) as functions of the temperature for $f = u,\ s$ and L, where L stands for light flavor

where a third quantity, w^f, was introduced because it is more sensitive to the flavor content. In particular, in the hadronic phase it only receives contributions from hadrons containing more than one quark of flavor f. The temperatures at which v_1^f and v_2^f deviate from the HRG model define the onset of deconfinement for flavor f. For observables that are most sensitive to multi-strange content, a pronouncedly higher temperature of deviation from HRG than for the analogous quantity in the light sector was observed (see Fig. 6.4).

6.3.3 The Resonance Spectrum

The only freedom available in the non-interacting Hadron Resonance Gas model, is the number and type of hadronic resonances over which the sums in Eq. (6.6) are taken. In the early versions of the model an exponential spectrum for the hadronic resonances was used, to be replaced in more recent versions with the discrete sum over particles listed in the Particle Data Book. This list is continuously evolving.

It was suggested in Refs. [17–19] to use the precise lattice QCD results on specific observables and their possible discrepancy with the HRG model predictions, to infer the existence of higher mass states, either through an exponential spectrum or the predictions of the Quark Model [20,21] and lattice QCD spectroscopy [22]. In

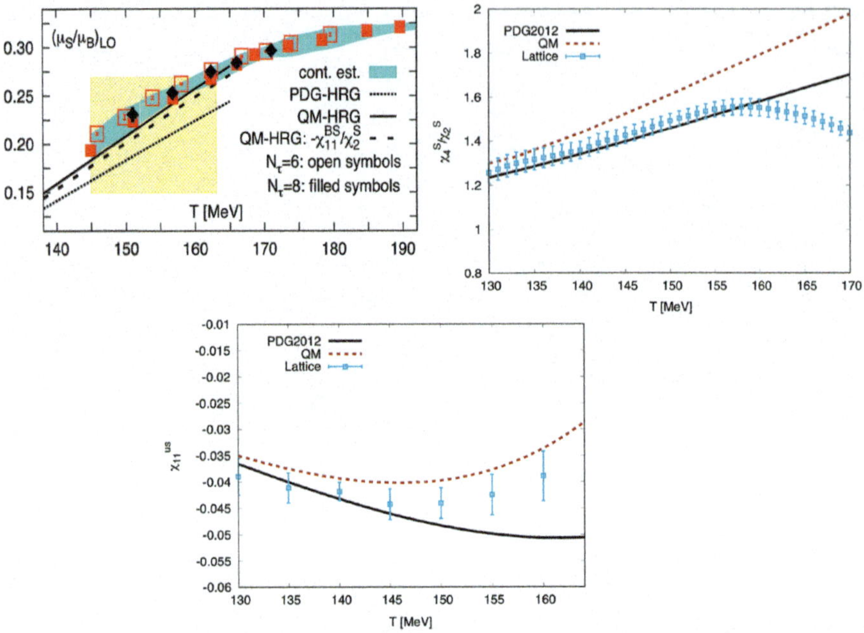

Fig. 6.5 Top left: from Ref. [19]. Leading order result for the ratio of strangeness and baryon chemical potentials versus temperature. Data and HRG model results are for a strangeness neutral thermal system having a ratio of net-electric charge to net-baryon number density $n_Q/n_B = 0.4$. The dashed line shows the QM-HRG result for vanishing electric charge chemical potential. Top right: from Ref. [5]. Fourth-to-second order strangeness fluctuation as a function of the temperature. Bottom: from Ref. [5]. Up-strange correlator as a function of the temperature. In the top right and bottom panels, the lattice QCD results (cyan points) are compared to the predictions of the HRG model using the PDG2012 (black, solid line) or the Quark Model (red, dashed line) particle spectrum

particular, the authors of Ref. [19] pointed out the discrepancy between the leading order lattice QCD results for μ_S/μ_B for a strangeness neutral system and the predictions of the HRG model for the same quantity, based on the 2012 particle list of the PDG (see the top left panel of Fig. 6.5). They were able to restore the agreement between HRG model and lattice QCD by introducing in the particle spectrum all the strange states predicted by the Quark Model [20, 21] but not yet detected. It was also suggested that introducing these states will remove the flavor hierarchy observed in the thermal fit of STAR and LHC data (see the discussion in Chap. 7).

The authors of Ref. [5] pointed out that, while the agreement between lattice QCD results and HRG model predictions for μ_S/μ_B is certainly improved by the Quark Model resonance spectrum, this is not true for other observables such as χ_4^S/χ_2^S and χ_{11}^{us}. This is shown in the central and bottom panels of Fig. 6.5, and triggered a more in-depth analysis of the strange sectors of the PDG lists, to see how many and which states are actually needed.

By exploiting the Boltzmann approximation, the authors of Ref. [5] defined partial pressure contributions due to hadrons grouped according to their baryonic and strange quantum numbers. These quantities were defined in Eq. (6.17) and were obtained on the lattice by means of the simultaneous fit of observables defined in Eq. (6.18); by comparing the results from lattice QCD to the HRG model predictions for each one of them, it is possible to identify the families of hadrons that underestimate the lattice QCD results, and that therefore need to be complemented with additional states. Figure 6.6 shows the partial pressure due to non-strange baryons (top left), baryons with $|S| = 1$ (top right), $|S| = 2$ (bottom left) and $|S| = 3$ (bottom right). In the attempt of reducing the number of resonances with respect to the original Quark Model [20], additional versions have been considered: the hypercentral Quark Model (hQM) introduced in Ref. [23], together with a set of particles named "PDG2016+", which contains all hadronic particles listed in the 2016 version of the PDG [24], including those with just one star. From the Figure it is clear that, while the QM states are needed in some families (e.g. $|S| = 1$ baryons), they are redundant in other sectors (baryons with strangeness zero or $|S| = 2$). The authors of reference [5] concluded that the PDG2016+ spectrum is the one that

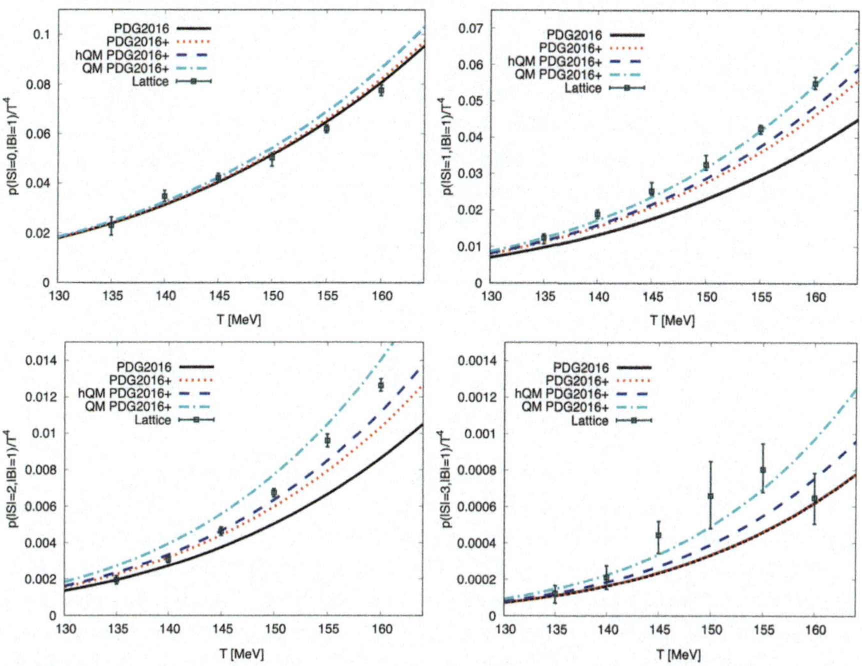

Fig. 6.6 From Ref. [5]. Contribution to the QCD pressure due to non-strange baryons (top left), baryons with $|S|$=1 (top right), baryons with $|S|$=2 (bottom left) and baryons with $|S|$=3 (bottom right). The dots are continuum extrapolated lattice QCD results. The solid, black line is the result from the HRG model containing states with two, three and four stars from the PDG2016, the other curves correspond to PDG2016+ (including one star states, red dotted), PDG2016+ and additional states from the hQM (blue, dashed), PDG2016+ and additional states from the QM (cyan, dash-dotted)

allows a successful comparison between HRG model and lattice QCD results for the largest number of observables, including the problematic χ_4^S/χ_2^S and χ_{11}^{us} discussed earlier.

The effect of these additional resonances on multiparticle correlations in heavy-ion collisions was studied in Ref. [25], while in Ref. [26] it was shown that the flavor hierarchy in the freeze-out temperatures persists even when the thermal fits of particle yields and fluctuations are performed on the basis of the PDG2016+ particle list. We will discuss this in more detail in Chap. 7.

6.3.4 Off-Diagonal Correlators

Recently, comparison between the HRG model and lattice QCD was used as a guidance for the experimental program at RHIC, to compare lattice QCD simulations to experimental data [27]. While such comparison is relatively straightforward (given the caveats discussed in Chap. 7) for diagonal conserved charge fluctuations, it is not so trivial when considering off-diagonal correlators between different charges. Three correlators were considered; χ_{11}^{BQ}, χ_{11}^{BS} and χ_{11}^{QS}. The lattice QCD results were compared to HRG model predictions, separated into "measurable" and "non-measurable". The former contain the following particles: $\pi^{\pm}$, $K^{\pm}$, $p(\bar{p})$, $\Lambda(\bar{\Lambda})$, $\Xi^-(\bar{\Xi}^+)$, $\Omega^-(\bar{\Omega}^+)$, while the latter indicates neutral particles and those like the Σ baryons, that are potentially measurable but their yield and fluctuation results are not routinely performed at RHIC or at the LHC. It is straightforward to adapt the HRG model so that it is expressed in terms of stable hadronic states only. The sum over the whole hadronic sector is converted into a sum over both the whole hadronic spectrum and the list of states that are stable under strong interactions

$$\sum_R B_R^l Q_R^m S_R^n I_p^R \rightarrow \sum_{i \in stable} \sum_R (P_{R\rightarrow i}) B_i^l Q_i^m S_i^n I_p^R, \tag{6.22}$$

with $l + m + n = p$, the first sum on the r.h.s. runs only over the particle which are stable under strong interactions and $P_{R\rightarrow i} = \sum_\alpha N_{R\rightarrow i}^\alpha n_{i,\alpha}^R$ gives the average number of particle i produced by each particle R after the whole decay chain. Finally, $I_p^R = \partial^p(p_R/T^4)\partial(\mu_R/T)^p$. This comparison is shown in Fig. 6.7. Both the BQ and QS correlators are largely reproduced by the measured contribution, while the BS correlator is roughly split in half between measured and non-measured terms. This is because the BS correlator receives its main contribution from strange baryons, which are almost equally split between measurable and non-measurable.

In Ref. [27], it was observed that the main contribution to the off-diagonal correlators comes from self-correlations of hadrons, and not from correlations between different hadrons. Besides, exploiting the fact that also diagonal-fluctuations receive a contribution from non-measurable states, ratios of off-diagonal over diagonal correlators were constructed, which are entirely described by the measurable

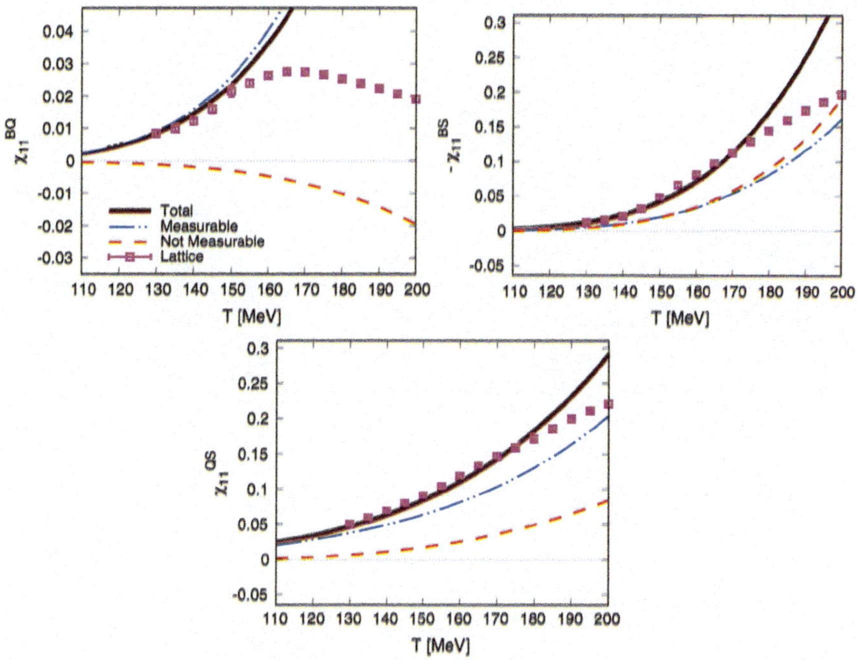

Fig. 6.7 From Ref. [27]. Second-order correlators of the conserved charges B, Q, S. The total contribution, the measurable and non-measurable parts evaluated in the HRG model are shown in solid black, dotted-dashed blue and dashed red, respectively. The lattice results are shown as magenta points

component. They are shown in Fig. 6.8. The comparison between the proxies and the experimental results is discussed in Chap. 7.

6.4 HRG Model Fits to Particle Yields

We spoke about the chemical freeze-out and the possible determination of its temperature and chemical potential by comparing experimental data for the cumulants of net-charge multiplicity distributions to the results for fluctuations of conserved charges from lattice QCD in Sect. 5.3. The advantage of that method is that, provided that a meaningful comparison of thermal fluctuations can actually be performed, it allows a model-independent determination of the freeze-out parameters by directly comparing experimental results to first principle calculations.

However, the easiest quantity to measure experimentally is the yield of each particle species. These quantities are difficult to calculate on the lattice, since the latter works in terms of conserved charges but it is not easy to identify the single particle contribution to the thermodynamics. The method that for many years has been used to determine the freeze-out parameters is to perform a combined fit

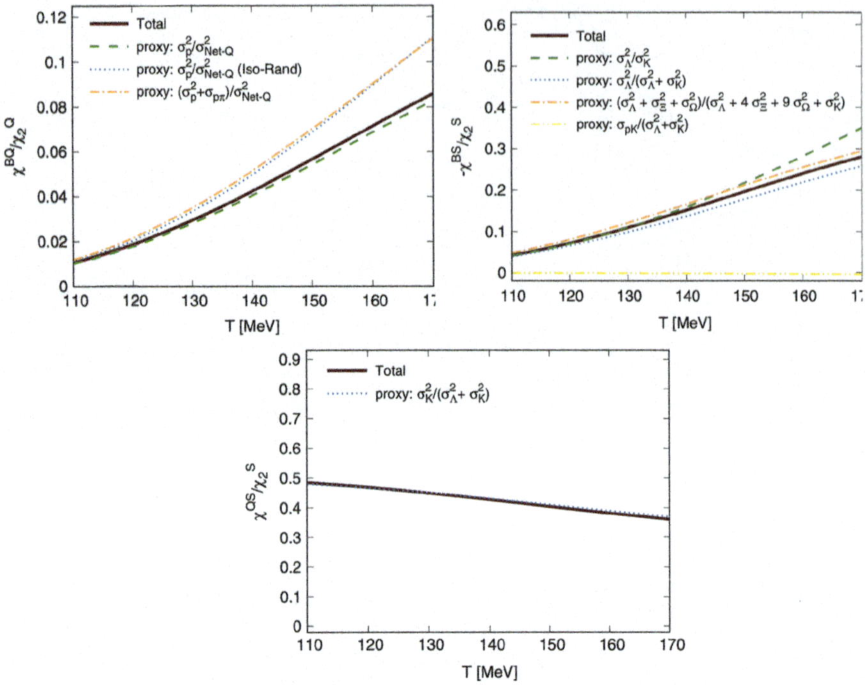

Fig. 6.8 From Ref. [27]. Temperature dependence of the ratio χ_{11}^{BQ}/χ_2^Q (top left), χ_{11}^{BS}/χ_2^S (top right), χ_{11}^{QS}/χ_2^S (bottom). In all cases the total contribution is shown as a black solid line, while the other curves correspond to the different proxies listed in the legends

of particle yields by means of the HRG model. This has been done over a broad range of energies, from the highest collision energy at the LHC [14], to the lowest ones at SIS/GSI [28]. If the QGP produced in a heavy-ion collision reaches thermal equilibrium, one expects that the thermal nature of the partonic medium could be preserved during hadronization. Consequently, the particle yields measured in the final state should resemble a thermal equilibrium population.

As we have seen, the HRG model with the full hadronic mass spectrum, yields a very good description of lattice QCD results. So, using it for the thermal statistical operator is a synonym to using the full QCD Hamiltonian. The only free parameters are T, μ_B and V, since μ_S and μ_Q can be fixed to impose the phenomenological net-charge conservation equations discussed in Sect. 4.4. The agreement between data and theoretical predictions implies statistical equilibrium at temperature T_{ch} and chemical potential μ_{Bch}. Of particular importance is to account for resonances and their decay into lighter particles: they decay strongly, so they have enough time to do so before they reach the detector.

We therefore get, for the average number $\langle N_i \rangle$ of particle species i that carries strangeness S_i, baryon number B_i and electric charge Q_i in volume V and temperature T:

$$\langle N_i\rangle(T,\mu) = \langle N_i\rangle^{th}(T,\mu) + \sum_j \Gamma_{j\to i}\langle N_j\rangle^{th}(T,\mu) \tag{6.23}$$

where the first term is the thermal average number of particles of species i and represents the primordial particles, namely those which were directly produced at hadronization, while the second term is the overall resonance contribution to the multiplicity of particle of species i, and it is given by the sum of all resonances j that decay into particle i, multiplied by the corresponding branching ratio for the process $j \to i$.

6.5 Interacting Hadron Resonance Gas Model

In contrast to the standard approach of simulating the high temperature hadron gas as a non-interacting hadron gas with the maximum amount of resonances in the spectrum, it was suggested to actually parametrize the interaction between hadrons based on lower energy measurements. These type of additions to the HRG presently fall into three categories: (a) the parametrization of repulsive interaction through an "excluded volume" HRG model, (b) the parametrization of repulsive and attractive forces measured at lower energies through the van der Waals (vdW) approach [29], and (c) the parametrization of attractive hadron interactions in resonance formation through the S-matrix approach of statistical mechanics [30, 31]. The shortcomings of the approaches are different. For the attractive and repulsive vdW forces one has to wonder whether the low energy measurements of vdW parameters a and b for ground state matter at the threshold of the nuclear liquid-gas phase transition can be applied at the high energies without any scaling corrections. A simplified version of the vdW is achieved when simply applying an excluded volume, due to the finite hadron eigenvolumes, to the non-interacting hadron gas [32]. Here an additional fit parameter arises from the scaling of the individual volume with particle species (meson/baryon) and flavor. Regarding the S-matrix approach, its constraints are largely based on the limitations of our knowledge of low energy particle interactions. Although the pion-nucleon interactions are well measured, the resulting phase shifts can be well applied to the Δ and N* resonances, which in the end impacts the proton yields, but any further application to, for example, strange particle yields requires a much better knowledge of the meson-baryon and baryon-baryon interactions in the strange sector. Ultimately, the optimization of the hadronic spectrum in the non-interacting HRG approach has shown reliable results in comparison to continuum extrapolated lattice QCD calculations for all relevant susceptibilities up to their respective pseudo-critical temperatures. In this sense, the historic approach of describing an interacting hadron gas through a non-interacting hadron resonance gas, first proposed by Hagedorn, is still alive and well [33]. Below we will present some more details on the three versions of the interacting HRG model discussed here.

6.5.1 Excluded Volume HRG Model

The non-interacting HRG model is based on the assumption of a relatively dilute system. However, hadron densities in heavy-ion collisions can be quite large, thus leading to deviations from the ideal gas picture. The short-range repulsion in multi-particle systems is often modeled by means of a hard-sphere interaction, which assumes that particles can move freely, unless the distance between their centers equals the sum of their hard-core radii [34]. The early versions of this model incorporate the same radius for all hadronic species [35–38], while more recent versions have attempted different radii for different particle species [39–43].

In the simple case of a one-component system, one can write the pressure of the system in the Boltzmann approximation as follows:

$$p = nTZ(\eta) \tag{6.24}$$

where n is the number density of particles, $\eta = nv$ indicates their packing fraction and $v = \frac{4}{3}\pi r^3$ is a single-particle volume, where r is the hard-core radius of the particle. $Z(\eta)$ is a dimensionless compressibility factor, which does not depend on temperature. Integrating the above equation for the pressure, one finds the trascendental equation for the particle density $n = n(T, \mu_B)$:

$$n = n^{id}\left[T, \mu_B - T\psi(vn)\right], \tag{6.25}$$

where $n^{id} = \frac{gm^2T}{2\pi^2}K_2\left(\frac{m}{T}\right)\exp\left(\frac{\mu_B}{T}\right)$ and the function $\psi(\eta)$ can be calculated analytically and yields the shift of the chemical potential for a one-component gas.

In the case of a multi-component system, the pressure can be parametrized as

$$p = T\sum_i \xi_i, \qquad \xi_i = \frac{n_i}{1 - \sum_j b_j n_j} \tag{6.26}$$

where $b_i = 16\pi r_i^3/3$ and n_i are the eigenvolume parameter and the density of the ith hadron respectively. One then finds the following equation for the chemical potential of the ith particle:

$$\mu_i = T\ln\frac{\xi_i}{\frac{g_i m_i^2 T}{2\pi^2}K_2\left(\frac{m_i}{T}\right)} + b_i p \tag{6.27}$$

which leads to the trascendental equation for the pressure

$$p = \sum_i p_i^{id}(T, \mu_i - b_i p) \tag{6.28}$$

and allows to calculate the particle number densities

$$n_i = \frac{\xi_i}{1 + \sum_j n_j \xi_j}. \tag{6.29}$$

In Ref. [44], thermal fits to particle yields were performed, using the excluded volume HRG model in which baryon-baryon and baryon-antibaryon interactions were considered, whereas for mesons the point-like approximation was used.

6.5.2 Van der Waals HRG Model

In the canonical ensemble, where the independent variables are the temperature T, volume V and number of particles N, the Van der Waals equation of state takes the form

$$p(T, n) = \frac{NT}{V - bN} - a\frac{N^2}{V^2} = \frac{nT}{1 - bn} - an^2 \tag{6.30}$$

where a and b are the parameters that describe the attractive and repulsive interaction, respectively. The first term on the rhs of the above equation corresponds to the excluded volume interaction seen in the previous Section. When considering the Van der Waals equation of state in the grand canonical ensemble, we get the following trascendental expression for the pressure

$$\begin{aligned} p(T, \mu) &= p^{id}(T, \mu^*) - an^2 \\ \mu^* &= \mu - bp(T, \mu) - abn^2 + 2an \end{aligned} \tag{6.31}$$

where n is the particle number density and can be defined as

$$n(T, \mu) = \frac{n^{id}(T, \mu^*)}{1 + bn^{id}(T, \mu^*)}. \tag{6.32}$$

The terms $p^{id}(T, \mu^*)$ and $n^{id}(T, \mu^*)$ indicate pressure and number densities calculated with the ideal HRG model, but with the shifted chemical potential μ^* defined above. In Ref. [45], this model was used to improve the agreement between lattice QCD and HRG model at $\mu_B = 0$ for several observables. Thus, the parameters a and b were obtained at $\mu_B = 0$. On the contrary, in Ref. [46] the authors fixed a and b to reproduce the saturation density and binding energy of the ground state of symmetric nuclear matter. They then used the model to extract freeze-out parameters and to study the beam energy dependence of the skewness and kurtosis.

6.5.3 S-Matrix Formulation

The S-matrix formalism allows to incorporate interactions into the description of a diluted system. Two-body interactions are included in the leading term of the S-matrix expansion of the grand canonical potential via the scattering phase shifts. The thermal yield of a πN interaction channel with spin J and isospin I is given by [47, 48]

$$\langle R_{J,I}\rangle = d_J \int_{m_{th}}^{\infty} dM \frac{1}{2\pi} B_{J,I}(M) \int \frac{d^3p}{(2\pi)^3} \frac{1}{e^{(\sqrt{p^2+M^2}-\mu)/T}+1} \tag{6.33}$$

where

$$B_{J,I}(M) = 2\frac{d\delta_J^I}{dM}, \tag{6.34}$$

T is the temperature, m_{th} the threshold mass, M the invariant mass and $\mu = \mu_B B + \mu_Q Q$ the chemical potential. Besides, $B_{J,I}$ is an effective spectral function derived from the scattering phase shift δ_J^I and $d_J = 2J+1$ is the degeneracy factor for spin J.

In Ref. [31], this formalism was selectively applied to the πN channel as a possible solution for the proton anomaly in the ALICE yields (see the discussion in Sect. 7.3.3). Before this can be unambiguously concluded, however, the same formalism would need to be applied to all other channels as well. Some steps in this direction, through a modeling of the unknown scattering phases, were performed in Ref. [49].

In Ref. [50], some guidance about the magnitude of the interaction in the other channels was obtained by calculating their contributions to the thermodynamics.

References

1. Hagedorn, R.: Nuovo Cim. Suppl. **3**, 147–186 (1965) CERN-TH-520
2. Venugopalan, R., Prakash, M.: Nucl. Phys. A **546**, 718–760 (1992)
3. Cabibbo, N., Parisi, G.: Phys. Lett. B **59**, 67–69 (1975)
4. Bazavov, A., Ding, H.T., Hegde, P., Kaczmarek, O., Karsch, F., Laermann, E., Maezawa, Y., Mukherjee, S., Ohno, H., Petreczky, P., Schmidt, C., Sharma, S., Soeldner, W., Wagner, M.: Phys. Rev. Lett. **111**, 082301 (2013)
5. Alba, P., Bellwied, R., Borsanyi, S., Fodor, Z., Günther, J., Katz, S.D., Mantovani Sarti, V., Noronha-Hostler, J., Parotto, P., Pasztor, A., Portillo Vazquez, I., Ratti, C.: Phys. Rev. D **96**, no.3, 034517 (2017)
6. Karsch, F., Redlich, K., Tawfik, A.: Eur. Phys. J. C **29**, 549–556 (2003)
7. Karsch, F., Redlich, K., Tawfik, A.: Phys. Lett. B **571**, 67–74 (2003)
8. Redlich, K., Karsch, F., Tawfik, A.: J. Phys. G **30**, S1271–S1274 (2004)
9. Huovinen, P., Petreczky, P.: Nucl. Phys. A **837**, 26–53 (2010)
10. Huovinen, P., Petreczky, P.: J. Phys. Conf. Ser. **230**, 012012 (2010)
11. Borsanyi, S., et al. [Wuppertal-Budapest] JHEP **09**, 073 (2010)
12. Bazavov, A., Bhattacharya, T., Cheng, M., Christ, N.H., DeTar, C., Ejiri, S., Gottlieb, S., Gupta, R., Heller, U.M., Huebner, K., Jung, C., Karsch, F., Laermann, E., Levkova, L., Miao, C., Mawhinney, R.D., Petreczky, P., Schmidt, C., Soltz, R.A., Soeldner, W., Sugar, R., Toussaint, D., Vranas, P.: Phys. Rev. D **80**, 014504 (2009)
13. Bazavov, A., et al.: [HotQCD] J. Phys. Conf. Ser. **230**, 012014 (2010)
14. Andronic, A., Braun-Munzinger, P., Redlich, K., Stachel, J.: Nature **561**(7723), 321–330 (2018)
15. Adamczyk, L., et al.: [STAR] Phys. Rev. C **96**, no.4, 044904 (2017)
16. Bellwied, R., Borsanyi, S., Fodor, Z., Katz, S.D., Ratti, C.: Phys. Rev. Lett. **111**, 202302 (2013)
17. Noronha-Hostler, J., Noronha J., Greiner, C.: Phys. Rev. Lett. **103**, 172302 (2009)

18. Majumder, A., Muller, B.: Phys. Rev. Lett. **105**, 252002 (2010)
19. Bazavov, A., Ding, H.T., Hegde, P., Kaczmarek, O., Karsch, F., Laermann, E., Maezawa, Y., Mukherjee, S., Ohno, H., Petreczky, P., Schmidt, C., Sharma, S., Soeldner, W., Wagner, M.: Phys. Rev. Lett. **113**(7), 072001 (2014)
20. Capstick, S., Isgur, N.: AIP Conf. Proc. **132**, 267–271 (1985)
21. Ebert, D., Faustov, R.N., Galkin, V.O.: Phys. Rev. D **79**, 114029 (2009)
22. Edwards, R.G., et al.: [Hadron Spectrum] Phys. Rev. D **87**(5), 054506 (2013)
23. Ferraris, M., Giannini, M.M., Pizzo, M., Santopinto, E., Tiator, L.: Phys. Lett. B **364**, 231–238 (1995)
24. Patrignani, C., et al.: [Particle Data Group] Chin. Phys. C **40**(10), 100001 (2016)
25. Alba, P., Mantovani Sarti, V., Noronha, J., Noronha-Hostler, J., Parotto, P., Portillo Vazquez, I., Ratti, C.: Phys. Rev. C **98**(3), 034909 (2018)
26. Alba, P., Sarti, V.M., Noronha-Hostler, J., Parotto, P., Portillo-Vazquez, I., Ratti, C., Stafford, J.M.: Phys. Rev. C **101**(5), 054905 (2020)
27. Bellwied, R., Borsanyi, S., Fodor, Z., Guenther, J.N., Noronha-Hostler, J., Parotto, P., Pasztor, A., Ratti, C., Stafford, J.M.: Phys. Rev. D **101**(3), 034506 (2020)
28. Redlich, K., Cleymans, J., Oeschler, H., Tounsi, A.: Acta Phys. Polon. B **33**, 1609–1628 (2002)
29. Vovchenko, V., Gorenstein, M.I., Stoecker, H.: Phys. Rev. Lett. **118**(18), 182301 (2017)
30. Lo, P.M., Friman, B., Redlich, K., Sasaki, C.: Phys. Lett. B **778**, 454–458 (2018)
31. Andronic, A., Braun-Munzinger, P., Friman, B., Lo, P.M., Redlich, K., Stachel, J.: Phys. Lett. B **792**, 304–309 (2019)
32. Vovchenko, V., Gorenstein, M.I., Stoecker, H.: Phys. Rev. C **98**(6), 064909 (2018)
33. Hagedorn, R.: Prog. Sci. Culture **1**, 395–411 (1976) CERN-TH-2272
34. Rischke, D.H., Gorenstein, M.I., Stoecker, H., Greiner, W.: Z. Phys. C **51**, 485–490 (1991)
35. Satarov, L.M., Dmitriev, M.N., Mishustin, I.N.: Phys. Atom. Nucl. **72**, 1390–1415 (2009)
36. Yen, G.D., Gorenstein, M.I., Greiner, W., Yang, S.N.: Phys. Rev. C **56**, 2210–2218 (1997)
37. Yen, G.D., Gorenstein, M.I.: Phys. Rev. C **59**, 2788–2791 (1999)
38. Wheaton, S., Cleymans, J.: Comput. Phys. Commun. **180**, 84–106 (2009)
39. Gorenstein, M.I., Kostyuk, A.P., Krivenko, Y.D.: J. Phys. G **25**, L75–L83 (1999)
40. Bugaev, K.A., Oliinychenko, D.R., Cleymans, J., Ivanytskyi, A.I., Mishustin, I.N., Nikonov, E.G., Sagun, V.V.: EPL **104**(2), 22002 (2013)
41. Bugaev, K.A., Oliinychenko, D.R., Sorin, A.S., Zinovjev, G.M.: Eur. Phys. J. A **49**, 30 (2013)
42. Vovchenko, V., Stöcker, H.: J. Phys. G **44**(5), 055103 (2017)
43. Vovchenko, V., Stoecker, H.: Phys. Rev. C **95**(4), 044904 (2017)
44. Satarov, L.M., Vovchenko, V., Alba, P., Gorenstein, M.I., Stoecker, H.: Phys. Rev. C **95**(2), 024902 (2017)
45. Samanta, S., Mohanty, B.: Phys. Rev. C **97**(1), 015201 (2018)
46. Poberezhnyuk, R., Vovchenko, V., Motornenko, A., Gorenstein, M.I., Stoecker, H.: Phys. Rev. C **100**(5), 054904 (2019)
47. Weinhold, W., Friman, B., Norenberg, W.: Phys. Lett. B **433**, 236–242 (1998)
48. Lo, P.M.: Eur. Phys. J. C **77**(8), 533 (2017)
49. Cleymans, J., Lo, P.M., Redlich, K., Sharma, N.: [arXiv:2009.04844 [hep-ph]]
50. Bellwied, R., et al.: 2102.06625 (2021)

Experimental Verification of Lattice QCD Predictions

7

Abstract

In this chapter we discuss the comparison between theoretical predictions and experimental measurements for several observables. In particular, we focus on fluctuations of conserved charges and how to perform a meaningful comparison between lattice QCD simulations and experimental results. We review several sources of non-thermal fluctuations and explain how to take them into account in the data or the theoretical results. We then discuss the details of the experimental analysis for fluctuations of net-protons, net-charge, net-kaons and net-Lambdas.

7.1 Introduction

Over the past decade the experimental relativistic heavy ion program has focused on event-by-event measurements of fluctuations of particle-identified multiplicity distributions to verify specific lattice QCD and other effective model predictions. These measurements can be directly related to the susceptibilities of conserved quantum numbers as functions of temperature and baryo-chemical potential on the lattice. The experimental motivation is three-fold.

The main focus of the experimental program at the GSI and RHIC has been to establish a critical point on the QCD phase diagram. Early lattice QCD calculations [1, 2], together with numerous effective theory predictions [3–23], hinted at the possibility that a critical point on the QCD phase diagram could be found by measuring critical fluctuations of conserved charges, and thus of their multiplicity distributions, near this point in T and μ_B.

An additional investigation of critical multiplicity fluctuations, even at low baryo-chemical potential, namely at the higher energies at RHIC and the LHC, was aimed at determining experimental signatures of the chiral transition for the first time, as well as its transition temperature.

C. Ratti, R. Bellwied, *The Deconfinement Transition of QCD*, Lecture Notes in Physics 981, https://doi.org/10.1007/978-3-030-67235-5_7

Finally, high precision measurements of the statistical multiplicity fluctuations away from the region of the critical point, have been used to investigate the processes of hadronization and chemical freeze-out in the QCD crossover region as a function of flavor and baryon number. Any viable comparison between multiplicity fluctuation measurements and susceptibility calculations on the lattice will yield a model-free determination of chemical freeze-out parameters in the thermally equilibrated system. Particle-identified fluctuation measurements turn out to lead to a more precise estimate of the hadronic freeze-out than particle yields, which have traditionally been used in statistical hadronization models. Lattice QCD has revealed that the freeze-out surface might be quantum number dependent, and thus could yield to a flavor hierarchy during the hadronization process. We have detailed the theoretical background for each of the three topics mentioned here in Chap. 5. Here we will show the most recent results of the past decade that relate to these predictions.

One should note, though, that lattice calculations follow a conserved quantum number (net-baryon number, net-strangeness or net-electric charge) into the crossover region, whereas the statistical model calculations follow a net-particle number through the crossover region from below. Comparison of experimental data to lattice QCD or HRG model predictions have both some caveats. For lattice calculations, the use of a net-partice number as a proxy for a net-quantum number behavior is hampered by several experimental caveats and the fact that single particle species will always be an incomplete proxy for the relevant quantum number. Numerous publications have dealt with this problem in the past decade and have arrived at the conclusion that a mapping, albeit provisionally, is applicable (see the discussion in Sects. 7.2.2 and 6.3.4). For the HRG model calculations, the problem has been all along to determine the proper spectrum of available states and their interactions (see the discussion in Sect. 6.3.3). The standard way of using a non-interacting resonance gas of PDG-established states with a particular mass cut-off has been recently summarized in a Nature article and applied to data from SPS to LHC [24]. Nevertheless, the past decade has seen improvements to that method either through parametrizing the interaction strength between the resonant states (van der Waals forces, S-matrix, excluded volume) [25–27] or by extending the spectrum of allowable non-interacting states using the latest resonance measurements or Quark Model calculations [28]. The advantage of these statistical models over first principle calculations is that any experimental constraints regarding acceptance and efficiency of particle identification can be folded into the calculation, which is not possible on the lattice. But in return, any confirmation of a lattice result through an HRG result can be used to establish the measurement as a viable proxy for the quantum behavior.

7.2 Experimental Methods

Prior to presenting and interpreting the relevant experimental results, we will give a detailed description of the several steps necessary in order to compare a particle measurement in a limited acceptance detector with a grand canonical quantum number based-calculation in an infinite volume.

7.2.1 Relation Between Moments, Cumulants and Susceptibilities

In this subsection, we describe how to calculate cumulants from experimental data and how these cumulants relate to quantum number susceptibilities on the lattice.

In order to determine a net-particle cumulant, consider an ensemble of events in each of which the number of particles and anti-particles of a particular species, which we shall denote x and y, are measured. In each case, the number is recorded in a specified rapidity window, which should be at least one unit wide, in order to avoid significant acceptance corrections, which are detailed further in the next section. The average value of x and y over the whole ensemble of events is defined as $\langle x\rangle$ and $\langle y\rangle$, and one can denote the deviation of x and y from their mean in a single event by

$$\delta x = x - \langle x\rangle$$
$$\delta y = y - \langle y\rangle. \tag{7.1}$$

The definition of the higher-order cumulants, and their relation with the moments and central moments of a distribution, was given in Sect. 5.2.

Cumulants are additive and thus any cumulant of extensive variables, such as particle multiplicities, is itself extensive, which makes them proportional to the volume of the system V in the thermodynamic limit. In order to cancel out the leading volume dependence, one can calculate simple cumulant ratios, as was detailed in Sect. 5.2.

Since the dependence on the correlation length increases with the moment of these distributions, the most obvious signals of criticality can be established in the highest moments. Presently, experiments are designed to reach reliable measurements up to the fourth moment (kurtosis), but the sixth order moment is considered the statistical limit for both BES-II energies at RHIC and ALICE measurements at the LHC in the future.

The higher moments can be derived from the cumulant generating function, which is

$$\ln(\Phi(t)) = \ln\langle e^{tx}\rangle \tag{7.2}$$

The nth cumulant is then given by

$$\kappa_n = \left. \frac{\mathrm{d}^n \ln(\Phi(t))}{\mathrm{d}t^n} \right|_{t=0} . \tag{7.3}$$

The derivations in Sect. 5.2 showed the relation between susceptibilities and cumulants of conserved quantum numbers calculated on the lattice. The relevance of the measured fluctuations now hinges on the question whether cumulants of single particle species or a sum of particle species in a limited acceptance can serve as a proxy for the behavior of the respective quantum number on the lattice and/or near a critical point. This question will be addressed in the next Section.

7.2.2 Caveats in the Comparison Between Theory and Experiment and How to Solve Them

Heavy ion collisions are highly dynamical, whereas lattice QCD calculates a static system in global equilibrium. In addition, all experiments measure particles only over a limited acceptance, which allows the generation of fluctuations of an otherwise globally conserved quantum number, but it also impacts the validity of determining non-statistical fluctuations. In other words, the strength of the non-statistical fluctuation will be acceptance-dependent. Several other, uniquely experimental effects, play a role when comparing to lattice QCD and we will describe them in the following:

1. Detector acceptance: Any particle yield obtained in a realistic detector will be measured over a limited geometrical space. Most detectors at collider experiments are cylindrical, which means they have full coverage in azimuth but they will be limited in the direction where the beams travel. This direction is generally denoted by the pseudo-rapidity, which is defined as a function of the Azimuthal angle θ (see the top panel of Fig. 7.1):

$$\eta = -\ln(\tan\frac{\theta}{2}) = \frac{1}{2}\ln\left(\frac{|\mathbf{p}| - p_L}{|\mathbf{p}| + p_L}\right), \tag{7.4}$$

where p_L is the component of the momentum along the beam axis, namely the longitudinal momentum. If the particle is identified and its mass is known, η can be replaced by the rapidity y, which is defined in momentum space, and reflects the longitudinal component of the momentum:

$$y = \frac{1}{2}\ln\left(\frac{E + p_L}{E - p_L}\right). \tag{7.5}$$

Notice that the rapidity or pseudo-rapidity are preferred over the Azimuthal angle θ, because differences in rapidity are Lorentz invariant under boosts along the longitudinal axis: they transform additively. The bottom panel of Fig. 7.1 shows the net-proton rapidity distributions for different collision energies. Although the detector is hermetic in azimuth, the detection capabilities of particles in the direction

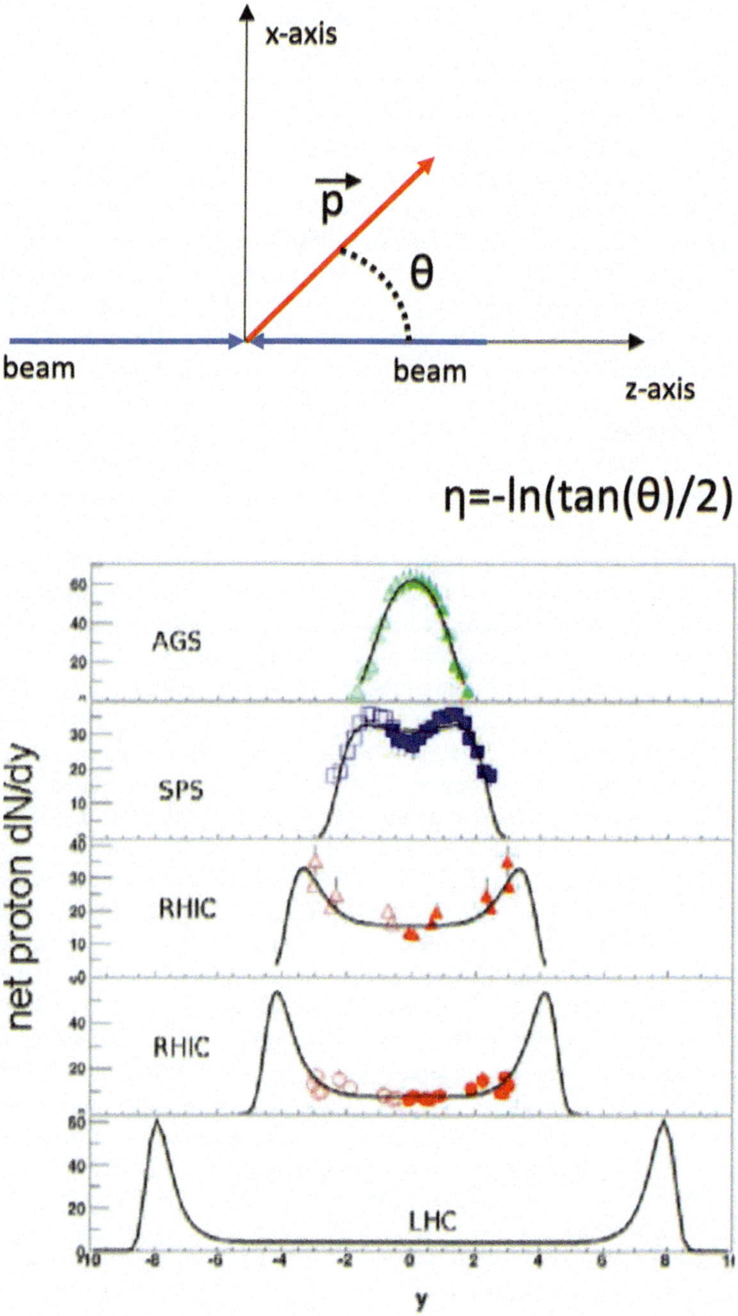

Fig. 7.1 Top: visualization of the Azimuthal angle θ and the particle momentum in the xz-plane. Bottom: net-proton rapidity distributions dN/dy for different collision energies

transverse to the beam direction are limited at low momentum due to a large deflecting magnetic field and at high momentum due to resolution limitations in the applied detector technology. In a particle yield determination, these limitations in the transverse momentum spectrum are generally corrected by extrapolating a well-known and scientifically motivated function from the measured part of the spectrum to the lower and higher momenta. The limited acceptance in pseudo-rapidity on the other hand is not corrected for, but rather used as a particular defining characteristic of the fluctuation measurement. Since all lattice results are calculated for a full 4π acceptance, a direct comparison between the measurement and lattice QCD results is not straightforward, but rather requires a mapping function, which is provided by a statistical model that gives a non-critical baseline, which on the one side has to match the lattice QCD results and on the other side takes into account acceptance cuts that mirror the acceptance of the actual detector. More about the baseline function can be found in one of the following sections. A statistical model that is based on a realistic spectrum of particle states will also allow us to correct for additional detector-specific effects which will be detailed in the following Section. For a recent result on correcting the lattice QCD results on fluctuations to account for limited detector acceptance, see Ref. [29].

2. *Efficiency and purity*: Any particle measured in a realistic detector can only be measured with a certain efficiency, which means that part of the produced particles will be lost or misidentified during the detection process. A loss can occur due to the limited acceptance or the insufficient resolution in identifying either the track of the particle itself or the momentum of the particle or the identity of the particle. All these effects are generally determined by propagating a simulated collision and its particle output through a so-called detector response program, which allows us to determine the efficiency of the detection by comparing the generated particles to the reconstructed particles. In most experiments, the resulting efficiency ϵ is multiplied with the limited acceptance factor A to give a total correction factor $A\epsilon$.

The simulated collision is also used to determine other interaction- or detector-specific contributions to the particle spectrum. The so-called feed-down contribution captures the number of non-primordial particles in the sample of interest. This means that, for example, not all measured protons are protons produced in the collisions but also protons that reach the detector because they were produced in a decay of a higher mass primordial particle. This feed-down contribution can be accounted for as long as the simulated collision has a realistic primordial particle distribution, which is generally true for most of the so-called Monte Carlo event generators presently available. Besides feed-down, two more unwanted contributions to the particle spectrum exist. First, one can show that additional protons can be generated through interactions of traversing particles with the material of the detector. Second, one can show that the detector sometimes misidentifies other particles as protons. The sum of feed-down, material and misidentified protons can be determined in the simulation and then treated as a contamination which can be subtracted on an event-by-event basis, leading to a purity factor. Therefore, an experimental sample is generally defined by its efficiency and its purity. The two are not independent, since the reconstruction of a particular particle species requires cuts on the detector response, and by tightening those cuts one can improve the purity, but at the same time will reduce the efficiency.

3. Volume fluctuations: The volume can fluctuate event by event, leading to a non-thermal contribution to fluctuations of conserved charges that needs to be investigated and understood in order to perform a meaningful comparison between theory and experiment. The effect of volume fluctuations was studied in Refs. [30, 31]. In Ref. [30], the contribution of volume fluctuations to the cumulants of net-baryon number distribution was obtained. The quantitative impact of this effect on the measured cumulants was recently addressed in Ref. [32].

4. Global conservation laws: The three quantum numbers that have been studied in detail in experiment are the electric charge, strangeness and baryon number. Lattice QCD calculations are done in a Grand Canonical ensemble, which allows for the exchange of conserved quantities with the heat bath, and thus, exhibits fluctuations. In experiment, the quantum numbers would be conserved globally with an infinite acceptance, i.e. there would be no fluctuation, statistical or non-statistical. Therefore, the experimental approach is to try to mimic a Grand Canonical ensemble by analyzing a subset of the particles in the final state in a limited acceptance. However, in order to extract actual non-statistical fluctuations, one has to account for the contribution of quantum number conservation in the limited acceptance. This can be done simply on a global basis by determining an acceptance factor based on the measured protons over all generated baryons in a simulated collision or on a more local basis if one can show that the correlation of the produced baryons does not extend over the full phase space. We will discuss this effect in more detail in Sect. 7.2.3.

5. Particles vs. quantum numbers: The main experimental caveat of fluctuation studies is the intrinsic requirement to compare the behavior of a quantum number on the lattice with the one of a particle in experiment. In other words, can a particular particle species or even a set of particle species be a good proxy for a quantum number?

Generally, the easiest example is the electric charge. Detectors are tuned to detect only charged particles and if the acceptance is properly corrected for, the charged particle yield will properly describe the behavior of electric charge. Unfortunately, electric charge is rather insensitive to critical fluctuations or flavor-specific freeze-out.

The focus over the past decade was on net-proton fluctuations, since the proton couples to the σ field and its yield can be rather easily corrected to reflect the baryon number. Although the other abundant baryon, namely the neutron, cannot be measured well, isospin symmetry allows us to simply double the yield of the protons to account for the neutrons. All higher mass baryons contribute only little to the total baryon yield, and thus the net-proton is considered a viable proxy for the net-baryon number [33].

Over the past few years, strangeness number studies have become a second focus of detailed fluctuation measurements. Although kaons are not expected to show any hint of critical fluctuations, the interest arose from the question of flavor-specific hadronization patterns in the QCD crossover region. Here, kaons together with first measurements of strange baryon fluctuations can serve as a viable proxy for the strangeness number. The kaon itself accounts for about 70% of the strangeness in a heavy ion event, but together with the Λ measurements this number increases to almost 90%. For a more detailed discussion on particle proxies for quantum number fluctuations, see the discussion in Sect. 6.3.4.

7.2.3 The Statistical Baseline and the Impact of Conservation Laws

Any multiplicity distribution that fluctuates in and out of the detector acceptance should be describable by a statistical probability function as long as no external circumstances causing thermal or critical fluctuations are relevant. The most basic probability distributions considered are the Poisson distribution and the Negative Binomial distribution (NBD). For those functions, the first two moments of the particle and anti-particle distributions are directly related to the second central moment of the net-particle multiplicity.

The Poisson distribution is a discrete probability distribution that expresses the probability of a given multiplicity in one event, given that the mean multiplicity is known and its likelihood is independent from event to event. However, there is the expectation that the number of particles from one event to the next will fluctuate around its mean due to limitations caused by the acceptance. Within Poisson statistics, if the average number of one kind of particle from one event to the next is λ, then for the number of occurrences: $k = 0, 1, 2, \ldots,$ its probability mass function is given by [34]:

$$P^{PD}(k) = \frac{\lambda^k}{k!} exp(-\lambda). \tag{7.6}$$

A Poisson distribution can be fully described by its mean, λ. As a result, the cumulant of a Poisson distribution is: $\kappa_n = \lambda$. In the case of convolution of two Poisson distributions with mean λ for the particle, and mean $\bar{\lambda}$ for the anti-particle, the net distribution is a so-called Skellam distribution with λ - $\bar{\lambda}$ and $\lambda + \bar{\lambda}$ for odd cumulants and even cumulants, respectively.

The Negative Binomial distribution (NBD) is also a discrete probability distribution and it very closely resembles the Poisson distribution. In fact, the NBD converges to the Poisson distribution when the sample size is large enough. The NBD is defined as:

$$P^{NBD}_{p,k}(n) = \binom{n+k-1}{n}(1-p)^n p^k. \tag{7.7}$$

The mean of the distribution $\langle n\rangle$ is related to p by $p^{-1} = 1+\langle n\rangle/k$. This leads to the form of NBD that is commonly used to describe multiplicity distributions [35]:

$$P_{p,k}^{NBD}(n) = \binom{n+k-1}{n} \left(\frac{\langle n \rangle / k}{1 + \langle n \rangle / k} \right)^n \frac{1}{(1 + \langle n \rangle / k)^k}. \tag{7.8}$$

The cumulants of the NBD distribution can be described by the mean, λ, and variance, σ^2. If two parameters: α and β are defined as [36]:

$$\alpha = \frac{\lambda}{\sigma^2}$$
$$\beta = \frac{\lambda\alpha}{1-\alpha}, \tag{7.9}$$

then the cumulants κ_1 and κ_2 are:

$$\kappa_1 = \frac{\beta(1-\alpha)}{\alpha}$$
$$\kappa_2 = \frac{\beta(1-\alpha)}{\alpha^2}. \tag{7.10}$$

In contrast to a Poissonian distribution, here the variance is larger than the mean and the cumulants of the net distribution are κ_n-(anti-κ_n) and κ_n+(anti-κ_n) for odd and even cumulants, respectively.

Whether Poisson or NBD distributions are more appropriate to set a baseline value will depend on the relative abundance of particles vs. anti-particles. For example, at collision energies that reflect a high chemical potential, the primordial baryons make the net-baryon number quite asymmetric and here the use of a NBD is more appropriate, whereas at higher collision energies such as at the LHC, where the matter and anti-matter yields are equal, the mean and the variance of the net-baryons are very similar and a Skellam distribution is most appropriate for comparison to net-proton and net-strange baryon data.

The impact of conservation laws on these statistical distributions can be easily addressed by simply assessing the number of measured baryons in a particular acceptance over all baryons produced in 4π. Since this number is not measurable, any correction factor relies on comparison to realistic MC event generators. The correction factor though is rather trivial, as long as it can be established that the full 4π multiplicity is thermally equilibrated, i.e. global thermalization = global conservation. In that case, and for a non-interacting system like the HRG model, the baseline second-order cumulant has to be simply multiplied by 1-α where $\alpha = N_{measured\ baryons}/N_{4\pi\ baryons}$ [37]. The formula also holds for a more local conservation approach by simply replacing the 4π number with a smaller acceptance. Local conservation is more appropriate for smaller systems or heavy ion systems at lower energies, where stopping generates an intrinsic matter over anti-matter imbalance.

More recently, these corrections were generalized to the case of a strongly correlated system in Ref. [38]. It was shown that the relevant cumulant ratios measured in experiment are related to the lattice QCD ones by the following relations:

$$\frac{\kappa_2}{\kappa_1} = (1-\alpha)\frac{\chi_2^B}{\chi_1^B}$$

$$\frac{\kappa_3}{\kappa_2} = (1-2\alpha)\frac{\chi_3^B}{\chi_2^B}$$

$$\frac{\kappa_4}{\kappa_2} = (1-3\alpha(1-\alpha))\frac{\chi_4^B}{\chi_2^B} - 3\alpha(1-\alpha)\left(\frac{\chi_3^B}{\chi_2^B}\right)^2, \tag{7.11}$$

where α is the ratio between the sub-volume and the whole volume of the system. Figure 7.2 shows the effect of these corrections on the lattice QCD results for κ_4/κ_2 (top panel) and κ_6/κ_2 (bottom panel) as functions of α.

Further insight in the underlying physics can be found by studying any deviations from Skellam or NBD in MC transport codes such as UrQMD, AMPT or EPOS, and these type of comparisons are often found in experimental papers where the data are discussed.

7.2.4 Experimental Procedures to Determine Particle-Identified Event by Event Multiplicity Distributions

The two major presently operating collider facilities, RHIC and LHC, are equipped with a series of multi-purpose experiments, many of them capable of measuring significant portions of the emission of particles from elementary and heavy-ion collisions. These experiments generally follow a common design, namely a cylindrical geometry of various detector components at different radii around the interaction point where the collision between the two particles occurs. This so-called primary vertex is achieved by crossing the two beams circulating (one clockwise, one counter clockwise) in the accelerator. The achievable highest collision energies are limited by the strength of the electromagnetic fields used to accelerate the ion beams and by the size of the accelerator. The larger the radius of the collider ring, the higher the final energy. The 26 km circumference LHC accelerates heavy ion up to 5.02 TeV, whereas the roughly five times smaller RHIC ring reaches a maximum energy of 200 GeV. The more recent program of mapping out the full QCD phase diagram in search of a critical point required to lower the beam energy, in particular at RHIC, in order to reach down to finite chemical potential. In the normal collider mode, the lowest collision energy available at RHIC is 7.7 GeV. Circulating lower energy beams proved impractical because the luminosity is proportional to the relativistic gamma factor of the individual ion beams to the third power. The implementation of a fixed-target (FXT) in STAR, to be hit by a single circulating beam, enables the possible beam energies to be extended downward to also cover

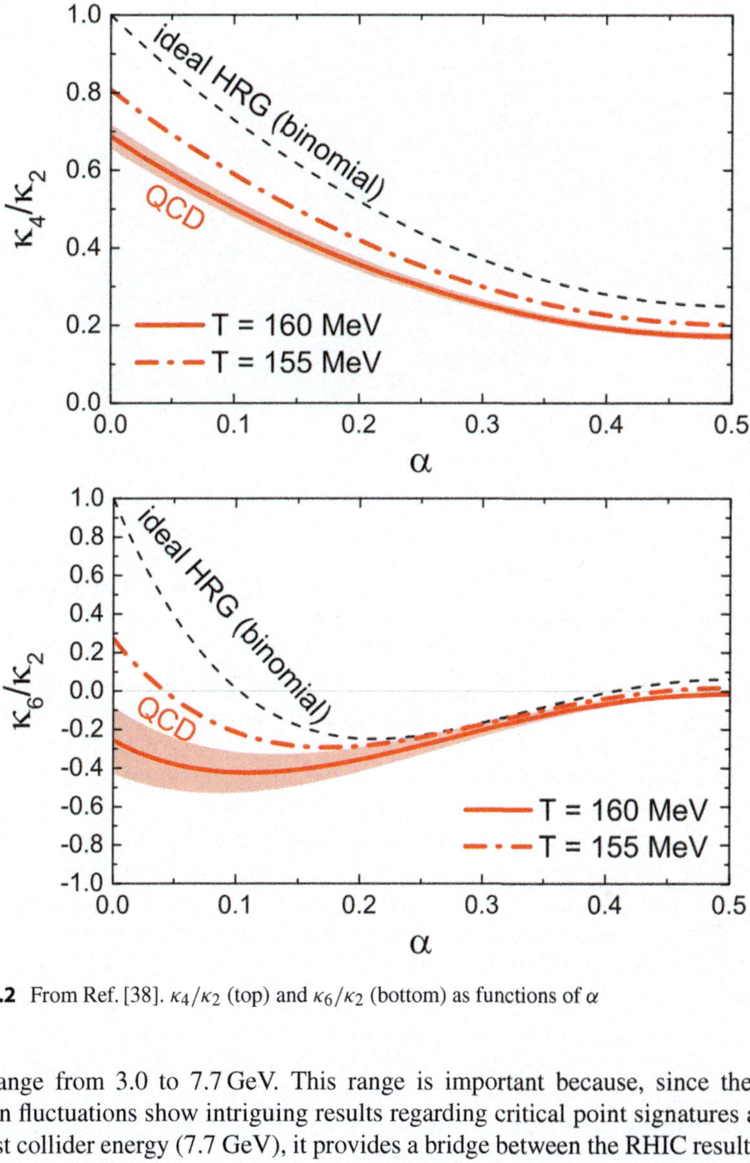

Fig. 7.2 From Ref. [38]. κ_4/κ_2 (top) and κ_6/κ_2 (bottom) as functions of α

the range from 3.0 to 7.7 GeV. This range is important because, since the net-proton fluctuations show intriguing results regarding critical point signatures at the lowest collider energy (7.7 GeV), it provides a bridge between the RHIC results and measurements from HADES at the GSI performed at 2.4 GeV (see Sect. 7.3.1.2). In order to conduct a FXT program at RHIC, a gold target, 1 cm in height, 6 cm in width, and 1 mm thick was installed inside the beam pipe at STAR. The target was positioned 2 cm below the beam axis and 2.1 meters to the West of the center of the detector. Several detector upgrades to the TPC and TOF components further improved the geometrical acceptance of STAR for the FXT.

A typical collider experiment, ALICE at the LHC, is shown in Fig. 7.3. The main cylindrical detector is augmented by 'forward' components, which capture a

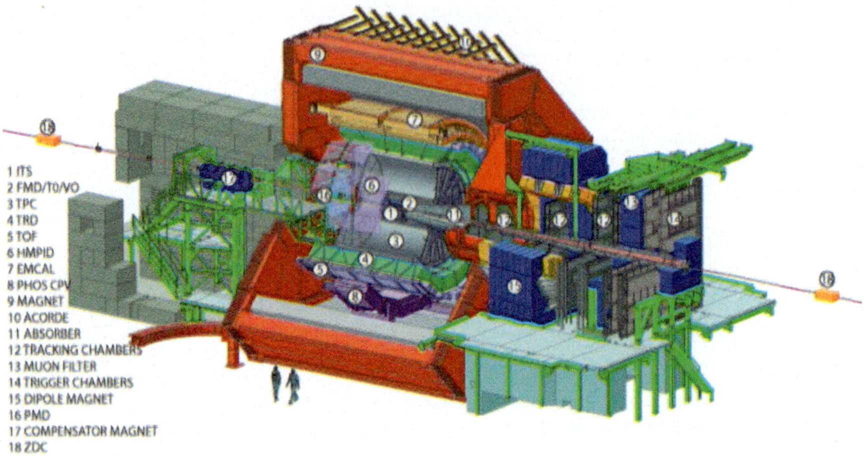

Fig. 7.3 Schematic diagram of the detector components of the ALICE experiment at the LHC

particular set of particles, namely muons, in addition to most of the particles which are emitted transversely, or perpendicular outward to the beam direction. Its overall dimensions are $16 \times 16 \times 26\,\text{m}^3$ and it weighs 10,000 tons. The central detector is embedded in a large 0.5T solenoid magnet, covering polar angles from 45 to 135 degrees. In this volume, any particle originating from the primary vertex will traverse several layers/components of the detector when moving radially outward. The initial layers have to be thin, if they are of solid state material or a low density gas volume, in order for the particle to be detected but not stopped or deflected on its path. After traversing most detector components, the final layer is of very dense solid material (often Lead or Tungsten) in order for the particle to stop and deposit all of its remaining energy. In general, the detector components that use minimal energy loss to determine a particle trajectory and identify the particle are called tracking detectors, the ones stopping the particle and measuring its full energy are called calorimeters. In the ALICE detectors the tracking detectors are the inner tracking system (ITS) and the time projection chamber (TPC). Another key particle identification detector is the time of flight detector (TOF) followed by the electromagnetic calorimeter (EMCal) as the outermost layer. All in all there are 18 different detector systems in ALICE. Here we focus only on those that are directly relevant for the measurements to be presented in this Chapter.

The ITS consists of seven concentric barrels of Silicon detectors located at radii between 2.3 and 39.3 cm from the primary vertex. The ITS is radially followed by the TPC, which is a 5 m long cylinder with a 1.65 m radial extension between r = 0.85 and 2.50 m. The $90\,\text{m}^3$ volume of the detector is filled with a gas mixture of 90% Ne and 10% CO_2/N_2. An electric field gradient of 400 V/cm is produced with a voltage of 100 kV on the central electrode. In both the ITS and the TPC any traversing charged particle will ionize the Silicon or the gas, respectively, and the released electrons will drift towards and electrode for readout. Both detectors thus collect position and energy loss information. Generally speaking, any event-by-event multiplicity distribution can be measured as long as the particle tracks can be sufficiently isolated and the event-by-event statistics is such that the measurement is significant. The easiest measurement is the net-charge measurement, since is does not require particle identification per se, but only the determination of positive and negative tracks in the magnetic field of the detector. It turns out that an important cross-check on the validity of the identified particle reconstruction approach is the agreement between a simple charged particle distribution and a distribution of the sum of identified charged pions, charged kaons and protons/anti-protons. This cross check has been successfully applied to the published STAR and ALICE data when accounted for weak decay and resonance contributions.

The main step for arriving at a valid particle distribution is the acceptance and efficiency correction based on the method and detectors employed to identify the individual particles. Certain detector components are optimized for specific momentum ranges of the particles. The top panel of Fig. 7.4 shows an example from ALICE for the pion spectrum. As can be seen, the low momentum part of the spectrum is determined by the tracking detectors, first the Inner Tracking Silicon (ITS) detectors and then the Time Projection Chamber (TPC). In both cases, the ionization energy loss per unit distance (dE/dx) is measured and used to identify the particles according to the Bethe-Bloch formalism [39] (see the bottom panel of Fig. 7.4). For higher momenta, both detectors employ a Time of Flight (TOF) measurement, before relying on the relativistic rise region of the Bethe-Bloch curve to identify even higher momenta. Generally the particle yield in the rise region is negligible and thus most experiments employ at most the TOF to reach out to high momenta [40, 41].

The dE/dx measurement can be analyzed in two different ways, the so-called 'traditional approach' and the 'identity approach' [42]. The identity approach leads to a more momentum-independent efficiency, but it is prone to a higher systematic uncertainty. ALICE investigated both approaches and the results are comparable, although the identity method might enable a larger statistical sample over a wider momentum range. Neutral weakly decaying particles, such as the Λ and the neutral Kaon, can also be investigated by reconstructing their decay and determining their yield through an invariant mass determination on an event-by-event basis (see e.g. [43, 44]). Since the efficiency of finding both decay daughters and properly reconstructing the original neutral particles is low, the correction factors are large, and the moments of these net-particle distributions carry a higher uncertainty. Nevertheless, the determination of e.g. the net-Λ cumulants turns out to be highly relevant for the question of hadronization and chemical freeze-out. Based on the theoretical modeling of chiral criticality and a critical point at finite baryon density, the key measurement at STAR, HADES and future lower energy experiments will be the net-proton fluctuations. It was also shown that a smaller event-by-event multiplicity leads to a better statistical error-bar, which experimentally favors the net-proton measurement over the net-charge measurement.

Generally, any ΔN_x measurement is carried out over a restricted rapidity and transverse momentum range. This leads to an acceptance which is determined

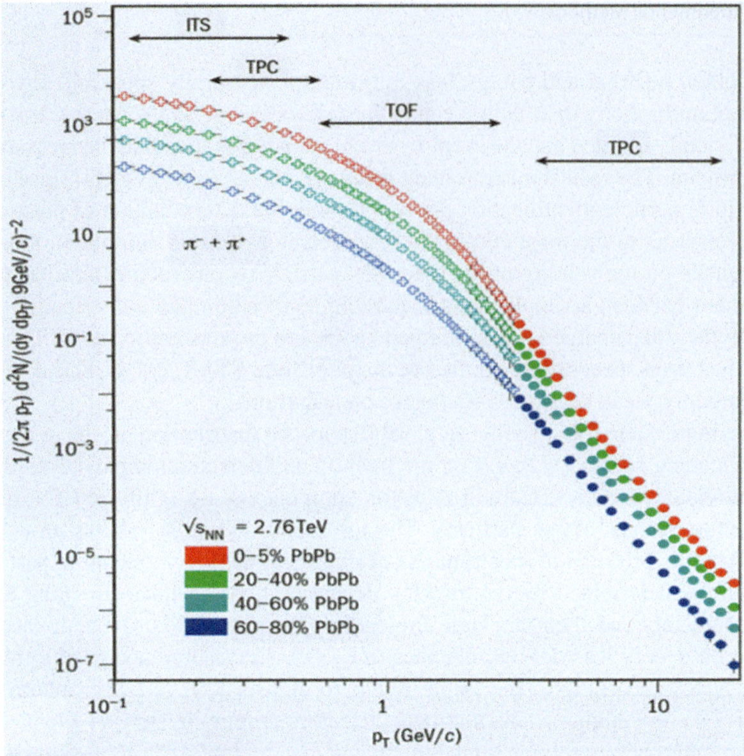

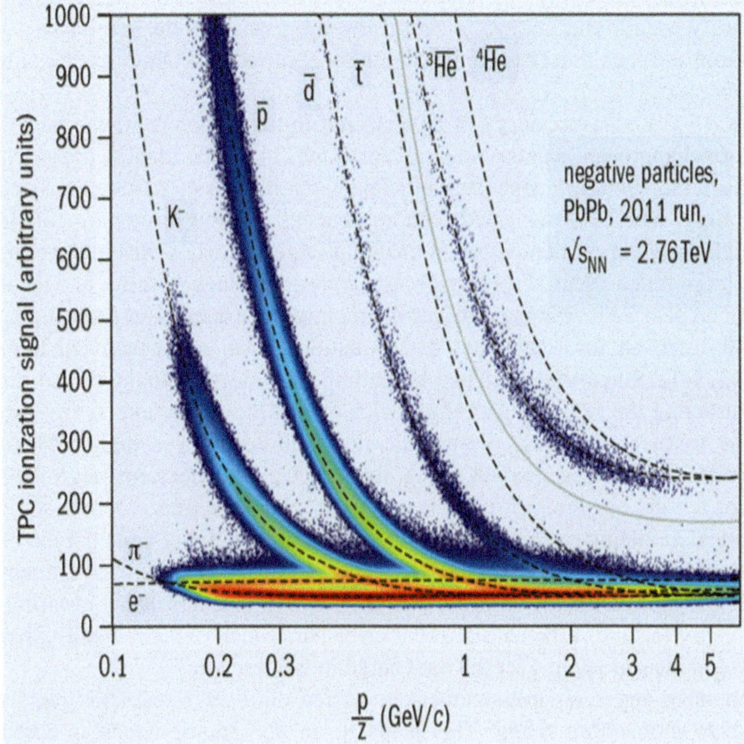

Fig. 7.4 (continued)

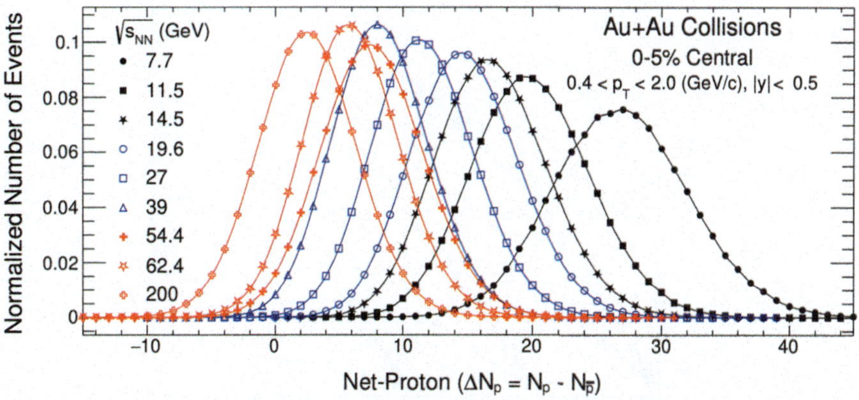

Fig. 7.5 From Ref. [45]. Event-by-event net-proton number distributions for head-on (0-5% central) Au+Au collisions for nine $\sqrt{s}$ values measured by the STAR detector at RHIC. The distributions are normalized to the total number of events at each $\sqrt{s}$. The statistical uncertainties are smaller than the symbol sizes and the lines are to guide the eye. The distributions in this figure are not corrected for proton and anti-proton detection efficiency

through comparisons to a 4π Monte Carlo simulation with a realistic event generator model such as HIJING, AMPT, UrQMD or EPOS [46–49]. Figure 7.5 shows typical net-proton multiplicity distributions for central Au+Au collisions at different collision energies. The centrality dependence of the resulting cumulants for different collision energies is shown in the top panel of Fig. 7.6. Particle production at any given centrality can be considered a superposition of several identically distributed independent sources, the number of which is proportional to N_{part}. This is the so-called central limit theorem, which leads to a linear increase of all cumulants as the volume of the system increases. Since skewness and kurtosis are cumulant ratios, their centrality dependence is not a linear increase but the curves still follow the central limit theorem as shown in the bottom panel of Fig. 7.6. All net-particle measurements obtained by STAR and ALICE follow these general trends. In order to now obtain a significant deviation from a statistical baseline, either Poissonian or NBD (see Sect. 7.2.3), one has to efficiency and acceptance correct these measurements.

It is difficult to correct moments for particle-reconstruction efficiencies on an event-by-event basis. The easiest way is to factorize the correction factors according

Fig. 7.4 Top: Pion yield from PbPb collisions at centre-of-mass energy $\sqrt{s}$= 2.76 TeV for different collision centralities measured in ALICE. To cover the whole momentum range, data from different ALICE subdetectors are combined - in this plot the particle identification information comes from ionization (ITS and TPC) and time-of-flight (TOF) measurements. Bottom: Ionization signals in the ALICE TPC as functions of the particle momentum for PbPb collisions at $\sqrt{s}$= 2.76 TeV. An offline trigger was applied to enhance track samples with charges $z < -1$. The dashed lines are parameterizations of the Bethe-Bloch curve

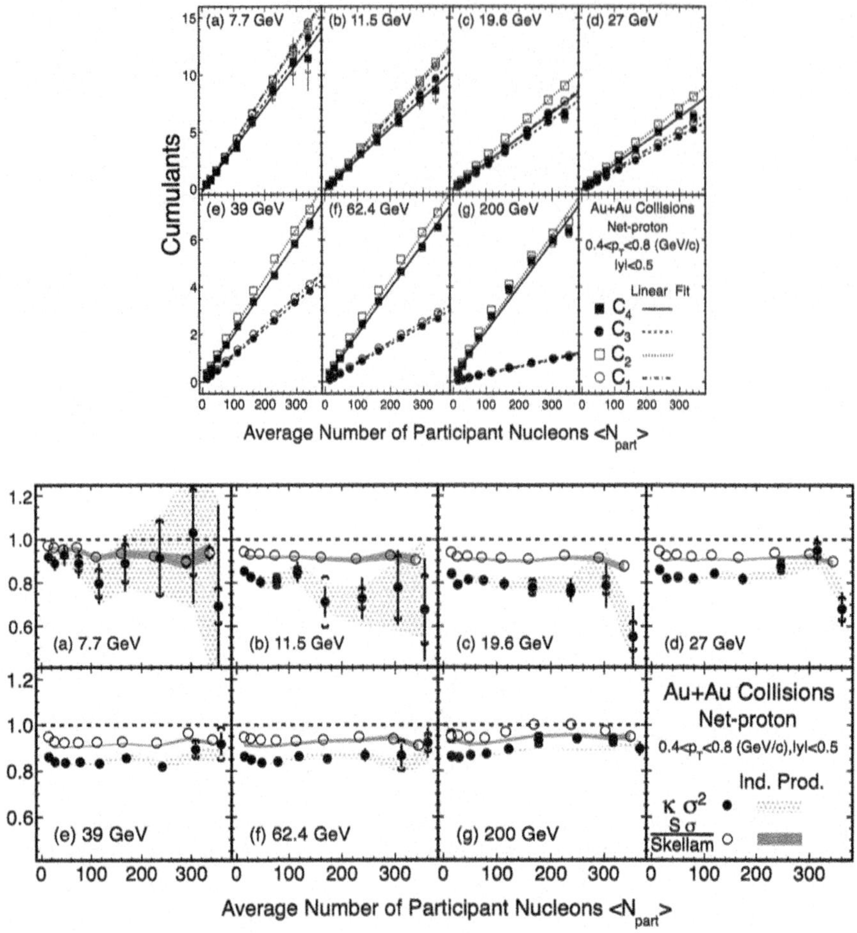

Fig. 7.6 From Ref. [51]. Top: centrality dependence of the cumulants of net-proton distributions for Au+Au collisions. Error bars are statistical and caps are systematic errors. Bottom: centrality dependence of $S\sigma$/Skellam and $\kappa\sigma^2$ for net-proton distributions in Au+Au collisions at different $\sqrt{s}$. The results are corrected for the $p(\bar{p})$ reconstruction efficiency. The error bars are statistical and caps are systematic errors. The shaded bands are expectations assuming the approach of independent proton and anti-proton production. The width of the bands represents statistical uncertainty

to the factorial moment under the assumption that the efficiency correction follows a binomial distribution. This method has been generally applied to the experimental data in the past for two different scenarios, namely a transverse momentum-independent efficiency measurement (small momentum range) [50] and a more involved procedure for a transverse momentum-dependent efficiency [52]. It can also be shown that the aforementioned identity method of particle identification allows for a large momentum range measurement with a momentum-independent efficiency. Lately, certain methods were also proposed based on non-binomial distributions or track-by-track efficiencies [53,54]. The efficiency corrections, together with a method to suppress volume fluctuation effects when the measurements are taken over a wide centrality range (Centrality Bin Width Correction (CBWC) [55]), and corrections based on the impact of conservation laws when the data are measured over a limited acceptance [37], lead to fully-corrected experimental data that can then be compared to statistical baseline values or HRG/lattice QCD predictions in order to determine relevant physics effects.

7.3 Results

7.3.1 Results on Searches for a Critical Point

7.3.1.1 Event-by-Event Net-Particle Multiplicities as a Proxy for Conserved Quantum Numbers

Since the early, coarse grain lattice QCD predictions on the location of the critical point were never verified in recent and more precise lattice studies, most of the guidance on the location of the critical point come from effective theories, and the predictions for relevant measurements are largely based on the early studies using the linear sigma model, as detailed in previous sections. Since the magnitude of critical fluctuations depends strongly on the order of the moments, due to their dependence on powers of the correlation length, higher moment measurements will exhibit the largest effects. As was shown, the baryon number couples to the σ field and the event-by-event net-proton multiplicity measurements are a quantitative proxy for the net-baryon number, since isospin randomization can account for the neutron contribution [33] and the remaining strange and charm baryon contributions are small.

A similar argument can be made for net-charge measurements when summing all the charged states for pion, kaon and proton and comparing them to unidentified charged particle measurements. Since the results matched, it was concluded that the net-charge fluctuations are also good proxies for the charge quantum number. Unfortunately, the higher statistics of charged particles made the measurement less reliable and the σ model also showed a much smaller coupling to the critical mode for net-charges than for net-protons.

Finally, it was suggested that net-kaon measurements could be a valid proxy for net-strangeness susceptibilities. This is a more complicated case, since strange baryons carry a substantial amount of the total strangeness in relativistic heavy ion collisions and cannot be ignored. Fortunately, early measurements on net-kaons were later on amended by net-Λ measurements. In general, net-kaons showed little or no sensitivity to the critical point in the theoretical models, but any flavor dependence in the fluctuation measurements could still be used to better understand the hadronization process in the QCD crossover region.

7.3.1.2 Net-Proton Measurements

The most recent results from the STAR collaboration for net-proton $S\sigma$ and $\kappa\sigma^2$ as functions of $\sqrt{s}$ are shown in the two panels of Fig. 7.7. Regarding the search for a critical point, none of the other identified net-particle measurements rises up to the quality of the net-proton measurements, and thus we focus here on the detailed results, in particular by the STAR and HADES experiments, on higher order net-proton fluctuations. The as of now only significant measurements outside of statistical and systematic error bars were for higher moments, in particular the lowest energy measurements of kurtosis ratio χ_4^B/χ_2^B. The STAR measurements, which were published first in 2013 [51] and then expanded in 2020 [45], exhibit two relevant features. First, they indicate a 'dip' and thus a change of sign in the energy dependence of χ_4^B/χ_2^B in the 10–20 A GeV energy range, which was one of the theoretical predictions [56]. This prediction was discussed in Sect. 5.4.1. In addition, they indicate a trend towards increasing values in χ_4^B/χ_2^B at the lowest measured collision energies, but still within statistical uncertainties of the baseline, as shown in the right panel of Fig. 7.7.

Other possible explanations for the dip emerged recently. It was pointed out in Refs. [32, 37] that the conservation of baryon number is expected to play a big role at low collision energies, where the system is smaller. The effect of baryon number conservation on the net-proton κ_4/κ_2 is shown in the left panel of Fig. 7.8. Lattice QCD predictions for χ_4^B/χ_2^B as functions of the chemical potential also show a

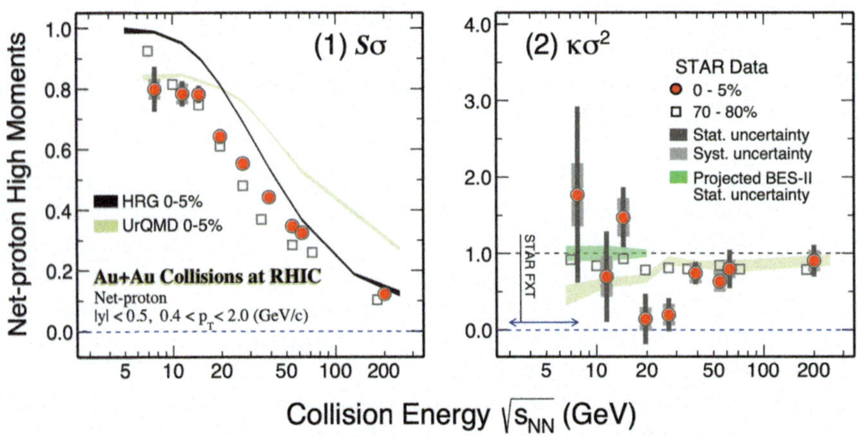

Fig. 7.7 From Ref. [45]. $S\sigma$ (left) and $\kappa\sigma^2$ (right) as functions of collision energy for net-proton distributions measured in Au+Au collisions. The results are shown for central (0-5%, filled circles) and peripheral (70–80%, open squares) collisions within $0.4 < p_T(GeV/c) < 2.0$ and $|y| < 0.5$. The vertical narrow and wide bars represent the statistical and systematic uncertainties, respectively. The shaded green band is the estimated statistical uncertainty for BES-II and the energy range for STAR fixed-target (FXT) program is shown as arrows in the right panel. The peripheral data points have been shifted along the x-axis for clarity of presentation. Results from a hadron resonance gas model [57] and a transport model calculation (UrQMD [58]) for central collisions (0–5%) are shown as black and gold bands, respectively

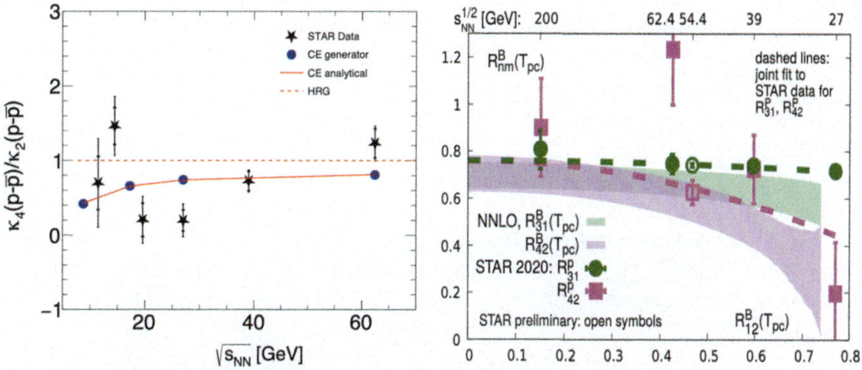

Fig. 7.8 Left: from Ref. [32]. Normalized fourth order cumulants of net protons in the collision energy range from $\sqrt{s} = 8.8$ to 62.4 GeV. A Kolmogorov-Smirnov test indicates that the deviations between the STAR data and the theoretical results, represented by the solid red line, are not statistically significant. The two-tailed p-value is about 0.3. Right: from Ref. [62]. $S_B\sigma_B^3/M_B$ (green) and $\kappa_B\sigma_B^2$ (pink) as functions of $M_B\sigma_B^2$ on the pseudo-critical line calculated from NNLO Taylor series. Data are results on cumulant ratios of net proton-number fluctuations obtained by the STAR Collaboration [45]

decrease of this ratio for increasing values of μ_B. This behavior is shown in the right panel of Figs. 7.8 and in the bottom panel of 7.9 and this suggests that at least some contribution to the experimentally observed dip comes from the equilibrium equation of state. Lattice QCD extrapolations to finite μ_B are also in line with the experimental results for χ_3^B/χ_1^B, as shown in the right panel of Fig. 7.8 and in the top panel of Fig. 7.9. It was recently pointed out that the original prediction for the dip [56] was based on the leading critical contribution to the fourth-order cumulant in the vicinity of a 3D Ising model critical point. However, when sub-leading contributions are taken into account, the dip along the chemical freeze-out line is no longer visible [59].

Recently, the HADES experiment contributed data to the same measurement at an even lower beam energy [60]. The data are again consistent with the statistical baseline at this lower energy, although the result is highly centrality dependent (see Fig. 7.10). At this point in time the data are, as of yet, inconclusive with respect to the discovery of a critical point and more high-precision data are necessary. It was pointed out that an extension of rapidity and p_T coverage would lead to an increase in the magnitude of the critical point fluctuation signatures [61].

With respect to connecting these measurements to lattice QCD one needs to clearly state that the latest state of the art lattice calculations up to around μ_B = 350 MeV show no indications of a critical point. This statement is hampered by the fact that all methods to expand to finite chemical potential are based on the extrapolation of a non-critical system at imaginary or zero chemical potential, and thus do not necessarily allow for critical behavior in their predictions. In that context, an extension of lattice QCD to higher μ_B using an AdS/CFT approach

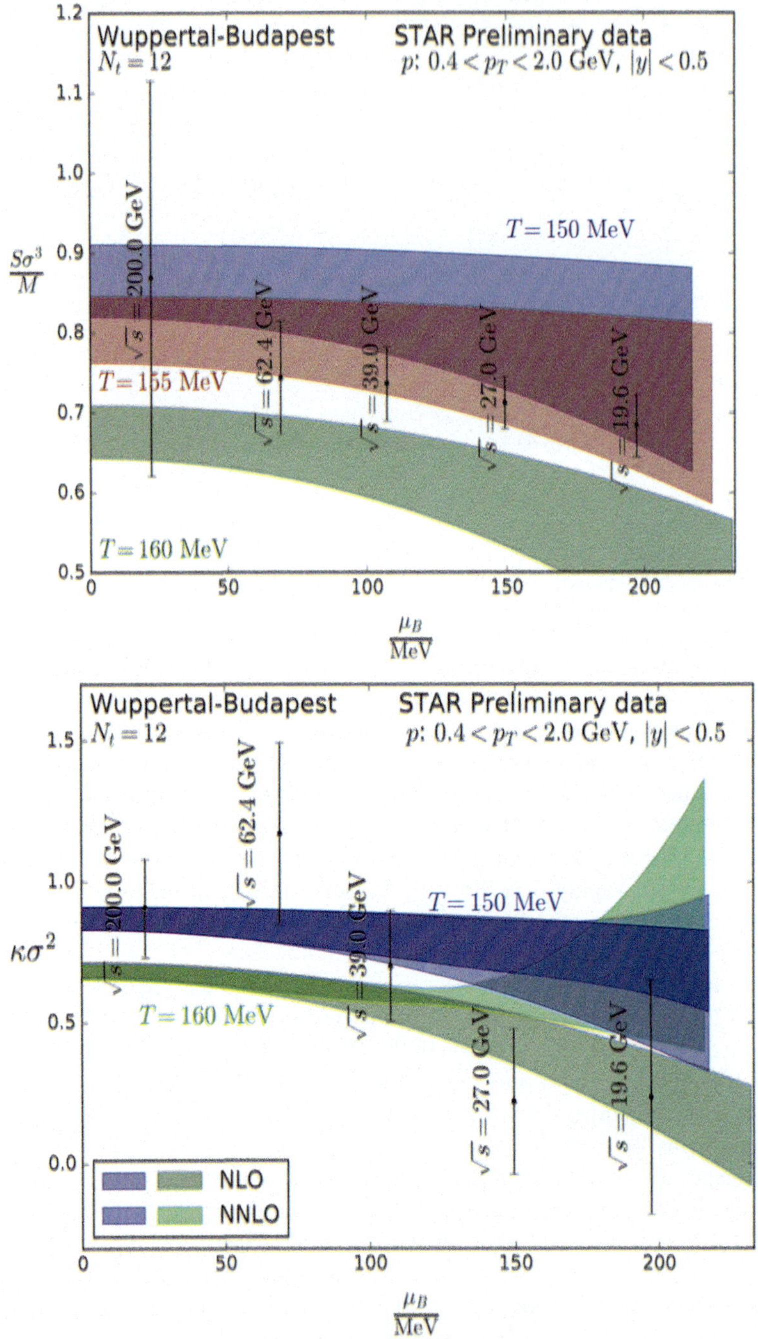

Fig. 7.9 From Ref. [63]. $S_B\sigma_B^3/M_B$ (top panel) and $\kappa_B\sigma_B^2$ (bottom panel) extrapolated to finite chemical potential. The top panel is extrapolated up to $\mathcal{O}(\mu_B^2)$. In the bottom panel, the darker bands correspond to the extrapolation up to $\mathcal{O}(\mu_B^2)$, whereas the lighter bands also include the $\mathcal{O}(\mu_B^4)$ term. In both panels, the black dots are experimental measurements on cumulant ratios of net proton-number fluctuations obtained by the STAR Collaboration [51]

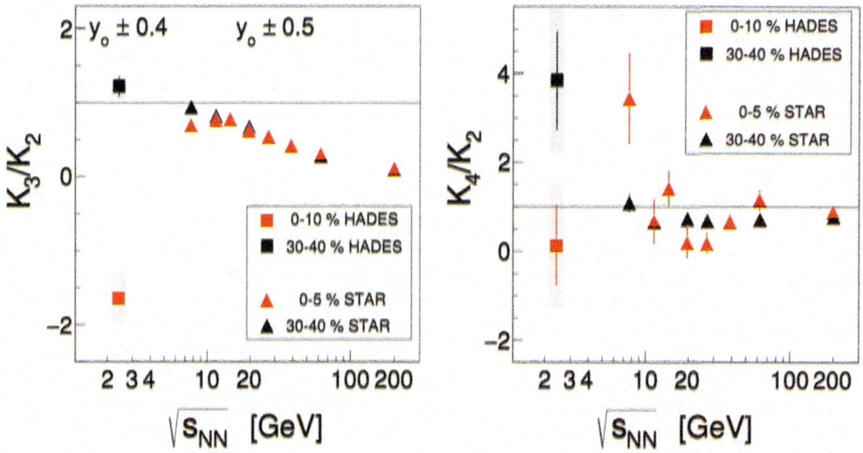

Fig. 7.10 From Ref. [60]. κ_3/κ_2 (left) and κ_4/κ_2 (right) as functions of $\sqrt{s}$. The triangles correspond to the preliminary STAR data presented in [65], while the squares are the HADES data at $\sqrt{s} = 2.4$ GeV

is of particular interest, since it matches the latest lattice results of higher order susceptibilities at lower μ_B and arrives at a specific prediction for higher μ_B, namely a critical point in the $\sqrt{s}$ range between 2.5 and 4.1 GeV [20].

Recently, ALICE also published its first results on net-protons at μ_B=0, i.e. outside the critical region in the phase diagram [64]. At this time, the measurements only reach out to κ_2, the variance, but they nevertheless turned out to be useful in determining the effect of baryon number conservation through variations in the detector acceptance (see the top panel of Fig. 7.11). The pseudo-rapidity dependence of the deviation of the variance from Skellam expectations revealed that net-baryon distributions are governed by long-range correlations over almost the complete fireball, at least in central Pb-Pb collisions. A comparable effect was also observed in net-Λ fluctuations at RHIC energies, taking into account not only baryon number but also strangeness conservation [44] (see the bottom panel of Fig. 7.11).

7.3.1.3 Other Net-Particle Measurements

Regarding other particle-identified measurements, it is important to note that, as of yet, no critical behavior has been detected in cumulant ratios for net-charges [67–69], net-kaons [70] and net-Λs [44]. Results from STAR are shown in Fig. 7.12. The net-charge and net-kaon measurements from STAR cover all cumulants up to κ_4 (note that, in the collaboration notation, cumulants are indicated as C_i and not κ_i as it is done here). The net-Λ measurements only reach up to κ_3 due to statistics restrictions. All of these measurements will receive a major statistics boost with the future runs at RHIC. Although there is no evidence for criticality in these measurements, they shed light on hadronization and freeze-out properties, which will be discussed in Sect. 7.3.3.

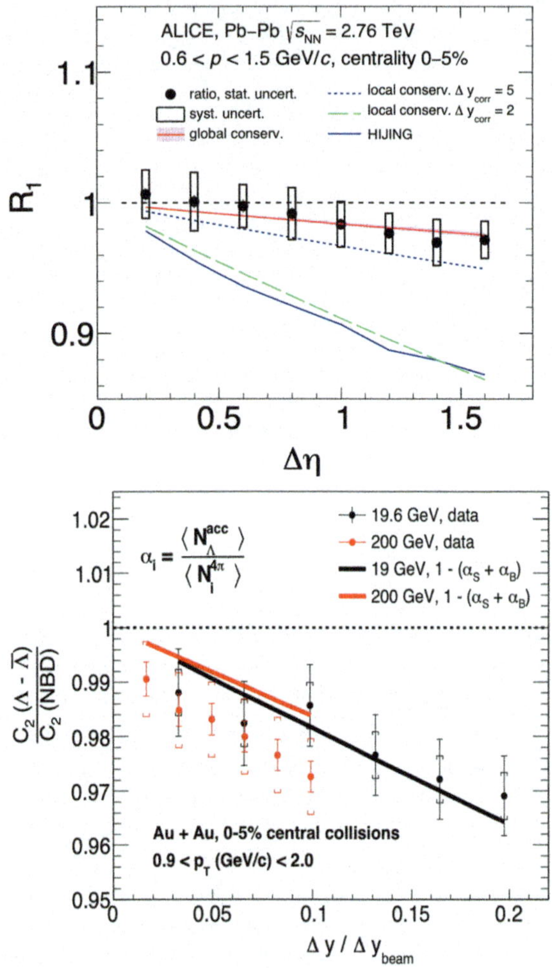

Fig. 7.11 Top: from Ref. [64]. Pseudo-rapidity dependence of the normalized second cumulant of net-protons $R_1 = \kappa_2(n_p - n_{\bar{p}})/\langle n_p + n_{\bar{p}}\rangle$. Global baryon number conservation is depicted as the pink band. The dashed lines represent the predictions from the model with local baryon number conservation [66]. The blue solid line represents the prediction using the HIJING generator. Bottom: from Ref. [44]. Rapidity dependence of the normalized $C_2(\Lambda - \bar{\Lambda})$ in most central (0-5%) collisions for 19.6 and 200 GeV collision energies. The solid lines show the expected effects from baryon number (B) and strangeness (S) conservation

7.3.2 Results on Searches for Chiral Criticality

Another type of critical fluctuations that has been predicted first by chiral effective theories [71] is based on the chiral transition at low μ_B, i.e. in the crossover region away from the critical point. Recent lattice QCD calculations seem to confirm this behavior for the higher order moments, which are potentially experimentally accessible (κ_6 and κ_8) [63, 72]. This is a region of the phase diagram that can be explored by RHIC and LHC experiments, but already sixth order cumulants require a very large number of recorded collisions in order to show a statistically significant result [73].

While ALICE is scheduled to accumulate the necessary statistics in the upcoming run period from 2022 to 2025, STAR has shown first preliminary results [74] that are

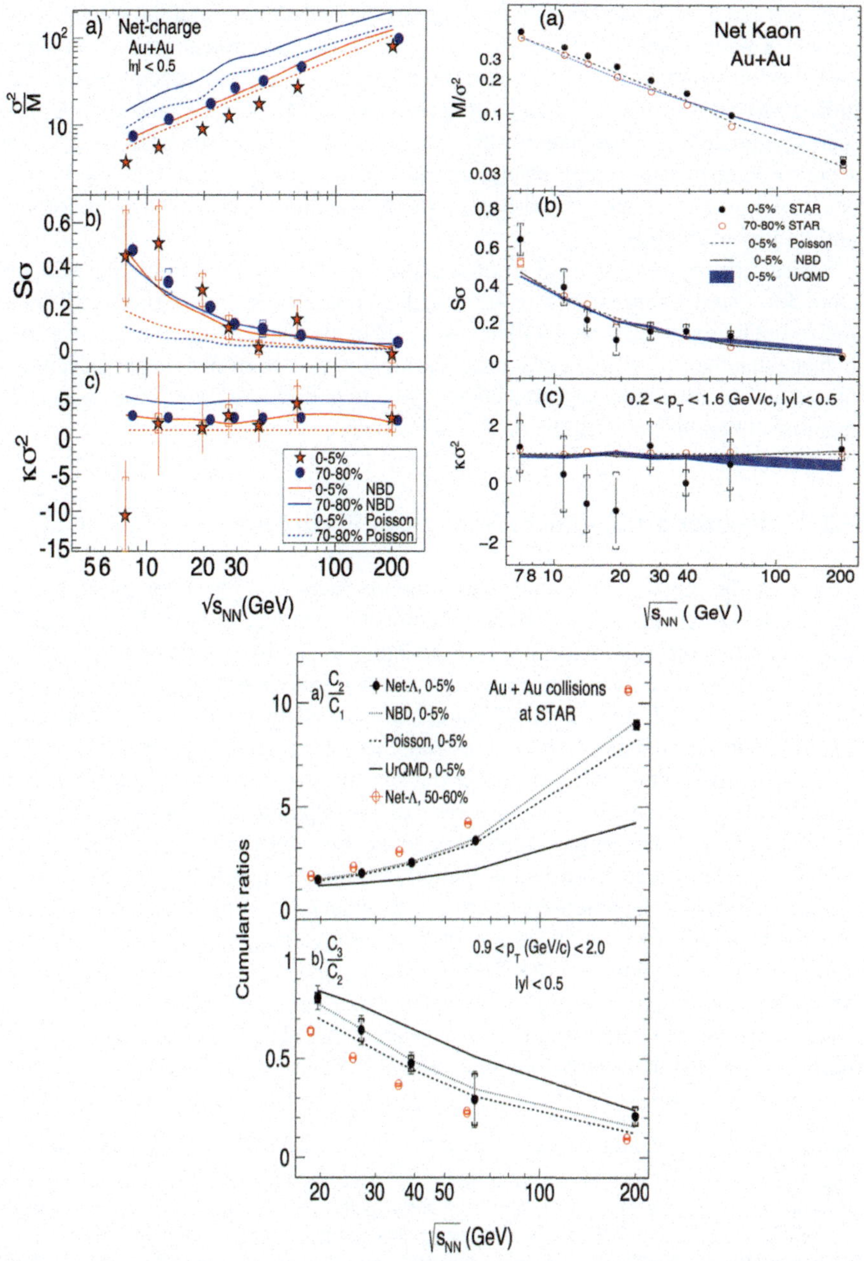

Fig. 7.12 Top left: from Ref. [68]. Beam-energy dependence of net-charge (**a**) σ^2/M, (**b**) $S\sigma$, and (**c**) $\kappa\sigma^2$, after all corrections, for most central (0–5%) and peripheral (70–80%) bins. The error bars are statistical and the caps represent systematic errors. Results from the Poisson and the NBD baselines are superimposed. The values of $\kappa\sigma^2$ for Poisson baseline are always unity. Top right: from Ref. [70]. Collision energy dependence of the values of M/σ^2, $S\sigma$, $\kappa\sigma^2$ for ΔN_K multiplicity distributions from 0-5% most central and 70-80% peripheral collisions in Au+Au collisions at $\sqrt{s} = 7.7,\ 11.5,\ 14.5,\ 19.6,\ 27,\ 39,\ 62.4$ and 200 GeV. The error bars are statistical uncertainties

quite intriguing inasmuch as they focus on one predicted feature of the calculations, namely a sign change of C_6/C_2 for net-protons as a function of collision energy (see the top panel of Fig. 7.13). The error bars are large and specific values are difficult to extract, but the change of sign between 54.4 A GeV and 200 A GeV seems significant. Unfortunately, very recent lattice QCD calculations [62] do not seem to confirm such a sign change in the relevant energy range (see the bottom panel of Fig. 7.13). One should note, though, that the experimental data are still preliminary at this time.

Chiral criticality is relevant to determine a model-free pseudo-critical temperature for chiral symmetry restoration and compare it to the deduced chemical freeze-out temperature which will be discussed in the next section. Most past results indicate that the two temperatures are close together, i.e. hadronization and chemical freeze-out occur almost simultaneously, which indicates a very short period of inelastic collisions between the produced hadrons.

7.3.3 Results on Chemical Freeze-Out Predictions

This is another highly relevant physics focus on fluctuation measurements which is independent of the existence of a critical point and relates directly to the susceptibilities of conserved quantum numbers on the lattice at any baryon density. In fact, the quantities derived here from an event-by-event net-particle measurement require the thermally equilibrated system that is a condition for lattice QCD to be applicable. The relationship between temperatures and chemical potentials derived from these measurements and the chemical freeze-out surface that they define, independent of any statistical hadronization model, has been first pointed out in Sect. 5.3.2 (see also Ref. [75]). This comparison between measurements and lattice QCD results therefore defines freeze-out/hadronization parameters based on first principles. Any additional agreement with Hadronic Resonance Gas models can be further utilized to confirm the applicability of both calculations, one based on hadrons and one on quark fields, in the QCD crossover region. Critical behavior though would be detrimental to the quantitative determination of a freeze-out surface in an equilibrated system and therefore the measurements need to be first examined for non-statistical fluctuations before being applied to thermal calculations.

Fig. 7.12 (continued) and the caps represent systematic uncertainties. The expectations from Poisson and NBD and the results of the UrQMD model calculations are all from the 0-5% centrality. Bottom: from Ref. [44]. Beam energy dependence of net-Λ cumulant ratios, C_2/C_1 (in our notation κ_2/κ_1) and C_3/C_2 (in our notation κ_3/κ_2) in most central (0–5%) and peripheral (50–60%) Au + Au collisions. NBD and Poisson baselines are presented by dashed lines. UrQMD predictions are shown in solid lines. Black vertical lines represent statistical uncertainties and caps represent systematic uncertainties. Results are corrected for the reconstruction efficiency and the CBWC is applied. The red data points are shifted left for clarity

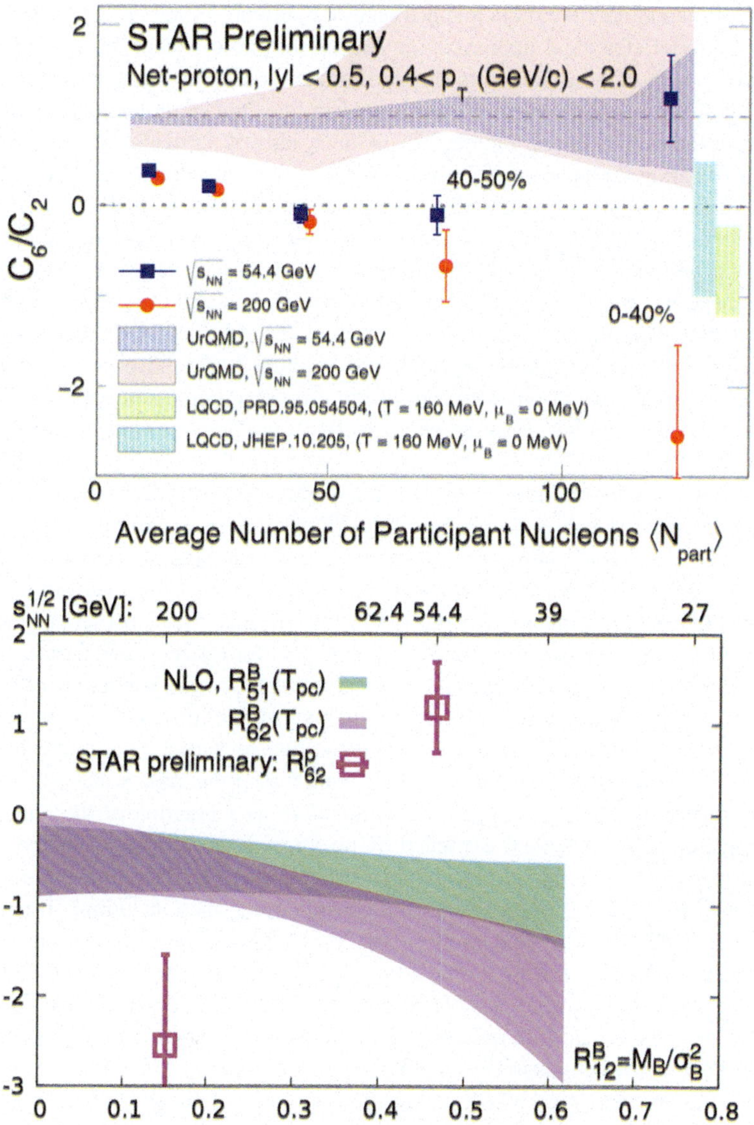

Fig. 7.13 Top: from Ref. [74]. Centrality dependence of C_6/C_2 of net-proton multiplicity distributions in Au+Au collisions at $\sqrt{s}$ = 54.4 and 200 GeV. The most central bin represents 0-40% centrality. The hadron transport model (UrQMD) calculations are shown in blue and red bands for two beam energies. The cyan and yellow bands show LQCD calculations taking T = 160 MeV. Bottom: from Ref. [62]. The cumulant ratios $R_{51}^{B}(T, \mu_B) = \chi_5^B(T, \mu_B)/\chi_1^B(T, \mu_B)$ and $R_{62}^{B}(T, \mu_B) = \chi_6^B(T, \mu_B)/\chi_2^B(T, \mu_B)$ vs $R_{12}^{B}(T, \mu_B) = \chi_1^B(T, \mu_B)/\chi_2^B(T, \mu_B)$ evaluated on the pseudo-critical line. Data are preliminary results for the cumulant ratio R_{62}^{p} of net-proton number fluctuations by STAR shown on the top panel

Experimental results on chemical freeze-out parameters are generally based on applications of statistical hadronization models to particle-identified yields under the assumption of emission from an equilibrated state. The same process can be applied to the event-by-event measurement of the higher moments, namely the variance, the skewness and the kurtosis of net yields of particular particle species. The justification for those measurements that go beyond the simple integrated yields is two-fold. First, they might lead to a higher precision in the deduced freeze-out parameters [76, 77], in particular if the well-measured event-by-event means and variances are used, and second, they link to a model-free determination of freeze-out parameters since lattice QCD enables the calculation of these parameters on the basis of the mapping between measured cumulants and calculated susceptibilities. Furthermore, quantum number-specific results from lattice have predicted subtle differences in the freeze-out between different quantum numbers based on higher moments, which can be measured exclusively on the basis of multiplicity fluctuations [78].

The most recent thermal fit to particle yields at the LHC was published in Ref. [24]. The fit yields a freeze-out temperature of $T_{ch} = 156.5 \pm 1.5$ MeV for $\sqrt{s} = 2.76$ TeV. The comparison between fit results and data is shown in the left panel of Fig. 7.14. It is evident that this freeze-out temperature overestimates the protons and underestimates the (multi)-strange baryons. The most recent fit from the STAR collaboration was performed in Ref. [79] for different collision energies. The extracted freeze-out parameters are shown in the top right panel of Fig. 7.14. In the same reference it was pointed out that the inclusion of different sets of particles in the fit yields different freeze-out temperatures. In particular, strange hadrons systematically increase T_{ch}. This is evident from the bottom right panel of Fig. 7.14, which shows the chemical freeze-out temperature as a function of the number of participants at $\sqrt{s}$ =39 GeV, extracted by performing thermal fits of different sets of particle yields. The flavor hierarchy in the freeze-out temperature was recently confirmed in Ref. [28], where it was shown that, even when including all particles from the PDG2016+ list and even Quark Model states, the difference in freeze-out temperatures between light and strange hadrons persists (see the two panels of Fig. 7.15), contrary to what was predicted in Ref. [80]. Other possible explanations for this discrepancy include rescattering in the hadronic phase [81, 82], in-medium effects [83] or corrections to the ideal HRG model [26, 84].

The first results from STAR on net-proton and net-charge fluctuations were used to show that indeed a model-free, i.e. lattice QCD-based, determination of a common freeze-out surface for the two related quantum numbers (charge and baryon number) is possible and slightly below the latest pseudo-critical temperature from the lattice in the baryo-chemical potential range of μ_B=0–200 MeV [86, 87]. An upper limit for the transition temperature was obtained by comparing results for the lattice curve for χ_3^B/χ_1^B, which is independent of the chemical potential to leading order as shown in Eq. (5.24), to the experimental value of net-proton $S\sigma^3/M$ from Ref. [51]. This comparison is shown in the left panel of Fig. 7.16. The right panel of Fig. 7.16 shows the determination of the baryonic chemical potential, through a

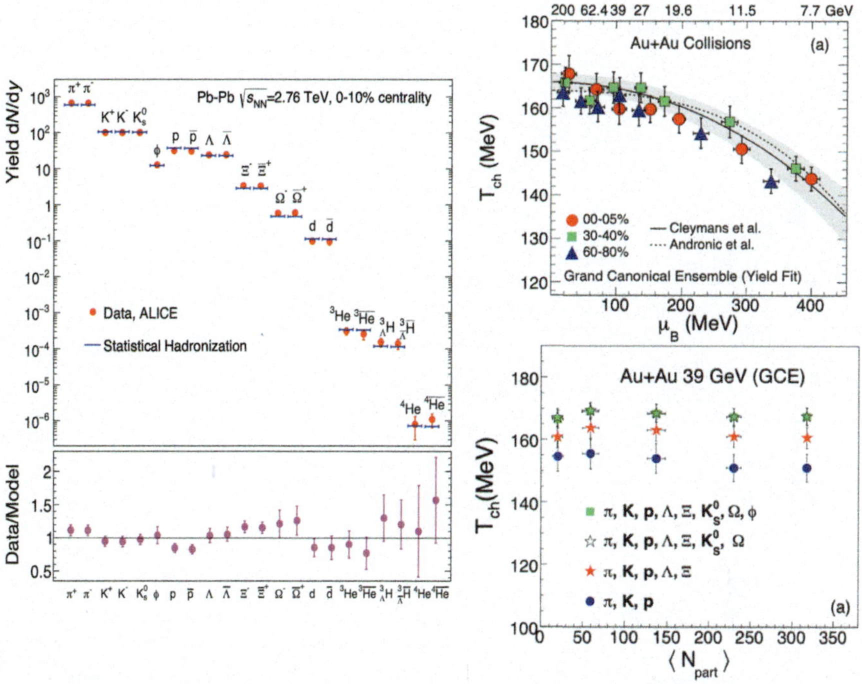

Fig. 7.14 Left: from Ref. [24]. Hadron abundances and statistical hadronization model predictions. Here dN/dy values for different hadrons measured at mid-rapidity are compared with the statistical hadronization analysis. The data are from the ALICE collaboration for central Pb-Pb collisions at the LHC. The lower panel shows the ratio of data and statistical hadronization predictions with uncertainties determined only from the data. Right: from Ref. [79]. Extracted chemical freeze-out temperature versus baryon chemical potential (top). Extracted chemical freeze-out temperature using particle yields as inputs in the fit. Results are compared for Au+Au collisions at $\sqrt{s}$=39 GeV for four different sets of particle yields (bottom). In bot panels on the right, uncertainties represent systematic errors

comparison of the ratio χ_1^B/χ_2^B from lattice QCD extrapolated to finite μ_B in the range of temperatures extracted from χ_3^B/χ_1^B, and the net-proton M/σ^2 at the four highest collision energies at RHIC from Ref. [51]. The left panel of Fig. 7.17 shows an independent determination of μ_{Bch}, through a comparison of the ratio χ_1^Q/χ_2^Q from lattice QCD extrapolated to finite μ_B in the range of temperatures extracted from χ_3^B/χ_1^B, and the net-electric charge M/σ^2 at the four highest collision energies at RHIC from Ref. [68]. The right panel of Fig. 7.17 shows that the two sets of chemical potentials obtained from these two independent analyses are consistent with each other [86].

Once the consistency between the electric charge and baryon number freeze-out parameters was established, a combined fit of the lowest-order ratio χ_1/χ_2 for electric charge and baryon number was performed, to obtain a more precise

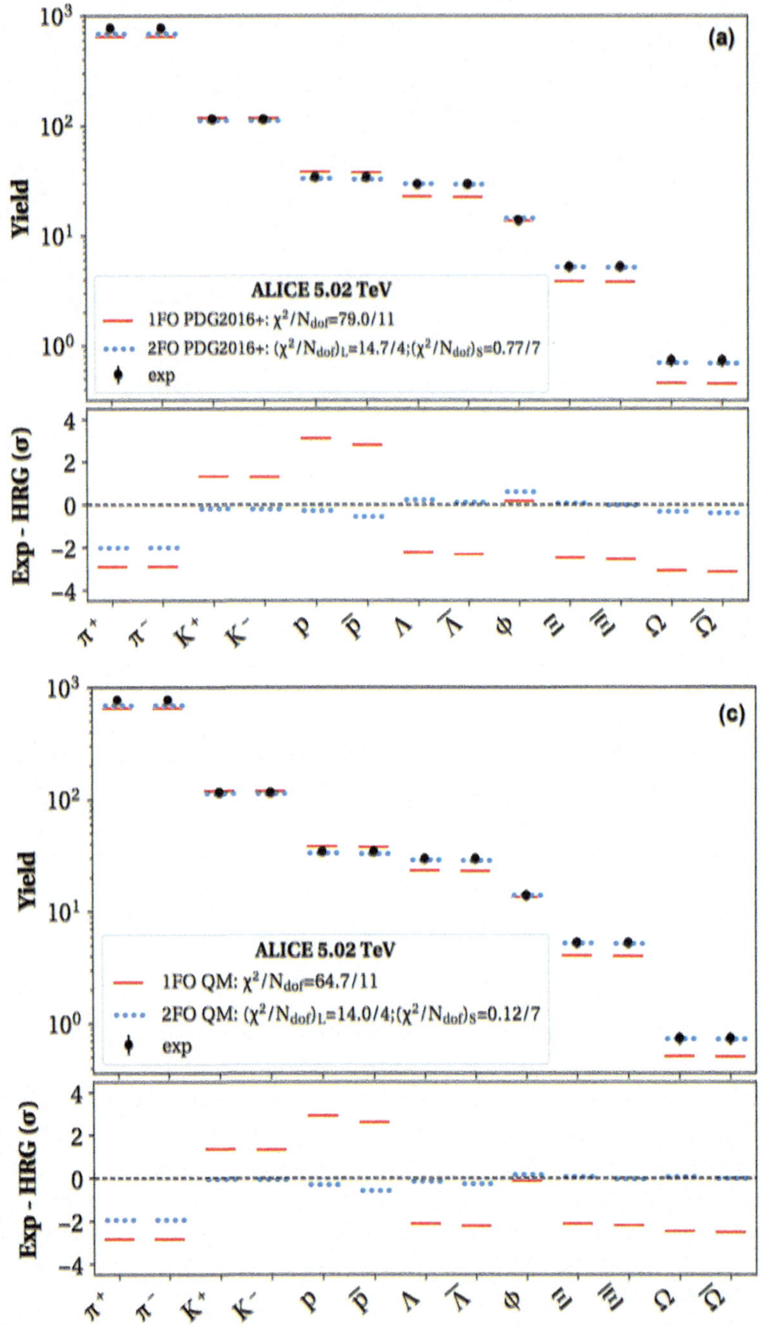

Fig. 7.15 From Ref. [28]. ALICE PbPb $\sqrt{s}$= 5.02 TeV data for particle yields [85] in 0-10% collisions, in comparison to HRG model calculations with the PDG2016+ (top) and QM (bottom) lists; deviations in units of experimental errors σ are shown below each panel

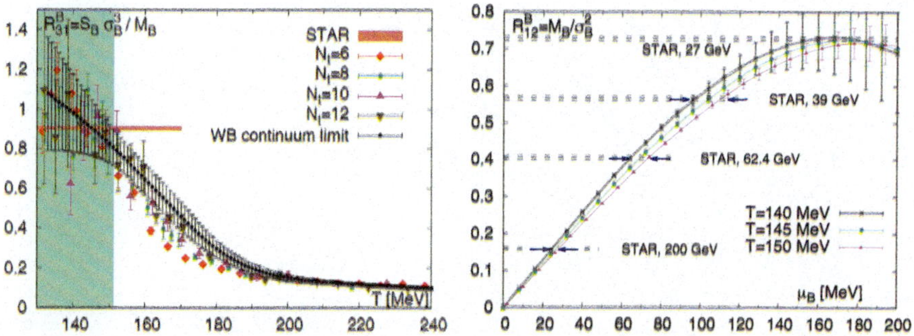

Fig. 7.16 From Ref. [86]. Left: χ_3^B/χ_1^B as a function of the temperature at $\mu_B = 0$. The colored symbols show finite-N_t lattice QCD results. The continuum extrapolation is represented by black points (lattice results from Ref. [87]). The dark-orange band shows the experimental measurement by the STAR collaboration [51]: it was obtained by averaging the 0-5% and 5-10% data at the four highest energies ($\sqrt{s}$ = 27, 39, 62.4, 200 GeV). The green-shaded area shows the valid temperature range. Right: χ_1^B/χ_2^B as a function of μ_B. The three points (dashed lines) correspond to the STAR data for M_p/σ_p^2 at collision energies $\sqrt{s} = 39, 62.4, 200$ GeV and centrality 0-10%, from Ref. [51] (the $\sqrt{s}$ = 27 GeV point is also shown, but the non- monotonicity of the lattice results at $\mu_B \geq$ 130 MeV does not allow a determination of μ_B from it). The colored symbols correspond to the lattice QCD results in the continuum limit, for the range $(140 \leq T_{ch} \leq 150)$ MeV. The arrows show the extracted values for μ_B at freeze-out

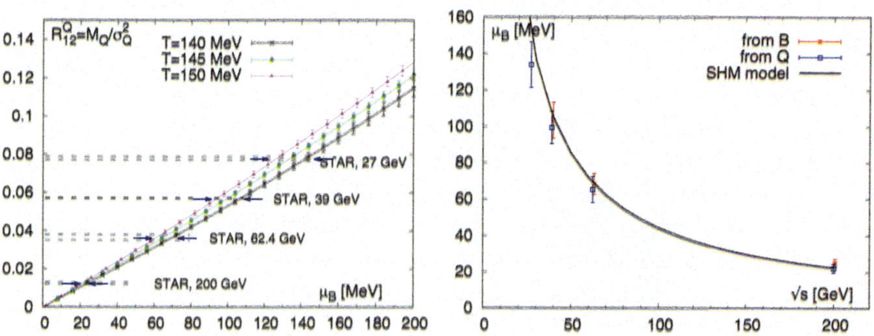

Fig. 7.17 From Ref. [86]. Left: χ_1^Q/χ_2^Q as a function of μ_B. The four points (dashed lines) correspond to the STAR data for M_Q/σ_Q^2 at collision energies $\sqrt{s} = 27, 39, 62.4, 200$ GeV and centrality 0-10%, from Ref. [68]. The colored symbols correspond to the lattice QCD results in the continuum limit, for the range $(140 \leq T_{ch} \leq 150)$ MeV. The arrows show the extracted values for μ_B at freeze-out. Right: freeze-out chemical potential as a function of the collision energy. The red stars show the result obtained by fitting χ_1^B/χ_2^B, the blue squares have been obtained by fitting χ_1^Q/χ_2^Q. The black curve comes from the statistical hadronization model analysis of Refs. [90,91]

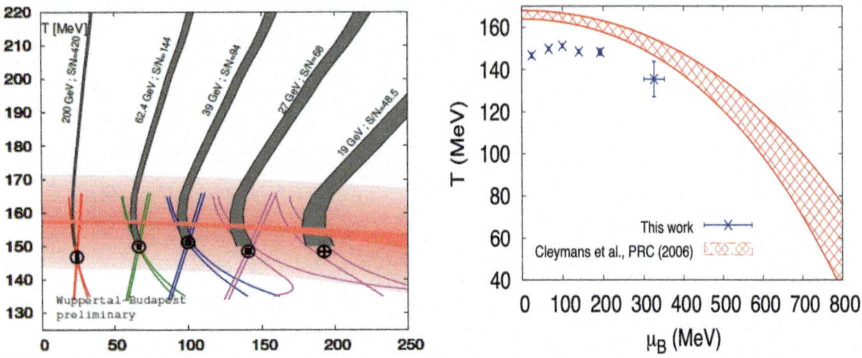

Fig. 7.18 Left: from Ref. [94]. Preliminary results of the WB collaboration. The colored lines are the contours at constant mean/variance ratios of the net electric charge and net-baryon number from lattice simulations. The contours that correspond to STAR data intersect in the freeze-out points of Ref. [88]. The red band is the QCD phase diagram. Also shown are the isentropic contours that match the chemical freeze-out data. Right: from Ref. [88]. Freeze-out parameters in the (T, μ_B)-plane: comparison between the curve obtained in Ref. [89] (red band) and the values obtained from the combined analysis of σ^2/M for net-electric charge and net-protons (blue symbols) from the HRG model

determination of the freeze-out parameters. These results are shown in the left panel of Fig. 7.18. They are in agreement with hadronic resonance gas calculations, which were able to extend the determination of the freeze-out temperature to chemical potential ranges beyond the present applicability limit of lattice QCD (see the right panel of Fig. 7.18) [88]. This figure shows a comparison between the freeze-out parameters obtained in Ref. [88] and the freeze-out curve from Ref. [89]. While the chemical potentials are consistent with each other, the temperature from the fit of net-baryon number and net-electric charge χ_2/χ_1 is about 15 MeV lower than the one previously obtained in the literature. This is a further confirmation of the flavor hierarchy, since net-B and net-Q fluctuations are dominated by protons and pions, respectively, while the thermal fit of particle yields performed in Ref. [89] contains (multi-)strange baryons as well.

Since the lattice has predicted a higher transition temperature for the strange quantum number (see the discussion in Sect. 6.3.2), the measurement of net-kaons was of significant importance. Although the net-kaons carry only slightly more than half of all strange quarks, the net-kaon fluctuations serve as an early indication of the hadronization features of particles carrying strangeness. The central panel of Fig. 7.12 shows the collision energy dependence of the net-Kaon cumulants measured by STAR and published in 2018 [70]. These data were again modeled in different ways, first using the HRG model with latest hadronic spectrum as defined by the particle data group [28, 92] and then also using a partial pressure approach applied to lattice QCD predictions for the μ_B=0 case [93]. In both cases, the results confirm the separation of proton and kaon freeze-out over a large range of collision energies, which can be interpreted as a difference between light and strange quark hadronization (see the left panel of Fig. 7.19).

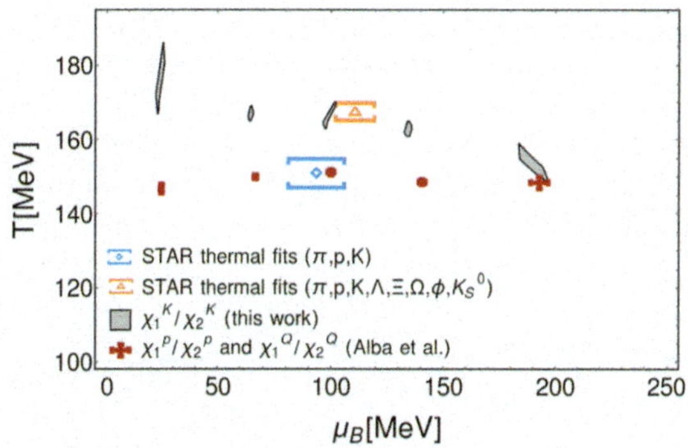

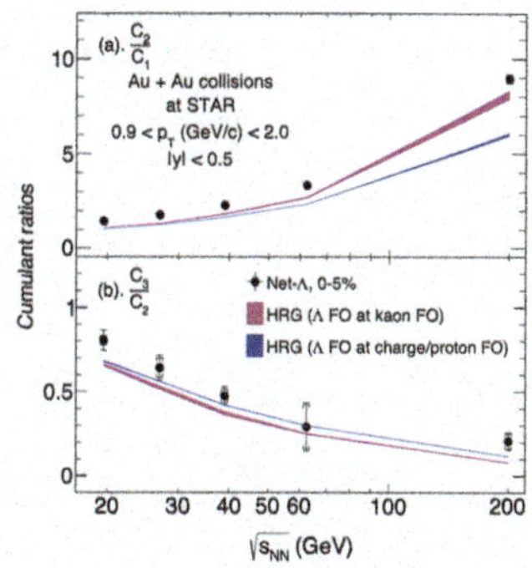

Fig. 7.19 Left: from Ref. [92]. Freeze-out parameters across the highest five energies from the Beam Energy Scan. The red points were obtained from the combined fit of χ_1^p/χ_2^p and χ_1^Q/χ_2^Q [88], while the gray bands are obtained from the fit of χ_1^K/χ_2^K in Ref, [92]. Also shown are the freeze-out parameters obtained by the STAR collaboration at $\sqrt{s}$ =39 GeV [79] from thermal fits to all measured ground-state yields (orange triangle) and only to protons, pions, and kaons (blue diamond-shaped symbol). Right: from Ref. [44]. Black markers show the beam energy dependence of the measured net-Λ cumulant ratios, (**a**) C_2/C_1 and (**b**) C_3/C_2 in most central (0–5%) Au + Au collisions. Magenta bands show the net-Λ cumulant ratios from a HRG calculation [92], assuming Λs freeze-out under the same FO conditions as the kaons. Blue bands show the net-Λ cumulant ratios from the same HRG calculation, assuming Λs freeze out under the same FO conditions as the charged particles/protons. Results are corrected for the reconstruction efficiency, and CBWC is applied. The vertical bars and the caps represent the statistical and systematic uncertainties, respectively. Uncertainties in the HRG calculations are shown by the width of the bands

In addition to the net-kaons, STAR has recently completed and published also the net-Λ fluctuations, which confirmed the conclusions drawn from the net-kaons [44]. Both particle species exhibit a freeze-out surface that is about 15-20 MeV higher than that of the net-protons and of the net-charges, which are dominated by net-protons and net-pions, respectively. These measurements link nicely to the predictions from the lattice. One should point out that the measurements are based on the lower cumulant ratios (C_2/C_1, see the right panel of Fig. 7.19), whereas the unambiguous predictions from the lattice show the effect most prominently in the higher cumulant ratios (χ_4/χ_2). The statistics to significantly determine these moments experimentally for all relevant particle species will require datasets from RHIC and the LHC, which should become available in the next five years.

Based on all accumulated net-particle measurements, one can also try to determine the off-diagonal cumulant components. Of particular interest is the baryon-strangeness correlator (BS), which was early suggested as a good measure for the level of deconfinement in the produced medium [95, 96]. STAR has recently published off-diagonal correlations by forming first and second order correlators between pions and protons, pions and kaons, and protons and kaons [97]. The latter should have a distinct contribution to the BS correlator. However it turns out that, based on HRG and lattice studies [98], the self-correlation term of the Λs is even stronger. We have extensively discussed this issue in Sect. 6.3.4. By constructing

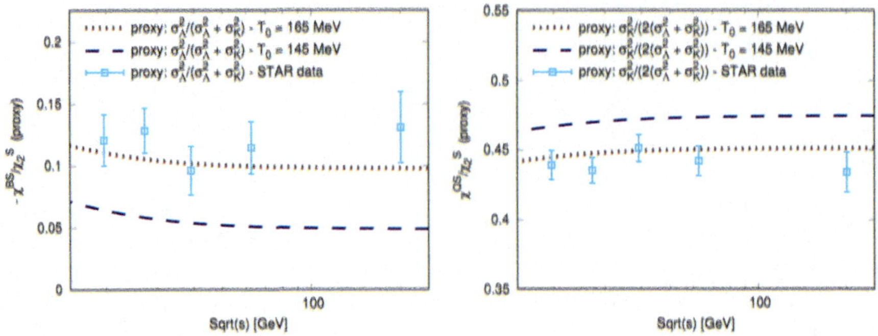

Fig. 7.20 From Ref. [98]. HRG model proxy for the χ_{11}^{BS}/χ_2^S correlator (left) and χ_{11}^{QS}/χ_2^S correlator (right) along freeze-out lines with $T_{ch} = 145$ MeV (blue dashed line) and T_{ch}= 165 MeV (red dotted line) compared to the experimental results [44, 70] (light blue points)

a ratio between Λ and Kaon moments it is possible to determine the freeze-out features of the BS correlator. Figure 7.20 shows a comparison between the HRG model proxy for the χ_{11}^{BS}/χ_2^S correlator (left) and χ_{11}^{QS}/χ_2^S correlator (right) along freeze-out lines with $T_{ch} = 145$ MeV (blue dashed line) and T_{ch}= 165 MeV (red dotted line) and the experimental results [44,70] (light blue points). This comparison again points at a slightly higher freeze-out temperature than for the protons.

7.3.4 Expectations for Near Term Future Measurements

Both RHIC [99] and the LHC [100] have now specific run plans for the near future where fluctuation measurements are prominently featured. STAR and RHIC have extended their capabilities to enable fixed target measurements in the STAR detector. The Beam Energy Scan (BES-II) program, which is scheduled to run until 2022, is supposed to collect sufficient statistics to make definitive measurements up to the fourth order cumulant in the collision energy range of $\sqrt{s}$ = 3–27 GeV. Here the fixed target runs cover nine different energies between 3 and 7.7 A GeV and the collider runs add another seven energies between 7.7 and 27 A GeV. Fixed target running is necessary since the collider requires a minimum energy of 7.7 A GeV for simultaneous injection. Sixth and potentially eighth order cumulants might be possible to measure in STAR at the higher energies to a degree that would allow at least a better determination of a sign change as a function of energy. This signal of chiral criticality at lower baryon densities prior to a critical point on the phase diagram is also one of the goals of the ALICE program. Here, simulations show that 200 Million central Pb-Pb events would be sufficient to determine deviations in the χ_4/χ_2 ratio, whereas χ_6/χ_2 will require the full luminosity of 13 nb^{-1} from Run-3 and Run-4 at the LHC. In other words, this measurement needs to accumulate data over the next decade in order to yield a significant result. One caveat is that the assumed strength of the fluctuation is based on the chiral effective theory [71], which might be an underestimate of the anticipated effect of chiral criticality.

References

1. Fodor, Z., Katz, S.D.: JHEP **04**, 050 (2004)
2. Gavai, R.V., Gupta, S.: Phys. Rev. D **71**, 114014 (2005)
3. Asakawa, M., Yazaki, K.: Nucl. Phys. A **504**, 668–684 (1989)
4. Barducci, A., Casalbuoni, R., De Curtis, S., Gatto, R., Pettini, G.: Phys. Lett. B **231**, 463–470 (1989)
5. Barducci, A., Casalbuoni, R., De Curtis, S., Gatto, R., Pettini, G.: Phys. Rev. D **41**, 1610 (1990)
6. Barducci, A., Casalbuoni, R., Pettini, G., Gatto, R.: Phys. Rev. D **49**, 426–436 (1994)
7. Berges, J., Rajagopal, K.: Nucl. Phys. B **538**, 215–232 (1999)
8. Halasz, A.M., Jackson, A.D., Shrock, R.E., Stephanov, M.A., Verbaarschot, J.J.M.: Phys. Rev. D **58**, 096007 (1998)
9. Scavenius, O., Mocsy, A., Mishustin, I.N., Rischke, D.H.: Phys. Rev. C **64**, 045202 (2001)
10. Antoniou, N.G., Kapoyannis, A.S.: Phys. Lett. B **563**, 165–172 (2003)
11. Hatta, Y., Ikeda, T.: Phys. Rev. D **67**, 014028 (2003)
12. Roessner, S., Ratti, C., Weise, W.: Phys. Rev. D **75**, 034007 (2007)
13. Roessner, S., Hell, T., Ratti, C., Weise, W.: Nucl. Phys. A **814**, 118–143 (2008)
14. Sasaki, C., Friman, B., Redlich, K.: Phys. Rev. D **75**, 074013 (2007)
15. Fukushima, K.: Phys. Rev. D **78**, 114019 (2008)
16. Zhang, Z., Fukushima, K., Kunihiro, T.: Phys. Rev. D **79**, 014004 (2009)
17. Fan, W., Luo, X., Zong, H.: Chin. Phys. C **43**(3), 033103 (2019)
18. Herbst, T.K., Pawlowski, J.M., Schaefer, B.J.: Phys. Rev. D **88**(1), 014007 (2013)
19. Eichmann, G., Fischer, C.S., Welzbacher, C.A.: Phys. Rev. D **93**(3), 034013 (2016)
20. Critelli, R., Noronha, J., Noronha-Hostler, J., Portillo, I., Ratti, C., Rougemont, R.: Phys. Rev. D **96**(9), 096026 (2017)
21. Stephanov, M.A., Rajagopal, K., Shuryak, E.V.: Phys. Rev. Lett. **81**, 4816–4819 (1998)
22. Athanasiou, C., Rajagopal, K., Stephanov, M.: Phys. Rev. D **82**, 074008 (2010)
23. Stephanov, M.A.: Phys. Rev. Lett. **102**, 032301 (2009)
24. Andronic, A., Braun-Munzinger, P., Redlich, K., Stachel, J.: Nature **561**(7723), 321–330 (2018)
25. Poberezhnyuk, R., Vovchenko, V., Motornenko, A., Gorenstein, M.I., Stoecker, H.: Phys. Rev. C **100**(5), 054904 (2019)
26. Andronic, A., Braun-Munzinger, P., Friman, B., Lo, P.M., Redlich, K., Stachel, J.: Phys. Lett. B **792**, 304–309 (2019)
27. Satarov, L.M., Vovchenko, V., Alba, P., Gorenstein, M.I., Stoecker, H.: Phys. Rev. C **95**(2), 024902 (2017)
28. Alba, P., Sarti, V.M., Noronha-Hostler, J., Parotto, P., Portillo-Vazquez, I., Ratti, C., Stafford, J.M.: Phys. Rev. C **101**(5), 054905 (2020)
29. Vovchenko, V., Poberezhnyuk, R.V., Koch, V.: [arXiv:2007.03850 [hep-ph]]
30. Skokov, V., Friman, B., Redlich, K.: Phys. Rev. C **88**, 034911 (2013)
31. Luo, X., Xu, J., Mohanty, B., Xu, N.: J. Phys. G **40**, 105104 (2013)
32. Braun-Munzinger, P., Friman, B., Redlich, K., Rustamov, A., Stachel, J.: [arXiv:2007.02463 [nucl-th]]
33. Kitazawa, M., Asakawa, M.: Phys. Rev. C **86**, 024904 (2012)
34. Yates, R.D., Goodman, D.J.: Probability and Stochastic Processes: A Friendly Introduction for Electrical and Computer Engineers, 2nd edn. Wiley, Hoboken (2014). ISBN 978-0-471-45259-1
35. Alner, G.J., et al.: [UA5] Phys. Lett. B **160**, 193–198 (1985)
36. Tarnowsky, T.J., Westfall, G.D.: Phys. Lett. B **724**, 51–55 (2013)
37. Braun-Munzinger, P., Rustamov, A., Stachel, J.: Nucl. Phys. A **960**, 114–130 (2017)
38. Vovchenko, V., Savchuk, O., Poberezhnyuk, R.V., Gorenstein, M.I., Koch, V.: [arXiv:2003.13905 [hep-ph]]

39. Patrignani, C., et al.: [Particle Data Group], Chin. Phys. C **40**(10), 100001 (2016)
40. Aamodt, K., et al.: [ALICE] JINST **3**, S08002 (2008)
41. Alessandro, B., et al.: [ALICE] J. Phys. G **32**, 1295–2040 (2006)
42. Rustamov, A., Gorenstein, M.I.: Phys. Rev. C **86**, 044906 (2012)
43. Aggarwal, M.M., et al.: [STAR] Phys. Rev. C **83**, 024901 (2011)
44. Adam, J., et al.: [STAR] Phys. Rev. C **102**(2), 024903 (2020)
45. Adam, J., et al.: [STAR] [arXiv:2001.02852 [nucl-ex]]
46. Wang, X.N., Gyulassy, M.: Phys. Rev. D **44**, 3501–3516 (1991)
47. Lin, Z.W., Ko, C.M., Li, B.A., Zhang, B., Pal, S.: Phys. Rev. C **72**, 064901 (2005)
48. Petersen, H., Bleicher, M., Bass, S.A., Stocker, H.: [arXiv:0805.0567 [hep-ph]]
49. Pierog, T., Karpenko, I., Katzy, J.M., Yatsenko, E., Werner, K.: Phys. Rev. C **92**(3), 034906 (2015)
50. Bzdak, A., Koch, V.: Phys. Rev. C **86**, 044904 (2012)
51. Adamczyk, L., et al.: [STAR] Phys. Rev. Lett. **112**, 032302 (2014)
52. Bzdak, A., Koch, V.: Phys. Rev. C **91**(2), 027901 (2015)
53. Nonaka, T., Kitazawa, M., Esumi, S.: Phys. Rev. C **95**(6), 064912 (2017)
54. Luo, X., Nonaka, T.: Phys. Rev. C **99**(4), 044917 (2019)
55. Gonzalez, V., Marin, A., Ladron De Guevara, P., Pan, J., Basu, S., Pruneau, C.: Phys. Rev. C **99**(3), 034907 (2019)
56. Stephanov, M.A.: Phys. Rev. Lett. **107**, 052301 (2011)
57. Garg, P., Mishra, D.K., Netrakanti, P.K., Mohanty, B., Mohanty, A.K., Singh, B.K., Xu, N.: Phys. Lett. B **726**, 691–696 (2013)
58. Bleicher, M., Zabrodin, E., Spieles, C., Bass, S.A., Ernst, C., Soff, S., Bravina, L., Belkacem, M., Weber, H., Stoecker, H., Greiner, W.: J. Phys. G **25**, 1859–1896 (1999)
59. Mroczek, D., Noronha-Hostler, J., Acuna, A., R.N., Ratti, C., Parotto, P., Stephanov, M.A.: [arXiv:2008.04022 [nucl-th]]
60. Adamczewski-Musch, J., et al.: [HADES] Phys. Rev. C **102**(2), 024914 (2020)
61. Ling, B., Stephanov, M.A.: Phys. Rev. C **93**(3), 034915 (2016)
62. Bazavov, A., Bollweg, D., Ding, H.T., Enns, P., Goswami, J., Hegde, P., Kaczmarek, O., Karsch, F., Larsen, R., Mukherjee, S., Ohno, H., Petreczky, P., Schmidt, C., Sharma, S., Steinbrecher, P.: Phys. Rev. D **101**(7), 074502 (2020)
63. Borsanyi, S., Fodor, Z., Guenther, J.N., Katz, S.K., Szabo, K.K., Pasztor, A., Portillo, I., Ratti, C.: JHEP **10**, 205 (2018)
64. Acharya, S., et al.: [ALICE] Phys. Lett. B **807**, 135564 (2020)
65. Luo, X.: [STAR Collaboration] PoS CPOD **2014**, 019 (2015)
66. Braun-Munzinger, P., Rustamov, A., Stachel, J.: [arXiv:1907.03032 [nucl-th]]
67. Abelev, B., et al.: [ALICE Collaboration] Phys. Rev. Lett. **110**(15), 152301 (2013)
68. Adamczyk, L., et al.: [STAR Collaboration] Phys. Rev. Lett. **113**, 092301 (2014)
69. Adare, A., et al.: [PHENIX Collaboration] Phys. Rev. C **93**(1), 011901 (2016)
70. Adamczyk, L., et al.: [STAR Collaboration] Phys. Lett. B **785**, 551 (2018)
71. Friman, B., Karsch, F., Redlich, K., Skokov, V.: Eur. Phys. J. C **71**, 1694 (2011)
72. Bazavov, A., Ding, H.T., Hegde, P., Kaczmarek, O., Karsch, F., Laermann, E., Maezawa, Y., Mukherjee, S., Ohno, H., Petreczky, P., Sandmeyer, H., Steinbrecher, P., Schmidt, C., Sharma, S., Soeldner, W., Wagner, M.: Phys. Rev. D **95**(5), 054504 (2017)
73. Luo, X.: J. Phys. G **39**, 025008 (2012)
74. Nonaka, T.: [STAR Collaboration] arXiv:2002.12505 [nucl-ex]
75. Karsch, F.: Central Eur. J. Phys. **10**, 1234–1237 (2012)
76. Nahrgang, M., Bluhm, M., Alba, P., Bellwied, R., Ratti, C.: Eur. Phys. J. C **75**(12), 573 (2015)
77. Alba, P., Bellwied, R., Bluhm, M., Mantovani Sarti, V., Nahrgang, M., Ratti, C.: Phys. Rev. C **92**(6), 064910 (2015)
78. Bellwied, R., Borsanyi, S., Fodor, Z., Katz, S.D., Ratti, C.: Phys. Rev. Lett. **111**, 202302 (2013)
79. Adamczyk, L., et al.: [STAR] Phys. Rev. C **96**(4), 044904 (2017)

80. Bazavov, A., Ding, H.T., Hegde, P., Kaczmarek, O., Karsch, F., Laermann, E., Maezawa, Y., Mukherjee, S., Ohno, H., Petreczky, P., Schmidt, C., Sharma, S., Soeldner, W., Wagner, M.: Phys. Rev. Lett. **111**, 082301 (2013)
81. Rapp, R., Shuryak, E.V.: Phys. Rev. Lett. **86**, 2980–2983 (2001)
82. Steinheimer, J., Aichelin, J., Bleicher, M.: Phys. Rev. Lett. **110**(4), 042501 (2013)
83. Aarts, G., Allton, C., De Boni, D., Jäger, B.: Phys. Rev. D **99**(7), 074503 (2019)
84. Lo, P.M., Friman, B., Redlich, K., Sasaki, C.: Phys. Lett. B **778**, 454–458 (2018)
85. Bellini, F.: [ALICE] Nucl. Phys. A **982**, 427–430 (2019)
86. Borsanyi, S., Fodor, Z., Katz, S.D., Krieg, S., Ratti, C., Szabo, K.K.: Phys. Rev. Lett. **113**, 052301 (2014)
87. Borsanyi, S., Fodor, Z., Katz, S.D., Krieg, S., Ratti, C., Szabo, K.K.: Phys. Rev. Lett. **111**, 062005 (2013)
88. Alba, P., Alberico, W., Bellwied, R., Bluhm, M., Mantovani Sarti, V., Nahrgang, M., Ratti, C.: Phys. Lett. B **738**, 305–310 (2014)
89. Cleymans, J., Oeschler, H., Redlich, K., Wheaton, S.: Phys. Rev. C **73**, 034905 (2006)
90. Andronic, A., Braun-Munzinger, P., Stachel, J.: Phys. Lett. B **673**, 142–145 (2009)
91. Andronic, A., Braun-Munzinger, P., Stachel, J.: Nucl. Phys. A **772**, 167–199 (2006)
92. Bellwied, R., Noronha-Hostler, J., Parotto, P., Portillo Vazquez, I., Ratti, C., Stafford, J.M.: Phys. Rev. C **99**(3), 034912 (2019)
93. Noronha-Hostler, J., Bellwied, R., Gunther, J., Parotto, P., Pasztor, A., Portillo Vazquez, I., Ratti, C.: [arXiv:1607.02527 [hep-ph]]
94. Ratti, C.: Nucl. Phys. A **956**, 51–58 (2016)
95. Koch, V., Majumder, A., Randrup, J.: Phys. Rev. Lett. **95**, 182301 (2005)
96. Majumder, A., Muller, B.: Phys. Rev. C **74**, 054901 (2006)
97. Adam, J., et al.: [STAR] Phys. Rev. C **100**(1), 014902 (2019)
98. Bellwied, R., Borsanyi, S., Fodor, Z., Guenther, J.N., Noronha-Hostler, J., Parotto, P., Pasztor, A., Ratti, C., Stafford, J.M.: Phys. Rev. D **101**(3), 034506 (2020)
99. Aprahamian, A., Robert, A., Caines, H., Cates, G., Cizewski, J.A., Cirigliano, V., Dean, D.J., Deshpande, A., Ent, R., Fahey, F., Fatemi, R., Filippone, B., Fuller, G., Gagliardi, C., Gao, H., Geesaman, D., Greene, G., Hardy, J., Heeger, K., Hertzog, D., Hobart, D., Howell, C., Hutton, A., Jacobs, P., Ji, X., Kruecken, R., Kuhn, S., Kumar, K., Lapi, S., Lisa, M., Lu, Z.T., Macchiavelli, A., Makins, N.C., R., Paul, M., Meziani, Z.E., Milner, R., Mueller, B., Nagle, J., Nazarewicz, W., Nunes, F., Ormand, E., Piekarewicz, J., Radford, D., Rajagopal, K., Musolf, M., , R., Robertson, H., Roser, T., Rossi, P., Savage, M., Savard, G., Schellman, H., Scholberg, K., Schukraft, J., Shepherd, M., Velkovska, J., Venugopalan, R., Vigdor, S., Wiescher, M., Wilkerson, J., Zajc, W.: Reaching for the Horizon: The 2015 Long Range Plan for Nuclear Science. United States. https://www.osti.gov/servlets/purl/1296778
100. Citron, Z., et al.: CERN Yellow Rep. Monogr. **7**, 1159–1410 (2019)

Part II

Transport Properties of Strongly Interacting Matter

Transport Properties of QCD Matter

8

Abstract

In this chapter we review the state of the art of transport properties of strongly interacting matter from lattice QCD. All these quantities have a common problem: they require the inversion of an integral, calculated on a discrete set of points. We begin the chapter by reviewing the most common inversion methods used in the literature. We then focus on three main topics: electric diffusion coefficient, heavy quark diffusion coefficient and viscosity.

8.1 Introduction

Transport properties of QCD matter are expected to be severely modified in the vicinity of the phase transition, due to the strongly coupled nature of the system in that regime. Lattice QCD is unfortunately limited in studying dynamical quantities: simulations of real-time properties would lead to oscillatory behavior like the one encountered in finite-chemical potential simulations. Current-current correlators can be investigated on a discrete set of points. All these correlators G have spectral representations involving integrals of spectral functions ρ weighted by appropriate integration kernels K:

$$G(\tau, \mathbf{p}) = \int_0^\infty \frac{d\omega}{2\pi} \rho(\omega, \mathbf{p}, T) K(\omega, \tau, T), \text{ with } K(\tau, \omega) = \frac{\cosh[\omega(\tau - 1/2T)]}{\sinh[\omega/2T]}. \tag{8.1}$$

The Euclidean correlator above, therefore, depends on the temperature both through the Kernel K and through the spectral function. The two effects need to be disentangled, as the observables of interest are the low-frequency and low-momentum limits of the spectral functions ρ. This can be achieved by comparing the actual correlator

C. Ratti, R. Bellwied, *The Deconfinement Transition of QCD*, Lecture Notes in Physics 981, https://doi.org/10.1007/978-3-030-67235-5_8

above to a reconstructed one (we set $\mathbf{p} = 0$)

$$G_r(\tau, T; T_0) = \int_0^\infty \frac{d\omega}{2\pi} K(\omega, \tau, T)\rho(\omega, T_0). \tag{8.2}$$

Thus, a difference between the actual and reconstructed correlators must be due to a change in the spectral function.

Spectral functions can only be extracted through inversion methods or a modeling of ρ at low frequencies in order to integrate over a discrete set of lattice points. This is often referred-to as an "ill-posed" problem [1]. In the case of lattice QCD, it is also a discrete problem, as only the values of the correlators on a finite set of points in τT can be accessed. In fact, the numerical temporal correlator dataset contains $O(10)$ points, while ideally a solution should be continuous. This lack of information means that the problem is under-determined; besides, the few available points are the results of a lattice QCD simulation, and as such they are subject to uncertainties. Finally, the lattice QCD results have ideally to be extrapolated to the continuum in order to remove cut-off effects due to the lattice regularization.

The spectral function at finite temperature is constrained through sum rules. In particular, $\Delta\rho(\omega, T) = \rho_{ii}(\omega, T) - \rho_{ii}(\omega, 0)$ has to obey the following

$$\int_{-\infty}^\infty \frac{d\omega}{\omega} \Delta\rho(\omega, T) = 0. \tag{8.3}$$

in the thermodynamic limit [2]. The above integral converges, as the spectral function decreases as $(T/\omega)^2$ at large ω [3].

Any approach in solving an ill-posed problem needs to add information to render it "better-posed". Effective methods to this purpose are the Tikhonov regularization [4] and the Maximum Entropy Method (MEM) [5, 6], or related Bayesian approaches [7]. More recently, stochastic approaches [8, 9] and the Backus-Gilbert method [10] have also been investigated. As we will see, in spite of the difficulties due to such procedures, several results have been obtained on the transport properties of matter: we will summarize the most recent ones in this Chapter.

8.2 Reconstruction Methods

8.2.1 Physics-Based Ansätze

In order to reduce the ill-posedness of the problem discussed above, one possibility is to provide an ansatz for the behavior of the spectral function, containing fewer parameters than data points. This, of course, introduces an unavoidable bias; besides, the ansatz needs to capture both the low- and high-temperature behavior of ρ. In particular, some features that must be included are: a peak at small frequencies, with a linear slope in ω, a continuum contribution at high ω and at least one peak in the low-T phase.

A common ansatz [11–13] includes a transport peak and the expected continuum behavior in the deconfined phase

$$\rho(\omega) = A_{trans} \frac{\gamma\omega}{\omega^2 + \gamma^2} + \frac{3}{2\pi} A_{pert} \omega^2 \left[1 - 2n_F\left(\frac{\omega}{2}\right)\right], \tag{8.4}$$

where $A_{trans,pert}$ and γ (the width of the transport peak) are temperature-dependent parameters.

A different ansatz has been used in Ref. [2] for $\Delta\rho(\omega, T) = \rho_{ii}(\omega, T) - \rho_{ii}(\omega, 0)$:

$$\Delta\rho(\omega) = \rho_{trans}(\omega) + \Delta\rho_{pert}(\omega) + \rho_{bound}(\omega) \tag{8.5}$$

where

$$\rho_{bound}(\omega) = A_{bound} \frac{2g_B \tanh(\omega/T)^3}{4(\omega - m_B)^2 + g_B^2} \tag{8.6}$$

is the ansatz for the bound-state peak with mass m_B, width g_B and strength A_{bound}. The tanh function ensures a smooth fading of the bound state contribution at small and large ω. It was noted in Ref. [2] that the transport peak introduced above violates this condition, which was then modified as follows

$$\rho_{trans,mod}(\omega) = A_{trans} \frac{T \tanh(\omega/T)_\gamma}{\omega^2 + \gamma^2} \tag{8.7}$$

which now decays as $1/\omega^2$ at large ω. Finally, the subtraction in $\Delta\rho_{pert}$ eliminates the zero-temperature ω^2 contribution.

8.2.2 Maximum Entropy Method

The theoretical basis of the Maximum Entropy Method (MEM) is Bayes' theorem in probability theory [6]

$$P[X|Y] = \frac{P[Y|X]P[X]}{P[Y]}, \tag{8.8}$$

where $P[X|Y]$ is the conditional probability of X given Y. Using the ideas of Bayesian probability theory, one can construct the most probable spectral function by maximizing the conditional probability $P[\rho|G, D]$, where G indicates the data and D some additional prior knowledge. According to Bayes' theorem, the conditional probability of having ρ with given G and D is given by

$$P[\rho|G, D] = \frac{P[G|\rho, D]P[\rho|D]}{P[G|D]}, \tag{8.9}$$

where $P[G|\rho, D]$ and $P[\rho|D]$ are the likelihood function and the prior probability, respectively, while $P[G|D]$ is a $\rho-$independent normalization. The most probable ρ is the one that satisfies the condition

$$\frac{\delta P[\rho|G, D]}{\delta \rho} = 0. \tag{8.10}$$

Besides, one can estimate the reliability of the solution through $\delta^2 P[\rho|G, D]/\delta\rho\delta\rho$. The probability $P[\rho|G, D]$ can be updated once more data become available, which is evident from the following chain rule

$$P[\rho|G_1, G_2, D] = P[G_2|G_1, \rho, D]P[G_1|\rho, D]P[\rho|D]/P[G_1, G_2|D]. \tag{8.11}$$

In the Maximum Entropy Method, the prior knowledge is encoded in an entropy term

$$S = \int_0^{\infty} \frac{d\omega}{2\pi} \left[\rho(\omega) - m(\omega) - \rho(\omega) \ln \frac{\rho(\omega)}{m(\omega)} \right], \tag{8.12}$$

where $m(\omega)$ is the default model containing the additional information. A popular choice is $m(\omega) = m_0\omega^2$, with m_0 constant. This is motivated by the large-ω behavior of meson spectral functions in the continuum, which is accessible in perturbation theory. One then extremizes the conditional probability

$$P[\rho|G, D] = \exp\left(-\frac{1}{2}\chi^2 + \alpha S\right), \tag{8.13}$$

where χ^2 is the standard likelihood function and α is a parameter balancing the relative importance of the data and the prior knowledge. $\rho(\omega)$ is non-negative for positive ω, and it is therefore written as

$$\rho(\omega) = m(\omega) \exp[f(\omega)], \tag{8.14}$$

where $f(\omega)$ is a function to be determined. The arbitrary function $f(\omega)$ has to be reduced to one containing at most N parameters for the problem to be well-defined, where N is typically $O(10)$. The subspace can then be determined using a singular value decomposition. One discretizes the ω integral, after which the Kernel can be viewed as an $N_\omega \times N$ matrix $K(\omega_n, \tau_i)$, where N_ω is typically $O(10^3)$.

It follows from the extremum condition, that $f(\omega)$ can be parametrized as

$$f(\omega) = \sum_{i=1}^{N} c_i u_i(\omega), \tag{8.15}$$

which reduces the problem to finding the N coefficients c_i. The functions u_i are to be considered as basis functions. Due to the following limit

$$\lim_{\omega\to 0} K(\omega, \tau) = \frac{2T}{\omega} + \frac{\omega}{T}\left[\frac{1}{6} - \tau T(1-\tau T)\right] + O(\frac{\omega^3}{T^3}) \tag{8.16}$$

the kernel diverges for small ω, which can lead to a numerically unstable algorithm. However, this can be avoided by writing

$$\bar{K}(\omega, \tau) = \frac{\omega}{2T} K(\omega, \tau), \qquad \bar{\rho}(\omega) = \frac{2T}{\omega}\rho(\omega) \tag{8.17}$$

so that $\bar{K}(\omega, \tau)\bar{\rho}(\omega) = K(\omega, \tau)\rho(\omega)$. This fixes the problem and allows one to include the $\omega = 0$ point in the analysis. One can then parametrize the redifined spectral function as

$$\bar{\rho}(\omega) = \bar{m}(\omega) \exp \sum_{i=1}^{N} \bar{c}_i \bar{u}_i(\omega), \tag{8.18}$$

where the default model becomes $\bar{m}(\omega) \sim m(\omega)/\omega$. The previous results are recovered if one imposes $\bar{m}(\omega) \sim \omega$ at large ω, whereas at small ω one reconstructs $\bar{\rho} \sim \rho/\omega$, so that the intercept at $\omega = 0$ is proportional to the corresponding transport coefficient.

8.2.3 Bayesian Approaches

A novel Bayesian approach was introduced in [7], to address some key issues affecting the MEM: the slow convergence of the underlying optimization task, the high computational cost for extended search space, the scale dependence in the prior functional and, finally, the Gaussian approximation that is required in the parameter estimate.

As discussed before, the Bayesian strategy relies on an application of the multiplication law for the probability distribution of the spectral function ρ, the measured data G and prior information D:

$$P[\rho|G, D] = \frac{P[G|\rho, D]P[\rho|D]}{P[G|D]}. \tag{8.19}$$

In this case, the prior probability $P[\rho|D] \propto \exp[S]$ is constructed by the four axioms detailed below. The prior information is contained in a function $m(\omega_l) = m_l$

which is the correct spectral function in the absence of data, by definition. We now introduce the four axioms.

- Consider two different subsets Ω_1 and Ω_2 along the frequency axis. If constraints are imposed on ρ from prior information within each of these subsets, the result of the reconstruction should not depend on whether one treats these domains separately or combined. This is true if S is written as

$$S \propto \int d\omega s(\rho(\omega), m(\omega), \omega). \tag{8.20}$$

  This axiom coincides with one of the four axioms used in the MEM.
- In general, $\rho(\omega)$ does not need to be a probability distribution, namely its scaling can be different from $1/\omega$. Therefore, one requires that the choice of units for ρ and m does not change the result, namely S can only depend on the ratio ρ/m

$$S = \tilde{\alpha} \int d\omega s(\rho(\omega)/m(\omega)). \tag{8.21}$$

  Therefore, s does not carry a dimension, which is instead absorbed in the parameter $\tilde{\alpha}$ to make the argument of the exponential dimensionless as well.
- The spectral function has to be positive definite and smooth. This is achieved by penalizing spectra that differ between two adjacent frequencies ω_1 and ω_2. This yields the following functional form for S

$$S = \tilde{\alpha} \int d\omega \left(C_0 - C_1 \frac{\rho}{m} + C_2 \ln\left(\frac{\rho}{m}\right)\right), \tag{8.22}$$

  where C_i are constants to be fixed.
- When data are not available, S must have a maximum at $\rho = m$, with the convention that $S(\rho/m = 1) = 0$. Therefore

$$S'(\rho/m = 1) = 0, \qquad S''(\rho/m = 1) < 0. \tag{8.23}$$

  These conditions fix all the C_i up to an overall constant that can be absorbed in $\tilde{\alpha}$. Therefore

$$S = \alpha \int d\omega \left(1 - \frac{\rho}{m} + \ln\left(\frac{\rho}{m}\right)\right). \tag{8.24}$$

The authors of Ref. [7] proceed as follows. They consider the joint probability distribution

$$\begin{aligned} P[\rho, G, \alpha, m] &= P[G|\rho, \alpha, m]P[\rho|\alpha, m]P[\alpha, m] \\ &= P[\alpha|\rho, G, m]P[\rho|G, m]P[G, m] \end{aligned} \tag{8.25}$$

where

$$P[\rho|G,m] = \frac{P[G|\rho,D]}{P[G|m]} \int d\alpha P[\rho|\alpha,m]. \tag{8.26}$$

Here, $P[G|\rho, D] = \exp[-L - \gamma(L - N_\tau)^2]$, with $L/N_\tau = 1$ and $P[G|m]$ is an irrelevant constant. After numerical integration, the dependence on α is eliminated. Therefore, $P[\rho|G,m]$ does not contain any free parameters and one can calculate its maximum numerically.

8.2.4 Stochastic Approaches

While in the Maximum Entropy Method approach the most probable spectral function can be proven to be unique, if it exists, in stochastic approaches a sequence of possible spectra is generated stochastically and then the average is taken [8,9].

Suppose we are trying to reconstruct a spectral image ρ from correlator function data G, with a given default model D which contains some information about the spectral function. Let us also introduce a parameter α, which controls the contribution to the reconstructed image from the prior information relative to the data. According to Bayes' theorem, the conditional probability of having ρ with given G, D and α is given by

$$P[\rho|G,D,\alpha] = \frac{P[G|\rho,D,\alpha]P[\rho|D,\alpha]}{P[G|D,\alpha]}, \tag{8.27}$$

where $P[G|\rho, D, \alpha]$ and $P[\rho|D, \alpha]$ are the likelihood function and the prior probability, respectively, while $P[G|D, \alpha]$ is a $\rho-$independent normalization.

One usually calculates $P[\rho|G, D, \alpha]$ and then takes the average over all possible spectra weighted by it:

$$\langle\rho\rangle_\alpha = \int \mathcal{D}\rho\rho P[\rho|G,D,\alpha]. \tag{8.28}$$

Finally, the α dependence is eliminated by taking a weighted average over it

$$\langle\langle\rho\rangle\rangle = \int d\alpha \langle\rho\rangle_\alpha P[\alpha|G,D], \tag{8.29}$$

where we can write

$$P[\alpha|G,D] = \frac{P[G|D,\alpha]P[\alpha|D]}{P[G|D]} = \frac{P[\alpha|D]}{P[G|D]} \int \mathcal{D}\rho P[G|\rho,D,\alpha]P[\rho|D,\alpha]. \tag{8.30}$$

The statistical uncertainties of the reconstructed image can be introduced by considering the spectral function averaged over a certain frequency range I

$$\langle\bar{\rho}_I\rangle_\alpha = \frac{\int_I d\omega \langle\rho(\omega)\rangle_\alpha}{\int_I d\omega} \tag{8.31}$$

and the variance

$$\langle(\delta\bar{\rho}_I)^2\rangle_\alpha = \frac{\int_{I\times I} d\omega d\omega' \langle\delta\rho(\omega)\delta\rho(\omega')\rangle_\alpha}{\int_{I\times I} d\omega d\omega'}, \tag{8.32}$$

where $\delta\rho(\omega) = \rho(\omega) - \langle\rho(\omega)\rangle_\alpha$. The dependence on α can be then eliminated as

$$\begin{aligned}\langle\langle\bar{\rho}_I\rangle\rangle &= \int d\alpha \langle\bar{\rho}_I\rangle_\alpha P[\alpha|G, D],\\ \langle\langle(\delta\bar{\rho}_I)^2\rangle\rangle &= \int d\alpha \langle(\delta\bar{\rho}_I)^2\rangle_\alpha P[\alpha|G, D].\end{aligned} \tag{8.33}$$

In MEM, one assumes that the probability $P[\rho|G, D, \alpha]$ is a sharp Gaussian distribution. Therefore, MEM is a special limit of stochastic analytic continuation [14].

8.2.5 Backus-Gilbert Method

This method represents a paradigm change, compared to the usual goal of reconstructing the entire spectral function from the lattice correlation function data [15]. One defines a rescaled kernel

$$K(\omega, \tau) = f(\omega/T)\frac{\cosh[\omega(\beta/2 - \tau)]}{\sinh[\omega\beta/2]}, \tag{8.34}$$

where $f(x) \propto x$ for $x \to 0$ and $f(x) > 0$ for $x > 0$. One then considers a resolution function $\delta(\omega, \omega')$, such that the filtered spectral function $\hat{\rho}$ yields an average value of ρ around ω:

$$\hat{\rho} = f(\omega/T)\int_0^\infty d\omega' \delta(\omega, \omega')\frac{\rho(\omega', T)}{f(\omega', T)}. \tag{8.35}$$

The goal then becomes to have a narrowly concentrated function $\delta(\omega, \omega')$ around ω, viewed as a function of ω' for a given ω. The filtered spectral function is then necessarily given by a linear combination

$$\hat{\rho}(\omega) = f(\omega/T) \sum_{i=1}^{n} g_i(\omega) G(\tau_i), \tag{8.36}$$

where τ_i are the Euclidean times at which the data for the correlator are available. One then chooses the coefficients $g_i(\omega)$ to realize this goal, and then concludes that the resolution function can be written as

$$\delta(\omega, \omega') = \sum_{i=1}^{n} g_i(\omega) K(\omega', \tau_i). \tag{8.37}$$

Therefore, the coefficients $g_i(\omega)$ completely define the resolution function for a given ω. The method of Backus and Gilbert provides a recipe to construct the $g_i(\omega)$. One gets

$$g_i(\omega) = \frac{(W^{-1})_{ij} R_j}{R_k (W^{-1})_{kl} R_l}, \tag{8.38}$$

where

$$W_{ij}(\omega) = \int_0^\infty d\omega' K(\omega', \tau_i) K(\omega', \tau_j)(\omega - \omega')^2 + (1 - \lambda) S_{ij}$$

$$\text{and } R_i = \int_0^\infty d\omega K(\omega, \tau_i). \tag{8.39}$$

In the above formula, S_{ij} is the covariance matrix for the correlator $G(\tau_i)$, and λ is a tunable parameter, determined by comparing the behaviour of $\delta(\omega, \omega')$ for different values of λ. As for the choice of the reweighting function $f(x)$, it is better to choose it such that the reconstructed function $\rho(\omega, T)/f(\omega/T)$ does not show a global trend. In general, this choice depends on the channel under consideration.

8.2.6 Tikhonov Regularization

Both the Backus-Gilbert and Tikhonov regularization methods are non-parametric approaches that can be used to study the spectral function. They aim at solving the following equation

$$C(\tau) = \int_0^\infty \frac{d\omega}{2\pi} \frac{\rho(\omega)}{f(\omega)} K(\omega, \tau) \tag{8.40}$$

where

$$K(\omega, \tau) = \frac{\cosh \omega(\tau - \beta/2)}{\sinh \omega\beta/2} f(\omega) \tag{8.41}$$

was introduced in Eq. (8.34) and $f(\omega)$ is an arbitrary function. Similarly to what was done in the Backus-Gilbert method, also here one reconstructs a filtered spectral function $\hat{\rho}$. While in the Backus-Gilbert method the regularization is performed as

$$W_{ij} \to \lambda S_{ij} + (1-\lambda) W_{ij}, \tag{8.42}$$

in the Tikhonov regularization scheme the single value decomposition of $W^{-1} = VDU^T$ is regularized. The entries of the diagonal matrix D are the susceptibilities of the data to noise, and they might be very large. The regularization is performed by adding the regularizer γ to all entries, which corresponds to a smooth cutoff. In both methods, the resolution function cannot be chosen a priori, but it is an outcome of the method itself. This makes it difficult to continuum extrapolate the lattice data. To perform a continuum extrapolation, a fixed resolution function will be needed [16].

8.3 Charge Diffusion and Electromagnetic Probes

Consider the electromagnetic current in the case of three quark flavors

$$j_\mu^{em} = \frac{2e}{3} j_\mu^u - \frac{e}{3} j_\mu^d - \frac{e}{3} j_\mu^s \tag{8.43}$$

where j_μ^f are the vector currents for each flavor. We can define the Euclidean current-current correlator $G_{\mu\nu}^{em}(\tau)$ at zero spatial momentum

$$G_{\mu\nu}^{em}(\tau) = \int d^3x \langle j_\mu^{em}(\tau, \mathbf{x}) j_\nu^{em}(0, \mathbf{0})^\dagger \rangle. \tag{8.44}$$

We can write for the above correlator the following spectral representation

$$G_{\mu\nu}^{em}(\tau) = \int_0^\infty \frac{d\omega}{2\pi} K(\tau, \omega) \rho_{\mu\nu}^{em}(\omega) \tag{8.45}$$

where

$$K(\tau, \omega) = \frac{\cosh[\omega(\tau - 1/2T)]}{\sinh[\omega/2T]} \tag{8.46}$$

is the kernel and $\rho_{\mu\nu}^{em}(\omega)$ is the spectral function. The electrical conductivity σ can be obtained using Kubo formula from linear response theory [17]

$$\frac{\sigma}{T} = \frac{1}{6T} \lim_{\omega \to 0} \frac{\rho^{em}(\omega)}{\omega}, \qquad \rho^{em}(\omega) = \sum_{i=1}^{3} \rho_{ii}^{em}(\omega). \tag{8.47}$$

The practical problem with the above equation is that the spectral function $\rho^{em}(\omega)$ has to be obtained from the Euclidean correlator $G^{em}(\tau) = \sum_i G^{em}_{ii}(\tau)$ by inverting Eq. (8.45). As already discussed, this is not an easy task, when the Euclidean correlator is only available on a limited set of lattice points.

Usually, the electromagnetic observables are normalized by the sum of the squares of the individual quark charges:

$$C_{em} = e^2 \sum_f q_f^2 \tag{8.48}$$

where eq_f is the electric charge for flavor f. For three flavors we have $C_{em} = 2e^2/3$. One can then define

$$G^{em}(\tau) = C_{em} G(\tau), \qquad \rho^{em}(\omega) = C_{em} \rho(\omega), \tag{8.49}$$

and work in terms of $G(\tau)$ and $\rho(\omega)$ instead.

The exactly conserved vector current was used in Ref. [18] as an interpolator for j^{em}_μ, since it is protected from renormalization under quantum corrections. It is defined as

$$V^C_\mu(x) = c_\mu \left[\bar{\psi}(x+\hat{\mu})(1+\gamma_\mu)U^\dagger_\mu(x)\psi(x) - \bar{\psi}(x)(1-\gamma_\mu)U_\mu(x)\psi(x+\hat{\mu})\right], \tag{8.50}$$

where $c_4 = 1/2$, $c_i = 1/(2\gamma_i)$ and $U_\mu(x)$ are the gauge links. There are two connected diagrams contributing to the correlator (8.44):

$$\begin{aligned}\langle V^C_\mu(x) V^C_\nu(x)^\dagger\rangle = 2c_\mu c_\nu \Re\mathrm{tr}\Big[&S(y+\hat{\nu}, x+\hat{\mu})U^\dagger_\mu(x)\Gamma^+_\mu S(x,y)U_\nu(y)\tilde{\Gamma}^+_\nu + \\ &- S(y, x+\hat{\mu})U^\dagger_\mu(x)\Gamma^+_\mu S(x, y+\hat{\nu})U^\dagger_\nu(n)\tilde{\Gamma}^-_\nu\Big],\end{aligned} \tag{8.51}$$

where $S(x,y) = \langle\psi(x)\bar{\psi}(y)\rangle$ is the fermion propagator, $\Gamma^\pm_\mu = 1\pm\gamma_\mu$, $\tilde{\Gamma}^\pm_\mu = 1\pm\tilde{\gamma}_\mu$ and $\tilde{\gamma}_\mu = \gamma_4\gamma_\mu\gamma_4$. One usually neglects the disconnected contribution, since it is supposed to be small [2, 11, 19–21].

Figure 8.1 shows the conserved vector correlator $G(\tau)$, normalized by the corresponding correlator for free lattice fermions: the left panel is the result for light quarks, while the right panel is the result for strange quarks. They are plotted for different temperatures, as functions of τT. This results in their separation, even though they have identical decay. This result for the correlator was then used to invert Eq. (8.44) and obtain the electric conductivity. This was done in Ref. [18] using the MEM. The results are shown in Fig. 8.2 in the case of light and strange quarks separately (left) and combined (right). These results correspond to a system of 2+1 quark flavors, with smeared stout staggered fermions with an anisotropy parameter $\xi = 3.5$, physical mass for strange quarks and a light quark

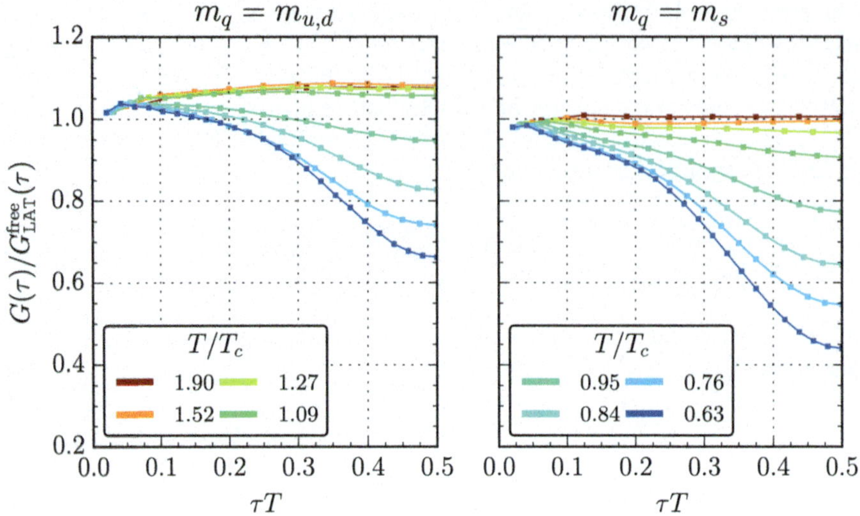

Fig. 8.1 From Ref [18]. Conserved vector current correlator, $G(\tau)$, for light (left) and strange (right) quarks. In both cases, the correlator is normalized by the one for free lattice fermions and is plotted as a function of τT for different values of the temperature. Plotting the correlators as functions of τT has the effect of separating them, even when they have identical decay

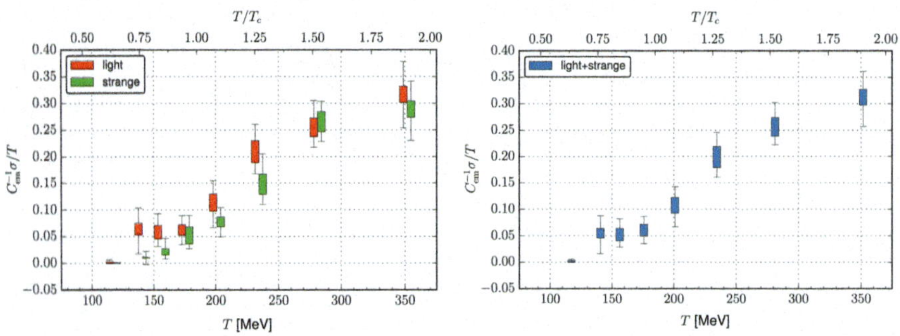

Fig. 8.2 From Ref. [18]. $C_{em}^{-1}\sigma/T$ as a function of the temperature for light and strange quarks separately (left) and combined (right). Statistical and systematic errors are shown as lines and rectangles, respectively

mass leading to $M_\pi = 384(4)$ MeV. Notice that the values of σ increase by a factor of 6 in the range of temperatures between 140 and 350 MeV. Results in the quenched approximation were obtained in Refs. [11–13, 20], while $N_f = 2$ and other $N_f = 2+1$ results were obtained in Refs. [2, 15] and [22], respectively.

In particular, the authors of Ref. [2] used two degenerate, light Wilson quarks with $O(a)$ improvement, corresponding to a pion mass $M_\pi = 270$ MeV and generated one value for σ at $N_\tau = 16$, corresponding to $T \simeq 250$ MeV. The

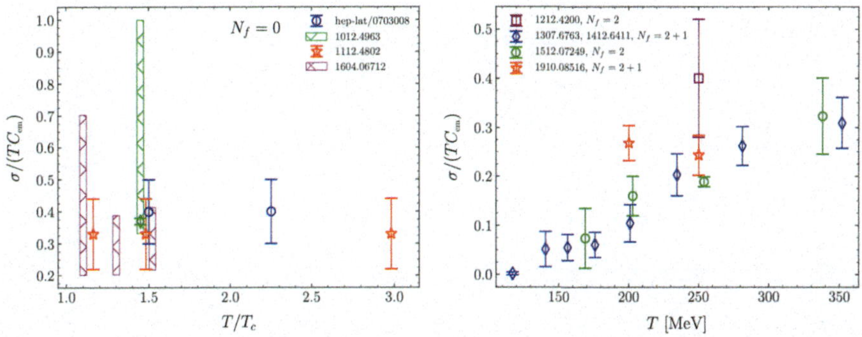

Fig. 8.3 From Ref. [23]. Compilation of results for the electrical conductivity, normalized as $\sigma/(TC_{em})$. In the left panel, results are calculated in the quenched approximation and they are shown as functions of T/T_c. In the right panel, results are calculated in a system of $N_f = 2$ and $N_f = 2 + 1$ quark flavors, and plotted as a function of the temperature in MeV

inversion was performed by using an operator product expansion-based ansatz for the spectral function.

In Ref. [15], these results have been extended to several values of the temperature and, besides the phenomenologically-motivated Ansatz on the spectral function, the Backus-Gilbert method has been used for the inversion, as a model-independent check.

Finally, in Ref. [22], $N_f = 2 + 1$ flavors at the physical point have been used for the first time. The electromagnetic conductivity has been obtained in this case at $T = 200, 250$ MeV and the Tikhonov approach was chosen for the inversion. It is worth pointing out that a compilation of all these results shows that they are consistent within error-bars: see Fig. 8.3.

Another interesting quantity is the charge diffusion coefficient D, which can be obtained as $D = \sigma/\chi_2^Q$. The result for D from Ref. [18] is shown in the left panel of Fig. 8.4. It exhibits a dip in the vicinity of T_c, which is consistent with the expectations for a strongly coupled system.

These calculations are phenomenologically very relevant, since the vector channel spectral function ρ_V is directly related to the dilepton and photon rates $(dW)/(d\omega d^3p)$ and dR/d^3p, respectively, and to the electrical conductivity σ (see Eqs. (8.45) and (8.47) for the electrical conductivity)

$$\frac{dW}{d\omega d^3p} = \frac{C_{em}\alpha_{em}^2 \rho_V(\omega, \mathbf{p}, T)}{6\pi^2(\omega^2 - \mathbf{p}^2)(e^{\omega/T} - 1)},$$
$$\omega \frac{dR_\gamma}{d^3p} = \frac{C_{em}\alpha_{em}\rho^T(\omega = |\mathbf{p}|, T)}{4\pi^2(e^{\omega/T} - 1)}, \tag{8.52}$$

where ρ^T is the transversally-polarized spectral function with respect to the direction of $\mathbf{p}$. Thermal photons are a very interesting observable, since they can

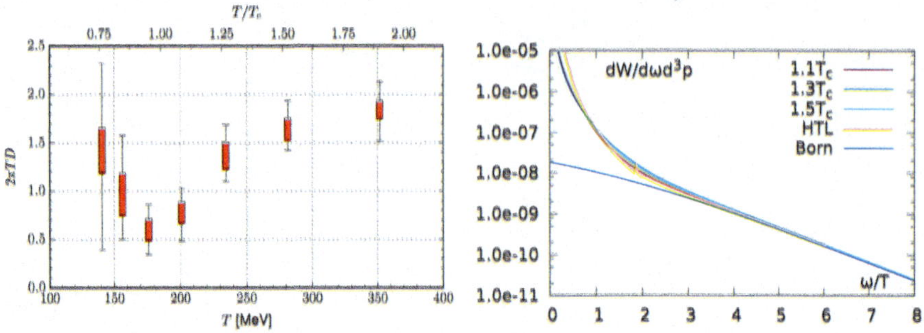

Fig. 8.4 Left: from Ref. [18]. Normalized charge diffusion coefficient $2\pi T D$ as a function of the temperature. Right: from Ref. [13]. Thermal dilepton rate obtained from ρ_V as a function of ω/T for different values of the temperature. Also shown are the Hard Thermal Loop (HTL) and the non-interacting Born results

provide information about the interaction that partons experience in the QGP. Dilepton spectra can instead be related to chiral symmetry restoration. The authors of Ref. [13] have obtained the thermal dilepton rate from ρ_V, as a function of ω/T for three different values of the temperature. It is shown in the right panel of Fig. 8.4. These results have been normalized by $C_{em} = 5/9$, corresponding to two valence quark flavors. They are qualitatively compatible with the rates obtained in HTL perturbation theory, and also with the leading order Born rate. However, these results show an enhancement around $\omega/T \sim 2$ and a qualitatively different behavior for small frequencies.

The low-mass dilepton emissivity is directly proportional to ρ_V, as we have seen. When chiral symmetry is broken, the masses of the chiral partners ρ and a_1 split, and this dominates the dilepton signal. Besides the result presented in the right panel of Fig. 8.4, several approaches to this problem exist, based e.g. on QCD and Weinberg sum rules and using input for the chiral condensate from lattice QCD, or on hadronic chiral lagrangians, which are used to evaluate the medium effects (for recent reviews see e.g. [24–26]). In the case of the ρ meson, the QCD sum rules provide constraints on model calculations [27–29], while the Weinberg sum rule relates the chiral condensate to the energy moments of the difference between the ρ and a_1 spectral functions [30]. It was found in Ref. [31] that, by combining these two approaches, it is possible to show that the ρ meson spectral function that successfully describes the experimental spectra is compatible with the temperature dependence of the chiral condensate obtained from lattice QCD simulations. Through the chiral lagrangian approach, chiral order parameters can be calculated as functions of the temperature. The temperature dependence of the ρ and a_1 meson spectral functions was studied in Ref. [32], where it was shown that the chiral condensate drops by about 15–20% at $T \simeq 160\,\text{MeV}$ as the two spectral functions become degenerate. This corroborates the idea that the degeneracy in these spectral functions is related to chiral symmetry restoration.

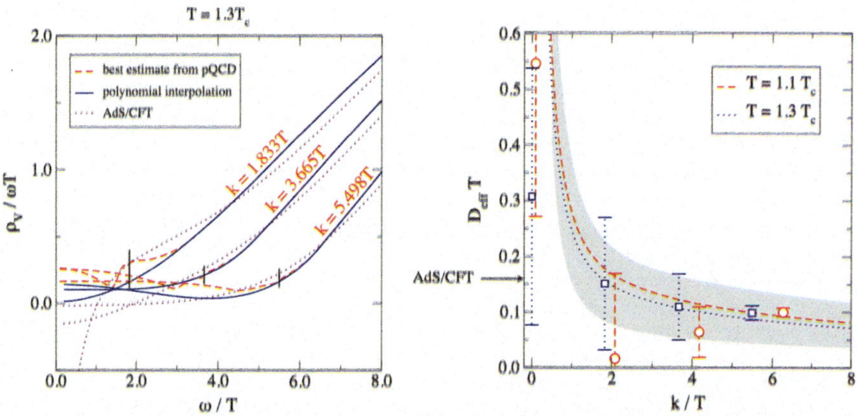

Fig. 8.5 From Ref. [33]. Left: Vector spectral functions as functions of ω/T, calculated at $T = 1.3T_c$. The vertical bars are placed at the light cone. The perturbative QCD results (in red) are based on Refs. [34–36], while the AdS/CFT curves (dotted, black curves) are rescaled from Ref. [38]. Right: lattice QCD results for D_{eff} as a function of $|\mathbf{p}|/T$. The lattice data points are compared to the NLO perturbative prediction [34] (curves)

It is possible to rewrite the thermal photon rate in terms of the charge diffusion coefficient D as

$$\frac{dR_\gamma}{d^3p} = \frac{2\alpha_{em}\chi_2^Q}{3\pi^2} n_B(\mathbf{p}) D_{eff}(\mathbf{p}) \tag{8.53}$$

where n_B is the Bose distribution and $D = \lim_{|\mathbf{p}|\to 0} D_{eff}(\mathbf{p})$. Results for vector meson correlation functions at non-vanishing momenta have been obtained in Ref. [33] by combining lattice QCD and perturbative techniques [34–36]. A polynomial description of the spectral shape was used in the regime in which lattice results deviate from the perturbative ones. These results are shown in the left panel of Fig. 8.5 at $T = 1.3T_c$. They agree with perturbative QCD predictions for $|\mathbf{p}| > 3T$, while the discrepancy at low momenta, with pQCD results overestimating the lattice QCD ones, is in apparent agreement with phenomenology [37]. The right panel of Fig. 8.5 shows the behavior of $D_{eff}(\mathbf{p})$ as a function of $|\mathbf{p}|/T$.

8.4 Heavy Quark Diffusion Coefficient

The heavy quark diffusion coefficient characterizes the movement of a heavy quark with momentum of at most the order of the temperature, with respect to the rest frame of the medium. It is defined as follows

$$D = \frac{1}{6\chi^{00}} \sum_{i=1}^{3} \frac{\rho_{ii}^{V}(\omega, \mathbf{p} = 0, T)}{\omega}, \tag{8.54}$$

where χ^{00} is the quark number susceptibility defined through the zeroth component of the temporal correlator in the vector channel, and ρ_V is the spectral function in the vector channel.

The Euclidean temporal correlation function in channel H, $G_H(\tau, \mathbf{p})$, can be defined as

$$G_H(\tau, \mathbf{p}) = \int d^3\mathbf{x} e^{-i\mathbf{p}\cdot\mathbf{x}} \langle J_H(\tau, \mathbf{p}) J_H(0, \mathbf{0}) \rangle, \tag{8.55}$$

where $G_H(\tau, \mathbf{p})$ is the analytical continuation of $D^+(t, \mathbf{p})$ from real to imaginary time

$$G_H(\tau, \mathbf{p}) = D^+(-i\tau, \mathbf{p}). \tag{8.56}$$

Also in this case, one can relate the correlation function to the spectral function

$$G_H(\tau, \mathbf{p}) = \int \frac{d\omega}{2\pi} \rho_H(\omega, \mathbf{p}) K(\omega, \tau), \tag{8.57}$$

where the integration kernel is the same as in Eq. (8.1). Due to asymptotic freedom, at high energy the spectral function is supposed to describe the propagation of a free quark-antiquark pair, which is known analytically [39, 40].

Spectral functions studied in lattice QCD at finite lattice spacing suffer from discretization effects, which are more severe on anisotropic lattices. For this reason, the authors of Ref. [41] decided to use isotropic lattices in their simulations. They obtained their results in the quenched approximation, but for a finite value of the heavy quark mass; these results are not continuum extrapolated. Earlier studies obtained results in the quenched and static approximation at finite lattice spacing [42] The spectral functions in this manuscript were reconstructed from the Euclidean correlators by means of the MEM. The left panel of Fig. 8.6 shows the charm transport coefficient obtained in Ref. [41] for three values of the temperature above T_c.

A related quantity is the heavy quark momentum diffusion coefficient, defined as

$$\kappa = \lim_{\omega \to 0} \frac{2T\rho_E(\omega)}{\omega}, \tag{8.58}$$

where ρ_E is the spectral function that can be reconstructed from the "color-electric correlator", related to the force that the heavy quark feels as it propagates through a gluon plasma:

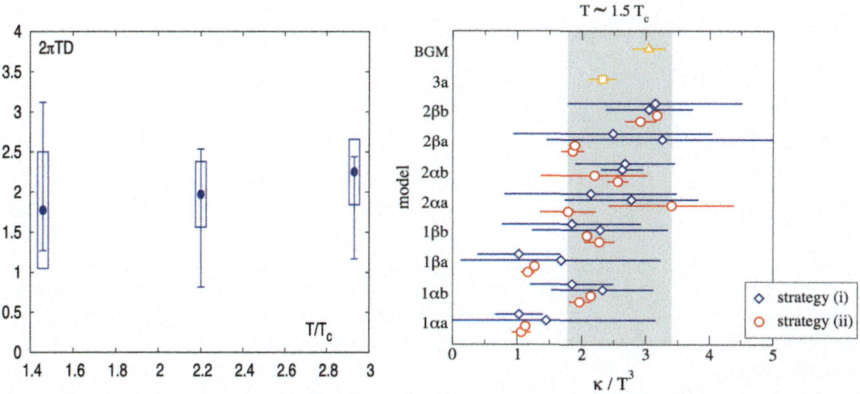

Fig. 8.6 Left: from Ref. [41]. Charm quark diffusion coefficient as a function of T/T_c, calculated in the quenched approximation and at finite lattice spacing. The boxes indicate the statistical error, while the lines show the systematic one due to the MEM analysis. Right: from Ref. [43]. Fit results for κ based on different models and different fit strategies

$$G_E(\tau) = -\frac{1}{3}\sum_{i=1}^{3}\frac{\langle \Re \mathrm{tr}[U(\beta;\tau)gE_i(\tau,\mathbf{0})U(\tau,0)gE_i(0,\mathbf{0})]\rangle}{\langle \Re \mathrm{tr}[U(\beta;0)]\rangle}, \quad \beta = \frac{1}{T}, \tag{8.59}$$

where E_i represents the color-electric field, T the temperature, and $U(\tau_2;\tau_1)$ is a Wilson line in the Euclidean time direction. In the non-relativistic limit, the following relationship between D and κ holds

$$D = \frac{2T^2}{\kappa}, \tag{8.60}$$

or, in terms of the drag coefficients η_D:

$$\eta_D = \frac{\kappa}{2MT}. \tag{8.61}$$

The drag coefficient is related to the kinetic equilibration time scale associated with heavy quarks: $\tau_{kin} = \eta_D^{-1}$. The authors of Ref. [43] calculated κ in the quenched approximation, in the continuum limit for the first time, but with an infinite mass for the heavy quark. They find $\kappa/T^3 = 1.8 - 3.4$, which leads to an equilibration time $\tau_{kin} = (1.8\ldots3.4)(T_c/T)^2(\frac{M}{1.5\ \mathrm{GeV}})$ fm/c, which close to T_c is of the order of 1 fm/c. For the diffusion coefficient they obtain a value $DT = 0.59\ldots1.1$. The right panel of Fig. 8.6 shows a compilation of fit results for κ, based on different models and different fitting strategies (see Ref. [43] for details).

More recently, Ref. [44] presented a progress report on calculating the heavy quark momentum diffusion coefficient from the correlator of two chromo-electric fields attached to a Polyakov loop in $SU(3)$ pure gauge theory. The left panel of

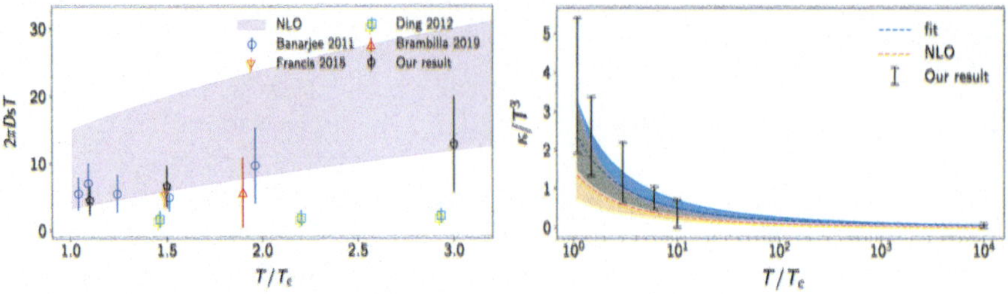

Fig. 8.7 From Ref. [44]. Left: Heavy quark diffusion coefficient D as a function of the temperature: lattice QCD results are compared to the perturbative QCD purple band. Right: κ/T^3 as a function of T/T_c over a broad temperature range. Also in this case, results are compared to NLO perturbation theory

Fig. 8.7 shows D as a function of the temperature: lattice QCD results are compared to the purple band from perturbation theory. The right panel of Fig. 8.7 shows κ/T^3 as a function of T/T_c over a broad temperature range. Also in this case, results are compared to NLO perturbation theory.

8.5 Viscosity

One of the most impressive characteristics of the Quark-Gluon Plasma is that it behaves like the most ideal fluid ever observed, namely with a very small shear viscosity over entropy density ratio η/s [45–48]. This feature emerged from the fact that experimental measurements of elliptic flow reveal an unexpected collective behavior, which can be described in terms of hydrodynamics with very small η/s. This collective behavior is reflected by the fact that the initial anisotropic pressure gradients in the fireball drive the system to develop a momentum anisotropy. The shear viscosity is therefore one of the most important transport coefficients for heavy-ion physics.

While perturbative QCD results based on a weakly interacting gas of quarks and gluons predict a viscosity value which is an order of magnitude larger than the "experimental" one described above, the gauge-string duality between Anti-de-Sitter space and conformal field theory provided a paradigm shift in our understanding of strongly interacting matter. For any field theory at infinite coupling, a value of $\eta/s = 1/(4\pi)$ was predicted [49], which led to the idea that the quark-gluon plasma, in the regime of temperatures reached in heavy-ion collision experiments, is strongly coupled.

In classical transport theory, the ratio η/s for a dilute gas at temperature T is given by

$$\frac{\eta}{s} \sim T l_{mfp} \bar{v} \sim \frac{T \bar{v}}{n\sigma}, \tag{8.62}$$

where l_{mfp} is the mean free path, n is the particle number density, $\bar{v}$ is the mean speed and σ the cross section. Since for a weakly interacting system σ is small, the viscosity is supposed to be large: it is in fact infinitely large for a free gas. One possible way of determining the shear viscosity is through the Kubo formula

$$\eta(T) = \pi \lim_{\omega\to 0} \lim_{\mathbf{k}\to 0} \frac{\rho_{ijij}(\omega, \mathbf{k}, T)}{\omega}, \tag{8.63}$$

where $\rho_{ijij}(\omega, \mathbf{k}, T)$ is the spectral function corresponding to the energy-momentum tensor, for spatial indices $i \neq j$, and with momentum $\mathbf{k}$ in the j direction.

The correlator of the energy-momentum tensor is defined as

$$C_{\mu\nu,\rho\sigma}(\tau, \mathbf{x}) = \int d\tau' d\mathbf{x}' \langle T_{\mu\nu}(\tau', \mathbf{x}') T_{\rho\sigma}(\tau' + \tau, \mathbf{x}' + \mathbf{x})\rangle, \tag{8.64}$$

and it can be given the following spectral representation

$$C_{\mu\nu,\rho\sigma}(\tau, \mathbf{q}) = \int_0^\infty d\omega \rho_{\mu\nu\rho\sigma}(\omega, \boldsymbol{\rho}, T) K(\omega, \tau; T), \tag{8.65}$$

where the kernel is the same as in Eq. (8.1). Early [50–52] and more recent [53] lattice QCD simulations used the above Kubo formula, extracted the correlator in Eq. (8.64) from lattice QCD simulations and inverted Eq. (8.65) to obtain the spectral function. This strategy is affected by two main problems

- the inversion of Eq. (8.65) is an ill-posed problem, as we have already discussed for other inversion problems in this Chapter;
- for the specific case of the viscosity, the signal in the stress-energy tensor is dominated by the high-ω part of the spectral function [54]. This makes the inversion of Eq. (8.65) even harder. In the case of ρ_{1313}, the asymptotic UV behavior is $\sim \omega^4$.

The situation is much better for the correlators of conserved charges or heavy flavors encountered in the previous chapters. In those cases, the spectral function at large ω only grows like ω^2, making the problem of UV contamination less severe. It was noticed in Refs. [1,55], that the problem can be mitigated to $\sim \omega^2$ through the Ward identity

$$-\omega^2 \rho_{0101} = \mathbf{q}^2 \rho_{1313} \tag{8.66}$$

and therefore calculating ρ_{0101} instead. This improves the UV behavior. However, in the continuum limit the following identity holds

$$\frac{\langle T_{01}T_{01}\rangle(\tau, \mathbf{q}=0)}{T^5} = \frac{s}{T^3} \tag{8.67}$$

which means that one needs non-zero momenta to obtain information about the shear viscosity from this correlator.

Finally, the stress-energy tensor correlator itself has a quickly degrading signal as τ is increased beyond a few lattice spacings. Therefore, the signal is evaluated as an average of widely fluctuating contributions, which is analogous to a sign problem. This further complication can luckily be mitigated in the quenched case, by using the multilevel algorithm [56, 57]. This algorithm crucially depends on the locality of the action, which is the reason why it is not applicable to full QCD simulations with dynamical quarks. Recent attempts to extend this algorithm to full QCD are described in Ref. [58].

The first continuum extrapolated results for this observable in pure gauge QCD were presented in Ref. [59] at $T = 1.5T_c$ and $T = 2T_c$. These results are compatible with previous ones. It is worth pointing out that the focus of Ref. [59] was mainly to determine the stress energy tensor correlator, and the estimate of the viscosity is provided through a hydrodynamics-motivated ansatz which was used previously in the literature [1]. The left panel of Fig. 8.8 shows a compilation of all results on η/s from lattice QCD simulations and a lattice-based approach [60].

All of these results have been obtained on lattices with relatively small temporal extent, compared to the ones used e.g. in the electric conductivity calculations presented in Sect. 8.3. The magnitude of these systematic errors needs to be better estimated.

The right panel of Fig. 8.8 shows an estimate of the temperature dependence of η/s obtained through a Bayesian analysis of heavy-ion collision data [65]. The model was calibrated to multiplicity, transverse momentum and flow data,

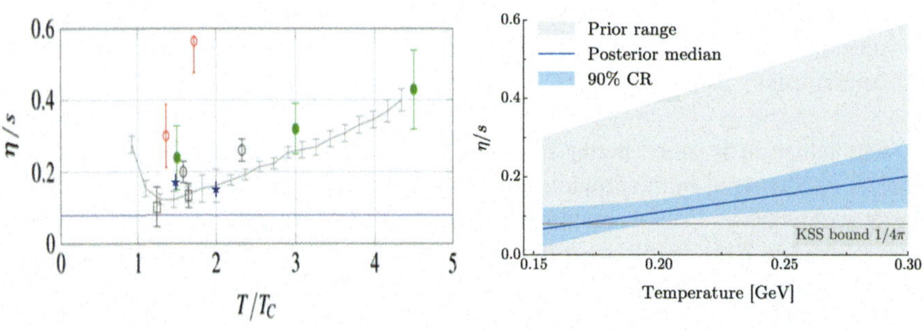

Fig. 8.8 Left: From Ref. [61]. Compilation of pure gauge lattice QCD results and selected lattice-based approaches for η/s as a function of the temperature: the black empty squares and circles are from [62, 63], the vertical lines are from [60], the empty red circles from [51], the green full circles from Ref. [64] and the blue stars from Ref. [59]. Right: From Ref. [65]. Bayesian analysis prediction for the temperature dependence of η/s above T_c. The prior range is indicated by the gray band, the blue band is a 90% credible region around the posterior distribution median (blue line)

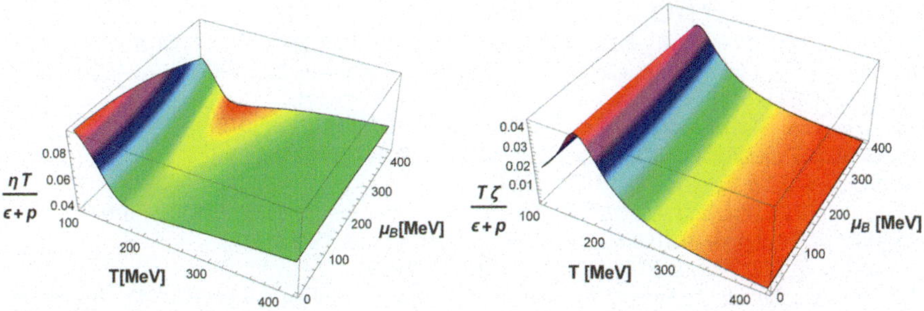

Fig. 8.9 From Ref. [75]. Shear (left) and bulk (right) viscosities, multiplied by T and divided by $\epsilon + p$ as functions of temperature and baryonic chemical potential

and predicted constraints on the initial conditions and T−dependent transport coefficients in the quark-gluon plasma. The behavior of the shear viscosity over entropy ratio is in reasonable agreement with the lattice QCD results discussed above.

Several other estimates of η/s exist in the literature (for a review see e.g. [66]). In particular, the constraint $\frac{1}{4\pi} < \frac{\eta}{s} < \frac{5}{8\pi}$ was obtained from the experimental data for the second-order anisotropic flow coefficient of charged hadrons measured at RHIC [67]. The more recent, very precise measurements of higher order anisotropic flow coefficients [68] have led to a more precise determination of η/s, which can now be obtained phenomenologically with a precision of 5−10% [69, 70]. Even the temperature dependence of this quantity, as well as of other transport coefficients [71–74], is now within reach thanks to the precise experimental result, and the sensitivity of the phenomenological approaches. It is worth mentioning that the AdS/CFT-based approach introduced in Sect. 4.5.2 is the only method currently available, which is able to reproduce both lattice QCD thermodynamics and provide a realistic estimate of the temperature and chemical potential dependence of transport coefficients. Figure 8.9 shows the shear (left) and bulk (right) viscosities multiplied by T and divided by $\epsilon + p$ as functions of temperature and baryonic chemical potential from Ref. [75].

References

1. Meyer, H.B.: Eur. Phys. J. A **47** (2011), 86
2. Brandt, B.B., Francis, A., Meyer, H.B., Wittig, H.: JHEP **03**, 100 (2013)
3. Caron-Huot, S., Phys. Rev. D **79**, 125009 (2009)
4. Hansen, P.: Numer. Algorithms **46**, 189 (2007)
5. Tarantola, A.: Inverse Problem Theory and Methods for Model Parameter Estimation. Society for Industrial and Applied Mathematics, Philadelphia (2004)
6. Asakawa, M., Hatsuda, T., Nakahara, Y.: Prog. Part. Nucl. Phys. **46**, 459 (2001)
7. Burnier, Y., Rothkopf, A.: Phys. Rev. Lett. **111**, 182003 (2013)
8. Shu, H.-T., Ding, H.-T., Kaczmarek, O., Mukherjee, S., Ohno, H.: **PoSLATTICE2015**, 180 (2016)

9. Ohno, H.: **PoSLATTICE2015**, 175 (2016)
10. Brandt, B.B., Francis, A., Jäger, B., Meyer, H.B.: Phys. Rev. **D93**, 054510 (2016)
11. Ding, H.T., Francis, A., Kaczmarek, O., Karsch, F., Laermann, E., Soeldner, W.: Phys. Rev. D **83**, 034504 (2011)
12. Francis, A., Kaczmarek, O.: Prog. Part. Nucl. Phys. **67**, 212–217 (2012)
13. Ding, H.T., Kaczmarek, O., Meyer, F.: Phys. Rev. D **94**(3), 034504 (2016)
14. Beach, K.S.D.: cond-mat/0403055 (2004)
15. Brandt, B.B., Francis, A., Jäger, B., Meyer, H.B.: Phys. Rev. D **93**(5), 054510 (2016)
16. Hansen, M., Lupo, A., Tantalo, N.: Phys. Rev. D **99**(9), 094508 (2019)
17. Kadanoff, L.P., Martin, P.C.: Ann. Phys. **24**, 419 (1963)
18. Aarts, G., Allton, C., Amato, A., Giudice, P., Hands, S., Skullerud, J.I.: JHEP **02**, 186 (2015)
19. Gupta, S.: Phys. Lett. B **597**, 57–62 (2004)
20. Aarts, G., Allton, C., Foley, J., Hands, S., Kim, S.: Phys. Rev. Lett. **99**, 022002 (2007)
21. Kaczmarek, O., Müller, M.: PoS **LATTICE2013**, 175 (2014)
22. Astrakhantsev, N.Y.: Braguta, V.V., D'Elia, M., Kotov, A.Y., Nikolaev, A.A., Sanfilippo, F.: Phys. Rev. D **102**(5), 054516 (2020)
23. Aarts, G., Nikolaev, A.: [arXiv:2008.12326 [hep-lat]]
24. van Hees, H., Rapp, R.: Nucl. Phys. A **806**, 339–387 (2008)
25. Rapp, R., Wambach, J., van Hees, H.: Landolt-Bornstein **23**, 134 (2010)
26. Rapp, R.: Adv. High Energy Phys. **2013**, 148253 (2013)
27. Hatsuda, T., Koike, Y., Lee, S.H.: Nucl. Phys. B **394**, 221–266 (1993)
28. Leupold, S., Peters, W., Mosel, U.: Nucl. Phys. A **628**, 311–324 (1998)
29. Zschocke, S., Pavlenko, O.P., Kampfer, B.: Eur. Phys. J. A **15**, 529–537 (2002)
30. Weinberg, S.: Phys. Rev. Lett. **18**, 507–509 (1967)
31. Hohler, P.M., Rapp, R.: Phys. Lett. B **731**, 103–109 (2014)
32. Hohler, P.M., Rapp, R.: Ann. Phys. **368**, 70-s-109 (2016)
33. Ghiglieri, J., Kaczmarek, O., Laine, M., Meyer, F.: Phys. Rev. D **94**(1), 016005 (2016)
34. Ghiglieri, J., Moore, G.D.: JHEP **12**, 029 (2014)
35. Ghisoiu, I., Laine, M.: JHEP **10**, 083 (2014)
36. Laine, M.: JHEP **11**, 120 (2013)
37. Burnier, Y., Gastaldi, C.: Phys. Rev. C **93**(4), 044902 (2016)
38. Caron-Huot, S., Kovtun, P., Moore, G.D., Starinets, A., Yaffe, L.G.: JHEP **12**, 015 (2006)
39. Karsch, F., Laermann, E., Petreczky, P., Stickan, S.: Phys. Rev. D **68**, 014504 (2003)
40. Aarts, G., Martinez Resco, J.M.: Nucl. Phys. B **726**, 93–108 (2005)
41. Ding, H.T., Francis, A., Kaczmarek, O., Karsch, F., Satz, H., Soeldner, W.: Phys. Rev. D **86**, 014509 (2012)
42. Banerjee, D., Datta, S., Gavai, R., Majumdar, P.: Phys. Rev. D **85**, 014510 (2012)
43. Francis, A., Kaczmarek, O., Laine, M., Neuhaus, T., Ohno, H.: Phys. Rev. D **92**(11), 116003 (2015)
44. Brambilla, N., Leino, V., Petreczky, P., Vairo, A.: [arXiv:2007.10078 [hep-lat]]
45. Teaney, D., Lauret, J., Shuryak, E.V.: [arXiv:nucl-th/0110037 [nucl-th]]
46. Teaney, D.: Phys. Rev. C **68**, 034913 (2003)
47. Romatschke, P., Romatschke, U.: Phys. Rev. Lett. **99**, 172301 (2007)
48. Song, H., Heinz, U.W.: J. Phys. G **36**, 064033 (2009)
49. Kovtun, P., Son, D.T., Starinets, A.O.: Phys. Rev. Lett. **94**, 111601 (2005)
50. Karsch, F., Wyld, H.W.: Phys. Rev. D **35**, 2518 (1987)
51. Nakamura, A., Sakai, S.: Phys. Rev. Lett. **94**, 072305 (2005)
52. Meyer, H.B.: Phys. Rev. D **76**, 101701 (2007)
53. Astrakhantsev, N., Braguta, V., Kotov, A.: EPJ Web Conf. **137**, 07003 (2017)
54. Aarts, G., Martinez Resco, J.M.: JHEP **04**, 053 (2002)
55. Meyer, H.B.: JHEP **08**, 031 (2008)
56. Luscher, M., Weisz, P.: JHEP **09**, 010 (2001)
57. Meyer, H.B.: JHEP **01**, 048 (2003)
58. Cè, M., Giusti, L., Schaefer, S.: PoS **LATTICE2016**, 263 (2016)

59. Borsányi, S., Fodor, Z., Giordano, M., Katz, S.D., Pasztor, A., Ratti, C., Schäfer, A., Szabo, K.K., Tóth, B.C.: Phys. Rev. D **98**(1), 014512 (2018)
60. Christiansen, N., Haas, M., Pawlowski, J.M., Strodthoff, N.: Phys. Rev. Lett. **115**(11), 112002 (2015)
61. Ratti, C.: Rept. Prog. Phys. **81**(8), 084301 (2018)
62. Meyer, H.B.: Phys. Rev. Lett. **100**, 162001 (2008)
63. Meyer, H.B.: Nucl. Phys. A **830**, 641C–648C (2009)
64. Mages, S.W., Borsányi, S., Fodor, Z., Schäfer, A., Szabó, K.: PoS **LATTICE2014**, 232 (2015)
65. Bernhard, J.E., Moreland, J.S., Bass, S.A., Liu, J., Heinz, U.: Phys. Rev. C **94**(2), 024907 (2016)
66. Shen, C., Heinz, U.: Nucl. Phys. News **25**(2), 6–11 (2015)
67. Song, H., Bass, S.A., Heinz, U., Hirano, T., Shen, C.: Phys. Rev. Lett. **106**, 192301 (2011)
68. Aamodt, K., et al.: [ALICE] Phys. Rev. Lett. **107**, 032301 (2011)
69. Gale, C., Jeon, S., Schenke, B., Tribedy, P., Venugopalan, R.: Phys. Rev. Lett. **110**(1), 012302 (2013)
70. Schenke, B., Venugopalan, R.: Phys. Rev. Lett. **113**, 102301 (2014)
71. Shen, C., Heinz, U., Huovinen, P., Song, H.: Phys. Rev. C **84**, 044903 (2011)
72. Niemi, H., Denicol, G.S., Huovinen, P., Molnar, E., Rischke, D.H.: Phys. Rev. Lett. **106**, 212302 (2011)
73. Shen, C., Heinz, U.: Phys. Rev. C **83**, 044909 (2011)
74. Noronha-Hostler, J., Denicol, G.S., Noronha, J., Andrade, R.P.G., Grassi, F.: Phys. Rev. C **88**(4), 044916 (2013)
75. Rougemont, R., Critelli, R., Noronha-Hostler, J., Noronha, J., Ratti, C.: Phys. Rev. D **96**(1), 014032 (2017)

Heavy Flavors and Quarkonia

9

Abstract

In this chapter we review recent progress in the study of heavy flavors and quarkonia. These are considered a thermometer of the system produced in heavy-ion collisions. We will discuss the in-medium quark-antiquark potential, the free-energy, the euclidean temporal and spatial correlators and their relationships with the quarkonium spectral function. We also discuss effective theories for heavy quarkonia, besides lattice QCD results.

9.1 Introduction

Heavy flavor quark are very interesting probes to understand the properties of the QGP, since they are produced in the early stage of the collision and thus enable us to learn about the entire evolution of the system. They are considered as a thermometer of the system produced in a heavy-ion collision, in which their melting is a dynamical process that filters out quarkonium states over time, depending on their binding energies. The physics of heavy quarks and quarkonia has recently reached a new exciting stage, in which it is possible for the first time to compare results of lattice QCD simulations to experimental measurements [1–3].

Usually, the in-medium behavior of quarkonia is studied by means of three different approaches

1. by simulating a $q\bar{q}$ potential on the lattice, plugging it into Schrödinger's equation for the quarkonium two-point function and solving it;
2. by simulating the quarkonium Euclidean temporal correlators on the lattice, inverting them and reconstructing their spectral function, analogously to what was discussed in Chap. 8;
3. by studying the in-medium screening properties of spatial correlators.

For a recent comprehensive review on the physics of heavy flavors see Ref. [2].

C. Ratti, R. Bellwied, *The Deconfinement Transition of QCD*, Lecture Notes in Physics 981, https://doi.org/10.1007/978-3-030-67235-5_9

9.2 In-Medium $q\bar{q}$ Potential

The first approach listed above is based on the idea that the interaction between the static quarks within a bound state can be described by means of an instantaneous, temperature-dependent potential [4–6], which can be obtained from effective theories or lattice QCD simulations.

The real-time $q\bar{q}$ potential in thermal equilibrium has been evaluated for the first time in Ref. [5] using Hard Thermal Loop perturbation theory. Such a potential has a real and an imaginary part, which are given by the following expressions

$$\Re[V_{S,HTL}](\mathbf{r}) = -\frac{g^2 C_F}{4\pi}\left[m_D + \frac{\exp(-m_D r)}{r}\right] = -\alpha_S\left[m_D + \frac{\exp(-m_D r)}{r}\right],$$
$$\Im[V_{S,HTL}](\mathbf{r}) = -\alpha_S T 2\int_0^\infty \frac{dzz}{(z^2+1)^2}\left[1 - \frac{\sin(zx)}{zx}\right], \tag{9.1}$$

where $\alpha_S = g^2 C_F/(4\pi)$ is the strong coupling constant and m_D is the Debye screening mass. The real part of the potential is related to Debye screening effects. At intermediate distance, where the r−dependence of α_S is weak, the temperature-dependence of the real part is entirely determined by m_D, while the imaginary part explicitly depends on temperature. The real part of the potential from HTL is shown in the top panel of Fig. 9.1.

The real part of the potential shows an asymptotic behavior at high distance, that tends to twice the value of the in-medium heavy quark mass [8]. It turns out that this lowest-order HTL result coincides with the color singlet free energy in the Coulomb gauge. This is no longer true at higher orders in the perturbative expansion, even though lattice QCD results show that the two are still quite close. The imaginary part of the potential arises from the phenomenon of Landau damping [5, 6] at this scale. At other scales it relates to gluo-dissociation and in-medium inelastic parton scattering.

Notice that the potential we discussed so far governs the evolution in time of the medium-averaged, color-singlet correlator of point split mesonic operators, and not of the heavy quarkonium wavefunction. In fact, the two heavy-quarks are separated in space and can never meet and therefore annihilate. As a consequence, the imaginary part of the potential represents decoherence between the quarkonium state at initial and later times, and therefore relates to the thermalization of the quarkonium system.

In the non-perturbative regime, one needs to calculate the potential from lattice QCD simulations. This has been done in a system of $N_f = 2 + 1$ dynamical quarks for the first time in Refs. [7, 9], in which spectral functions were extracted using a Bayesian inference prescription. In fact, the real-time definition of the static $q\bar{q}$ potential is formulated in Minkowski space, and is thus affected by the same reconstruction problems discussed in Chap. 8. These results are shown in the bottom panel of Fig. 9.1. At low temperatures, the results exhibit a Cornell-like behavior, but

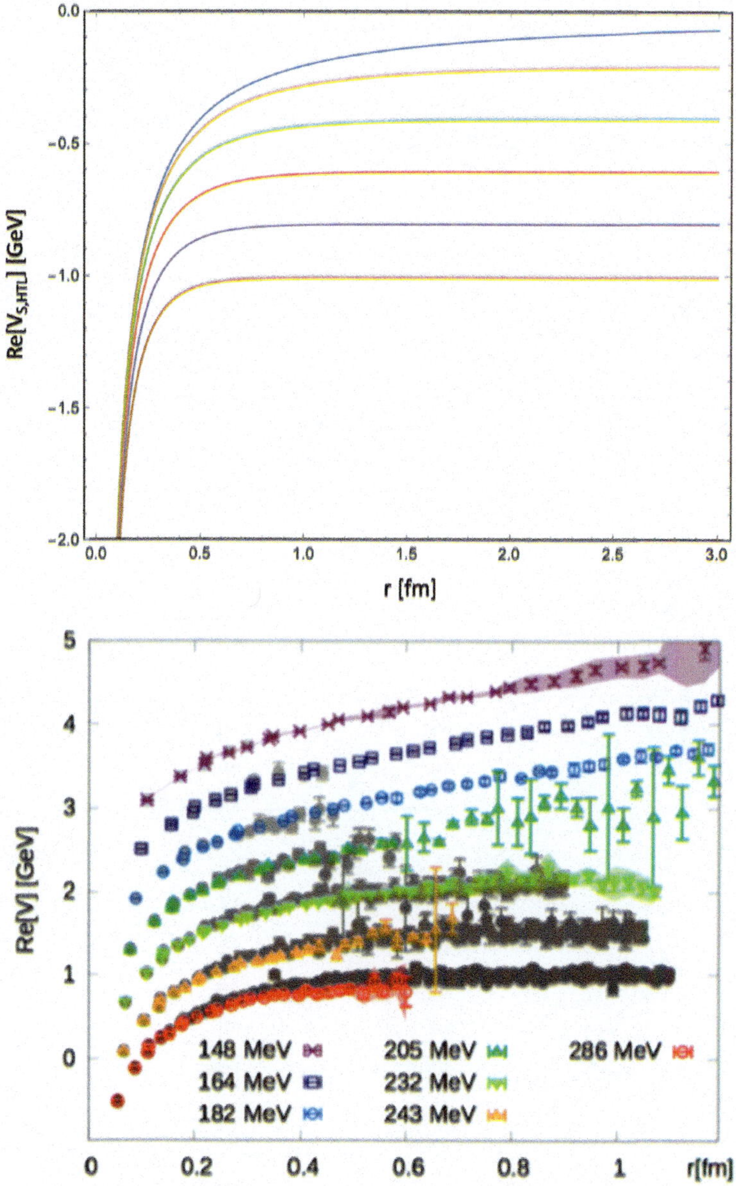

Fig. 9.1 Top: The real part of the static $q\bar{q}$ potential from HTL perturbation theory, calculated for $\alpha_S = 1$ and different values of the Debye mass between 0 (highest, blue line) and 1 GeV (lowest, brown line). Bottom: from Ref. [7]. Results for the real part of the $q\bar{q}$ potential in a system of 2+1 dynamical quark flavors (colored points) compared to the color singlet free energies in Coulomb gauge (gray bands)

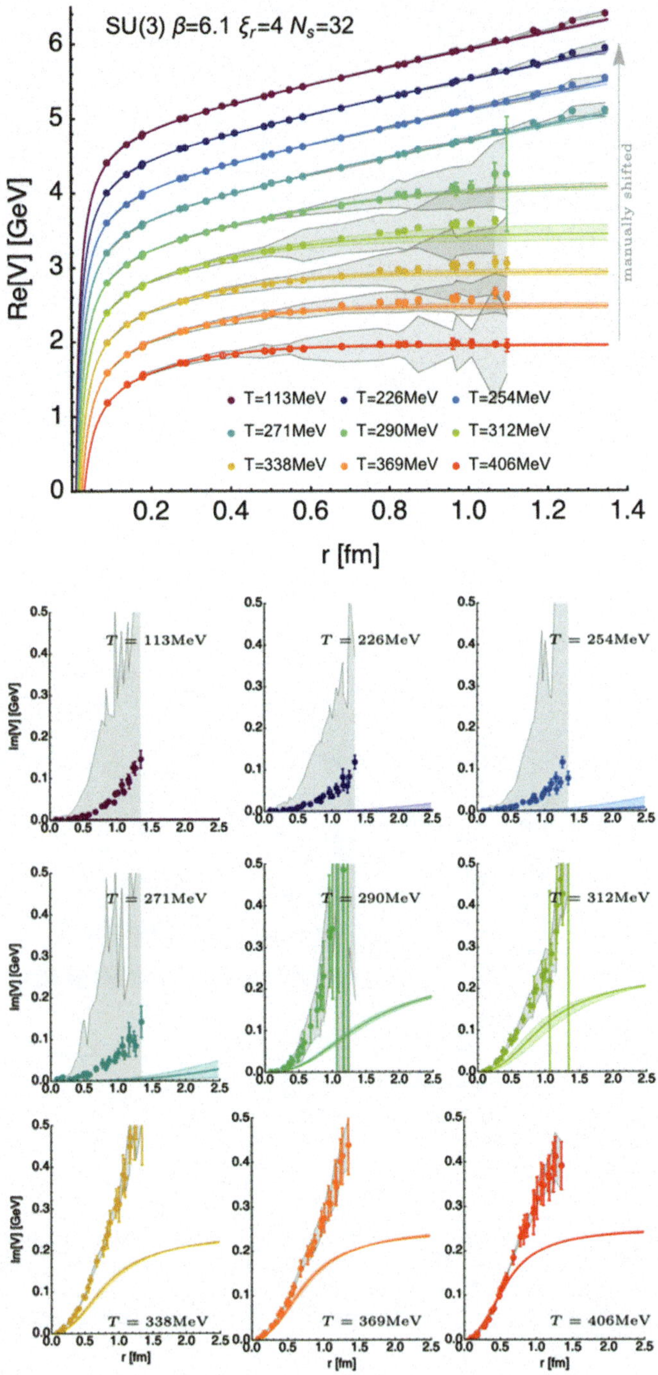

Fig. 9.2 From Ref. [10]. Top: The real part of the static $q\bar{q}$ potential from pure gauge, lattice QCD simulations at different temperature values. Bottom: The imaginary part of the static $q\bar{q}$ potential. Error-bars indicate statistical uncertainties, while gray bands indicate systematic uncertainties

no indications of string breaking up to $r \sim 1.2$ fm. This relatively large value is due to the heavier-than-physical dynamical quark masses used in the simulations.

In Ref. [10], these results in pure gauge QCD were considerably improved, by considering significantly larger physical volumes, thus clarifying the effect of finite-volume artifacts. Results for the real and imaginary parts of V from this study are shown in Fig. 9.2.

It is evident from comparing the bottom panel of Fig. 9.1 to the top panel of Fig. 9.2 that the results in the presence of dynamical quarks show a smooth decrease of the string-like part of the potential, compared to the abrupt onset observed in the pure gauge case.

Recently, more realistic results have been obtained in Ref. [11], in which the HISQ action was used to implement $N_f = 2 + 1$ dynamical flavors with a pion mass $m_\pi = 160\,\text{MeV}$ and $N_s^3 \times N_\tau = 48^3 \times 12$ and $48^3 \times 16$. These results for the real and imaginary parts of the potential are shown in the top and bottom panels of Fig. 9.3, respectively. They have been obtained through the Pade reconstruction method.

9.3 Quark-Antiquark Free Energy

The free-energy of a quark-antiquark pair can be related to the partition function as follows

$$\exp[-\beta F] = \frac{1}{N_c^2} \sum_s \langle s | e^{-\beta H} | s \rangle, \tag{9.2}$$

where the states $|s\rangle$ contain the medium degrees of freedom and also the static quarks. One can rewrite the above expression in terms of the color trace of the Polyakov loop

$$\exp[-\beta F] = \frac{1}{N_c^2} \mathrm{tr}_{medium}[\mathrm{tr}_c(L(\mathbf{x}_1))\mathrm{tr}_c(L^\dagger(\mathbf{x}_2))] \tag{9.3}$$

where

$$L(\mathbf{x}) = \mathcal{T} \exp[i \int_0^\beta D\tau A_4(\mathbf{x}, \tau)]. \tag{9.4}$$

Usually, the definition above gets re-written by normalizing with respect to the partition function of the medium without heavy quarks: $\Delta F = F - F_0$. One thus gets

$$\exp[-\beta \Delta F] = \frac{1}{N_c^2} [\mathrm{tr}_c(L(\mathbf{x}_1))\mathrm{tr}_c(L^\dagger(\mathbf{x}_2))]. \tag{9.5}$$

As we saw in Chap. 2, the Polyakov loop itself is related to the free energy of a single heavy quark in the medium.

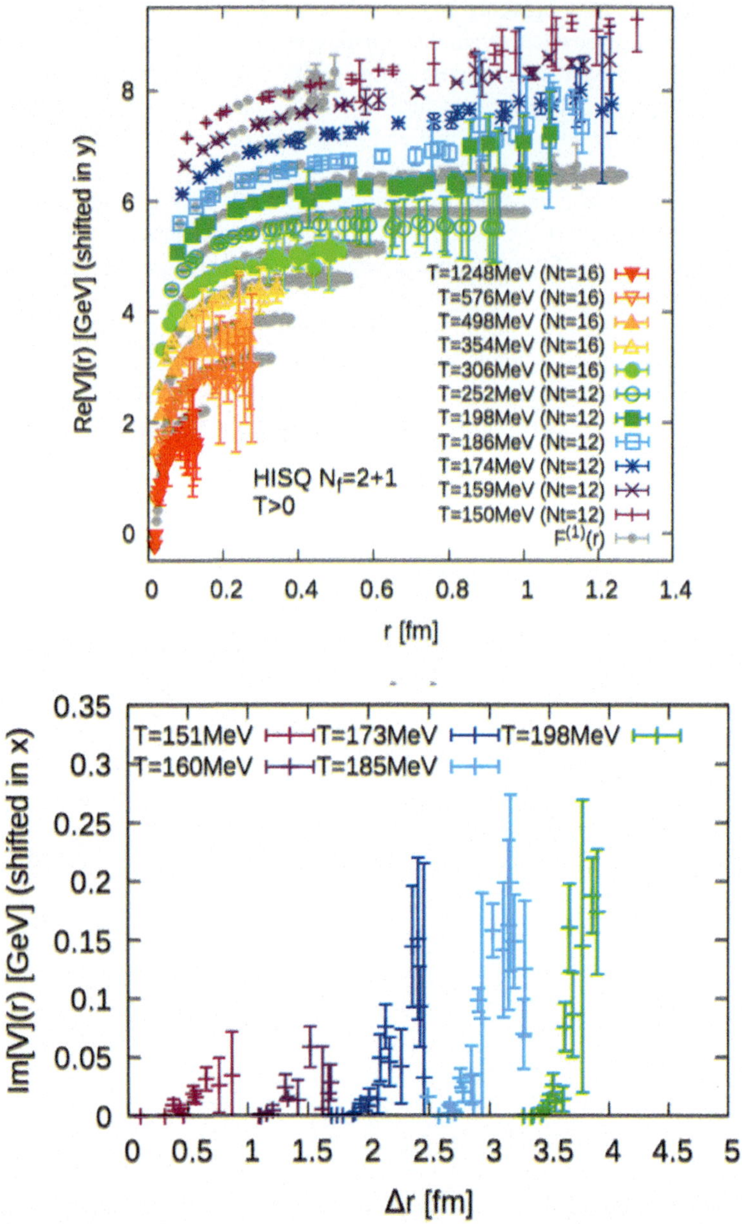

Fig. 9.3 From Ref. [11]. Top: The real part of the static $q\bar{q}$ potential from $N_f = 2+1$ dynamical lattice QCD simulations at different temperature values. Bottom: The imaginary part of the static $q\bar{q}$ potential. The gray bands in the top panel indicate the color singlet free energies

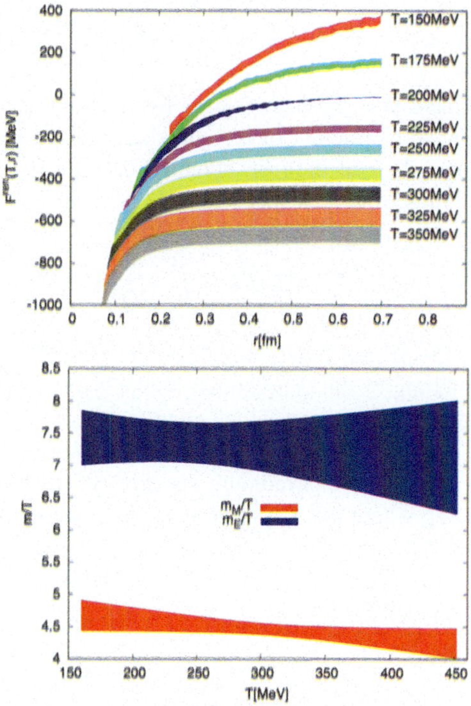

Fig. 9.4 From Ref. [13]. Top: Continuum extrapolation of the static quark-antiquark free energy at different temperatures. Bottom: Continuum extrapolation of electric and magnetic screening masses as functions of the temperature

The Polyakov loop defined above mixes color electric and color magnetic contributions [12]; besides, since it diverges in the continuum limit, so does the free energy. One therefore needs to renormalize this divergence. The authors of Ref. [13] chose the following renormalization: $\Delta F^{ren} = \Delta F(r, T) - \Delta F(\infty, T_0)$, where T_0 is a fixed reference temperature. The corresponding continuum extrapolated results for this renormalized free energy are shown in the top panel of Fig. 9.4. These results were then used to obtain the electric and magnetic screening masses through the following equations:

$$C_{M^+}(r) = \langle \sum_x \mathrm{tr}_c[L_{M^+}(\mathbf{x})]\mathrm{tr}_c[L_{M^+}(\mathbf{x}+\mathbf{r})]\rangle - \left|\sum_x \mathrm{tr}_c[L(\mathbf{x})]\right|^2 \sim_{r\to\infty} \gamma_M \frac{e^{-m_M r}}{rT}$$

$$C_{E^-}(r) = \langle \sum_x \mathrm{tr}_c[L_{E^-}(\mathbf{x})]\mathrm{tr}_c[L_{E^-}(\mathbf{x}+\mathbf{r})]\rangle \sim_{r\to\infty} \gamma_E \frac{e^{-m_E r}}{rT}. \tag{9.6}$$

Results for the masses as functions of the temperature are shown in the bottom panel of Fig. 9.4.

9.4 Euclidean Temporal Correlators and Spectral Functions

Lattice QCD simulations can provide the two-point correlation functions $G(\tau, T)$ through this definition:

$$G(\tau, T) = \int d^3x \langle (\bar{\psi}(\tau, \mathbf{x})\Gamma\psi(\tau, \mathbf{x}))(\bar{\psi}(0, \mathbf{0})\Gamma\psi(0, \mathbf{0}))^\dagger \rangle, \tag{9.7}$$

where ψ is the heavy quark field. The matrices Γ are Dirac matrices which define the spin structure of the quarkonium bound state of interest. The spectral function $\rho(\omega, T)$ can then be obtained through the inversion of the following equation, similarly to what was discussed in Chap. 8

$$G(\tau, T) = \int_0^\infty \frac{d\omega}{2\pi} \rho(\omega, T) \frac{\cosh(\omega(\tau - 1/(2T)))}{\sinh(\omega/(2T))}. \tag{9.8}$$

The spectral function embeds the information about in-medium modification and melting of the quarkonium bound state. As we discussed in Chap. 8, this is an ill-posed problem as lattice simulations only provide a limited amount of points for $G(\tau, T)$, based on which the above integral should be inverted.

The first charmonium spectral function reconstruction was performed in Ref. [14], where the J/ψ and η_c states were studied. Many more results were obtained in the quenched approximation [15–20]. Dynamical quarks on isotropic and anisotropic lattices have been studied in Refs. [21–23]. In the case of charmonium, all results yield a dissociation temperature $T \simeq 1.5T_c$, as it is evident from Fig. 9.5.

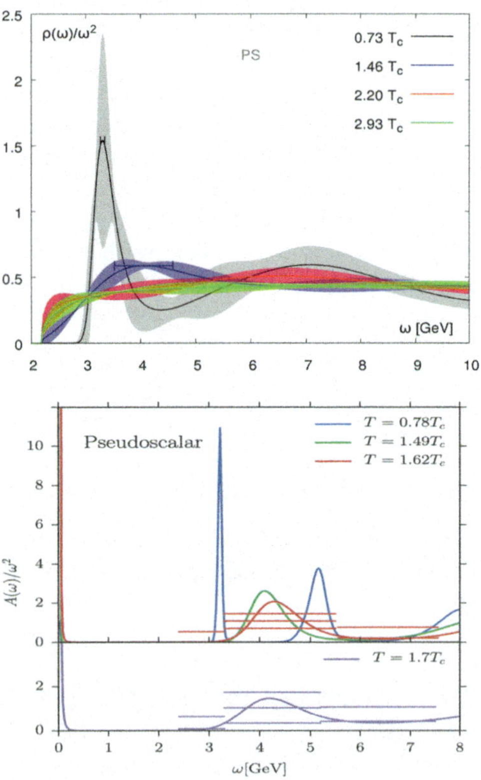

Fig. 9.5 Top: from Ref. [18]. Pseudoscalar charmonium spectral function on an isotropic lattice at different temperatures. Bottom: from Ref. [19]. Pseudoscalar charmonium spectral function obtained on an anisotropic lattice at different temperatures. In both cases, the spectral functions have been reconstructed using the MEM method

It is interesting to notice that this is true both in quenched simulations and in the case in which the $c\bar{c}$ state sits in a medium in which dynamical quarks are present.

Most of these studies used the MEM technique in order to reconstruct the spectral function from the Euclidean correlator, while the Bayesian Reconstruction (BR) approach was used in Ref. [24]. The MEM technique is prone to over-smoothing, in particular when few correlator datapoints are available. The FASTSUM collaboration has tested these limitations by changing the search space size through the number of included datapoints [25, 26]. On the other hand, the BR method leads to ringing artifacts that lead to artificial peak structures in the case of limited amount of datapoints.

The issue with studying heavy quarks on the lattice is that the discretization artifacts grow as $a_s m_Q$, where a_s is the lattice spacing in the space-like direction and m_Q is the heavy-quark mass. For this reason, one needs extremely fine lattices to study e.g. bottomonium. Results have been obtained in the quenched approximation, on a lattice with $N_s = 192$ and $N_t = 96$ and 48, corresponding to $T = 0.7T_c$ and $T = 1.4T_c$ for $a = 0.01$ fm [27]. The conclusion of this paper is that, at $T = 1.4T_c$, only the vector channel correlation function shows smaller thermal modification than the corresponding charmonium one in the same channel.

Some more results about bottomonium have been obtained through NRQCD, which allow to describe heavy-quark physics in a kinematic regime in which the lattice cut-off can take moderately small values. Within this framework, there have been discrepancies in the $P-$wave channel between Ref. [26], which reported a dissociation temperature right above T_c using the MEM method, and Ref. [28], which reported $T \simeq 250\,\text{MeV}$, based on the BR method. This tension has been settled recently, leading to a dissociation temperature in agreement with the one of Ref. [26], due to a deeper understanding of the ringing and smoothing features of the reconstruction methods. More recently, NRQCD has been used to study excited bottomonium states up to 3s and 2p, on $48^3 \times 12$ lattices with $m_\pi \simeq 161\,\text{MeV}$ [29]. The results on the width of these states are compatible with the quarkonia sequential dissociation picture. The thermal widths Γ_α of the χ_{b0} states as functions of the temperature are shown in the top panel of Fig. 9.6. The authors of Ref. [30] found a thermally broadened resonance peak for bottomonium in the vector channel, that is only melted at $T = 2.25T_c$. This was obtained by combining vacuum asymptotics with pNRQCD (NRQCD and pNRQCD are described in Sect. 9.6 below).

9.5 Spatial Correlation Functions

This method is somewhat complementary to the ones presented above, as it is not limited by the finite temporal extent of the lattice. Several excitations can be studied, but the connection to phenomenology is less straightforward.

The idea is to study the spatial correlation functions of mesons

$$G(z, T) = \int_0^{1/T} d\tau \int dxdy \langle J_H(\tau, x, y, z) J_H(0, 0, 0, 0)\rangle, \qquad (9.9)$$

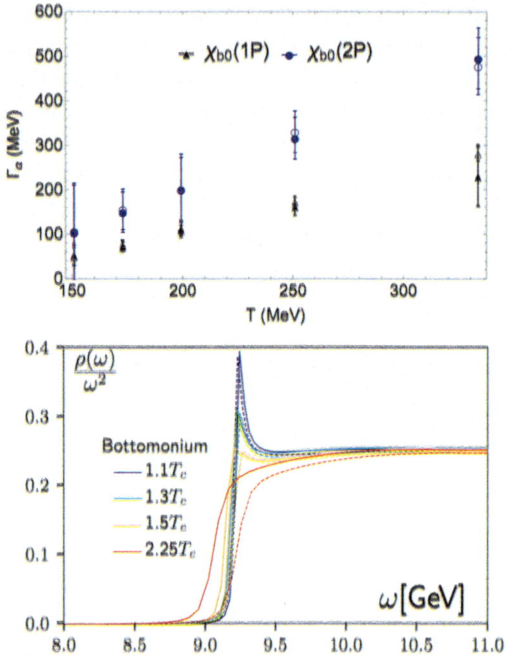

Fig. 9.6 Top: from Ref. [29]. Thermal widths Γ_α of the χ_{b0} states as functions of the temperature. Bottom: from Ref. [30]. Spectral functions in the high-ω region for bottomonium in the vector channel. The dashed lines show the perturbative spectral functions, while the solid lines are the modified ones obtained in the analysis

which are related to the spectral function $\rho(\omega, p_z, T)$ as

$$G(z, T) = \int_0^\infty d\tau \frac{2d\omega}{\omega} \int_{-\infty}^{\infty} dp_z e^{ip_z z} \rho(\omega, p_z, T). \tag{9.10}$$

The advantage of this approach is that the spatial correlator can be studied at separations larger than $1/T$, as the spatial separation is not limited by $1/T$. For this reason, it might be more sensitive to the meson in-medium modification. Besides, as we discussed in Chap. 8, the temperature dependence of the spectral function can be partially masked by the one of the Kernel, while this complication is not there for the spatial correlator.

As evident from Eq. (9.10), the relationship between this correlator and the spectral function is more complicated. There are however some simple limiting cases. At large distances we have $G(z, T) \sim \exp(-M(T)z)$, where $M(T)$ is the temperature-dependent screening mass. At small temperatures, where the mesonic bound state is well-defined, we get $\rho(\omega, 0, 0, p_z, T) \sim \delta(\omega^2 - p_z^2 - m_0^2)$ and $M(T)$ becomes the pole mass of the meson m_0. Finally, at very large temperatures, the spatial correlation function describes the propagation of a free quark-antiquark pair.

The spatial correlation functions of mesonic excitations with strange and charm quarks ($s\bar{s}$, $s\bar{c}$, $c\bar{c}$) have been studied in lattice QCD using the HISQ action in Ref. [31], with physical strange mass and almost physical up and down quark

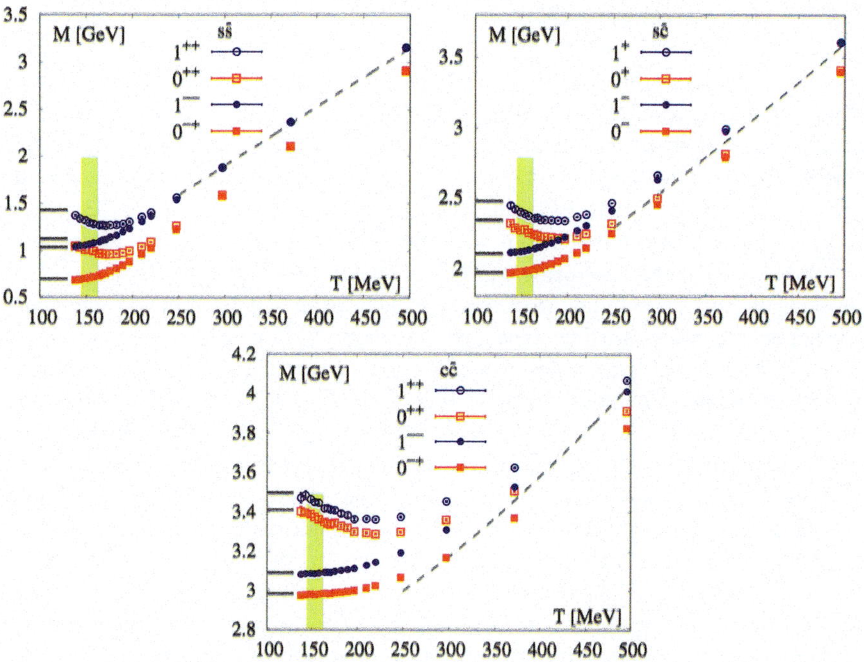

Fig. 9.7 From Ref. [31]. Screening masses for the $s\bar{s}$ (top left), $s\bar{c}$ (top right) and $c\bar{c}$ (bottom) states in different channels, as functions of the temperature. The solid, black lines indicate the zero-temperature meson mass, the yellow band is the QCD transition temperature and the dashed lines indicate the free field theory results

masses, corresponding to $m_\pi = 160\,\text{MeV}$. The lattice size is $48^3 \times 12$ and the temperature is in the range $T = 138 \ldots 248\,\text{MeV}$.

The screening masses for the $s\bar{s}$, $s\bar{c}$ and $c\bar{c}$ bound states are shown in the three panels of Fig. 9.7 as functions of the temperature.

9.6 Effective Theories for Heavy Quarkonium

If the quark mass m is large but finite, quarkonium bound states are expected to survive up to temperatures that are much smaller than m. Since the quark mass is much larger than the relevant temperature scale, one can construct effective theories in which m is integrated out, and quantities are expanded in powers of $1/m$. The static limit represents the first term in such expansion, whereas higher order corrections can be incorporated by means of effective field theories such as non-relativistic QCD (NRQCD) and potential non-relativistic QCD (pNRQCD) described below. For a review on the topic see Ref. [32].

9.6.1 Non-relativistic QCD

We start by considering a meson which contains one light and one heavy flavor. There are two relevant scales in such a system: the mass m of the heavy quark, which is large (perturbative scale) and the scale Λ_{QCD} which is low (non-perturbative scale). The other dimensional quantities (energy E and momentum p) reduce to the Λ_{QCD} scale. Heavy Quark Effective Theories (HQET, for a review see [33]) allow us to distinguish hard momenta ($p \sim m$) and soft ones ($p \sim \Lambda_{QCD}$). Phenomena that take place at the hard scale can be calculated in perturbative QCD, whereas those at the soft scale follow from the spin-flavor symmetry that makes the theory predictive. Quantities are then expanded in powers of $\alpha_s(m)$ and Λ_{QCD}/m.

If we want to describe a meson which contains two heavy quarks, the situation is more complicated. There are now four scales: m, $p \sim mv$, $E \sim mv^2$ and Λ_{QCD}. In general, $v = v(\alpha_s, \Lambda_{QCD}) \ll 1$. Therefore, the different scales are well separated. The fact that the scales are entangled makes both perturbative and non-perturbative calculations very difficult. In particular, lattice QCD simulations would need a lattice of volume $V \sim 100^4$, as the spacetime grid should be much larger than $1/mv^2$ but the lattice spacing should be small compared to $1/m$.

Non-Relativistic QCD (NRQCD) is the first effective theory that was introduced to disentangle these physical scales. In this theory, the mass m is factorized but the scale hierarchy $mv \gg mv^2$ is not resolved. Ambiguities arise beyond leading order in the perturbative expansion, as the scales p, E, Λ_{QCD} are entangled.

The theory is constructed by introducing an ultraviolet cutoff Λ of the order of the mass m. This excludes relativistic heavy quarks from the theory, which is compensated by adding new coefficients and new interaction terms in the effective Lagrangian. To order $1/\Lambda$, these terms are already present in the theory; beyond leading order one has to include non-renormalizable terms multiplied by constant parameters to be determined. It is convenient to choose $\Lambda = 1/m$, so that the Lagrangian is organized in powers of $1/m$. Usually, these parameters are chosen by calculating correlation functions of heavy quark fields both in QCD and NRQCD and by requiring that they agree at energies below Λ. Notice that, if one changes the value of the cutoff Λ, in principle all these coefficients have to be re-evaluated.

We should remember that [34]

- The matching of NRQCD to QCD is perturbative.
- In NRQCD we still have two scales, mv and mv^2. For this reason, the size of each term in the effective NRQCD Lagrangian is not unique. This makes the calculation of Feynman diagrams very difficult.
- As a consequence, the NRQCD Lagrangian contains both soft and ultrasoft scales.
- Since the rest mass has been removed from the theory, much coarser lattices are allowed. The quark Green function satisfies a Schrödinger-like equation that can easily be solved numerically.

In this effective theory, one gets explicit analytic expressions for mesonic correlation functions and spectral functions [6, 35]. There are several lattice groups which use this theory to describe bottom quarks propagating through a system of $N_f = 2 + 1$ light quarks [25, 26, 28]. These results are obtained in different lattice setups but they are in approximate agreement in the case of the $\Upsilon(1S)$ state, and now also for $P-$wave states.

9.6.2 Potential Non-relativistic QCD

The difference to NRQCD is that in this case the ultrasoft degrees of freedom remain dynamical, whereas all others are integrated out. In order to achieve this, the approach needs two ultraviolet cutoffs Λ_1 and Λ_2. Their respective conditions are $mv^2 \ll \Lambda_1 \ll mv$ and $mv \ll \Lambda_2 \ll mv^2$: Λ_1 is the cutoff of the energy of quarks and of the energy and momentum of gluons, whereas Λ_2 is the cutoff of the heavy quark bound system three-momentum $\mathbf{p}$.

Two situations are possible: in the first scenario, the matching is perturbative and $mv \ll \Lambda_{QCD}$, in the second one $mv \sim \Lambda_{QCD}$ and the matching cannot be obtained through an expansion in α_s. The lowest quarkonium excitations belong to the first scenario, the excited states to the second.

At leading order in the multipole expansion (i.e. expansion in $r \sim (mv)^{-1}$), the equations of motion of the singlet field coincide with Schrödinger's equation; for this reason, we can say that pNRQCD highlights the dominant role of the potential and the quantum mechanical nature of the $q\bar{q}$ bound state.

The generalization of pNRQCD to a thermal environment has been presented in Ref. [36]. It can be used to obtain in-medium properties of tightly bound states like the $\Upsilon(1S)$ [37]; the thermal width induced by the imaginary part of the in-medium potential can also be obtained [5, 6, 36].

The dissociation temperature of the $\Upsilon(1S)$ is found from pQCD to be $T_{diss} \sim$ 450 MeV [38, 39]. One can argue that pNRQCD can be generalized to strong coupling, because when the binding energy is small, the potential coincides with the energy of a static $q\bar{q}$ pair that can be calculated on the lattice [40]. Besides, the authors of Ref. [41] showed that, as long as a well defined Lorentzian peak can be observed in the spectrum of the Wilson loop, its late-time behavior can be obtained through a Schrödinger-like equation with a time-independent potential. The real and imaginary parts of this potential are related to the position and width of the peak. However, more work is needed to understand the relation between the energy of the static $q\bar{q}$ pair and the potential defined in pNRQCD.

The bottomonium and charmonium states in the $S-$ and $P-$wave channels have been computed based on this potential and its in-medium modifications from lattice QCD [9, 42]. These results show that the hierarchy of the narrow vacuum states is modified, due to their transition into open heavy-flavor states or excited bound states. This modification shifts the states to lower masses and increases their widths. This broadening could be due to gluo-dissociation and inelastic parton scattering, or to transitions to a color octet state.

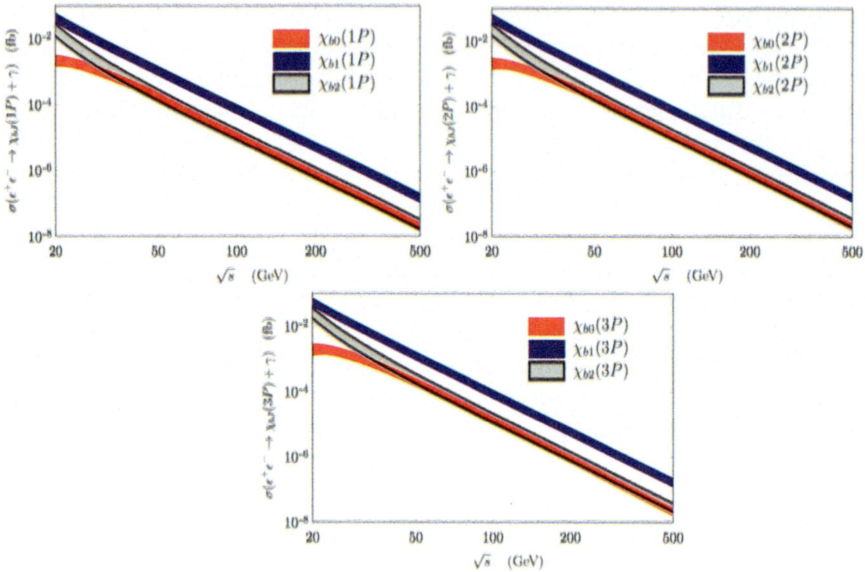

Fig. 9.8 From Ref. [43]. Cross sections $\sigma(e^+e^-) \rightarrow \chi_{bJ}(nP) + \gamma$ (top left: $n = 1$, top right: $n = 2$, bottom: $n = 3$), for $J = 0$ (red), $J = 1$ (blue) and $J = 2$ (gray)

More recently, strongly coupled quarkonia such as excited charmonium and bottomonium states have been studied in pNRQCD: their exclusive electromagnetic production has been obtained in Ref. [43]. The cross sections for production of $1P$, $2P$ and $3P$ states are shown in the three panels of Fig. 9.8. In the same reference, the electromagnetic decay widths, exclusive production cross sections and inclusive decay widths into light hadrons were obtained for $P-$wave states, together with the decay widths into lepton pairs and their ratios with the inclusive widths into light hadrons for $2S$ and $3S$ bottomonium.

9.7 Other Approaches

The T-matrix approach has extensively been used to obtain spectral functions for open heavy-flavors and quarkonia [44–47]. This method is based on a three-dimensional Bethe-Salpeter equation containing a local interaction kernel; a real-valued interaction potential needs to be calculated and entered into the Bethe-Salpeter kernel. This has been done self-consistently in Ref. [48], in which it was shown how finite-width effects in the potential and heavy-quark propagator can affect the extraction of the interaction kernel. The authors of Ref. [49] have constructed a renormalized potential that is tuned to reproduce lattice QCD data for the Equation of State, quarkonium correlators and static $q\bar{q}$ free energies. However, it turns out that this set of data does not univocally constrain the potential. Its possible range is consequently classified by a weakly- and a strongly-coupled solutions. The strongly coupled solution seems to be favored, based on preliminary studies of heavy quark transport coefficients.

One of the main challenges in relating the results for heavy flavors and quarkonia to experimental measurements is to make the connection between phenomenological concepts and theoretically well-defined quantities. For example, in order to describe heavy quark diffusion and jets, one needs the heavy quark diffusion coefficient (described in Sect. 8.4) and the jet quenching parameters $\hat{q}$ and $\hat{q}_L$. The two jet quenching parameters enter a Fokker-Plank equation, which is based on the assumption that the number of hard particles is conserved. However, this assumption is violated already at the order α_s by some processes [50]. While on the lattice it is not possible to isolate such processes, this can in principle be done in perturbative approaches, which resum the contribution of certain "soft" momentum scales to all orders. This effective theory can then be simulated on the lattice, which leads to the determination of $\hat{q}$ [51–55].

Other attempts to relate theoretical observables to phenomenological quantities include the definition of a temperature-dependent Debye screening mass $m_D(T)$ [56, 57] and effective strong coupling constant $\alpha_s(r, T)$ [58–60], which can be obtained from lattice QCD simulations of the heavy-quark potential and used in transport approaches [61–67].

The concept of open quantum systems has been introduced, to take into account the possibility that quarkonia can melt already in the regime $rm_D < 1$ [68–76]. This approach is based on the possibility of being able to separate the constituent quarks from the thermal medium and leads to a relationship between the real and imaginary parts of the heavy-quark potential and the stochastic evolution of the bound state wave function.

References

1. Aarts, G., Aichelin, J., Allton, C., Arnaldi, R., Bass, S.A., Bedda, C., Brambilla, N., Bratkovskaya, E., Braun-Munzinger, P., Bruno, G.E., Dahms, T., Das, S.K., Dembinski, H., Djordjevic, M., Ferreiro, E., Frawley, A., Gossiaux, P.B., Granier de Cassagnac, R., Grelli, A., He, M., Horowitz, W., Innocenti, G.M., Jo, M., Kaczmarek, O., Kuijer, P.G., Laine, M., Lombardo, M.P., Mischke, A., Munhoz, M., Nahrgang, M., Nguyen, M., Oliveira da Silva, A.C., Petreczky, P., Rothkopf, A., Schmelling, M., Scomparin, E., Song, T., Stachel, J., Suaide, A., Tolos, L., Trzeciak, B., Uras, A., L. van Doremalen, Vermunt, L., Vigolo, S., Xu, N., Ye, Z., Zanoli, H., Zhuang, P.: Eur. Phys. J. A **53**(5), 93 (2017)
2. Rothkopf, A.: Phys. Rept. **858**, 1–117 (2020)
3. Rothkopf, A.: [arXiv:2002.04938 [hep-ph]]
4. Mocsy, A.: Eur. Phys. J. C **61**, 705–710 (2009)
5. Laine, M., Philipsen, O., Romatschke, P., Tassler, M.: JHEP **03**, 054 (2007)
6. Beraudo, A., Blaizot, J.P., Ratti, C.: Nucl. Phys. A **806**, 312–338 (2008)
7. Burnier, Y., Kaczmarek, O., Rothkopf, A.: Phys. Rev. Lett. **114**(8), 082001 (2015)
8. Gava, E., Jengo, R.: Phys. Lett. **105B**, 285 (1981)
9. Burnier, Y., Kaczmarek, O., Rothkopf, A.: JHEP **12**, 101 (2015)
10. Burnier, Y., Rothkopf, A.: Phys. Rev. D **95**(5), 054511 (2017)
11. Petreczky, P., Rothkopf, A., Weber, J.: Nucl. Phys. A **982**, 735–738 (2019)
12. Arnold, P.B., Yaffe, L.G.: Phys. Rev. D **52**, 7208–7219 (1995)

13. Borsányi, S., Fodor, Z., Katz, S.D., Pásztor, A., Szabó, K.K., Török, C.: JHEP **04**, 138 (2015)
14. Asakawa, M., Hatsuda, T.: Phys. Rev. Lett. **92**, 012001 (2004)
15. Umeda, T., Matsufuru, H., Miyamura, O., Nomura, K.: Nucl. Phys. A **721**, 922–925 (2003)
16. Datta, S., Karsch, F., Petreczky, P., Wetzorke, I.: Phys. Rev. D **69**, 094507 (2004)
17. Jakovac, A., Petreczky, P., Petrov, K., Velytsky, A.: Phys. Rev. D **75**, 014506 (2007)
18. Ding, H.T., Francis, A., Kaczmarek, O., Karsch, F., Satz, H., Soeldner, W.: Phys. Rev. D **86**, 014509 (2012)
19. Ikeda, A., Asakawa, M., Kitazawa, M.: Phys. Rev. D **95**(1), 014504 (2017)
20. Burnier, Y., Ding, H.T., Kaczmarek, O., Kruse, A.L., Laine, M., Ohno, H., Sandmeyer, H.: JHEP **11**, 206 (2017)
21. Aarts, G., Allton, C., Oktay, M.B., Peardon, M., Skullerud, J.I.: Phys. Rev. D **76**, 094513 (2007)
22. Borsanyi, S., Dürr, S., Fodor, Z., Hoelbling, C., Katz, S.D., Krieg, S., Mages, S., Nogradi, D., Pasztor, A., Schäfer, A., et al.: JHEP **04**, 132 (2014)
23. Kelly, A., Rothkopf, A., Skullerud, J.I.: Phys. Rev. D **97**(11), 114509 (2018)
24. Burnier, Y., Rothkopf, A.: Phys. Rev. Lett. **111**, 182003 (2013)
25. Aarts, G., Allton, C., Amato, A., Evans, W., Giudice, P., Harris, T., Kelly, A., Kim, S.Y., Lombardo, M.P., Praki, K., Ryan, S.M., Skullerud, J.I.: Nucl. Phys. A **956**, 717–720 (2016)
26. Aarts, G., Allton, C., Harris, T., Kim, S., Lombardo, M.P., Ryan, S.M., Skullerud, J.I.: JHEP **07**, 097 (2014)
27. Ohno, H., Ding, H.T., Kaczmarek, O.: PoS **LATTICE2014**, 219 (2014)
28. Kim, S., Petreczky, P., Rothkopf, A.: Phys. Rev. D **91**, 054511 (2015)
29. Larsen, R., Meinel, S., Mukherjee, S., Petreczky, P.: Phys. Lett. B **800**, 135119 (2020)
30. Lorenz, A.L., Ding, H.T., Kaczmarek, O., Ohno, H., Sandmeyer, H., Shu, H.T.: PoS **LATTICE2019**, 207 (2020)
31. Bazavov, A., Karsch, F., Maezawa, Y., Mukherjee, S., Petreczky, P.: Phys. Rev. D **91**(5), 054503 (2015)
32. Brambilla, N.: [arXiv:hep-ph/0012026 [hep-ph]]
33. Grozin, A.G.: Springer Tracts Mod. Phys. **201**, 1–213 (2004)
34. Brambilla, N.: [arXiv:hep-ph/0202215 [hep-ph]]
35. Aarts, G., Allton, C., Kim, S., Lombardo, M.P., Oktay, M.B., Ryan, S.M., Sinclair, D.K., Skullerud, J.I.: JHEP **11**, 103 (2011)
36. Brambilla, N., Ghiglieri, J., Vairo, A., Petreczky, P.: Phys. Rev. D **78**, 014017 (2008)
37. Brambilla, N., Escobedo, M.A., Ghiglieri, J., Soto, J., Vairo, A.: JHEP **09**, 038 (2010)
38. Laine, M.: JHEP **05**, 028 (2007)
39. Escobedo, M.A., Soto, J.: Phys. Rev. A **78**, 032520 (2008)
40. Petreczky, P., Miao, C., Mocsy, A.: Nucl. Phys. A **855**, 125–132 (2011)
41. Burnier, Y., Rothkopf, A.: Phys. Rev. D **86**, 051503 (2012)
42. Burnier, Y., Kaczmarek, O., Rothkopf, A.: JHEP **10**, 032 (2016)
43. Brambilla, N., Chung, H.S., Müller, D., Vairo, A.: JHEP **04**, 095 (2020)
44. Mannarelli, M., Rapp, R.: Phys. Rev. C **72**, 064905 (2005)
45. Cabrera, D., Rapp, R.: Phys. Rev. D **76**, 114506 (2007)
46. Riek, F., Rapp, R.: Phys. Rev. C **82**, 035201 (2010)
47. Riek, F., Rapp, R.: New J. Phys. **13**, 045007 (2011)
48. Liu, S.Y.F., Rapp, R.: Nucl. Phys. A **941**, 179–187 (2015)
49. Liu, S.Y.F., Rapp, R.: Phys. Rev. C **97**(3), 034918 (2018)
50. Ghiglieri, J., Moore, G.D., Teaney, D.: JHEP **03**, 095 (2016)
51. Laine, M.: Eur. Phys. J. C **72**, 2233 (2012)
52. Benzke, M., Brambilla, N., Escobedo, M.A., Vairo, A.: JHEP **02**, 129 (2013)
53. Panero, M., Rummukainen, K., Schäfer, A.: Phys. Rev. Lett. **112**(16), 162001 (2014)
54. Laine, M., Rothkopf, A.: PoS **LATTICE2013**, 174 (2014)
55. D'Onofrio, M., Kurkela, A., Moore, G.D.: JHEP **03**, 125 (2014)
56. Burnier, Y., Rothkopf, A.: Phys. Lett. B **753**, 232–236 (2016)
57. Lafferty, D., Rothkopf, A.: Phys. Rev. D **101**(5), 056010 (2020)
58. Kaczmarek, O., Karsch, F., Zantow, F., Petreczky, P.: Phys. Rev. D **70**, 074505 (2004)

59. Kaczmarek, O., Zantow, F.: Phys. Rev. D **71**, 114510 (2005)
60. Bazavov, A., Brambilla, N., Tormo, X.G.I, Petreczky, P., Soto, J., Vairo, A.: Phys. Rev. D **90**(7), 074038 (2014)
61. Song, T., Berrehrah, H., Torres-Rincon, J.M., Tolos, L., Cabrera, D., Cassing, W., Bratkovskaya, E.: Phys. Rev. C **96**(1), 014905 (2017)
62. Song, T., Berrehrah, H., Cabrera, D., Cassing, W., Bratkovskaya, E.: Phys. Rev. C **93**(3), 034906 (2016)
63. Song, T., Berrehrah, H., Cabrera, D., Torres-Rincon, J.M., Tolos, L., Cassing, W., Bratkovskaya, E.: Phys. Rev. C **92**(1), 014910 (2015)
64. Song, T., Cassing, W., Moreau, P., Bratkovskaya, E.: Phys. Rev. C **97**(6), 064907 (2018)
65. Das, S.K., Scardina, F., Plumari, S., Greco, V.: Phys. Rev. C **90**, 044901 (2014)
66. Das, S.K., Plumari, S., Chatterjee, S., Alam, J., Scardina, F., Greco, V.: Phys. Lett. B **768**, 260–264 (2017)
67. Das, S.K., Scardina, F., Plumari, S., Greco, V.: Phys. Lett. B **747**, 260–264 (2015)
68. Borghini, N., Gombeaud, C.: Eur. Phys. J. C **72**, 2000 (2012)
69. Young, C., Dusling, K.: Phys. Rev. C **87**(6), 065206 (2013)
70. Akamatsu, Y., Rothkopf, A.: Phys. Rev. D **85**, 105011 (2012)
71. Akamatsu, Y.: Phys. Rev. D **87**(4), 045016 (2013)
72. Rothkopf, A.: JHEP **04**, 085 (2014)
73. Akamatsu, Y.: Phys. Rev. D **91**(5), 056002 (2015)
74. Akamatsu, Y.: Phys. Rev. C **92**(4), 044911 (2015) [erratum: Phys. Rev. C **101**(5), 059901 (2020)]
75. Blaizot, J.P., De Boni, D., Faccioli, P., Garberoglio, G.: Nucl. Phys. A **946**, 49–88 (2016)
76. Brambilla, N., Escobedo, M.A., Soto, J., Vairo, A.: Phys. Rev. D **96**(3), 034021 (2017)

Index

C. Ratti, R. Bellwied, *The Deconfinement Transition of QCD*, Lecture Notes in Physics 981, https://doi.org/10.1007/978-3-030-67235-5

The manufacturer's authorised representative in the EU is Springer Nature Customer Service Centre GmbH, Europaplatz 3, 69115 Heidelberg, Germany. If you have any concerns regarding our products, please contact ProductSafety@springernature.com

Printed and bound by CPI Group (UK) Ltd, Croydon, CR0 4YY

30/07/2026

02181732-0001